VOLUME EIGHTY FIVE

CURRENT TOPICS IN
DEVELOPMENTAL BIOLOGY

Ciliary Function in Mammalian Development

Series Editor

Paul M. Wassarman
Department of Developmental and Regenerative Biology
Mount Sinai School of Medicine
New York, NY 10029-6574
USA

Olivier Pourquié
Investigator Howard Hughes Medical Institute
Stowers Institute for Medical Research
Kansas City, MO, USA

Editorial Board

Blanche Capel
Duke University Medical Center
Durham, USA

B. Denis Duboule
Department of Zoology and Animal Biology
NCCR 'Frontiers in Genetics'
Geneva, Switzerland

Anne Ephrussi
European Molecular Biology Laboratory
Heidelberg, Germany

Janet Heasman
Cincinnati Children's Hospital Medical Center
Department of Pediatrics
Cincinnati, USA

Julian Lewis
Vertebrate Development Laboratory
Cancer Research UK London Research Institute
London WC2A 3PX, UK

Yoshiki Sasai
Director of the Neurogenesis and Organogenesis Group
RIKEN Center for Developmental Biology
Chuo, Japan

Cliff Tabin
Harvard Medical School
Department of Genetics
Boston, USA

Founding Editors

A. A. Moscona
Alberto Monroy

VOLUME EIGHTY FIVE

CURRENT TOPICS IN
DEVELOPMENTAL BIOLOGY

Ciliary Function in Mammalian Development

Edited by

BRADLEY K. YODER
Department of Cell Biology,
University of Alabama at Birmingham,
School of Medicine,
Birmingham, Alabama

AMSTERDAM • BOSTON • HEIDELBERG • LONDON
NEW YORK • OXFORD • PARIS • SAN DIEGO
SAN FRANCISCO • SINGAPORE • SYDNEY • TOKYO
Academic Press is an imprint of Elsevier

ELSEVIER

Academic Press is an imprint of Elsevier
525 B Street, Suite 1900, San Diego, CA 92101-4495, USA
30 Corporate Drive, Suite 400, Burlington, MA 01803, USA
32, Jamestown Road, London NW1 7BY, UK
Linacre House, Jordan Hill, Oxford OX2 8DP, UK

First edition 2008

Copyright © 2008 Elsevier Inc. All rights reserved.

No part of this publication may be reproduced, stored in a retrieval system
or transmitted in any form or by any means electronic, mechanical, photocopying,
recording or otherwise without the prior written permission of the publisher

Permissions may be sought directly from Elsevier's Science & Technology Rights
Department in Oxford, UK: phone (+44) (0) 1865 843830; fax (+44) (0) 1865 853333;
email: permissions@elsevier.com. Alternatively you can submit your request online by
visiting the Elsevier web site at http://elsevier.com/locate/permissions, and selecting
Obtaining permission to use Elsevier material

Notice

No responsibility is assumed by the publisher for any injury and/or damage to persons
or property as a matter of products liability, negligence or otherwise, or from any use
or operation of any methods, products, instructions or ideas contained in the material
herein. Because of rapid advances in the medical sciences, in particular, independent
verification of diagnoses and drug dosages should be made

ISBN: 978-0-12-374453-1
ISSN: 0070-2153

For information on all Academic Press publications
visit our website at elsevierdirect.com

Printed and bound in USA
08 09 10 11   9 8 7 6 5 4 3 2 1

Working together to grow
libraries in developing countries

www.elsevier.com | www.bookaid.org | www.sabre.org

ELSEVIER   BOOK AID International   Sabre Foundation

Contents

Contributors	xi
Preface	xv

1. Basal Bodies: Platforms for Building Cilia   1
Wallace F. Marshall

1. Introduction	2
2. Basal Body Architecture and Assembly	3
3. Basal Body Functions in Ciliogenesis	7
4. Approaches to the Study of Basal Bodies	14
5. Conclusions	17
References	17

2. Intraflagellar Transport (IFT): Role in Ciliary Assembly, Resorption and Signalling   23
Lotte B. Pedersen and Joel L. Rosenbaum

1. Introduction	24
2. Early Events in Ciliogenesis	28
3. General Description of Intraflagellar Transport	29
4. Discovery and Evolution of IFT	30
5. The Canonical Anterograde IFT Motor	32
6. Additional Kinesin Motors Involved in Ciliogenesis	35
7. The Retrograde IFT Motor	37
8. IFT Particle Polypeptides	38
9. Functional Analysis of IFT Particle Polypeptides: Distinct Roles of IFT Complexes A and B	39
10. Regulation of IFT	41
11. Targeting of Proteins to the Ciliary Compartment: Clues from Ciliary Disease Genes	43
12. IFT and Cilia-Mediated Signaling	47
13. Conclusions and Perspectives	49
Acknowledgments	50
References	50

3. How Did the Cilium Evolve?   63
Peter Satir, David R. Mitchell, and Gáspár Jékely

1. Introduction   64
2. The Link Between Ciliary Evolution and Eukaryotic Divergence   66
3. Hypotheses of the Origin of Cilia   67
4. The Viral Hypothesis of Ciliary Origin   68
5. The Autogenous Model for the Origin of Cilia   71
6. Origin of Intraflagellar Transport and the Sensory Function of Cilia   76
7. The Evolution of Ciliary Motility   78
Acknowledgments   79
References   79

4. Ciliary Tubulin and Its and Post-Translational Modifications   83
Jacek Gaertig and Dorota Wloga

1. Introduction   83
2. Production and Turnover of Ciliary Tubulin   84
3. Post-Translational Modifications of Tubulin   89
4. Conclusions   103
Acknowledgments   103
References   104

5. Targeting Proteins to the Ciliary Membrane   115
Gregory J. Pazour and Robert A. Bloodgood

1. Introduction   116
2. Structure of the Ciliary Membrane   118
3. Functions of the Ciliary Membrane   121
4. Protein Machinery Involved in Trafficking to the Ciliary Membrane   127
5. Ciliary Targeting Sequences   134
References   139

6. Cilia: Multifunctional Organelles at the Center of Vertebrate Left–Right Asymmetry   151
Basudha Basu and Martina Brueckner

1. Introduction   152
2. Vertebrate LR Asymmetry is Chiral Asymmetry   154
3. The Structure and Distribution of Cilia in the Vertebrate Embryo   155
4. Motile Primary Cilia in LR Development   156
5. Mouse and Human Mutations Affecting Cilia Have a Wide-Ranging Effect on Left–Right and Cardiac Development   158

6. A Structure with Prominent Primary Cilia is Essential to the
　　　 Development of LR Asymmetry in Most Vertebrates　　　　　　　160
　　7. Cilia at the LR Organizer Generate Directional Fluid Flow　　 162
　　8. A Conserved Asymmetric Calcium Signal Is Found at the
　　　 LR Organizer　　　　　　　　　　　　　　　　　　　　　　　 164
　　9. Do Cilia Lie at the Root of All Vertebrate LR Asymmetry,
　　　 and What Leads from Asymmetric Calcium to Asymmetric
　　　 Organogenesis?　　　　　　　　　　　　　　　　　　　　　　168
　　References　　　　　　　　　　　　　　　　　　　　　　　　　 169

7. Ciliary Function and Wnt Signal Modulation　　　　　　　　　175
Jantje M. Gerdes and Nicholas Katsanis

　　1. Introduction　　　　　　　　　　　　　　　　　　　　　　　176
　　2. Overview of Wnt Signaling　　　　　　　　　　　　　　　　　176
　　3. Emerging Roles of the Primary Cilium and Basal Body
　　　 in Wnt Signaling Modulation　　　　　　　　　　　　　　　　179
　　4. Wnt and Ciliogenesis　　　　　　　　　　　　　　　　　　　184
　　5. Wnt Dysfunction and Ciliopathy Phenotypes　　　　　　　　　185
　　6. What Drives the Phenotype: Gain of Canonical Wnt
　　　 Signaling or Loss of Noncanonical Wnt Signaling?　　　　　　　185
　　7. Other Phenotypes　　　　　　　　　　　　　　　　　　　　　187
　　8. Discussion　　　　　　　　　　　　　　　　　　　　　　　　189
　　Acknowledgments　　　　　　　　　　　　　　　　　　　　　　 190
　　References　　　　　　　　　　　　　　　　　　　　　　　　　 191

8. Primary Cilia in Planar Cell Polarity Regulation of the Inner Ear　　　　　　　　　　　　　　　　　　　　　　　　197
Chonnettia Jones and Ping Chen

　　1. Introduction　　　　　　　　　　　　　　　　　　　　　　　198
　　2. Planar Cell Polarity and Ciliogenesis of the Inner Ear Sensory Organs　200
　　3. Planar Cell Polarity Regulation for the Development of the Inner Ear　206
　　4. Cilia and PCP Regulation　　　　　　　　　　　　　　　　　 209
　　5. Cilia and Determination of the Intrinsic Cellular
　　　 Polarity of Inner Ear Cells　　　　　　　　　　　　　　　　　214
　　6. Conclusions and Perspectives　　　　　　　　　　　　　　　　217
　　References　　　　　　　　　　　　　　　　　　　　　　　　　 219

9. The Primary Cilium: At the Crossroads of Mammalian Hedgehog Signaling   225

Sunny Y. Wong and Jeremy F. Reiter

1. Hh Signals Pattern Diverse Developing Tissues   226
2. Hh Signal Transduction: Activating an Activator and Repressing a Repressor   229
3. Functions of Gli Proteins in Development   233
4. Defective Intraflagellar Transport Disrupts Mammalian Hh Signaling   236
5. The Primary Cilium as a Cellular Signaling Center   239
6. Mammalian Smo Activates the Hh Pathway at the Primary Cilium   241
7. Proper Ciliary Function is Required for Processing of Gli Proteins   242
8. Additional Ciliary Components Function in Hh Signal Transduction   243
9. Are Cilia Involved in Hh Pathway-Mediated Tumorigenesis?   246
10. Lingering Questions   248
Acknowledgments   249
References   249

10. The Primary Cilium Coordinates Signaling Pathways in Cell Cycle Control and Migration During Development and Tissue Repair   261

Søren T. Christensen, Stine F. Pedersen, Peter Satir, Iben R. Veland, and Linda Schneider

1. Introduction   262
2. Cell Cycle Entry Regulated by PDGFR$\alpha\alpha$ Signaling in the Primary Cilium   267
3. Directional Cell Migration is Regulated by PDGFRα in the Primary Cilium   272
4. The Extracellular Matrix and the Primary Cilium   280
5. Polarization, Cell Migration, Cell Cycle Control and Wnt Signaling in the Primary Cilium   282
6. Conclusions and Perspectives   286
Acknowledgments   288
References   288

11. Cilia Involvement in Patterning and Maintenance of the Skeleton   303

Courtney J. Haycraft and Rosa Serra

1. Introduction   304
2. Cilia are Required for Anterior–Posterior Limb Patterning   305
3. A Role for IFT/Primary Cilia in Endochondral Bone Formation and Development of the Postnatal Growth Plate   312
4. Primary Cilia in Articular Cartilage   318

5. A Role for Primary Cilia in the Development of the Bone Collar 319
  6. Primary Cilia in the Maintenance of Bone 321
  7. Primary Cilia in Craniofacial Development 323
  8. Primary Cilia in Tooth Development 325
  9. Summary 326
  Acknowledgments 327
  References 327

12. Olfactory Cilia: Our Direct Neuronal Connection to the External World   333

Dyke P. McEwen, Paul M. Jenkins, and Jeffrey R. Martens

  1. Olfaction as a Sensory Modality 334
  2. Anatomy and Organization of the Olfactory Epithelium 335
  3. Structure of Olfactory Cilia 338
  4. Formation of Olfactory Cilia 344
  5. Intraflagellar Transport 347
  6. Ciliary Proteome 349
  7. Olfactory Cilia and Disease 355
  8. Summary 358
  References 359

13. Ciliary Dysfunction in Developmental Abnormalities and Diseases   371

Neeraj Sharma, Nicolas F. Berbari, and Bradley K. Yoder

  1. Introduction 372
  2. The Oak Ridge Polycystic Kidney (*Orpk*) Mouse: A Model For Human Ciliopathies 374
  3. Functions and Phenotypes Associated with Abnormal Motile Cilia 376
  4. Functions and Diseases Associated with Immotile Cilium 383
  5. Oligogenic Inheritance and Clinical Variability in the Ciliopathies 409
  6. Cilia Proteome and Genome Databases and the Identification of Human Ciliopathy Genes 410
  Acknowledgments 412
  References 413

Index 429
Contents of Previous Volumes 437

Contributors

Basudha Basu
Department of Pediatrics, Yale University School of Medicine, New Haven, Connecticut

Nicolas F. Berbari
Department of Cell Biology, University of Alabama at Birmingham, School of Medicine, Birmingham, Alabama

Robert A. Bloodgood
Department of Cell Biology, University of Virginia School of Medicine, Charlottesville, Virginia

Martina Brueckner
Department of Genetics, Yale University School of Medicine, New Haven, Connecticut, and Department of Pediatrics, Yale University School of Medicine, New Haven, Connecticut

Ping Chen
Department of Cell Biology, Emory University School of Medicine, Atlanta, Georgia

Søren T. Christensen
Department of Biology, Section of Cell and Developmental Biology, The August Krogh Building, University of Copenhagen, Universitetsparken 13, DK-2100 Copenhagen OE, Denmark

Jacek Gaertig
Department of Cellular Biology, University of Georgia, Athens, Georgia

Jantje M. Gerdes
McKusick-Nathans Institute of Genetic Medicine, Johns Hopkins University School of Medicine, Baltimore, Maryland

Courtney J. Haycraft
Department of Medicine/Division of Nephrology, Medical University of South Carolina, Charleston, SC 29425

Gáspár Jékely
Max Planck Institute for Developmental Biology, Spemannstrasse 35, 72076 Tübingen, Germany

Paul M. Jenkins
Department of Pharmacology, University of Michigan, Ann Arbor, MI 48109-5632

Chonnettia Jones
Department of Cell Biology, Emory University School of Medicine, Atlanta, Georgia

Nicholas Katsanis
McKusick-Nathans Institute of Genetic Medicine, Johns Hopkins University School of Medicine, Baltimore, Maryland

Wallace F. Marshall
Department of Biochemistry and Biophysics, University of California, San Francisco, California

Jeffrey R. Martens
Department of Pharmacology, University of Michigan, Ann Arbor, MI 48109-5632

Dyke P. McEwen
Department of Pharmacology, University of Michigan, Ann Arbor, MI 48109-5632

David R. Mitchell
Department of Cell and Developmental Biology, SUNY Upstate Medical University, Syracuse, New York

Gregory J. Pazour
Program in Molecular Medicine, University of Massachusetts Medical School, Biotech II, Worcester, Massachusetts

Lotte B. Pedersen
Department of Biology, Section of Cell and Developmental Biology, University of Copenhagen, Universitetsparken 13, DK-2100 Copenhagen OE, Denmark

Stine F. Pedersen
Department of Biology, Section of Cell and Developmental Biology, The August Krogh Building, University of Copenhagen, Universitetsparken 13, DK-2100 Copenhagen OE, Denmark

Jeremy F. Reiter
Department of Biochemistry and Biophysics, Cardiovascular Research Institute, University of California, San Francisco, California

Joel L. Rosenbaum
Department of Molecular, Cellular, and Developmental Biology, Yale University, New Haven, Connecticut

Peter Satir
Department of Anatomy and Structural Biology, Albert Einstein College of Medicine of yeshiva university, Bronx, NY 10461

Linda Schneider
Department of Biology, Section of Cell and Developmental Biology, The August Krogh Building, University of Copenhagen, Universitetsparken 13, DK-2100 Copenhagen OE, Denmark

Rosa Serra
Department of Cell Biology, University of Alabama at Birmingham, 1918 University Blvd, Birmingham, AL 35294-0005

Neeraj Sharma
Department of Cell Biology, University of Alabama at Birmingham, School of Medicine, Birmingham, Alabama

Iben R. Veland
Department of Biology, Section of Cell and Developmental Biology, The August Krogh Building, University of Copenhagen, Universitetsparken 13, DK-2100 Copenhagen OE, Denmark

Dorota Wloga
Department of Cellular Biology, University of Georgia, Athens, Georgia

Sunny Y. Wong
Department of Biochemistry and Biophysics, Cardiovascular Research Institute, University of California, San Francisco, California

Bradley K. Yoder
Department of Cell Biology, University of Alabama at Birmingham, School of Medicine, Birmingham, Alabama

Preface

The recent rise of the cilium as an important cellular organelle is a remarkable tale. Although identified nearly a century ago, in comparison to the nucleus, Golgi, or endoplasmic reticulum (ER), the cilium has been relatively ignored by most scientists and clinicians. The main research interests in cilia were with regards to the basic cellular and molecular mechanisms regulating its assembly and motility rather than its possible role as a signaling center. Clinically, it was recognized that defects in the motile cilia were associated with respiratory dysfunction; however, the general consensus was that the cilium had minimal importance for human development and health. This was particularly evident in the case of the primary or nonmotile form of the cilium which is present on most mammalian cells. Until recently, references in text books and the opinion of most scientists and clinicians was that the primary cilia were likely evolutionary detritus with no significant functional roles.

During these dark ages for cilium, times were tough for scientists whose research interests centered on this organelle. We are fortunate that we had several visionaries, such as Joel Rosenbaum and Bjorn Afzelius, who were able to maintain interest and research programs focused on the cilium despite their status as the red-headed step *children* of cellular organelles. Their research programs are largely responsible for building the foundation upon which our recent advance have been made. For reasons discussed in this volume the cilium has rapidly emerged from its obscurity to an area of prominence as a major signaling hub involved in multiple pathways important for normal development and tissue functions in adult.

The origin of the mammalian cilium remains a matter of debate as described by third chapter of this volume and range from an invasion of a centriolar-like virus to functional modification of an existing microtubule organizing center. In addition to its evolution, the cilium has undergone significant morphogenic diversification adapted for specific functions in different tissues or cell types. This can be seen best in the case of the nonmotile $9 + 0$ axoneme cilia on rods and cone photoreceptor cells in the retina where the cilia have extensive membrane folds; whereas, cilia on most epithelial cells, neurons, and fibroblasts have a relatively simple linear morphology. Yet another form of the cilium is found on olfactory neurons where they are required for odor detection. As described in Chapter 12, these cilia are immotile despite having $9 + 2$ axoneme architecture and are present in small groups on each cell. Further, there are multiple forms of motile cilia. Motile cilia with the prototypical 9 + 2 microtubule arrangement of the axoneme are exemplified by the hundreds of cilia that are found on a typical tracheal epithelial cell. There are also examples where cells have single motile cilia such as the flagella of the sperm or on the cells of the embryonic node. Each of these forms of cilia has a unique beat pattern required for the specific function of the cells. Thus, the cell has altered its cilia number, length, and morphology to specialize in fluid or cell movement, and for sensory reception needed for these diverse different cell types. Exploring how these cilia function to mediate or regulate these events is the focus of intense research activity.

The cilia axoneme extends from basal body, which are attached to the cell membrane. As discussed in the initial chapter, the basal body is a turning out to be a very complex and interesting structure that has important functions in regulating the number of cilia that will form and the direction or angle in which they will extend from the cell. Defects in the angle of the cilium appear to have important implications for the direction and rate of fluid movement and cell division. Furthermore, many cilia proteins initially localize to the basal body suggesting that another function of the basal body is as the docking site for protein destined to enter the cilium. In addition, the basal body is generated from the centrioles which have important roles in mitotic spindle formation. In order for cells to divide, the cilium must be resorbed to free up the centrioles for the mitotic spindle. Once the division is complete the basal bodies again attach to the membrane and re-initiate cilia extension. Thus, cells are not able to divide while they maintain the cilium raising the possibility that cilia function as a mitotic check points and that loss of this organelle may be associated with increase rates of proliferation and cancer.

The structural complexity and diversity of the cilium puts significant demands on the cell since neither ribosomes or membrane vesicles are present in the cilium. Thus all components of the cilium must be generated in the cytosol and subsequently imported into the cilium. The mechanism of by which this occurs is known as intraflagellar transport (IFT) and was uncovered by ground breaking studies in *Chlamydomonas* flagella. This process is discussed in detail in the second chapter of this volume.

The proteins in the cilium axoneme are subject to a number of post-translational modifications. Among the targets modified is tubulin, the most abundant protein in the axoneme. As discussed in the fourth chapter, data suggest these modifications are not benign but have significant influences on ciliary function, protein transport properties along the microtubule axoneme, and may regulate cilia stability and motility.

It is important to note that protein entry into the cilium appears to be under tight control. How proteins become targeted to the basal body and are then are loaded onto the IFT particle for transport into the cilium is not well understood and remains a main research question. This is particularly interesting with regard to transmembrane protein since membrane vesicles that would normally transport these proteins are not present in the cilium. The sequences and machinery involved in ciliary targeting and Golgi to cilia transport are now beginning to emerge and are discussed in the fifth chapter. This is a critical issue since several proteins involved in human disease and developmental abnormalities are now thought to require localization in the cilium for normal function and regulation.

The renaissance that resulted in the emergence of the cilium from a rudimentary organelle to one of clinical and developmental importance began with several manuscripts around the turn of the century from research groups in diverse model organisms looking at what was thought to be unrelated phenomena. It should be noted that many of the initial breakthroughs leading to our appreciation of cilia as being important in human health and develop came from studies conducted in simple model organisms which interestingly lack most mammalian tissues. This demonstrates the need for continued support of basic science as a means of uncovering unexpected and clinically relevant connections. Who would have predicted that proteomic analysis of *Chlamydomonas* flagella and genetic screens to identify genes involved in male mating responses in *C. elegans* would have such a major impact on the direction of research in polycystic kidney disease (PKD), one of the most common human disorders.

The initial data demonstrating a critical role for the cilium in mammalian development came from Hirokawa's laboratory working on mice lacking Kif3a, the kinesin needed for IFT. These mutants exhibited a random specification of the left-right body axis. This was found to involve motile cilia on cells of an early developmental structure called the node. Despite there being a single cilium on each of the node cells, elegant studies from this group demonstrated that these cilia rotate and generate fluid flow across the node surface. The direction of this flow was important in determining axis speciation. As described in the sixth chapter, the mechanism by which the cells respond to this flow to break left-right symmetry is a matter of discussion. One model proposes that the node cells secrete membrane vesicles containing important morphogenic factors that then release their content on the left side of the node as a consequence of the cilia generated fluid movement. The other model involves mechanosensory function of a second group of nonmotile cilia located at the lateral edge of the node that elicit a transient calcium signal in response the fluid mediated deflection of the cilia axoneme. Further studies are needed to address whether either or a combination of these models are key features for breaking symmetry in the early mammalian embryo and whether similar mechanism may be involved in axis formation in additional nonmammalian metazoans.

The next major advance came from a genetic screen in *C. elegans* to identify genes involved in male mating response, from proteomic analysis of the *Chlamydomonas* flagella, and from work on a mouse model of cystic kidney disease. As described in chapter 13, these seemingly unrelated studies revealed a connection between polycystins (the major proteins involved in human PKD), cilia, and IFT. This connection had a remarkable impact of the direction of research in cystogenesis. Since then there have been multiple other genes identified that are involved in different human cystic disorders, most of which encode proteins that localize to the cilium. This led to the flow-mechanosensory model of cyst formation and subsequently to the hypothesis that cilia/basal body function in regulating mitotic spindle orientation, defects in which result in tubule enlargement, rather than elongation, and cyst formation.

Shortly after these seminal studies, it was found that cilia also function as important regulators of multiple signaling pathways needed for normal development and tissue homeostasis. As discussed in several chapters within this volume, the pathways affected by cilia dysfunction now include hedgehog (Hh), wingless (Wnt), and platelet derived growth factor (PDGF). The cilia proteomes have been evaluated for multiple mammalian tissues and model organism such as *Chlamydomonas* and *C. elegans* and the overall result has been that we greatly underestimated the cilium in its complexity as it contains an abundance of receptors, transcriptional regulatory factors, and signaling molecules. Based on the proteome data it is likely that many pathways in addition to those mentioned above will also be under ciliary control as well. It is currently not known how disruption of cilia alters signaling; however, many components of these pathways localize to the cilium or to the basal body at the base of the cilium and it seem likely that this localization is needed for either reception of the ligand or subsequent activation of downstream effectors of the pathway.

As discussed in several of the chapters in this volume, abnormal regulation of these pathways due to loss of cilia function is responsible for a wide range of phenotypes in mice and is now associated with a large number of human syndromes collectively called the ciliopathies. The phenotypic presentation depends on the severity of the genetic mutation and its effect on cilia function. These phenotypes range from the severe gestational lethality, neural tube abnormalities, and heterotaxia to more mild phenotypes such as inner ear patterning defects, anosmia and blindness, cystic lesions in the liver, kidney, and pancreas, hydrocephalus, mental deficits, skeletal abnormalities, and obesity. In some cases, the phenotypes can be explained by abnormal regulation of one of the pathways mentioned above. This would include Hh signaling defects in neural tube patterning/closure and polydactyly or possible effects on the canonical and noncanonical Wnt pathways with the inner ear abnormalities. However, most of the other phenotypes such as abnormal bone homeostasis, mental retardation, and obesity likely involve additional yet to be identified signaling pathways. Although I believe there has been amazing progress over the past decade in demonstrating the importance of the cilium in human health and development, there remains a lot that needs to be discovered, in particular with how the cilium regulates the activities of these pathways.

The cilium has had a rapid trajectory to stardom over the past decade, and can now stand proudly among its compatriots the Golgi, mitochondria, and the endoplasmic reticulum as a critically important cellular organelle.

Finally, I would like to thank all of the authors who have taken precious time from their very busy schedules to make contributions to this important volume. Their efforts along with the comments of the reviewers of each chapter have been essential for completion of this project. I also would also like to recognize the outstanding efforts of Kirsten Funk and Tara Hoey on the editorial staff who have had extraordinary patience (and persistence) with us scientists and clinicians who worked on this project and who have made sure that it came to fruition.

CHAPTER ONE

Basal Bodies: Platforms for Building Cilia

Wallace F. Marshall

Contents

1. Introduction	2
2. Basal Body Architecture and Assembly	3
2.1. Basal body structure	3
2.2. Basal body assembly	4
2.3. Basal body genome concept	6
3. Basal Body Functions in Ciliogenesis	7
3.1. Basal body function in templating axoneme	7
3.2. Basal body functions in attaching and orienting cilium at cortex	9
3.3. Basal body functions in regulating protein import into cilium	11
3.4. Basal body function in spindle orientation	12
4. Approaches to the Study of Basal Bodies	14
4.1. Proteomics	14
4.2. Comparative genomics	14
4.3. RNAi-mediated knockdown of basal body-related genes	15
4.4. Genetic analysis of basal bodies	15
4.5. Imaging	16
5. Conclusions	17
References	17

Abstract

Basal bodies are modified centrioles that give rise to cilia and flagella. The basal body is a complex structure that can form through at least two distinct pathways depending on the cell type. Corresponding to this structural complexity, the basal body proteome contains a large number of proteins, many of which correspond to cilia-related disease genes, especially genes involved in nephronophthisis and cone-rod dystrophy. Basal bodies appear to play several roles in the cell. First, they provide a ninefold symmetric template on which the ninefold symmetry axonemal structure of the cilium can be built. Second, they dictate

Department of Biochemistry and Biophysics, University of California, San Francisco, California the position and orientation of the cilium, which is especially critical for ensuring that cilia-driven fluid flows move in the correct direction. Third, they are the point at which entry of proteins into the cilium is regulated. Finally, recent evidence suggests that basal body position may be involved in coupling planar cell polarity cues with the axis of cell division. Defects in any of these functions could lead to disease symptoms. Current studies of basal body biology include both proteomic and genetic approaches, relying on ciliated cell culture lines as well as genetically tractable systems such as *Chlamydomonas reinhardtii*. The "parts list" of basal body proteins and genes is rapidly being completed, opening the way to more mechanistic studies in the future.

1. INTRODUCTION

Basal bodies are protein-based structures located at the base of cilia which are thought to provide a platform on which the cilium is constructed. The basal body is a modified form of the centriole, an organelle that is found at the core of the mitotic spindle pole. In dividing cells, the centriole moves to the cell surface during G1 where it acts as a basal body to template ciliogenesis. Then, during mitosis, the cilia resorb and the basal body moves to the spindle pole, at which point it is called a centriole. The function of centrioles has long been mysterious and controversial. Much of this controversy arose from a desire to assign a mitotic role to this organelle despite an ever increasing list of experiments showing that mitosis can proceed rather well when centrioles are removed from cells (for a recent, and extremely convincing, example, see Uetake et al., 2007), not to mention the fact that many organisms, such as higher plants, lack centrioles altogether. Nevertheless centrioles have continued to be discussed in terms of mitotic roles, brushing aside the obvious fact that the centriole is also involved in making cilia. The central link between centrioles and cilia is highlighted by the fact that centrioles are only found in species that have cilia at some point in their life cycle. Why, then, was this ever considered mysterious or enigmatic? The obsessive focus on a mitotic function for centrioles was probably due, in large part, to the old idea that primary cilia were nonfunctional vestiges, hence nobody wanted to believe that centrioles evolved purely for the purpose of making cilia. With the new appreciation of the ubiquitous importance of cilia in physiology and development, it no longer seems odd to think that the centriole exists primarily for the purpose of driving ciliogenesis. The majority of this chapter will therefore focus on the functions that basal bodies perform in support of ciliogenesis. We first begin with an overview of basal body structure and composition, and after a discussion of known and possible basal body functions, we will close with an overview of approaches and model systems that are currently being used to understand these fascinating organelles.

2. Basal Body Architecture and Assembly

2.1. Basal body structure

Basal bodies are specialized forms of centrioles, which in turn are cylinders composed of nine triplet microtubule "blades" arranged in the shape of a barrel (Ringo, 1967). Each of the blades is composed of one complete microtubule, with a second partial microtubule grown off the side of the first, and then a third partial microtubule grown off of the side of the second. These tubules are named A, B, and C. The microtubule blades are arranged roughly parallel to the long axis of the cylinder, and the end of the centriole/basal body that contains the plus ends of the microtubules is called the distal end. The other end is called the proximal end. The term centrosome refers to a composite structure consisting of a centriole surrounded by a cloud of microtubule-nucleating material. Although it is thought that the centriole may help recruit this material to form the mitotic spindle poles, the actual function of centrioles in mitosis is poorly understood. When a centriole structure gives rise to a cilium, it is called a basal body. Attached to the basal bodies are a number of fibrous structures that probably act to link the basal body to the rest of the cytoskeleton. A structure called the basal foot protrudes in a direction correlated with the direction of ciliary beating (Satir and Dirksen, 1985), while a prominent fiber called the striated rootlet protrudes inward toward the cell interior (Hagiwara et al., 1997). Distal fibers called distal and subdistal appendages probably act to attach the basal body to the cell cortex (Ringo, 1967).

Although most eukaryotes, including vertebrates as well as ciliates and green algae, contain centrioles and basal bodies with the canonical nine triplet microtubule blades, some more highly specialized organisms have divergent centriole/basal body structures. Unusual centriole morphology is a characteristic of nematodes such as *Caenorhabditis elegans* as well as insects such as *Drosophila*. In the case of nematodes, the centriole is reduced to a thin disk of singlet microtubules, while in *Drosophila* it is a short ring of doublets. The strange ultrastructures of centrioles in these two organisms is reflected by a highly unusual molecular composition and may be related to the lack of motile cilia in all nematode cells and most *Drosophila* cells. Confirming this idea, when *Drosophila* forms sperm with motile flagella, the basal bodies elongate and acquire a third microtubule in each blade, becoming more like the canonical structures seen in humans or *Chlamydomonas*. Another unusual feature of centrioles in worms and flies is the presence of a central tube running down the middle of the centriole—this is not seen in other species and may represent a specialization of the ecdysozoa.

The complex ultrastructure of the basal body is reflected in a complex protein composition, which is still being determined. A significant number of ciliary disease genes have been found to encode proteins that localize to the basal body. These include oral–facial–digital syndrome gene *OFD1* (Ferrante *et al.*, 2006; Romio *et al.*, 2004), the nephronophthisis protein NPHP-4 (Winklebauer *et al.*, 2005), the cone-rod dystrophy-related genes *RPGR* and *HRG4* (Kobayashi *et al.*, 2000; Shu *et al.*, 2005), the Joubert syndrome protein RPGRIP1L (Arts *et al.*, 2007), and the Meckel syndrome protein MKS1 (Kyttala *et al.*, 2006). Many Bardet-Biedl syndrome proteins accumulate around the basal body as well (Ansley *et al.*, 2003). In most cases, the functional role that these proteins play in basal body assembly and/or function is still a mystery.

As we gradually complete the "parts list," with the aid of systematic proteomic analyses discussed below (Keller *et al.*, 2005; Kilburn *et al.*, 2007), it will become increasingly important to start more mechanistic studies using genetic and biochemical approaches. Perhaps one important first step will be to determine which ultrastructural components of the basal body correspond to which proteins. In this regard, basal body proteins can be classified into three categories. First, there are the core structural components of the basal body, which would include of course tubulin as well as other components of the centriole triplet microtubule blades, for example, tektin (Hinchcliffe and Linck, 1998). The second class would be proteins that are recruited transiently to the basal bodies prior to their transport elsewhere. This would include IFT and BBS proteins, which do not copurify with basal bodies on sucrose gradients and therefore are probably only loosely associated (Keller *et al.*, 2005). The third class are proteins that are permanently associated with the basal body, but that form associated fibrous structures. Basal bodies in different cell types contain a range of different associated fibers and protrusions, mostly of unknown function. Several protein constituents of basal body-associated fibers have been identified using purified algal basal bodies (Geimer *et al.*, 1998a,b; Lechtreck and Melkonian, 1991; Lechtreck *et al.*, 1999). Other important basal body fiber proteins include rootletin, a component of the striated rootlet (Yang *et al.*, 2002) and ODF2/cenexin, a component of the distal appendages (Ishikawa *et al.*, 2005). The high protein complexity and morphological detail of centrioles raises the obvious question of how such a complicated structure is assembled, which will next be addressed.

2.2. Basal body assembly

Basal bodies can arise through two distinct pathways. In the first pathway, a centriole undergoes a maturation process that allows it to dock on the plasma membrane and nucleate formation of cilia. The centrioles arise in this case via the same duplication process that gives rise to new centrioles during division. This apparent duplication of centrioles to produce new centrioles is one of their most interesting biological features. Every cell cycle, each centriole gives rise to a new centriole, which is referred to as its daughter. Normally, new centrioles do not arise *de novo*. However, when pre-existing centrioles are removed from virtually any cell type, they are able to produce new ones *de novo* (Marshall et al., 2001; Uetake et al., 2007) which indicates that the old centriole does not perform any essential function in new centriole formation, but rather exerts an inhibitory influence on the *de novo* assembly pathway. In the absence of *de novo* assembly, however, the centriole duplication mechanism is able to maintain the number of centrioles in a cell at the desired number of two per cell. This is because in a cell that starts with two, each will produce one new one by the duplication pathway, and then one pair of centrioles will associate with each spindle pole. The result is that when a cell divides, each daughter cell inherits two centrioles. Cells have additional control mechanisms to ensure that centrioles only duplicate once per cell cycle (Tsou and Stearns, 2006; Wong and Stearns, 2003), as well as an error correction mechanism that can restore the correct number of two per cell if the number is transiently perturbed (Marshall, 2007). In terms of the basic process of centriole duplication, however, it remains unknown how the mother centriole biases formation of new centrioles. In cells that rely on the canonical duplication pathway, cells generally contain two centrioles. In vertebrate cells, only the older of the two, that is, the "mother" centriole, is capable of acting as a basal body to nucleate cilia, and for this reason these cells have at most one cilium. In the key model system *Chlamydomonas*, both centrioles are capable of acting as basal bodies, so that cells have two flagella, although the molecular basis for this difference is not known. A key outstanding question in basal body biology is to understand the molecular differences between the mother and daughter centrioles, and how these determine the ability to become a basal body.

In the second pathway of basal body formation, a large number of basal bodies form all at once in a large intracellular structure called the deuterosome or generative complex (Dirksen, 1991; Dirksen and Crocker, 1966). A similar process appears to give rise to the basal bodies seen on the sperm cells of lower plants, but in plants the corresponding assembly structure is called a blepharoplast (Hepler, 1976; Klink and Wolniak, 2001; Mizukami and Gall, 1966). Because these structures have mostly been studied at the ultrastructural level, there is not yet much known about their molecular composition. It is therefore hard to say, at this point, whether the underlying mechanism that induces basal body assembly in deuterosomes/blepharoplasts is the same or different from that involved in formation of daughter centrioles from mothers via the normal duplication process (which is also, by the way, not really understood at a molecular level). It has been proposed that the generative complex arises from material produced by a pre-existing centriole (Dirksen, 1991), implying perhaps that mother centrioles still play a key role as they do in normal duplication. There is also likely to be at least some molecular similarity in the two pathways. For example, the EF-hand protein centrin, shown to be required for centriole duplication (Salisbury *et al.*, 2002) is also required for basal body production by blepharoplasts in the fern *Marsilea* (Klink and Wolniak, 2001). Differences in the two pathways would be exceedingly important from a disease perspective, because a gene required for one pathway but not the other could give rise to a ciliary disease syndrome in which motile cilia of multiciliated epithelia were defective, but primary sensory cilia were normal, or vice versa. Understanding the molecular basis of basal body production in multiciliated cells is thus of the very highest priority, and has prompted a number of researchers to revisit this process in mammalian cells (Vladar and Stearns, 2007). It is also interesting to consider why this alternative pathway should exist at all. Why not simply use the normal centriole duplication pathway? One possible explanation could be speed—a multiciliated cell such as those of the trachea or oviduct can have up to 200–300 cilia, requiring formation of an equal number of basal bodies. Since each round of centriole duplication doubles the number of centrioles, to go from 2 to, say, 256, would require seven cycles of duplication, which might simply take too long. Alternatively, it is possible that the basal bodies produced by the generative complex/deuterosome/blepharoplast are structurally different from those produced by normal duplication, a possibility that will only be resolved by comparative proteomic analysis, which has not yet been done.

2.3. Basal body genome concept

For cells with primary cilia, the centrioles that become basal bodies always arise by duplication of pre-existing centrioles. Although centrioles and basal bodies can arise *de novo*, the duplication that is normally seen suggests some propagation of information from one organelle to another. One might be tempted to ask whether these organelles have a genome that is passed on to daughter centrioles during cell division. Several independent lines of evidence had, for a time, suggested that basal bodies might, like mitochondria or chloroplasts, contain their own independent organelle genomes. Staining of ciliates with DNA-binding fluorescent dyes showed rows of dots on the cell surface, suggesting possible association of DNA with basal bodies (Randall and Disbrey, 1965), but because mitochondria are also found on the cortex in a similar arrangement, these analyses were probably just staining the mitochondrial nucleoids. Biochemical analysis showed that cilia cortex preparations enriched for basal bodies contained substantial quantities of DNA, but more careful preparations in which electron microscopy was used to verify loss of mitochondria showed that DNA was also lost, even though basal bodies were retained, thus arguing the ciliate basal bodies probably did not contain DNA (Argetsinger, 1965). In *Chlamydomonas*, genetic studies suggested that an unusually large fraction of cilia-related genes were concentrated on a single, circular chromosome (Hall *et al.*, 1989). Because circular chromosomes are a feature of organelle genomes but not of nuclear chromosomes, this seemed to provide strong genetic evidence for a basal body genome, which was confirmed by *in situ* hybridization showing the chromosome, known as the UNI linkage group, localized at the basal bodies (Hall *et al.*, 1989). Despite this strong set of evidence, it now seems highly unlikely that such a basal body genome exists. First of all, highly sensitive direct visualization experiments using fluorescent dyes, radioactive probes, and DNA-specific antibodies, all failed to detect any DNA in basal bodies (Johnson and Rosenbaum, 1990; Kuroiwa *et al.*, 1990; Pyne, 1968). Moreover, the circularity of the UNI linkage group was an artifact that resulted from mismapping of several key markers and in fact the chromosome is linear (Holmes *et al.*, 1993). The apparent concentration of cilia-related genes on that chromosome was also apparently a mistake, and current genetic maps show no particular bias for cilia-related genes to be on one particular chromosome. The *in situ* hybridization result showing that this one chromosome localized to the basal bodies was also an artifact, resulting from protease digestion of the cells prior to fixation, because *in situ* experiments done without cell digestion show the chromosome is within the nucleus (Hall and Luck, 1995). We are thus forced to conclude that while the basal body has many interesting features, a self-contained genome is not one of them.

3. Basal Body Functions in Ciliogenesis

3.1. Basal body function in templating axoneme

Basal bodies are strictly required for formation of cilia—there are no known cases in which cilia can form without a basal body being present. Although there is one documented case of a cilium that lacks a basal body, that is, in the green alga *Chlorogonium*, in this case the basal body is present during ciliogenesis, and then subsequently detaches from the cilium which continues to beat (Hoops and Witman, 1985). The requirement of basal bodies for ciliogenesis thus appears to be absolute. Why is this? The key function of the basal body in ciliogenesis is most likely to provide the template for the formation of the axoneme—the array of nine microtubule doublets that gives structural form and rigidity to the cilium. In electron micrographs, it is clearly seen that the outer doublet microtubules of the axoneme are contiguous with the A and B tubules of the triplets of the basal body (Ringo, 1967). Isolated basal bodies can nucleate formation of microtubule growth off of the ends of their triplets (Snell *et al.*, 1974). Presumably, the ability of the basal body triplets to directly template the axonemal doublets allows the ninefold symmetry of the basal body to be propagated into the axoneme. The propagation of geometry from the basal body to the cilium is supported by recent genetic experiments in *Chlamydomonas*. The *Chlamydomonas* bld12 mutant has a deletion in the conserved centriole/basal body protein SAS6, and results in basal bodies forming with variable numbers of triplets. Corresponding to the variability in triplet number in the basal bodies of bld12 mutants, variability is also seen in the number of doublets in the flagellar axonemes, which, given the fact that SAS6 protein localizes to the basal bodies but not the flagella, strongly implies numerical control of doublets by the basal body (Nakazawa et al., 2007). The distribution of doublet numbers does not exactly match the distribution of triplet numbers, which complicates the interpretation somewhat. One possible explanation is that the axoneme has a statistical preference to have ninefold symmetry, so that the distribution of doublet numbers is constrained compared to the distribution of triplet numbers. A second *Chlamydomonas* genetic experiment, using the *BLD10* gene, gives a similar result. The BLD10 protein localizes to the cartwheel spoke, and a bld10 null mutation mostly eliminates recognizable basal bodies. Rescue of the bld10 null mutation with a truncated construct of the *BLD10* gene that produces a shorter protein, leads to formation of centrioles with eight, instead of nine, triplets. These cells produce some flagella with eight doublets (Hiraki et al., 2007). These data hint that the ninefold symmetry of the axoneme is imposed by the ninefold symmetry of the basal body. To really establish a correlation, however, it would be desirable to use serial section microscopy or tomography to determine the basal body triplet number and the axoneme doublet number for the same basal body, and ask whether there is a strict correlation.

It is also a formal possibility that the axoneme can self-organize a ninefold symmetry and then propagate this to the basal bodies. This could be ruled out by examining the distribution of triplet number in doublet mutants of bld12 and a second mutation that prevents flagellar assembly. This alternative possibility seems relatively far-fetched, although it is worth pointing out that expression of alternative tubulin isoforms in *Drosophila* spermatogenesis can produce sperm flagella with extra doublets without having any effect on triplet number in the basal bodies (Raff et al., 2000). In these experiments it was noted that the doublet number only becomes abnormal distally from the transition zone, and that near the basal body, the axoneme still has the normal ninefold symmetry. These results taken together do seem to imply that the axoneme can have a self-organizing propensity to form a particular number of doublets, nine in wild-type and ten in tubulin isoform altered mutations, and that the basal body can propagate symmetry information to the axoneme to influence the doublet number, but this influence only extends for a certain distance along the length of the axoneme. Another case of an axoneme with abnormal numbers of doublets is in the Gregarine protozoan *Lecudina tuzetae*, whose flagellum contains an axoneme with only six microtubule doublets (Schrevel and Besse, 1975). In this case, the basal body was highly abnormal and did not show clear-cut triplet blades. Most regrettably, it was not possible to determine, from the electron micrographs, exactly what fold symmetry the reduced basal body in this organism has. It would be of great interest to know whether the *Lecudina* basal body retained ninefold symmetry or if it had a sixfold symmetry matching the axoneme it produces.

3.2. Basal body functions in attaching and orienting cilium at cortex

During ciliogenesis, the basal body moves to the cell surface and integrates itself into the cell cortex. The physical driving force for this movement is not entirely clear, in some cases an actin-rich structure has been reported trailing behind the basal body that is reminiscent of the actin "comet tails" that power *Listeria* motility (Tamm and Tamm, 1988). On the other hand, Brownian motion may be fast enough by itself. Estimates for the viscosity of cytoplasm range over several orders of magnitude, hence it is currently impossible to calculate a priori the effective diffusion constant of a basal body. In any case, as basal bodies approach the cell surface, they become associated with a membrane-bound vesicle (Sorokin, 1968) of unknown origin. As the basal body–vesicle complex reaches the surface, a hole is made in the actin cortex and the basal body-associated vesicle fuses with the plasma membrane, thus giving rise to the primordium of the ciliary membrane. The topology of this fusion event, based on electron micrographs (Sorokin, 1968), is such that the interior of the vesicle becomes the exterior of the ciliary membrane. During basal body attachment to the cortex, the distal connecting fibers/appendages make contact with the plasma membrane. How, or even whether, these fibers have a physical association with the lipid or with plasma membrane-associated proteins is not currently known, although the fact that one protein component of the fibers, p210, shows homology with a clathrin adaptor protein (Lechtreck *et al.*, 1999) is certainly suggestive of the idea that the basal body–membrane interaction is somehow related to clathrin-coated vesicle formation.

The site on the cell surface at which the basal body associates is probably not random, but instead can be dictated by intracellular polarity cues (Montcouquiol *et al.*, 2003). At least one factor involved in specifying basal body insertion site is a local enrichment of actin which is activated by the Foxj1 factor (Pan *et al.*, 2007), known to be required for basal body surface docking (Gomperts *et al.*, 2004). The precise site of surface insertion can have profound effect on the function of the cilia that the basal body will nucleate. For example, sensory cilia in epithelial cells lining a duct must project into the lumen of the duct tube. This requires that the basal body associates with the portion of the plasma membrane facing the lumen. If the basal body were to associate with a nonlumenal portion of the plasma membrane, the cilium would not be able to sense conditions within the duct itself. Pathological conditions have been reported where cilia form within intracellular vacuoles rather than on the cell surface (Hagiwara et al., 2000), and again this would prevent the cilia from sensing the proper environment.

Proper placement in different regions of the cell surface is also important for motile cilia. A comparatively simple example is seen in the mouse node. The node is the site of left–right symmetry breaking during early embryonic development. The cells of the node have motile cilia, but the cilia do not beat back and forth like normal cilia but instead undergo a partially rotational motion, more like an eggbeater. This rotational motion drives a leftward flow of fluid over the surface of the node (Nonaka et al., 1998), and it has been demonstrated that this leftward flow is crucial for left–right symmetry breaking (Nonaka et al., 2002). Simple fluid mechanics predicts that a rotational motion will produce a circulation of fluid around the node rather than the linear flow across the node surface that has been observed. This puzzle can be resolved if the axis of cilia rotation is tilted toward the posterior, in which case surface effects will allow the clockwise rotation of the cilia to drive a leftward-directed linear flow (Nonaka et al., 2005). The posterior-directed tilt, in turn, is ultimately achieved by a posterior-directed displacement of the basal bodies (Nonaka et al., 2005). This works because the surface of the cells is curved into a roughly hemispherical shape, so that if the basal body were in the middle of the cell, the cilium would point straight up, but if the basal body is shifted toward the posterior, the cilium points out at an angle relative to the plane of the node (Nonaka et al., 2005). Thus, basal body position can have profound effects on even very large-scale features of development such as overall determination of body situs.

Basal bodies also dictate the direction in which motile cilia will beat (Hoops et al., 1984; Tamm et al., 1975), thus the rotational orientation of the basal body is critical to ensure that cilia-driven flows go in the correct direction. When the basal body first associates with the cortex, its rotational orientation is random, so that the ciliary doublets face in random directions. For motile cilia, this would mean that each cilium in a tissue would generate flow in a different direction, which would prevent formation of a coherent directed flow. About the time that ciliary motility begins, the basal bodies become rotationally aligned (Boisvieux-Ulrich and Sandoz, 1991; Boisvieux-Ulrich et al., 1985). The mechanisms that bring this alignment about are currently not known, but they must somehow act to couple planar cell polarity cues with forces exerted on the basal bodies to turn them in the right direction. In addition to a likely contribution of planar polarity pathways to basal body orientation, perhaps via actin/myosin-mediated force generation (Boisvieux-Ulrich et al., 1990; Lemullois et al., 1987) it has also been shown that cilia-driven fluid flows themselves can exert a powerful organizing influence on basal body orientation (Mitchell et al., 2007). In the case of nonmotile cilia, it is not known whether their basal bodies have a defined rotational orientation. They might, at least in the case of mechanosensory cilia, which could give the cilia the ability to discriminate forces applied in different directions, but this has yet to be tested experimentally.

Basal bodies thus perform two important tasks above and beyond simply initiating formation of the axonemal doublets. First, they bring the site of ciliary assembly to the cell surface, so that when the cilium forms it extends away from the surface of the cell, and second, they define the orientation of the cilium about its long axis. Both of these functions are critical for proper ciliary function, and defect in either process are predicted to lead to defects in physiology or development. It thus seems likely that a subclass of ciliary diseases might be caused by defects in basal body functions related to cortex association. This might be quite difficult to diagnose using current tests, especially in the case of a defect that results in randomization of rotational orientation. Many instances of randomly oriented cilia have been reported in patients with immotile cilia syndrome, however there remains a raging controversy over whether these defects are a cause or a consequence of the motility defects (Jorissen and Willems, 2004; Rayner *et al.*, 1996). In at least some cases, it seems clear that the disorientation is a secondary effect, but this by no means proves that this would be true in all reported cases.

3.3. Basal body functions in regulating protein import into cilium

The protein composition of the cilium is distinct from that of the cytoplasm, hence a mechanism is required to regulate selective import of ciliary proteins. This mechanism, whatever its molecular basis, should be functionally similar to the nuclear pore, and would have to be located at the basal body, since the basal body is the "last stop" in the cytoplasm before entering the cilium. The use of the basal body as a docking site to recruit proteins for import into the cilium also means that the basal body can exert control over the precise composition of the cilium that it nucleates. It is thus possible to have, in a single cell, more than one distinct type of cilium, determined by different basal bodies. This has been documented extensively in heterokont algae, in which basal bodies appear to go through a series of maturation steps synchronized with cell division, such that basal bodies of different ages produce flagella that are functionally and structurally different (Beech *et al.*, 1988; Heimann *et al.*, 1989). In these types of alga, several distinct types of flagella are seen. Some flagella contain long hair-like projections called mastigonemes projecting from their surface, while others are smooth and lack these projections. Some of the flagella execute large bending motions while others are either nonmotile or bend with a more symmetric waveform. The differences in motility and surface composition correlate strictly with the age of the basal body that nucleates the flagella. These studies have shown that basal bodies can go through a series of discrete maturation steps, with a basal body counting how many cell divisions since its initial assembly, and nucleating different types of flagella as a function of its age. This is one of the most clear-cut examples of cellular aging, but its molecular basis is entirely unknown.

From a thermodynamic perspective, the challenge of selective import is to create a barrier that is solid enough to prevent thermally driven nonselective crossing of random proteins while fluid enough to allow crossing of desired proteins. One possible way to regulate protein import into the cilium would be to exploit the IFT machinery to provide a driving force to push ciliary proteins across a barrier, such that proteins that cannot interact with the IFT system are left behind. Consistent with this idea, many protein components of the IFT machinery, as well as proteins involved in handoff of cargo the IFT system, accumulate around the basal bodies (Cole et al., 1998; Stephan et al., 2007), and IFT52 protein has been specifically localized to the transitional fibers by immunoelectron microscopy (Deane et al., 2001). This type of model would suggest that IFT exists not just to move proteins along the cilium, but also to get proteins into the cilium in the first place. The main apparent problem with this model is the fact that when IFT is inactivated using conditional mutants in *Chlamydomonas*, resulting in disassembly of the flagellum, the flagellar proteins are not left behind, but are returned to the cytoplasm. This demonstrates that IFT is not required to provide a driving force to allow proteins to cross the ciliary import diffusional barrier. This is an indirect argument, however, and it would be highly desirable to develop imaging approaches to monitor individual protein import events at the ciliary base.

3.4. Basal body function in spindle orientation

During mitosis, basal bodies usually detach from the cell surface and move to the interior of the cell, where they act as nucleating centers to form the centrosome. In this capacity, the basal bodies are called centrioles due to their central position in the spindle pole. The equivalence between basal bodies and centrioles was recognized by Henneguy and van Lenhossek but the teleological purpose in using the same structure in both contexts remains obscure to this day. Indeed, the function of centrioles in mitosis is still not entirely clear and there has been considerable debate about whether they play any role in mitosis at all. Much of the confusion arose because nobody was able to imagine that a structure as complex as the spindle could form by itself, and so the idea arose early on that the centrioles would act as nucleators to initiate spindle pole formation. When it was subsequently found that spindle could form in cells lacking centrioles, the pendulum of public opinion swung the other way, and it was generally inferred that centrioles had no role in spindle formation.

The current data can probably best be summarized in two points. The first point is that we now know bipolar spindles can self-organize when centrioles or centrosomes are absent. This is based on many experimental studies but is most dramatically demonstrated using DNA-coated beads in Xenopus extracts, which robustly form bipolar spindle structures (Heald et al., 1996). This self-organization can be recapitulated in computer simulations and seems to be an intrinsic result of the combined action of opposing motor proteins that drive microtubule rearrangements until a stable configuration is reached. This stable configuration happens to be the bipolar spindle geometry. As a result, cells from which centrioles are removed can in many cases still organize microtubule-based bipolar spindles (e.g., de Saint Phalle and Sullivan, 1998).

The second key point is that, as with many other self-organizing systems, spindle assembly is subject to spatial regulation by biasing inputs. It is a common theme in developmental biology that spontaneous symmetry breaking systems, which left to their own devices will organize along a random axis, are usually biased by some simple upstream cue to ensure that they break symmetry in the desired direction in all embryos. In the case of the spindle, it appears that centrioles, when present, exert a biasing input that drives spindle self-organizing in such a way that the poles form near the position of the centrioles.

The net result of these two points is that the main contribution centrioles make to mitosis is not to allow spindle formation, but rather to specify WHERE the spindle will form. The effect of the centrioles is not permissive, but instructive. Defects in centrioles are thus expected to give rise to randomization of spindle orientation, and this has been reported in analysis of basal body mutants in Chlamydomonas (Ehler et al., 1995). This, in turn, can lead to abnormal development of tissues. For example, oriented mitosis appears to play a key role in kidney tubule morphogenesis, and alteration of spindle orientation may directly cause formation of kidney cysts (Fischer et al., 2006; Simons and Walz, 2006). One would thus expect that defects in centriole positioning or function could cause cystic kidney diseases that have nothing to do with sensory cilia function. This may explain the prevalence of nephronophthisis gene products within the centriole/basal body proteome (Keller et al., 2005). One key question is whether sensory inputs to the basal body during interphase, when it is attached to a cilium, can bias the subsequent position of the spindle. The fact that the EB1 protein, which tracks the plus ends of growing microtubules, is found at basal bodies (Pedersen et al., 2003), suggests a possible mechanism for linking basal body position and/or activity to the organization of the microtubule-based cytoskeleton.

4. Approaches to the Study of Basal Bodies

4.1. Proteomics

Basal bodies are among the most complex structures in all of cell biology, and are of indisputably great importance for understanding ciliogenesis and cilia-related diseases, but our understanding of their composition, assembly, and function currently lags behind that of most other organelles. Probably the most fundamental question one can ask about an organelle is what proteins it contains. Although a number of proteins have been found to localize to the basal body, there is a clear need for a systematic cataloging of all basal body-associated proteins. Purification of organelles for proteomic analysis is always a challenging problem, and it is particularly difficult for centrioles that, in vertebrate cell culture systems, are embedded within a large mass of microtubule-nucleating material including gamma tubulin ring complexes, pericentrin (Dictenberg *et al.*, 1998), etc. As a result, proteomic analysis of isolated centrosomes reveals a huge list of proteins, only a fraction of which are bona fide centriole or basal body components (Andersen *et al.*, 2003). Selective proteomic analysis therefore requires cell types in which the basal body exists in a naked state, free from pericentriolar material. One example of such a cell is *Chlamydomonas reinhardtii*, and this particular advantage of *Chlamydomonas* was exploited to obtain the first direct proteome of isolated basal bodies (Keller *et al.*, 2005). Another cell type that provides excellent starting material is the cilia *Tetrahymena*, and a careful analysis of the *Tetrahymena* basal body proteome has been published (Kilburn *et al.*, 2007). These two studies identified dozens of basal body-specific proteins.

4.2. Comparative genomics

Another approach to identifying the basal body "parts list" is to use comparisons between completed genomes to identify genes conserved in species that have basal bodies but missing from species that do not. Several labs have presented comparisons of this type (Avidor-Reiss *et al.*, 2004; Li *et al.*, 2004). Of course, since the set of species that have basal bodies is exactly the same as the set of species that have cilia, the resulting set of conserved genes will include both basal body and cilia-specific protein-encoding genes. Nevertheless, this type of approach is quite interesting and harnesses the incredible explosion of genomic sequence data appearing in recent years. One point that must be clarified is that while one of these studies called the set of conserved genes the "flagellar and basal body proteome" (FABP), it must be emphasized that the use of the term "proteome" is potentially misleading—the studies did not in any way shape or form identify a proteome, but rather a set of genes. In fact, the comparative genomics studies and proteomics studies are complementary, rather than equivalent, and this is a good thing. Proteomic analysis directly shows the components of the basal body. Comparative genomics can reveal genes whose function may be required for basal body assembly or activity but which do not encode components of the basal body itself. Therefore, both types of data are of use.

4.3. RNAi-mediated knockdown of basal body-related genes

With lists of candidate basal body proteins gradually growing, interesting results have been achieved by knocking down expression of the corresponding genes using RNAi. RNAi-mediated knockdown of basal body protein-encoding genes requires careful choice of model system. RNAi is a standard tool in *C. elegans* but unfortunately, the ciliated cells in this organism are, for some reason refractory to RNAi-mediated knockdown. This has prevented the powerful genome-wide RNAi technologies possible in nematodes from being brought to bear on questions of cilia or basal bodies. In *Drosophila*, genome-wide RNAi is conveniently accomplished in S2 tissue culture cells, however these cells lack cilia and their centrioles do not form basal bodies, thus this other powerful genome-wide RNAi system also has not been applicable to the study of basal bodies. RNAi has mostly been applied in vertebrate tissue culture cells that have nonmotile primary cilia (e.g., IMCD-3 or RPE-1 cells). Such cells also have the advantage that, as vertebrate cells, their basal bodies have the canonical triplet microtubule structure as opposed to the more aberrant structures of *Drosophila* or *C. elegans*. However, these cells all form their basal bodies from pre-existing centrioles, so while they are excellent systems to study the basal bodies of primary cilia, they cannot address the special assembly pathways seen in multiciliated cells. Nor would they allow testing of any basal body components related to ciliary motility, since these cells only have nonmotile cilia. It was recently shown to be possible to perform lentiviral-mediated RNAi in mouse tracheal epithelial cells recently (Vladar and Stearns, 2007). This is an exceedingly important step forward because these cells form basal bodies via the deuterosome-mediated pathway seen in other multiciliated cells.

4.4. Genetic analysis of basal bodies

However RNAi often fails to achieve complete knockdown, and can give highly variable results between cells. Ultimately it can be much more convenient for mechanistic studies to have bona fide mutants. Forward genetic analysis of basal bodies has been carried out primarily in the green alga *Chlamydomonas* (Lefebvre and Silflow, 1999), which has many of the same genetic advantages as budding yeast, especially growth in the haploid state which allows phenotypes to be immediately revealed without the need for backcrossing. Due to these advantages, an entire screen can be completed in *Chlamydomonas* in a tiny fraction of the time needed to perform a similar screen in more complicated, slow growing organisms such as flies or worms. Probably for this reason, most all of the known mutations affecting basal body structure and function have been first identified in *Chlamydomonas*. Because *Chlamydomonas* uses its flagella to swim, and swimming motility is an easily scored phenotype, it has been possible to conduct large-scale screens with minimal effort, and these have revealed many types of mutant phenotypes involving basal body defects. These include mutants that alter basal body length (Goodenough and StClair, 1975), symmetry (Hiraki *et al.*, 2007; Nakazawa *et al.*, 2007), number (Kuchka and Jarvik, 1982; Wright *et al.*, 1983), and positioning (Feldman *et al.*, 2007). The mutants are named after the phenotype observed in the motility screens, and include the BLD mutants, which lack flagella entirely, and the VFL mutants which have variable numbers of flagella due to defects in basal body number regulation.

One particularly interesting class of mutants affects the maturation rate of centrioles into basal bodies, so that instead of a newly formed centriole being able to make a flagellum in the next cycle, an additional delay of one cell cycle is imposed, such that cells have one flagellum rather than two, despite having two basal bodies (Dutcher and Trabuco, 1998; Huang *et al.*, 1982; Piasecki *et al.*, 2008). This type of mutation converts the *Chlamydomonas* pattern of basal body differentiation, in which each centriole acts as a basal body, to the pattern seen in vertebrate cells, where only the mother centriole acts as a basal body. This so-called UNI phenotype may thus point the way to understanding the cellular aging of basal bodies that was commented on above. In addition to providing useful genetic tools for probing the pathways that regulate basal bodies, these mutant screens, because they are conducted in an unbiased way, have proven to be an important way to identify key basal body proteins. Examples of genes and proteins first identified via genetic screens in *Chlamydomonas* include delta and epsilon tubulin (products of the *UNI3* and *BLD2* genes, respectively).

Although most studies in mammalian cells have relied on RNAi, it is also possible to create actual mutants in such cells. For example, somatic cell gene knockout methods were used to show that the *OFD2* gene, which encodes a basal body protein, is required for ciliogenesis in mouse cells (Ishikawa *et al.*, 2005).

4.5. Imaging

Basal bodies and centrioles have long been a favorite of electron microscopists, and indeed electron microscopy has been the key to understanding centriole/basal body structure (e.g., Johnson and Porter, 1968; Ringo, 1967). Recent advances in electron tomography have allowed three-dimensional images of basal bodies to be determined (O'Toole *et al.*, 2003). But while electron microscopy can reveal fine features, it suffers from two limitations. First, it cannot be used to study live cells since the sample has to be either chemically fixed or frozen in vitreous ice. Second, localization of proteins by EM is difficult, the usual method of immunogold localization cannot localize proteins precisely at high resolution due to the size of the gold beads and their distance from the actual protein. For both these reasons, fluorescence microscopy is used as an alternative to EM when *in vivo* imaging or protein localization is necessary. Unfortunately, the important features of this structure tend to be in the size scale of 10–100 nm, which is below the resolving power of traditional light microscopy. This means that light microscope images of centrioles almost always resemble round dots of light, with no real indication of substructure. New developments in optical technology have begun to circumvent the traditional resolution limits. Methods such as structured illumination (Gustafsson *et al.*, 2008) and STORM/PALM (Huang *et al.*, 2008) can bring the resolution of the light microscope down to the tens of nanometers range, ideal for studying basal bodies.

5. Conclusions

Although the basal body is one of the first subcellular structures to be observed by the early cytologists, it is only in the past decade that a combination of genetics, proteomics, and other experimental approaches have begun to shed light on how these complicated structures assemble and function. The crucial role that basal bodies play in so many aspects of ciliary biology means that these structures will continue to be the objects of intense study.

REFERENCES

Andersen, J. S., Wilkinson, C. J., Mayor, T., Mortensen, P., Nigg, E. A., and Mann, M. (2003). Proteomic characterization of the human centrosome by protein correlation profiling. *Nature* **426,** 570–574.

Ansley, S. J., Badano, J., Blacque, O. E., Hill, J., Hoskins, B. E., Leitch, C. C., Kim, J. C., Ross, A. J., Eihers, E. R., Teslovich, T. M., *et al.* (2003). Basal body dysfunction is a likely cause of pleiotropic Bardet-Biedl syndrome. *Nature* **425,** 628–633.

Argetsinger, J. (1965). The isolation of ciliary basal bodies (kinetosomes) from *Tetrahymena pyriformis*. *J. Cell Biol.* **24,** 154–157.

Arts, H. H., Doherty, D., van Beersum, S. E., Parisi, M. A., Lettboer, S. J., Gorden, N. T., Peters, T. A., Maerker, T., Voesenek, K., Kartono, A., Ozurek, H., Farin, F. M., *et al.* (2007). Mutations in the gene encoding the basal body protein RPGRIP1L, a nephrocystin-4 interactor, cause Joubert syndrome. *Nat. Genet.* **39,** 875–881.

Avidor-Reiss, T., Maer, A. M., Kouindakjian, E., Polyanovsky, A., Keil, T., Subramanian, S., and Zuker, C. S. (2004). Decoding cilia function: Defining specialized genes required for compartmentalized cilia biogenesis. *Cell* **117,** 527–539.

Beech, P. L., Wetherbee, R., and Pickett-Heaps, J. D. (1988). Transformation of the flagella and associated flagellar components during cell division in the coccolithophorid *Pleurochrysis carterae*. *Protoplasma* **145**, 37–46.

Boisvieux-Ulrich, E., and Sandoz, D. (1991). Determination of ciliary polarity precedes differentiation in the epithelial cells of quail oviduct. *Biol. Cell* **7**, 23–24.

Boisvieux-Ulrich, E., Laine, M. C., and Sandoz, D. (1985). The orientation of ciliary basal bodies in quail oviduct is related to the ciliary beating cycle commencement. *Biol. Cell* **55**, 147–150.

Boisvieux-Ulrich, E., Lainé, M. C., and Sandoz, D. (1990). Cytochalasin D inhibits basal body migration and ciliary elongation in quail oviduct epithelium. *Cell Tissue Res.* **259**, 443–454.

Cole, D. G., Diener, D. R., Himelblau, A. L., Beech, P. L., Fuster, J. C., and Rosenbaum, J. L. (1998). *Chlamydomonas* kinesin-II-dependent intraflagellar transport (IFT): IFT particles contain proteins required for ciliary assembly in *Caenorhabditis elegans* sensory neurons. *J. Cell Biol.* **141**, 993–1008.

Deane, J. A., Cole, D. G., Seeley, E. S., Diener, D. R., and Rosebaum, J. L. (2001). Localization of intraflagellar transport protein IFT52 identifies basal body transitional fibers as the docking site for IFT particles. *Curr. Biol.* **11**, 1586–1590.

de Saint Phalle, B., and Sullivan, W. (1998). Spindle assembly and mitosis without centrosomes in parthenogenetic Sciara embryos. *J. Cell Biol.* **141**, 1383–1391.

Dictenberg, J. B., Zimmerman, W., Sparks, C. A., Young, A., Vidair, C., Zheng, Y., Carrington, W., Fay, F. S., and Doxsey, S. J. (1998). Pericentrin and gamma-tubulin form a protein complex and are organized into a novel lattice at the centrosome. *J. Cell Biol.* **141**, 163–174.

Dirksen, E. R. (1991). Centriole and basal body formation during ciliogenesis revisited. *Biol. Cell* **72**, 31–38.

Dirksen, E. R., and Crocker, T. T. (1966). Centriole replication in differentiating ciliated cells of mammalian respiratory epithelium. An electron microscopic study. *J. Microsc.* **5**, 629–644.

Dutcher, S. K., and Trabuco, E. C. (1998). The UNI3 gene is required for assembly of basal bodies of *Chlamydomonas* and encodes δ-tubulin, a new member of the tubulin superfamily. *Mol. Biol. Cell* **9**, 1293–1308.

Ehler, L. L., Holmes, J. A., and Dutcher, S. K. (1995). Loss of spatial control of the mitotic spindle apparatus in a *Chlamydomonas reinhardtii* mutant strain lacking basal bodies. *Genetics* **141**, 945–960.

Feldman, J. L., Geimer, S., and Marshall, W. F. (2007). The mother centriole plays an instructive role in defining cell geometry. *PLoS Biol.* **5**, e149.

Ferrante, M. I., Zullo, Z., Barra, A., Bimonte, A., Messaddeq, N., Studer, M., Dolle, P., and Franco, B. (2006). Oral–facial–digital type I protein is required for primary cilia formation and left–right axis specification. *Nat. Genet.* **38**, 112–117.

Fischer, E., Legue, E., Doyen, A., Nato, F., Nicolas, J. F., Torres, V., Yaniv, M., and Pontoglio, M. (2006). Defective planar cell polarity in polycystic kidney disease. *Nat. Genet.* **38**, 21–23.

Geimer, A., Lechtreck, K. F., and Melkonian, M. (1998a). A novel basal apparatus protein of 90 kD (BAp90) from the flagellate green alga *Spermatozopsis similis* is a component of the proximal plates and identifies the d-(dexter) surface of the basal body. *Protist* **149**, 173–184.

Geimer, S., Clees, J., Melkonian, M., and Lechtreck, K. F. (1998b). A novel 95-kD protein is located in a linker between cytoplasmic microtubules and basal bodies in a green flagellate and forms striated filaments *in vitro*. *J. Cell Biol.* **140**, 1149–1158.

Gomperts, B. N., Gong-Cooper, X., and Hackett, B. P. (2004). Foxj1 regulates basal body anchoring to the cytoskeleton of ciliated pulmonary epithelial cells. *J. Cell Sci.* **117**, 1329–1337.

Goodenough, U. W., and StClair, H. S. (1975). BALD-2: A mutation affecting the formation of doublet and triplet sets of microtubules in *Chlamydomonas reinhardtii*. *J. Cell Biol.* **66**, 480–491.

Gustafsson, M. G., Shao, L., Carlton, P. M., Wang, C. J., Golubovskaya, I. N., Cande, W. Z., Agard, D. A., and Sedat, J. W. (2008). Three-dimensional resolution doubling in wide-field fluorescence microscopy by structured illumination. *Biophys. J.* **94**, 4957–4970.

Hagiwara, H., Aoki, T., Ohwada, N., and Fujimoto, T. (1997). Development of striated rootlets during ciliogenesis in the human oviduct epithelium. *Cell Tissue Res.* **290**, 39–42.

Hagiwara, H., Ohwada, N., Aoki, T., and Takata, K. (2000). Ciliogenesis and ciliary abnormalities. *Med. Electron Microsc.* **33**, 109–114.

Hall, J. L., and Luck, D. J. L. (1995). Basal body-associated DNA: *In situ* studies in *Chlamydomonas reinhardtii*. *Proc. Natl Acad. Sci. USA* **92**, 5129–5133.

Hall, J. L., Ramanis, Z., and Luck, D. J. L. (1989). Basal body/centriolar DNA: Molecular genetic studies in *Chlamydomonas*. *Cell* **59**, 121–132.

Heald, R., Tournebize, R., Blank, T., Sandaltzopoulos, R., Becker, P., Human, A., and karsenti, E. (1996). Self-organization of microtubules into bipolar spindles around artificial chromosomes in *Xenopus* egg extracts. *Nature* **382**, 420–425.

Heimann, K., Reize, I. B., and Melkonian, M. (1989). The flagellar developmental cycle in algae: Flagellar transformation in *Cyanophora paradoxa*. *Protoplasma* **148**, 106–110.

Hepler, P. K. (1976). The blepharoplast of *Marsilea*: Its *de novo* formation and spindle association. *J. Cell Sci.* **21**, 361–390.

Hinchcliffe, E. H., and Linck, R. W. (1998). Two proteins isolated from sea urchin sperm flagella: Structural components commons to the stable microtubules of axonemes and centrioles. *J. Cell Sci.* **111**, 585–595.

Hiraki, M., Nakazawa, Y., Kamiya, R., and Hirono, M. (2007). Bld10p constitutes the cartwheel-spoke tip and stabilizes the 9-fold symmetry of the centrioles. *Curr. Biol.* **17**, 1778–1783.

Holmes, J. A., Johnson, D. E., and Dutcher, S. K. (1993). Linkage group XIX of *Chlamydomonas reinhardtii* has a linear map. *Genetics* **133**, 865–874.

Hoops, H. J., and Witman, G. B. (1985). Basal bodies and associated structures are not required for normal flagellar motion or phototaxis in the green alga *Chlorogonium elongatum*. *J. Cell Biol.* **100**, 297–309.

Hoops, H. J., Wright, R. J., Jarvik, J. W., and Witman, G. B. (1984). Flagellar waveform and rotational orientation in a *Chlamydomonas* mutant lacking normal striated fibers. *J. Cell Biol.* **98**, 818–824.

Huang, B., Ramanis, Z., Dutcher, S. K., and Luck, D. J. (1982). Uniflagellar mutants of *Chlamydomonas*: Evidence for the role of basal bodies in transmission of positional information. *Cell* **29**, 745–753.

Huang, B., Wang, W., Bates, M., and Zhuang, X. (2008). Three-dimensional super-resolution imaging by stochastic optical reconstruction microscopy. *Science* **319**, 810–813.

Ishikawa, H., Kubo, A., Tsukita, S., and Tsukita, S. (2005). Odf2-deficient mother centrioles lack distal–subdistal appendages and the ability to generate cilia. *Nat. Cell Biol.* **7**, 517–524.

Johnson, U. G., and Porter, K. R. (1968). Fine structure of cell division in *Chlamydomonas reinhardi*. Basal bodies and microtubules. *J. Cell Biol.* **38**, 403–425.

Johnson, K. A., and Rosenbaum, J. L. (1990). The basal bodies of *Chlamydomonas reinhardtii* do not contain immunologically detectable DNA. *Cell* **62**, 615–619.

Jorissen, M., and Willems, T. (2004). The secondary nature of ciliary disorientation in secondary and primary ciliary dyskinesia. *Acta Otolaryngol.* **124**, 527–531.

Keller, L. C., Romijn, E. P., Zamora, I., Yates, J. R., and Marshall, W. F. (2005). Proteomic analysis of isolated *Chlamydomonas* centrioles reveals orthologs of ciliary disease genes. *Curr. Biol.* **15,** 1090–1098.

Kilburn, C. L., Pearson, C. G., Romijn, E. P., Meehl, J. B., Giddings, T. H., Culver, B. P., Yates, J. R., and Winey, M. (2007). New *Tetrahymena* basal body protein components identify basal body domain structure. *J. Cell Biol.* **178,** 905–912.

Klink, V. P., and Wolniak, S. M. (2001). Centrin is necessary for the formation of the motile apparatus in spermatids of *Marsilea*. *Mol. Biol. Cell* **12,** 761–776.

Kobayashi, A., Higashide, T., Hamasaki, D., Kubota, S., Sakuma, H., An, W., Tujimaki, T., McLaren, M. J., Weleber, R. G., and Inana, G. (2000). HRG (UNC119) mutation found in cone-rod dystrophy causes retinal degeneration in a transgenic model. *Invest. Ophthalmol. Vis. Sci.* **41,** 3268–3277.

Kuchka, M. R., and Jarvik, J. W. (1982). Analysis of flagellar size control using a mutant of *Chlamydomonas reinhardtii* with a variable number of flagella. *J. Cell Biol.* **92,** 170–175.

Kuroiwa, T., Yorihuzi, T., Yabe, N., Ohta, T., and Uchida, H. (1990). Absence of DNA in the basal body of *Chlamydomonas reinhardtii* by fluorimetry using a video-intensified microscope photon counting system. *Protoplasma* **158,** 155–164.

Kyttala, M., Tallila, J., Salonen, R., Kopra, O., Kohlschmidt, N., Paavola-Sakki, P., Peltonen, L., and Kestila, M. (2006). MKS1, encoding a component of the flagellar apparatus basal body proteome, is mutated in Meckel syndrome. *Nat. Genet.* **38,** 155–157.

Lechtreck, K. F., and Melkonian, M. (1991). Striated microtubule-associated fibers: Identification of assemblin, a novel 34-kD protein that forms paracrystals of 2-nm filaments in vitro. *J. Cell Biol.* **115,** 705–716.

Lechtreck, K. F., Teltenkoetter, A., and Grunow, A. (1999). A 210 kDa protein is located in a membrane–microtubule linker at the distal end of mature and nascent basal bodies. *J. Cell Sci.* **112,** 1633–1644.

Lefebvre, P. A., and Silflow, C. D. (1999). *Chlamydomonas*: The cell and its genomes. *Genetics* **151,** 9–14.

Lemullois, M., Klotz, C., and Sandoz, D. (1987). Immunocytochemical localization of myosin during ciliogenesis of quail oviduct. *Eur. J. Cell Biol.* **43,** 429–437.

Li, J. B., Gerdes, J. M., Haycraft, C. J., Fan, Y., Teslovich, T. M., May-Simera, H., Li, H., Blacque, O. E., Li, L., Leitch, C. C., Lewis, R. A., Green, J. S., *et al.* (2004). Comparative genomics identifies a flagellar and basal body proteome that includes the BBS5 human disease gene. *Cell* **117,** 541–552.

Marshall, W. F. (2007). Stability and robustness of an organelle number control system: Modeling and measuring homeostatic regulation of centriole abundance. *Biophys. J.* **93,** 1818–1833.

Marshall, W. F., Vucica, Y., and Rosenbaum, J. L. (2001). Kinetics and regulation of de novo centriole assembly. Implications for the mechanism of centriole duplication. *Curr. Biol.* **11,** 308–317.

Mitchell, B., Jacobs, R., Li, J., Chien, S., and Kintner, C. (2007). A positive feedback mechanism governs the polarity and motion of motile cilia. *Nature* **447,** 97–101.

Mizukami, I., and Gall, J. (1966). Centriole replication: II. Sperm formation in the fern, *Marsilea*, and the cycad, *Zamia*. *J. Cell Biol.* **29,** 97–111.

Montcouquiol, M., Rachel, R. A., Lanford, P. J., Copeland, N. G., Jenkins, N. A., and Kelley, M. W. (2003). Identification of Vanlg2 and Scrb1 as planar polarity genes in mammals. *Nature* **423,** 173–177.

Nakazawa, Y., Hiraki, M., Kamiya, R., and Hirono, M. (2007). SAS-6 is a cartwheel protein that establishes the 9-fold symmetry of the centriole. *Curr. Biol.* **17,** 2169–2174.

Nonaka, S., Tanaka, Y., Okada, Y., Takeda, S., Harada, A., Kanai, Y., Kido, M., and Hirokawa, N. (1998). Randomization of left–right asymmetry due to loss of nodal cilia generating leftward flow of extraembryonic fluid in mice lacking KIF3B motor protein. *Cell* **95**, 829–837.

Nonaka, S., Shiratori, H., Saijoh, Y., and Hamada, H. (2002). Determination of left–right patterning of the mouse embryo by artificial nodal flow. *Nature* **418**, 96–99.

Nonaka, S., Yoshiba, S., Watanabe, D., Ikeuchi, S., Goto, T., Marshall, W. F., and Hamada, H. (2005). De novo formation of left–right asymmetry by posterior tilt of nodal cilia. *PLoS Biol.* **3**, e268.

O'Toole, E. T., Giddings, T. H., McIntosh, J. R., and Dutcher, S. K. (2003). Three-dimensional organization of basal bodies from wild-type and delta-tubulin deletion strains of *Chlamydomonas reinhardtii*. *Mol. Biol. Cell* **14**, 2999–3012.

Pan, J., You, Y., Huan, T., and Brody, S. L. (2007). RhoA-mediated apical actin enrichment is required for ciliogenesis and promoted by Foxj1. *J. Cell Sci.* **120**, 1868–1876.

Pedersen, L. B., Geimer, S., Sloboda, R. D., and Rosenbuam, J. L. (2003). The microtubule plus end-tracking protein EB1 is localized to the flagellar tip and basal bodies in *Chlamydomonas reinhardtii*. *Curr. Biol.* **13**, 1969–1974.

Piasecki, B. P., LaVoie, M., Tam, L. W., Lefebvre, P. A., and Silflow, C. D. (2008). The Uni2 phosphoprotein is a cell cycle regulated component of the basal body maturation pathway in *Chlamydomonas reinhardtii*. *Mol. Biol. Cell* **19**, 262–273.

Pyne, C. K. (1968). Sur l'absence d'incorporation de la thymidine trite dans les cinetosomes de *Tetrahymena pyriformis* (Cilies Holotriches). *C. R. Acad. Sci. Paris D* **267**, 755–757.

Raff, E. C., Hutchsen, J. A., Hoyle, H. D., Nielsen, M. G., and Turner, F. R. (2000). Conserved axoneme symmetry altered by a component beta-tubulin. *Curr. Biol.* **10**, 1391–1394.

Randall, J., and Disbrey, C. (1965). Evidence for the presence of DNA at basal body sites in *Tetrahymena pyriformis*. *Proc. R. Soc. Lond. B* **162**, 473–491.

Rayner, C. F., Rutman, A., Dewar, A., Greenstone, M. A., Cole, P. J., and Wilson, R. (1996). Ciliary disorientation alone as a cause of primary ciliary dyskinesia syndrome. *Am. J. Respir. Crit. Care Med.* **153**, 1123–1129.

Ringo, D. L. (1967). Flagellar motion and fine structure of the flagellar apparatus in *Chlamydomonas*. *J. Cell Biol.* **33**, 543–571.

Romio, L., Fry, A. M., Winyard, P. J., Malcolm, S., Woolf, A. S., and Feather, S. A. (2004). OFD1 is a centrosomal/basal body protein expressed during mesenchymal–epithelial transition in human neurogenesis. *J. Am. Soc. Nephrol.* **15**, 2566–2568.

Salisbury, J. L., Suino, K. M., Busby, R., and Springett, M. (2002). Centrin-2 is required for centriole duplication in mammalian cells. *Curr. Biol.* **12**, 1287–1292.

Satir, P., and Dirksen, E. R. (1985). Function–structure correlations in cilia from mammalian respiratory tract. *In* "Handbook of Physiology—The Respiratory System" (A. P. Fishman, N. S. Cherniak, J. G. Widdicombe, and S. R. Gieger, Eds.), Vol. I, pp. 473–494. American Physiological Society, Bethesda.

Schrevel, J., and Besse, C. (1975). Un type flagellaire fonctionnel de base 6 + 0. *J. Cell Biol.* **66**, 492–507.

Shu, X., Fry, A. M., Tulloch, B., Manson, J. W., Crabb, J. W., Khanna, A., Faragher, A. J., Lennon, A., He, S., and Trojan, P. (2005). RPGR OFR15 isoform co-localizes with RPGRIP1 at centrioles and basal bodies and interacts with nucleophosmin. *Hum. Mol. Genet.* **14**, 1193–1197.

Simons, M., and Walz, G. (2006). Polycystic kidney disease: Cell division without a c(l)ue? *Kidney Int.* **70**, 854–864.

Snell, W. J., Dentler, W., Haimo, L. T., Binder, L. I., and Rosenbaum, J. L. (1974). Assembly of chick brain tubulin onto isolated basal bodies of *Chlamydomonas reinhardtii*. *Science* **185**, 33–38.

Sorokin, S. P. (1968). Reconstruction of centriole formation and ciliogenesis in mammalian lungs. *J. Cell Sci.* **3**, 207–230.

Stephan, A., Vaughan, S., Shaw, M. K., Gull, K., and McKean, P. G. (2007). An essential quality control mechanism at the eukaryotic basal body prior to intraflagellar transport. *Traffic* **8,** 1323–1330.

Tamm, S., and Tamm, S. L. (1988). Development of macrociliary cells in Beroe. I. Actin bundles and centriole migration. *J. Cell Sci.* **89,** 81–95.

Tamm, S. L., Sonneborn, T. M., and Dippell, R. V. (1975). The role of cortical orientation in the control of the direction of ciliary beat in Paramecium. *J. Cell Biol.* **64,** 98–112.

Tsou, M. F., and Stearns, T. (2006). Mechanism limiting centrosome duplication to once per cell cycle. *Nature* **442,** 947–951.

Uetake, Y., Loncarek, J., Nordberg, J. J., English, C. N., La Terra, S., Khodjakov, A., and Sluder, G. (2007). Cell cycle progression and *de novo* centriole assembly after centrosomal removal in untransformed human cells. *J. Cell Biol.* **176,** 173–182.

Vladar, E. K., and Stearns, T. (2007). Molecular characterization of centriole assembly in ciliated epithelial cells. *J. Cell Biol.* **178,** 31–42.

Winklebauer, M. E., Schafer, J. C., Haycraft, C. J., Swoboda, P., and Yoder, B. K. (2005). The *C. elegans* homologs of nephrocystin-1 and nephrocystin-4 are cilia transition zone proteins involved in chemosensory perception. *J. Cell Sci.* **118,** 5575–5587.

Wong, C., and Stearns, T. (2003). Centrosome number is controlled by a centrosome-intrinsic block to reduplication. *Nat. Cell Biol.* **5,** 539–544.

Wright, R. L., Chojnacki, B., and Jarvik, J. W. (1983). Abnormal basal-body number, location, and orientation in a striated fiber-defective mutant of *Chlamydomonas reinhardtii*. *J. Cell Biol.* **96,** 1697–1707.

Yang, J., Liu, X., Yue, G., Adamian, M., Bulgakov, O., and Li, T. (2002). Rootletin, a novel coiled-coil protein, is a structural component of the ciliary rootlet. *J. Cell Biol.* **159,** 431–440.

CHAPTER TWO

Intraflagellar Transport (IFT): Role in Ciliary Assembly, Resorption and Signalling

Lotte B. Pedersen[*] *and* Joel L. Rosenbaum[†]

Contents

1. Introduction	24
2. Early Events in Ciliogenesis	28
3. General Description of Intraflagellar Transport	29
4. Discovery and Evolution of IFT	30
5. The Canonical Anterograde IFT Motor	32
6. Additional Kinesin Motors Involved in Ciliogenesis	35
7. The Retrograde IFT Motor	37
8. IFT Particle Polypeptides	38
9. Functional Analysis of IFT Particle Polypeptides: Distinct Roles of IFT Complexes A and B	39
10. Regulation of IFT	41
11. Targeting of Proteins to the Ciliary Compartment: Clues from Ciliary Disease Genes	43
12. IFT and Cilia-Mediated Signaling	47
13. Conclusions and Perspectives	49
Acknowledgments	50
References	50

Abstract

Cilia and flagella have attracted tremendous attention in recent years as research demonstrated crucial roles for these organelles in coordinating a number of physiologically and developmentally important signaling pathways, including the platelet-derived growth factor receptor (PDGFR) α, Sonic hedgehog, polycystin, and Wnt pathways. In addition, the realization that defective assembly or function of cilia can cause a plethora of diseases and developmental defects ("ciliopathies") has increased focus on the mechanisms by which

[*] Department of Biology, Section of Cell and Developmental Biology, University of Copenhagen, Universitetsparken 13, DK-2100 Copenhagen OE, Denmark
[†] Department of Molecular, Cellular, and Developmental Biology, Yale University, New Haven, Connecticut

these antenna-like, microtubular structures assemble. Ciliogenesis is a complex, multistep process that is tightly coordinated with cell cycle progression and differentiation. The ciliary axoneme is extended from a modified centriole, the basal body, which migrates to and docks onto the apical plasma membrane early in ciliogenesis as cells enter growth arrest. The ciliary axoneme is elongated via intraflagellar transport (IFT), a bidirectional transport system that tracks along the polarized microtubules of the axoneme, and which is required for assembly of almost all cilia and flagella. Here, we provide an overview of ciliogenesis with particular emphasis on the molecular mechanisms and functions of IFT. In addition to a general, up-to-date description of IFT, we discuss mechanisms by which proteins are selectively targeted to the ciliary compartment, with special focus on the ciliary transition zone. Finally, we briefly review the role of IFT in cilia-mediated signaling, including how IFT is directly involved in moving signaling moieties into and out of the ciliary compartment.

1. INTRODUCTION

Cilia and flagella[1] are slender microtubule (MT)-based organelles with important motile and sensory functions. They consist of a highly organized MT-based axoneme that is extended from a modified centriole, the basal body, which anchors the axoneme in the cell. The ciliary axoneme projects out from the cell surface and is surrounded by a bilayer lipid membrane that is continuous with the plasma membrane of the cell body, but which contains a different complement of membrane receptors and ion channels (Fig. 2.1). Separating the two membrane compartments at the ciliary base is a region known as the "ciliary necklace" (Gilula and Satir, 1972), which is connected by fibers to the transition zone of the basal body (Fig. 2.1A, E, and H). It has been proposed that these transition zone fibers are part of a "ciliary pore complex," which, by analogy to the nuclear pore, functions as a regulated gate of entry where ciliary precursors and intraflagellar transport (IFT) proteins accumulate prior to entering the ciliary compartment via IFT (Rosenbaum and Witman, 2002).

In general, two types of cilia exist: those that are motile and those that are nonmotile. However, whether motile or not, all types of cilia have important sensory functions that are critical for controlling a large number of cellular and developmental processes (Christensen *et al.*, 2007; Eggenschwiler and Anderson, 2007). Motile cilia and flagella typically contain an axoneme with nine outer doublet MTs, composed of A and B subfibers, which are held together by nexin links, as well as a central pair of MTs. These cilia are

---

[1] Cilia and flagella are identical organelles, but for historical reasons, "cilia" is generally used when the organelle is present in multiple copies per cell while "flagella" is used when only one or two copies are present. Here, the terms will be used interchangeably.

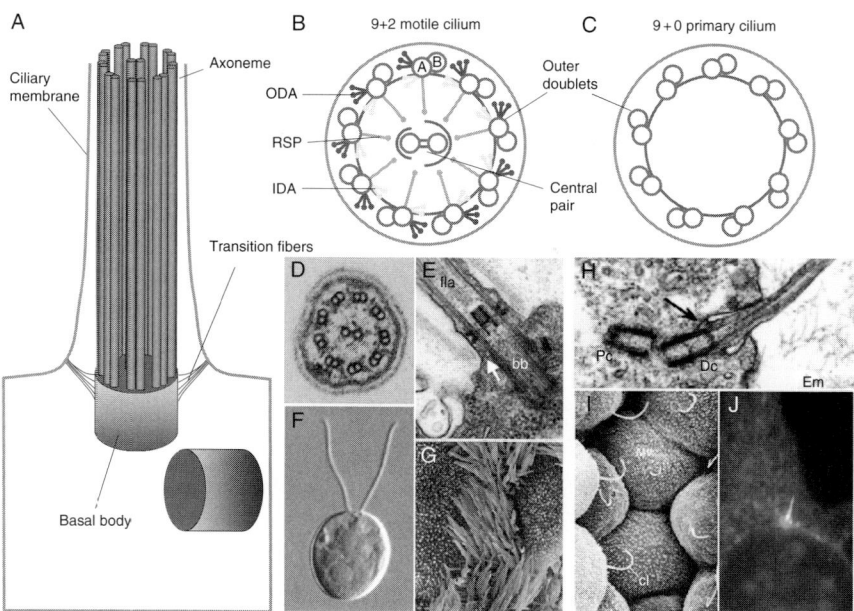

Figure 2.1 Diversity and structure of cilia and flagella. (A) Schematic showing the overall structure of a 9 + 2 cilium. (B) Schematic of a cross section of a 9 + 2 cilium highlighting the main axonemal components (ODA, outer dynein arm; IDA, inner dynein arm; RSP, radial spoke). (C) Schematic of a cross section of a 9 + 0 cilium that lacks motility-related structures. (D) TEM of a cross section of a *Chlamydomonas* flagellum (courtesy of Stefan Geimer, University of Bayreuth, Germany). (E) Longitudinal section of the base of a *Chlamydomonas* flagellum visualized by TEM. The arrow points to the flagellar base. *Abbreviations*: bb, basal body; fla, flagella. From Mitchell *et al.* (2005) with permission. (F) DIC of a *Chlamydomonas* cell with two 9 + 2 flagella. (G) Scanning EM of motile (9 + 2) tracheal cilia from a 4-week-old mouse (courtesy of Karl F. Lechtreck and George B. Witman, UMass Medical School, Worcester, MA). (H) Longitudinal section (TEM) of a primary cilium from a chicken chondrocyte showing the mother/distal centriole (Dc) and the daughter/proximal centriole (Pc). Arrow points to the transition fibers. *Abbreviation*: Em, extracellular matrix. Reprinted from Jensen *et al.* (2004) with permission. (I) SEM of ciliated renal epithelial cells from a kidney collecting tubule. *Abbreviations*: Ci, cilium; Mv, microvillus. From Kessel and Kardon (1979) with permission from Randy H. Kardon. (J) Immunofluorescence micrograph of a mouse NIH3T3 fibroblast stained with an antibody against detyrosinated α-tubulin to label the primary cilium (green) and with a p150^{Glued} antibody that labels the ciliary base (red). The nucleus is stained with DAPI (blue). Courtesy of Jacob M. Schrøder, University of Copenhagen, Denmark, who also generated the completed figure. (See Color Insert.)

hence designated "9 + 2" cilia. The axoneme of 9 + 2 cilia also contains accessory structures that are involved in motility, including outer and inner dynein arms, radial spokes, and central pair projections (Fig. 2.1A, B, and D). Sometimes 9 + 2 cilia are found in multiple copies per cell, for example, on the surface of epithelial cells lining the airways (Fig. 2.1G), oviducts, and brain ventricles, where the concerted action of the organelles mediates transport of fluids and substances across the epithelial surface. Single-celled organisms like *Tetrahymena* and *Paramecium* also contain multiple motile 9 + 2 cilia on their surface, while mammalian sperm cells and the green alga *Chlamydomonas* contain one or two motile 9 + 2 flagella, respectively (Fig. 2.1F). Motile cilia on single-celled organisms play an important role in cell motility (Ibanez-Tallon *et al.*, 2003; Rosenbaum and Witman, 2002; Satir and Christensen, 2006).

While 9 + 2 cilia are the most commonly encountered type of motile cilia, there are examples of motile cilia that lack the central pair of MTs, for example, motile 9 + 0 cilia on the nodal cells of developing mammalian embryos, which rotate vigorously in a manner that generates a directional flow across the node required for establishment of the left–right asymmetry axis (Hirokawa *et al.*, 2006). In addition, motile cilia with four central MTs (9 + 4 cilia) have been identified on the notochordal plate of rabbit embryos, suggesting that there is some degree of variation in axonemal structure of motile cilia (Feistel and Blum, 2006).

Nonmotile (primary) cilia typically contain a 9 + 0 axoneme, but lack outer and inner dynein arms and other accessory structures involved in motility (Fig. 2.1C and H–J). One exception to this canonical 9 + 0 structure is the mature kinocilium of hair cells in the inner ear, which contains a 9 + 2 axoneme, but is considered to be nonmotile (Dabdoub and Kelley, 2005). Specialized sensory 9 + 0 cilia in the vertebrate olfactory organs as well as the outer segment of vertebrate photoreceptor rod and cone cells, which develops as an extension of a primary cilium, also display an atypical axonemal structure in that the distal region of the axoneme consists of extended singlet A MT subfibers, similar to neuronal sensory cilia of nematodes (Insinna and Besharse, 2008; Reese, 1965).

In addition to specialized primary cilia of sensory organs, primary 9 + 0 cilia are present in a single copy of almost all other cell types in vertebrate organisms when these cells are in growth arrest (see http://www.bowserlab.org/primarycilia/ciliumpage2.html for a comprehensive listing of cells with primary cilia), but despite a near ubiquitous presence in vertebrates and although they were discovered over a century ago, primary cilia were, until recently, considered vestigial organelles with no important function. However, a number of key discoveries during the past decade have dramatically changed this view and it is now well established that primary cilia are indispensable sensory organelles involved in coordinating and regulating a variety of crucial cellular and developmental processes (Christensen *et al.*, 2007; Davenport and Yoder, 2005; Eggenschwiler and Anderson, 2007; Pazour and Witman, 2003). The first two key discoveries leading to this "renaissance" of the primary cilium were the discovery of IFT in *Chlamydomonas* by Kozminski *et al.* (1993) and a study by Cole *et al.* (1998), demonstrating that *Chlamydomonas* IFT particle polypeptides are homologous to proteins required for sensory cilia biogenesis in nematodes (Perkins *et al.*, 1986). These two studies provided the foundation for Pazour and coworkers who showed in 2000 that the mouse ortholog of one of the *Chlamydomonas* IFT particle polypeptides (IFT88/Tg737) is required for primary cilia formation in the kidney, and that failure to assemble these cilia leads to autosomal recessive polycystic kidney disease (Pazour et al., 2000). Finally, work by Barr, Pazour, Yoder, and colleagues on the autosomal dominant kidney disease-associated gene products polycystin-1 and polycystin-2 revealed that these proteins are localized to sensory cilia in *Caenorhabditis elegans* (Barr and Sternberg, 1999) as well as kidney primary cilia in the mouse (Pazour et al., 2002; Yoder et al., 2002) providing a link between IFT and polycystin-mediated signaling. While these studies were instrumental in defining a role for IFT and primary cilia in sensory signaling and mammalian health and disease, they also spurred an immense increase in primary cilia research, eventually leading to the realization that primary cilia are indispensable sensory organelles involved in coordinating and regulating a variety of crucial cellular and developmental signaling pathways, including the polycystin, Sonic hedgehog (Shh), platelet-derived growth factor receptor (PDGFR) α, and Wnt signaling pathways (Christensen et al., 2007; Davenport and Yoder, 2005; Eggenschwiler and Anderson, 2007; Pazour and Witman, 2003). Consequently, defects in the assembly or function of primary cilia can cause a plethora of diseases and developmental defects, which, along with disorders resulting from dysfunctional motile cilia, are now collectively known as "ciliopathies." The ciliopathies, which include cystic kidney and liver diseases, retinal degeneration, polydactyly, left–right patterning defects, obesity, airway disease, hydrocephalus, male and female infertility, and Kartagener, Joubert, Bardet-Biedl, Senior-Løken, and Meckel-Gruber syndromes (MKS), have been described in several recent reviews (Badano et al., 2006; Fliegauf et al., 2007; Ibanez-Tallon et al., 2003; Pan et al., 2005; Pazour and Rosenbaum, 2002) and will not be dwelled upon extensively here. Another exciting topic in cilia research, namely the role of cilia in cell cycle regulation and cancer, has also recently been reviewed in detail elsewhere (Pan and Snell, 2007; Plotnikova et al., 2008; Quarmby and Parker, 2005; Santos and Reiter, 2008) and will therefore not be covered extensively here.

In this review, we focus on the mechanisms by which cilia are assembled and maintained, with particular emphasis on the molecular mechanisms and functions of IFT (Kozminski et al., 1993), a process required for the assembly and maintenance of almost all eukaryotic cilia and flagella (Rosenbaum and Witman, 2002). We first provide a general description of IFT, from its initial discovery to the identification and characterization of individual IFT motor subunits and IFT particle polypeptides. Next, we discuss the mechanisms by which proteins are selectively targeted to the ciliary compartment with particular emphasis on the function and composition of the ciliary transition zone. Finally, we briefly review the role of IFT in cilia-mediated signaling, including how IFT is directly involved in moving signaling moieties within the cilium and between the ciliary compartment and the cytoplasm.

2. Early Events in Ciliogenesis

The ciliary axoneme is assembled onto a basal body, which is derived from a centriole, and ciliogenesis is therefore tightly coupled to centriole duplication and maturation. In multiciliated, differentiated mammalian epithelial cells, centrioles are formed en masse from a fibrogranular assembly site and centriole production is usually independent of pre-existing centrioles, but related to differentiation. These centrioles undergo a complex and poorly understood maturation process and migrate to the ciliary assembly site, possibly by interacting with the actin cytoskeleton and proteins of the planar cell polarity (PCP) pathway (Dawe et al., 2007a; Dirksen, 1991; Park et al., 2008).

In proliferating cells that will form a primary cilium, ciliogenesis is tightly coupled to the cell cycle (Fig. 2.2). Ciliogenesis is initiated in G1 by the addition of Golgi-derived vesicles to the distal end of the mother centriole, and the axoneme begins to elongate within this membrane-bound compartment (Sorokin, 1962). As ciliogenesis proceeds, the axoneme continues to elongate from its distal end (Johnson and Rosenbaum, 1992) and the mother centriole-associated membrane vesicle grows by recruitment of additional Golgi-derived vesicles (Sorokin, 1962). Concomitant with docking of the mature mother centriole/basal body onto the ciliary assembly site at the apical end of the cell, the mother centriole-associated membrane vesicle fuses with the plasma membrane of the cell thereby forming a membrane sheath around the emerging ciliary axoneme that is separated from the plasma membrane of the cell by the ciliary necklace (Sorokin, 1962). Elongation of the membrane-bound axoneme is mediated by IFT (Rosenbaum and Witman, 2002). However, in a few cases, for example, in *Drosophila* sperm cells, assembly of the ciliary axoneme takes place in the cytoplasm by an IFT-independent mechanism (Han et al., 2003), but this type of assembly will not be discussed further here.

New (daughter) centrioles are made from pre-existing (mother) centrioles in the S phase of the cell cycle and reach their mature length during late G2/M (Fig. 2.2; Azimzadeh and Bornens, 2007; Pan and Snell, 2007). When cells are in G1, the mother centriole is distinguished from the daughter centriole by the presence of two sets of nine appendages at its distal end termed distal and subdistal appendages, respectively. The subdistal appendages are thought to be required for anchoring of MTs at the mother centriole while the distal appendages are thought to be involved in the docking of the mother centriole/basal body to the apical cell membrane early in ciliogenesis. After basal body docking, these distal appendages constitute the transition fibers that link the basal body to the ciliary membrane/ciliary necklace (Paintrand et al., 1992), that is, the fibers proposed to be constituents of a ciliary pore complex (Rosenbaum and Witman, 2002).

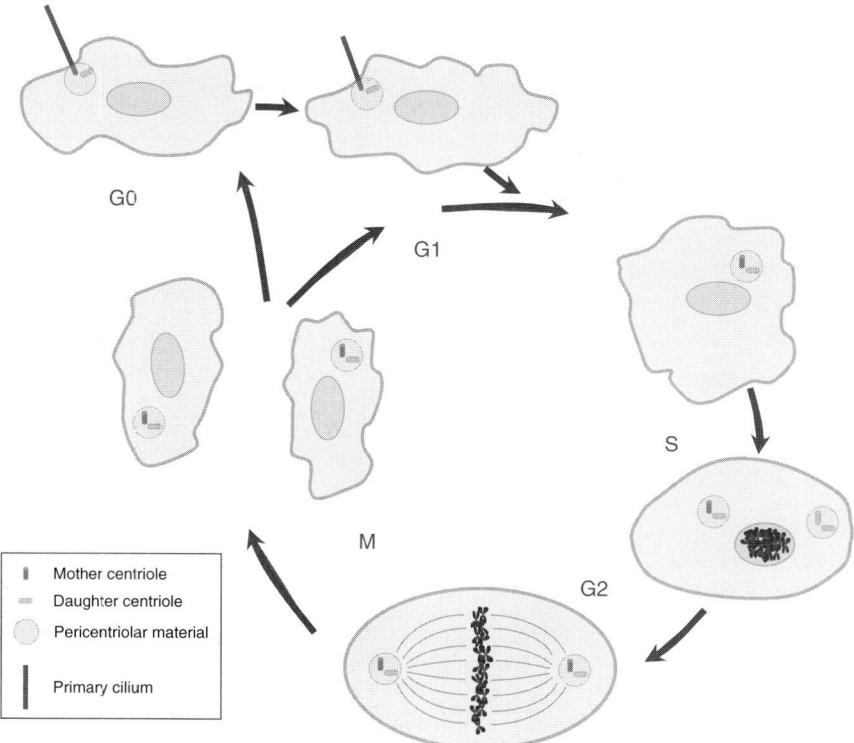

Figure 2.2 Assembly and disassembly of primary cilia are coordinated with the cell cycle. The primary cilium is formed in G1/G0 following docking of the mother centriole at the ciliary assembly site at the apical plasma membrane. Cilium elongation is mediated by IFT-dependent addition of ciliary precursors to the distal end of the mother centriole, which at this point is termed a basal body. Disassembly of the cilium prior to mitosis allows both centriole pairs to function in mitotic spindle formation. (See Color Insert.)

3. General Description of Intraflagellar Transport

Since cilia and flagella lack the machinery for *de novo* protein synthesis, and since the axoneme assembles from and continuously turns over at its distal tip (Johnson and Rosenbaum, 1992; Marshall and Rosenbaum, 2001), compartmentalized ciliary assembly and maintenance relies on the constant delivery of axonemal precursors from their site of synthesis in the cell body to the axonemal assembly site at the ciliary tip, a process that is mediated by IFT (Rosenbaum and Witman, 2002).

IFT is a bidirectional MT-based motility process during which groups of large protein complexes (IFT particles) are transported along the axonemal outer doublet MTs from the base of the cilium to its distal tip by kinesin-2 motors and then from the tip back to the cell body by cytoplasmic dynein 2 (Kozminski et al., 1993, 1995; Pazour et al., 1999; Porter et al., 1999). During anterograde (base to tip) IFT, axonemal precursor proteins associate with the IFT machinery, which ferries the precursors to the axonemal assembly site at the ciliary tip where precursors are unloaded for assembly. Following unloading of axonemal building blocks at the ciliary tip, the "empty" IFT particles bind to axonemal turn over products, which are then transported back to the cell body (retrogradely) for recycling (Qin et al., 2004; Fig. 2.3).

Much of our current knowledge about IFT stems from pioneering work in the biflagellate green alga *Chlamydomonas* (Cole et al., 1998; Kozminski et al., 1993, 1995), which remains an excellent model organism in which to study IFT, primarily because the flagella can be easily isolated and purified (Witman et al., 1972), the *Chlamydomonas* genome sequence (Merchant et al., 2007), flagellar proteome (Pazour et al., 2005), and flagellar transcriptome (Stolc et al., 2005) are known, a large number of IFT mutants are available (Cole, 2003), and IFT can be visualized *in vivo* without the aid of fluorescently tagged proteins (Kozminski et al., 1993). Nevertheless, since the discovery of IFT in *Chlamydomonas* in 1993 (Kozminski et al., 1993), studies in other ciliated model organisms such as *C. elegans*, *Tetrahymena*, trypanosomes, zebrafish, and the mouse have contributed significantly to our understanding of the molecular mechanisms and functions of IFT in various cellular and developmental contexts. These studies have shown that IFT is a highly conserved process required for the assembly and maintenance of almost all eukaryotic cilia and flagella, and that mutations leading to defective IFT can result in the development of severe diseases (ciliopathies) and developmental defects (Badano et al., 2006; Fliegauf et al., 2007; Pan et al., 2005; Pazour and Rosenbaum, 2002; Sloboda, 2002; see also Section 1). Since IFT is required for assembly of the cilium proper (Rosenbaum and Witman, 2002), which in turn functions as a repository for receptors and other signaling proteins involved in controlling important cellular and developmental processes (Christensen et al., 2007; Eggenschwiler and Anderson, 2007), some of the malignancies arising from defective IFT may result from the mere lack of an appropriate structure, the cilium, to harbor and coordinate specific signaling pathways. However, there is growing evidence indicating that IFT also participates *directly* in cilia-mediated signaling processes, for example, Shh signaling during vertebrate development (Eggenschwiler and Anderson, 2007) and cGMP-dependent signaling during mating in *Chlamydomonas* (Wang et al., 2006), a topic that will be discussed in Section 12.

4. Discovery and Evolution of IFT

IFT was first observed by enhanced digital interference contrast (DIC) microscopy of immobilized *Chlamydomonas* flagella as a continuous, nonsaltatory movement of granular particles (IFT particles), which were seen

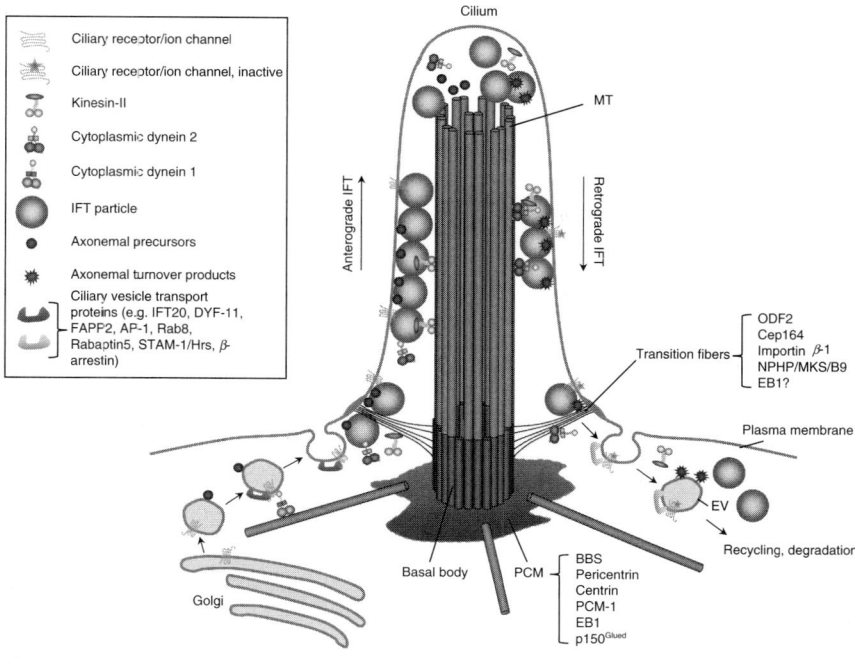

Figure 2.3 IFT and targeting of proteins to the ciliary compartment. Proteins destined for the ciliary compartment (membrane proteins as well as axonemal components) are transported in Golgi-derived vesicles to the base of the cilium where the vesicles are exocytosed and the ciliary proteins associate with IFT particles. This Golgi-to-cilium-mediated vesicle transport, which is proposed to involve cytoplasmic dynein 1 MT-based movement, depends on the IFT complex B proteins IFT20 and DYF-11, the small G protein Rab8 and associated GEFs (e.g., Rabin 8, Rabaptin 5), FAPP2, and adapter proteins such as AP-1. BBS proteins and other proteins localized in the pericentrosomal region (e.g., PCM-1, EB1, p150^{Glued}) may provide a link between the Golgi-derived vesicles and the transition fibers at the ciliary base and may also serve to anchor MTs at the basal body. At the ciliary base, only proteins (or protein complexes) containing specific ciliary targeting motifs are allowed access through the zone defined by the transition fibers. Selective entry of proteins into the ciliary compartment probably involves specific G proteins and GEFs that are associated with NPHPs, MKS, and B9 domain-containing proteins. Following entry into the ciliary compartment, these proteins, along with inactive cytoplasmic dynein 2, are transported anterogradely along the axoneme by kinesin-II-mediated IFT. At the ciliary tip IFT particles are remodeled, kinesin-II inactivated, and cytoplasmic dynein 2 activated. Ciliary turnover products (e.g., inactive receptors) are, in turn, transported retrogradely along ciliary axonemes by cytoplasmic dynein 2 for recycling or degradation in the cytoplasm. Recycling or turnover of ciliary membrane receptors may involve ubiquitination (e.g., via BBS proteins) and/or dephosphorylation of the receptors as well as binding to endosomal vesicle adapter proteins such as STAM-1/Hrs (Bae and Barr, 2008) or β-arrestin (Kovacs *et al.*, 2008). Figure based on references (Azimzadeh and Bornens, 2007; Leroux, 2007; Rosenbaum and Witman, 2002) as well as references cited in the text. *Abbreviations*: EV, endocytic vesicle; MT, microtubule; PCM, pericentriolar material. Figure generated by Jacob M. Schrøder, University of Copenhagen. (See Color Insert.)

to move bidirectionally underneath the flagellar membrane at *ca.* 2 μm/s in the anterograde direction and at *ca.* 3.5 μm/s in the retrograde direction (Kozminski *et al.*, 1993). Studies using GFP-tagged IFT particle polypeptides or motor subunits have demonstrated similar, but sometimes faster, movement of IFT particles in cilia from a variety of other cell types and organisms, including neuronal sensory cilia in *C. elegans* (Orozco *et al.*, 1999), primary cilia of cultured IMCD and LLC-PK1 kidney cells (Follit *et al.*, 2006; Tran *et al.*, 2008), and motile flagella of *Trypanosoma brucei* (Absalon *et al.*, 2008), indicating that IFT is conserved among ciliated organisms.

By using correlative light microscopy and transmission electron microscopy (TEM) of *Chlamydomonas* flagella, Kozminski *et al.* (1995) demonstrated that the IFT particles seen to move in the DIC consist of linear arrays of lollipop-shaped structures that are connected to both the axonemal outer doublet B MTs and the flagellar membrane (Fig. 2.4). Furthermore, immunogold EM using antibodies directed against IFT particle polypeptides and motor subunits provided conclusive evidence that these lollipop-shaped structures indeed are IFT particles (Pedersen *et al.*, 2006; Sloboda and Howard, 2007).

Consistent with a conserved role for IFT in ciliary assembly and maintenance (Cole *et al.*, 1998; Perkins *et al.*, 1986), comparative genomics studies have shown that the genes encoding IFT particle polypeptides or motor subunits are highly conserved among ciliated eukaryotes, but absent from nonciliated organisms such as higher plants and fungi (Avidor-Reiss *et al.*, 2004; Li *et al.*, 2004). In addition, bioinformatic analysis of IFT particle polypeptide sequences indicated that these polypeptides display similarities to components of coat protein I (COPI) and clathrin-coated vesicles, leading to the hypothesis that IFT evolved as a specialized form of coated vesicle transport from a protocoatomer complex (Jekely and Arendt, 2006).

Most of the genes coding for IFT particle polypeptides or motor subunits have now been cloned and characterized in several different organisms and, as will be discussed below, analyses of individual IFT components have provided significant insight into the molecular mechanisms and functions of IFT, although important questions still remain to be answered. For example, how individual IFT components interact and are regulated in different organisms and under diverse physiological conditions is only now beginning to be elucidated.

5. The Canonical Anterograde IFT Motor

The canonical motor for anterograde IFT is a heterotrimeric complex belonging to the kinesin-2 family. This kinesin-2 complex, which hereafter will be referred to as kinesin-II, was first purified from sea urchin eggs and

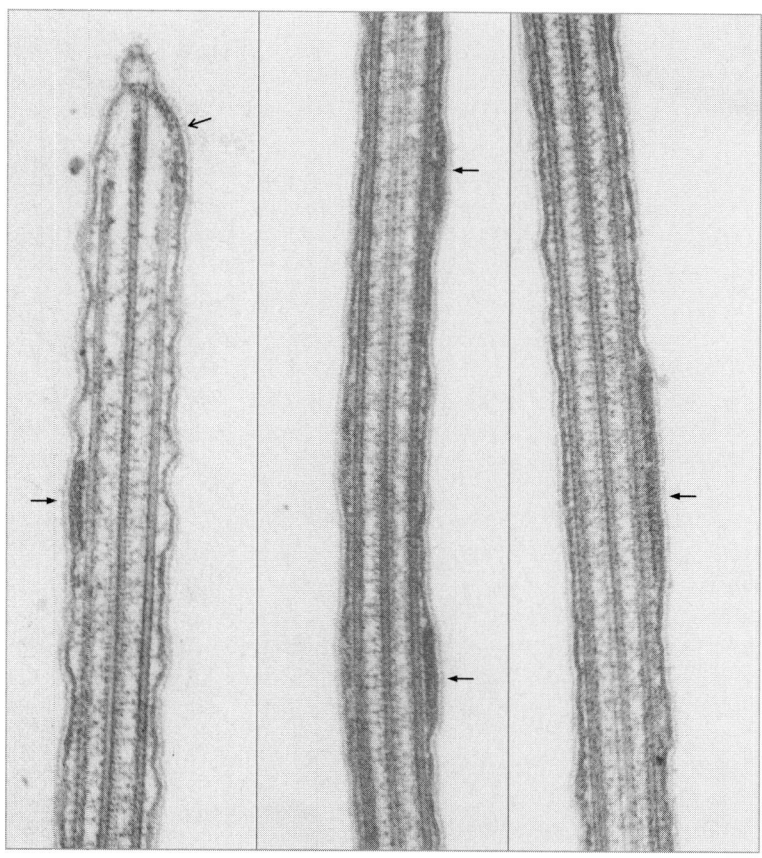

Figure 2.4 Images of IFT particles. The figure shows TEMs of longitudinal sections of *Chlamydomonas* flagella. Groups of IFT particles of variable size (closed arrows) are observed between the outer doublet B MTs and the flagellar membrane. Open arrow shows electron-dense IFT material near the flagellar tip. Courtesy of Stefan Geimer, University of Bayreuth, Germany.

consists of two motor subunits of 90 and 85 kDa, respectively, and a 100 kDa nonmotor subunit (Cole *et al.*, 1992, 1993; Scholey, 2003). In vertebrates, the 90 and 85 kDa motor subunits are termed KIF3A and KIF3B, respectively, whereas the 100 kDa nonmotor subunit is called kinesin-associated protein (KAP; Scholey, 2003). The *Chlamydomonas* orthologs of these subunits are termed FLA10 (KIF3A ortholog), FLA8 (KIF3B ortholog), and FLA3 (KAP ortholog), respectively (Table 2.1; Kozminski *et al.*, 1995; Miller *et al.*, 2005; Mueller *et al.*, 2005; Walther *et al.*, 1994). Analysis of a *Chlamydomonas* mutant, *fla10^{ts}*, which harbors a temperature-sensitive mutation in the *FLA10* gene (Adams *et al.*, 1982; Huang *et al.*, 1977; Walther *et al.*, 1994), first indicated that kinesin-II is

Table 2.1 Components of the IFT machinery

IFT polypeptide	Chlamydomonas	C. elegans	H. sapiens
Kinesin-2			
Heterotrimeric	FLA10	KRP85/KLP-20	KIF3A
	FLA8	KRP95/KLP-11	KIF3B
	FLA3	KAP/KAP-1	KAP
Homodimeric	?	OSM-3	KIF17
Cytoplasmic dynein 2			
	DHC1b	CHE-3	DYNC2H1/DHC2
	D1bLIC	D2LIC/XBX-1	DYNC2LI1/D2LIC
	FAP133	?	WD34
	LC8/FLA14	?	?
Complex A			
IFT144	FAP66	DYF-2	WDR19
IFT140	IFT140	CHE-11	IFT140
IFT139	FAP60	ZK328.7	THM-1
IFT122A	XP_001700201.1	DAF-10	IFT122/WDR10
IFT122B	IFT122B	IFTA-1	WDR35
IFT43	FAP118	?	?
Complex B			
IFT172	IFT172	OSM-1	IFT172
IFT88	IFT88	OSM-5	IFT88/Polaris
IFT81	IFT81	F32A6.2	IFT81
IFT80	IFT80	CHE-2	IFT80
IFT74/72	IFT74/72	C18H9.8	IFT74/72
IFT57	IFT57	CHE-13	IFT57/Hippi
IFT52	IFT52	OSM-6	IFT52/NGD5
IFT46	IFT46	DYF-6	C11orf60
IFT27	IFT27	?	RABL4
IFT20	IFT20	Y110A7A.20	IFT20
New putative IFT proteins			
	FAP259	DYF-1	TPR30A
	FAP22	DYF-3	CLUAP1
	XM_001698717.1	DYF-13	TTC26
	FAP116	DYF-11	MIP-T3

Data were obtained from GenBank and from the following references: Absalon et al. (2008), Blacque et al. (2006, 2008), Cole (2003), Li et al. (2008), Omori et al. (2008), Pazour et al. (2005), Pfister et al. (2005), Rompolas et al. (2007), Scholey (2003), Tran et al. (2008), and Tsao and Gorovsky (2008b).

required for anterograde IFT, because when mutant cells were placed at the restrictive temperature of 32 °C, IFT ceased and the flagella began to shorten. Furthermore, when the *fla10*^{ts} cells were placed at 32 °C and deflagellated, new flagella failed to form, indicating that kinesin-II and IFT are required for both assembly and maintenance of the flagella (Kozminski *et al.*, 1995). Subsequent studies in sea urchin, mouse, *Tetrahymena* and *Drosophila* showed that inactivation of kinesin-II similarly abolished ciliary assembly in these organisms (Brown *et al.*, 1999; Marszalek *et al.*, 1999; Morris and Scholey, 1997; Nonaka *et al.*, 1998; Sarpal *et al.*, 2003; Takeda *et al.*, 1999), whereas inactivation of kinesin-II in *C. elegans* prevented global ciliary assembly only when the homodimeric kinesin-2 motor OSM-3 was simultaneously inhibited, owing to partial redundancy between OSM-3 and kinesin-II in building specific subsets of sensory cilia of this organism (Scholey, 2008; Snow *et al.*, 2004). However, careful observations and measurements of motility rates of *C. elegans* kinesin-II subunits and IFT particle polypeptides, both *in vivo* and *in vitro*, demonstrated that kinesin-II indeed powers anterograde IFT in this organism (Orozco *et al.*, 1999; Ou *et al.*, 2005a; Pan *et al.*, 2006; Snow *et al.*, 2004). Such studies, along with genetic and biochemical studies in vertebrates, have also begun to shed light on the mechanisms by which kinesin-II is regulated (Burghoorn *et al.*, 2007; Pan *et al.*, 2006; Pedersen *et al.*, 2008; Scholey, 2008), a topic that will be dealt with in more detail in Section 10.

In summary, the available results suggest that heterotrimeric kinesin-II functions as the core anterograde IFT motor in virtually all ciliary systems studied to date. However, as will be discussed below, additional kinesins such as homodimeric kinesin-2 as well as other kinesin family members may cooperate with kinesin-II in the assembly of specific subsets of cilia to generate ciliary structural and functional diversity.

6. Additional Kinesin Motors Involved in Ciliogenesis

In *C. elegans* homodimeric kinesin-2 appears to cooperate with kinesin-II during anterograde IFT to mediate assembly of specific subsets of sensory cilia (Scholey, 2008; Snow *et al.*, 2004), and there is growing evidence suggesting that the deployment of "accessory" kinesins during ciliogenesis is not limited to *C. elegans* and homodimeric kinesin-2. In fact, numerous kinesins have been detected in cilia from diverse species, and some of these have been implicated in cilia assembly and maintenance, either by functioning cooperatively with kinesin-II during anterograde IFT to promote axonemal assembly, by affecting axonemal turnover and disassembly, or by transporting specific receptors and ion channels to the ciliary membrane (Scholey, 2008). Here, we will provide a brief description of "accessory" kinesins with putative roles in IFT and/or transport of ciliary membrane proteins, namely homodimeric kinesin-2, kinesin-3, and kinesin-16 family members. For detailed descriptions of homodimeric kinesin-2 and other ciliary kinesins, see Blacque et al. (2008) and Scholey (2008).

A role for homodimeric kinesin-2 in IFT and cilia biogenesis was first discovered in C. elegans (Shakir et al., 1993; Signor et al., 1999b), where genetic and motility studies later revealed that the homodimeric kinesin-2 motor OSM-3 functions redundantly with heterotrimeric kinesin-2 to build the middle axonemal segment of amphid channel cilia, which consists of doublet MTs, whereas OSM-3 alone mediates assembly of the singlet MTs that constitute the distal segment of the axoneme (Snow et al., 2004). In frogs, the olfactory cilia similarly contain a long segment of singlet MTs at their distal end (Reese, 1965), and biophysical measurements have indicated that cyclic nucleotide-gated channels (CNGs) are highly enriched at this site (Flannery et al., 2006). It is likely that this localization of CNGs is related to homodimeric kinesin-2-mediated assembly of distal singlet axonemal MTs, because in rodents the OSM-3 homolog KIF17 has been implicated in the transport of CNGs to the olfactory cilia (Jenkins et al., 2006). Furthermore, a recent study showed that morpholino-mediated knockdown of KIF17 in zebrafish inhibited the formation of photoreceptor outer segments, which essentially are modified primary cilia, and which also contain extended distal singlet axonemal MTs like C. elegans amphid channel and frog olfactory cilia. Importantly, the assembly of sensory cilia in the zebrafish pronephros was unaffected by KIF17 knockdown, suggesting that KIF17 plays a specific role in the assembly of cilia with extended distal singlet MTs (Insinna et al., 2008).

Members of the kinesin-3 and kinesin-16 families may also be involved in ciliary transport of specific membrane proteins. In C. elegans, the kinesin-3 family member KLP-6 is required for ciliary localization of the polycystin-2 calcium channel that controls male mating behavior in this organism (Peden and Barr, 2005). Furthermore, an extensive phylogenetic analysis of kinesins indicated that kinesin-16 family members, as well as other kinesin subfamilies such as kinesin-17, are specific to ciliated organisms (Wickstead and Gull, 2006). Humans contain six putative kinesin-3 family members and one kinesin-16 member (KIF12; Wickstead and Gull, 2006), and preliminary results indicate that at least one of the kinesin-3 proteins, KIF13A, as well as KIF12 localize to primary cilia in diverse cultured human cells and native human embryonic tissue (S. K. Nielsen, R. I. Thorsteinsson, S. T. Christensen, and L. B. Pedersen, unpublished data). While the functional implications of these observations are still to be determined, it is of interest that KIF12 was shown to be a polycystic kidney disease modifier in the cpk mouse hinting at a possible ciliary function for this protein (Mrug et al., 2005). Moreover, murine KIF13A associates directly with the AP-1 adaptor complex during transport of mannose-6-phosphate receptor-containing vesicles from the Golgi to the plasma membrane in various cultured cell types (Nakagawa *et al.*, 2000), and in *C. elegans* AP-1 has been implicated in ciliary transport of calcium channels and odorant receptors (Bae *et al.*, 2006; Dwyer *et al.*, 2001). Therefore we hypothesize that KIF13A and KIF12, possibly in conjunction with AP-1, are involved in targeting specific receptors and/or ion channels to the ciliary membrane in vertebrate cells. Whether such targeting occurs independently or in cooperation with kinesin-II, as observed for OSM-3 in *C. elegans* (Snow *et al.*, 2004), will be interesting to investigate.

7. THE RETROGRADE IFT MOTOR

The motor that powers retrograde IFT is an isoform of cytoplasmic dynein called cytoplasmic dynein 2 (previously known as cytoplasmic dynein 1b; Pfister *et al.*, 2005). This motor complex consists of at least four different subunits: a heavy chain termed DYNC2H1 that belongs to the AAA+ family of ATPases, a light intermediate chain (DYNC2LI1), a light chain (LC8), and a recently identified putative intermediate chain (IC)/WD repeat protein, which may be specific for motile cilia (Table 2.1; Pfister *et al.*, 2005; Rompolas *et al.*, 2007). The cytoplasmic dynein 2 motor subunit, DYNC2H1, was originally identified as a dynein heavy chain whose synthesis is induced by deciliation in sea urchin embryos (Gibbons *et al.*, 1994), and subsequent genetic and motility studies, primarily in *Chlamydomonas* and *C. elegans*, demonstrated that DYNC2H1 is essential for retrograde IFT. Thus null mutations in the gene encoding DYNC2H1 in these organisms led to cessation of retrograde IFT and formation of stunted cilia containing large accumulations of IFT particles, implying that transport of IFT particles out of the cilia was impaired (Pazour *et al.*, 1999; Porter *et al.*, 1999; Signor *et al.*, 1999a). In mammals, DYNC2H1 was found to localize to cilia in the brain and photoreceptors (Mikami *et al.*, 2002), and mutational inactivation of the corresponding gene led to the formation of stunted, bulbous cilia in the neuroectoderm, limb mesenchyme, and ventral node of developing mouse embryos (Huangfu and Anderson, 2005; May *et al.*, 2005). These mutant cilia are thus similar in appearance to the stunted cilia filled with IFT particles that were observed in *Chlamydomonas* mutants lacking DYNC2H1 (Pazour *et al.*, 1999; Porter *et al.*, 1999), indicating that the role of DYNC2H1 in retrograde IFT is conserved in mammals.

The DYNC2LI1 subunit of the cytoplasmic dynein 2 complex has been studied in several organisms, including *Chlamydomonas*, *C. elegans*, and mouse, and as for DYNC2H1, genetic and motility studies clearly demonstrated a requirement for this protein in retrograde IFT (Grissom *et al.*, 2002; Hou *et al.*, 2004; Perrone *et al.*, 2003; Rana *et al.*, 2004; Schafer *et al.*, 2003).

Furthermore, biochemical analyses of DYNC2LI1 from *Chlamydomonas* flagella demonstrated that DYNC2LI1 and DYNC2H1 are part of the same complex (Perrone *et al.*, 2003; Rompolas *et al.*, 2007), a complex that also contains LC8 and a recently identified WD repeat protein called FAP133 (Rompolas *et al.*, 2007). A *Chlamydomonas* mutant, *fla14*, that is defective in the gene encoding LC8 displays a phenotype consistent with defective retrograde IFT (Pazour *et al.*, 1998), and siRNA-mediated knockdown of the gene encoding the FAP133 homolog in trypanosomes resulted in flagellar dysfunction, suggesting that FAP133 may also be required for retrograde IFT (Baron *et al.*, 2007). However, the precise function of individual subunits of the cytoplasmic dynein 2 complex during retrograde IFT is still unclear. Moreover, it is possible that the complex contains additional subunits that have not yet been characterized in detail. For example, XBX-2, which contains a domain characteristic for dynein light chains of the Tctex-1 family, has been shown to undergo IFT in *C. elegans* sensory cilia (Efimenko *et al.*, 2005), but its ciliary function and relationship with the cytoplasmic dynein 2 complex are not known. However, given the large number of light chains that are involved in assembly and regulation of other dynein motor complexes, for example, outer arm dynein in *Chlamydomonas* flagella (King, 2003), it is plausible that cytoplasmic dynein 2 similarly contains multiple light chains involved in the assembly and function of the motor complex.

8. IFT Particle Polypeptides

By taking advantage of the *fla10*^{ts} mutant that harbors a temperature-sensitive mutation in the gene encoding the FLA10 motor subunit of kinesin-II (Adams *et al.*, 1982; Huang *et al.*, 1977; Walther *et al.*, 1994), IFT particle polypeptides were originally isolated from *Chlamydomonas* flagella using sucrose density gradient centrifugation and were found to consist of two separate complexes, termed A and B, containing six and ten subunits, respectively (Table 2.1; Cole *et al.*, 1998; Piperno and Mead, 1997). Genetic, motility, and bioinformatic studies in *C. elegans* subsequently led to the identification of three additional putative IFT particle polypeptides (DYF-1, DYF-3, DYF-13; Blacque *et al.*, 2005; Murayama *et al.*, 2005; Ou *et al.*, 2005a,b), and RNAi-based approaches in trypanosomes recently confirmed a role for these proteins in flagellar biogenesis (Absalon *et al.*, 2008). Homologues of DYF-1, DYF-3, and DYF-13 are also present in *Chlamydomonas* (Absalon *et al.*, 2008; Blacque *et al.*, 2008; Pazour *et al.*, 2005), but were not identified during initial purification of IFT particle polypeptides from *Chlamydomonas* flagella (Cole *et al.*, 1998; Piperno and Mead, 1997). However, genetic, motility, and RNAi-based studies in *C. elegans* and trypanosomes suggest that these three proteins are part of complex B (Absalon *et al.*, 2008; Blacque *et al.*, 2005; Murayama *et al.*, 2005; Ou *et al.*, 2005a,b), indicating that they may dissociate from the complex during sucrose density gradient centrifugation of *Chlamydomonas* flagella extracts, as has been observed for the complex B polypeptide IFT172 (Pedersen *et al.*, 2005). Yet another putative complex B polypeptide, DYF-11, was recently identified by two separate groups (Li *et al.*, 2008; Omori *et al.*, 2008), bringing the total number of likely complex B polypeptides to 14 (Table 2.1).

The genes for all known IFT particle polypeptides have now been cloned and sequenced in multiple organisms, including *Chlamydomonas*, *C. elegans*, *Tetrahymena*, *Drosophila*, trypanosomes, zebrafish, mouse, and human (Table 2.1; Absalon *et al.*, 2008; Blacque *et al.*, 2008; Cole, 2003; Li *et al.*, 2008; Omori *et al.*, 2008; Pedersen *et al.*, 2008; Tran *et al.*, 2008; Tsao and Gorovsky, 2008b). The corresponding amino acid sequences of IFT particle polypeptides are characterized by an abundance of motifs known to be involved in protein–protein interaction, including WD repeats, TPR repeats, and coiled coils (Cole, 2003). As mentioned previously, these features may reflect an evolutionary relationship of IFT particle polypeptides with components of COPI and clathrin-coated vesicles (Jekely and Arendt, 2006), and are also consistent with a role for IFT particle proteins in transport of ciliary precursors and turn over products (Qin *et al.*, 2004). The challenge now is to determine the function of individual IFT particle polypeptides, how they interact with each other as well as with motors and cargo proteins, and how these interactions are regulated. As will be described below, significant strides have been made in terms of elucidating the functions of individual IFT polypeptides, but much still remains to be learned about their interactions and regulation.

9. Functional Analysis of IFT Particle Polypeptides: Distinct Roles of IFT Complexes A and B

The first direct evidence demonstrating that IFT particle proteins are required for cilia assembly was provided by a combination of biochemical analyses of *Chlamydomonas* flagella and genetic studies in *C. elegans*. Thus, when IFT particle polypeptides were first isolated and sequenced from *Chlamydomonas* flagella, it was discovered that some of the IFT particle polypeptides are homologous to proteins encoded by genes that when mutated inhibit assembly of sensory cilia in *C. elegans* (Cole *et al.*, 1998; Perkins *et al.*, 1986). Using similar approaches, as well as motility- and RNAi-based methods, the function of almost all known IFT particle polypeptides has now been studied in multiple organisms. These studies have shown that in general, complete loss of any complex B polypeptide almost invariably leads to inhibition of cilia assembly, leading to the notion that complex B is required for anterograde transport of axonemal precursors into the ciliary compartment (Absalon et al., 2008; Beales et al., 2007; Brazelton et al., 2001; Brown et al., 2003; Cole, 2003; Deane et al., 2001; Follit et al., 2006; Fujiwara et al., 1999; Haycraft et al., 2001, 2003; Houde et al., 2006; Huangfu et al., 2003; Kobayashi et al., 2007; Li et al., 2008; Omori et al., 2008; Pathak et al., 2007; Pazour et al., 2000; Pedersen et al., 2005; Perkins et al., 1986; Qin et al., 2001, 2007; Sun et al., 2004; Tran et al., 2008; Tsao and Gorovsky, 2008a; Tsujikawa and Malicki, 2004). This notion is further supported by recent data from *Chlamydomonas* indicating that the complex B protein IFT46 is specifically involved in transporting outer dynein arms into flagella (Hou et al., 2007). In addition to IFT46, studies in which other IFT complex B proteins have been partially depleted or inactivated have led to assignment of specific functions for IFT172, IFT27, IFT20, and DYF-11/MIP-T3. Thus IFT172 appears to be required for regulating the transition from anterograde to retrograde IFT at the flagellar tip (Pedersen et al., 2005; Tsao and Gorovsky, 2008a), IFT27 is a small G protein involved in cell cycle control (Qin et al., 2007), whereas IFT20 and DYF-11/MIP-T3 play a role in transport of vesicles from the Golgi to the ciliary base (Follit et al., 2006; Li et al., 2008; Omori et al., 2008; see also Chapter 5). Furthermore, chemical crosslinking and yeast two- and three-hybrid analyses have shown that IFT81 and IFT74/72 interact directly, forming a tetrameric complex that serves as a scaffold for the intact IFT complex B (Lucker et al., 2005). The remaining subunit organization within complex B as well as interactions between complex B proteins and the remaining IFT constituents are not well characterized.

In contrast to complex B polypeptides, complex A polypeptides are not always essential for building the ciliary axoneme, but rather, seem to be important for retrograde IFT. For example, the *Chlamydomonas* ts mutants *fla15*, *fla16*, and *fla17*, which contain decreased amounts of complex A polypeptides in their flagella, are able to assemble flagella but display retrograde IFT phenotypes with accumulation of IFT complex B polypeptides at the flagellar tips (Iomini et al., 2001; Piperno et al., 1998). In addition, mutational inactivation of IFT140/CHE-11 or IFT122A/DAF-10 in *C. elegans* results in shortened cilia that accumulate complex B polypeptides at their distal end (Collet et al., 1998; Perkins et al., 1986; Qin et al., 2001; Schafer et al., 2003). Similar phenotypes were observed when these polypeptides were inactivated in *Tetrahymena* or trypanosomes (Absalon et al., 2008; Tsao and Gorovsky, 2008b), or when IFT139/THM-1 was inactivated in the mouse (Tran et al., 2008). In zebrafish auditory and olfactory organs, however, inactivation of IFT140 was reported to have little observable effect on cilia structure (Tsujikawa and Malicki, 2004). While the latter result might indicate that IFT140 is dispensable for IFT, it is also possible that there is some degree of functional overlap/redundancy among different IFT polypeptides, such that if IFT140 is not essential for stability of complex A, the remaining polypeptides may be able to compensate for its loss to some extent.

In summary, the available evidence strongly suggests that complex B is required for anterograde IFT while complex A is required for retrograde IFT. There is some evidence in the literature suggesting that complex A may also play a role in anterograde IFT, but whether this is limited to a few cellular systems is presently unclear. In *C. elegans* evidence from motility studies have suggested that complex A associates with kinesin-II during anterograde IFT and that the kinesin-II–IFT A complex in turn is linked to OSM-3–IFT B complex via Bardet-Biedl syndrome (BBS) proteins (Blacque et al., 2008; Ou et al., 2005a; Scholey, 2008; Snow et al., 2004). Furthermore, a *Chlamydomonas* IFT140 null mutant was reported to completely lack flagella, although the evidence was preliminary (Cole, 2003). Finally, gel filtration analysis of *Chlamydomonas* flagella extracts showed that complex A cofractionates with cytoplasmic dynein 2, and that this fraction also contains kinesin-II but not IFT complex B (Rompolas et al., 2007). While these findings suggest that complex A may play a role in anterograde IFT in addition to functioning in retrograde IFT, as we have previously hypothesized (Pedersen et al., 2006, 2008), it is also possible that IFT A complexes are simply cargoes during anterograde IFT, docked to the kinesin-II motor. More experimental data is required to distinguish between these possibilities.

10. Regulation of IFT

Immunolocalization studies in *Chlamydomonas* demonstrated that IFT particle polypeptides, motor subunits, and flagellar precursors are highly concentrated at the base of the flagella (Cole et al., 1998; Pazour et al., 1999; Qin et al., 2004), specifically near the site where transition fibers contact the flagellar membrane (Deane et al., 2001), suggesting that IFT is initiated by the gathering of IFT components and cargo proteins at this site. Active kinesin-II presumably associates with IFT complexes to which ciliary precursors and inactive cytoplasmic dynein 2 are bound, and transports these anterogradely along the ciliary axoneme toward the tip (Iomini et al., 2001; Kozminski et al., 1995; Pedersen et al., 2006; Qin et al., 2004; Walther et al., 1994). At the tip, significant remodeling of the IFT complexes takes place, precursors are unloaded for assembly, kinesin-II is inactivated, IFT complexes are remodeled, and cytoplasmic dynein 2 becomes active. Following binding of active cytoplasmic dynein 2 to complex A, complex B reassociates with A and, presumably, ciliary turn over products, whereafter cytoplasmic dynein 2 transports everything back to the cell body for recycling (Fig. 2.3; Iomini *et al.*, 2001; Pazour *et al.*, 1999; Pedersen *et al.*, 2006; Porter *et al.*, 1999; Qin *et al.*, 2004). Since both anterograde and retrograde IFT seemingly occur at a constant rate along the flagellar shaft (Kozminski *et al.*, 1993), with slight pauses at the base and tip (Iomini *et al.*, 2001), the main points of regulation of IFT are presumably at the ciliary base and tip.

Our knowledge of the mechanisms regulating IFT motor activity and loading and unloading of IFT cargo proteins at the ciliary base and tip is still very fragmentary, although some key players involved in various aspects of IFT regulation have been identified. First, regulation of ciliary length is influenced by the rates of anterograde IFT and ciliary disassembly (Marshall and Rosenbaum, 2001), and analysis of mutations or biochemical inhibitors that affect ciliary length have revealed a number of kinases as potential regulators of the kinesin-II activity. These include MAP kinases (Bengs *et al.*, 2005; Berman *et al.*, 2003; Burghoorn *et al.*, 2007), NIMA-related kinases (Shalom *et al.*, 2008; Surpili *et al.*, 2003; White and Quarmby, 2008), GSK-3β (Wilson and Lefebvre, 2004) and associated polarity proteins (Pedersen *et al.*, 2008), which may modulate kinesin-II motor activity via phosphorylation. Second, kinesin-II activity may be regulated via conformational changes in the KAP subunit since KAP is required for localization of kinesin-II at the flagellar base and for processive movement of the motor complex along flagella in *Chlamydomonas* (Mueller *et al.*, 2005; Scholey, 2008). Third, polyglutamylation of axonemal tubulin, mediated by the DYF-1 protein and an associated tubulin polyglutamylase (Ttll6), impairs the velocity of kinesin-2 motors in *C. elegans* amphid channel cilia, suggesting that tubulin polyglutamylation may regulate kinesin-2 motor processivity in the cilia (Ou *et al.*, 2005a; Pathak *et al.*, 2007; Scholey, 2008). Finally, in *C. elegans* amphid channel cilia, and potentially other types of cilia that deploy both heterotrimeric and homodimeric kinesin-2 motors during anterograde IFT, BBS proteins appear to play a critical role keeping the two kinesin-2 motors together (Blacque *et al.*, 2004; Ou *et al.*, 2005a; Pan *et al.*, 2006). Thus regulation of kinesin-2 motor activity appears to be quite complex involving a variety of different regulatory mechanisms (e.g., phosphorylation, conformational changes, tubulin modifications) and molecules, but how these different regulatory pathways and molecules interact and are coordinated during the IFT process is unclear. Further complicating the picture are findings suggesting that ciliary length may be regulated not only by changes in motor activity, but also via changes in the association between IFT particles and cargo proteins (Mukhopadhyay *et al.*, 2008; Pan and Snell, 2005; see also Section 12).

While the regulation of anterograde IFT and kinesin-II activity appears to be very complex and not yet fully understood, the regulation of cytoplasmic dynein 2 and retrograde IFT is even more obscure. This may in part be because the motor complex has been difficult to purify biochemically (Rompolas et al., 2007) and because the complex may contain additional components (e.g., LCs) that have not yet been identified (see Section 7). Nevertheless, it seems that in terms of docking of the dynein motor complex onto the anterograde IFT machinery, LC8 plays a critical role, because in a *Chlamydomonas* LC8 mutant, the dynein motor complex fails to enter the flagellar compartment (Rompolas et al., 2007). Furthermore, in terms of regulating the transition between anterograde and retrograde IFT at the flagellar tip, IFT172 seems to play an essential role (Pedersen et al., 2005; Tsao and Gorovsky, 2008a), possibly in conjunction with the small MT-associated protein EB1 (Pedersen et al., 2005), which localizes to the flagellar tip and basal bodies in *Chlamydomonas* (Pedersen et al., 2003). Interestingly, we recently found that EB1 is required for assembly of primary cilia in mouse fibroblasts, although in these cells, the effect of EB1 on ciliogenesis appeared to be mainly at the level of the basal body and not the ciliary tip (Schrøder et al., 2007). Consistent with these findings, EB1 was found to localize to the connecting cilia of rat photoreceptors (Schmitt and Wolfrum, 2001), which is a structure that is similar to the transition zone of other cilia and flagella (Insinna and Besharse, 2008). The EB1-binding partner p150^{Glued} interacts with BBS4 (Kim et al., 2004) as well as with the centrosomal protein Cep290/NPNP6 (Chang et al., 2006), suggesting that p150^{Glued} and EB1 may play a role in regulating ciliogenesis at the level of the transition zone/ciliary pore complex region (see Section 11), although further experiments are required to test this hypothesis. Since mammalian cells contain three EB1-related proteins (EB1, EB2, and EB3; Lansbergen and Akhmanova, 2006) it is also possible that EB2 and/or EB3 participate in some aspects of ciliogenesis, for example, IFT regulation at the ciliary tip. Indeed, preliminary results suggest that EB3, but not EB2, is required for ciliogenesis in cultured human fibroblasts, although the mechanism involved is still unknown (J. M. Schrøder and L. B. Pedersen, unpublished data). Identification of proteins that interact with IFT172 and EB proteins may provide more clues about the mechanisms by which these proteins affect ciliogenesis at the ciliary tip and base.

11. Targeting of Proteins to the Ciliary Compartment: Clues from Ciliary Disease Genes

Although the ciliary membrane is continuous with the plasma membrane of the cell, it is clear from numerous lines of evidence that the ciliary compartment is enriched for proteins, for example, specific membrane receptors and ion channels, that are absent or present in very low amounts in the remaining parts of the cell (Christensen *et al.*, 2007; Rosenbaum and Witman, 2002). Furthermore, specific ciliary targeting sequences/motifs have been identified in a number of ciliary membrane proteins (for reviews, see Christensen *et al.*, 2007 and Chapter 5), implying that cells possess mechanisms for selectively sorting and targeting proteins to the ciliary compartment. How is this specific sorting and targeting of ciliary proteins achieved? Based on EM observations in protists (Bouck, 1971) and cultures of vertebrate fibroblasts and smooth muscle cells (Sorokin, 1962), as well as studies in vertebrate photoreceptors (Deretic and Papermaster, 1991), it was proposed that specific ciliary membrane proteins synthesized on the rough endoplasmatic reticulum are trafficked through the Golgi apparatus and then transported in post-Golgi vesicles to the base of the cilium where the vesicles are exocytosed at the site where ciliary transition fibers contact the membrane, and ciliary proteins then become associated with the IFT machinery. Furthermore, it was proposed that axonemal proteins synthesized in the cytoplasm associate peripherically with post-Golgi vesicles and are transported to the ciliary compartment in a similar fashion (Rosenbaum and Witman, 2002). Consistent with this view, immunogold EM in *Chlamydomonas* revealed that IFT particle polypeptides are highly concentrated at the ciliary transition fiber-membrane contact site (Deane *et al.*, 2001), suggesting that this site indeed is where proteins destined for the ciliary compartment associate with the IFT machinery prior to being transported across the barrier formed by the transition fibers and into the cilium proper. Recent evidence from vertebrate systems as well as nematodes strongly support this scenario for ciliary protein targeting, and have also begun to shed light on the molecular mechanisms involved in different aspects of this process (summarized in Fig. 2.3).

In terms of delivery and fusion of Golgi-derived vesicles to the ciliary base, several key players involved have now been identified, including the IFT complex B proteins IFT20 and DYF-11/MIP-T3 (Follit *et al.*, 2006; Li *et al.*, 2008; Omori *et al.*, 2008), small GTPases such as Rab8 (Nachury *et al.*, 2007), the pleckstrin homology domain-containing protein FAPP2 (Vieira *et al.*, 2006), BBS proteins and their binding partners (Nachury *et al.*, 2007), as well as the AP-1 adaptor complex (Bae *et al.*, 2006; Dwyer *et al.*, 2001; for reviews, see Leroux, 2007; Reiter and Mostov, 2006 and Chapter 5). For example, subcellular localization and siRNA-based studies in cultured vertebrate cells indicated that IFT20 colocalizes with Golgi markers and is required for targeting of polycystin-2 to the ciliary compartment (Follit *et al.*, 2006). Furthermore, recent studies in *C. elegans* and zebrafish showed that IFT20 interacts directly with DYF-11/MIP-T3, which in turn was found to be essential for early stages of ciliogenesis, and which interacts, at least indirectly, with the endocytotic regulator Rabaptin 5 as well as with Rab8 (Kunitomo and Iino, 2008; Li *et al.*, 2008; Omori *et al.*, 2008). Rab8 localizes to cilia and has been implicated in ciliary membrane transport in vertebrates and *C. elegans* (Moritz et al., 2001; Mukhopadhyay et al., 2008; Nachury et al., 2007). In the latter organism, it was suggested that Rab8-mediated vesicle trafficking to the cilium might be controlled by sensory input from specific receptors and ion channels located in the ciliary membrane (Mukhopadhyay et al., 2008). Rab8 also interacts with a complex of at least seven BBS proteins, termed the BBSome, which was proposed to function in ciliary membrane biogenesis via a mechanism that might also include the pericentriolar matrix protein PCM-1 (Nachury et al., 2007).

BBS is a genetically heterogeneous disorder characterized by cognitive impairment, obesity, polydactyly, renal cystic disease, and retinal degeneration. So far a total of 12 BBS genes have been identified, most of which have been characterized in detail at the molecular level (for reviews, see Blacque and Leroux, 2006; Tobin and Beales, 2007). Of particular interest here is a recent study demonstrating that mice lacking the genes encoding BBS2 or BBS4 exhibit a lack of ciliary localization of somatostatin receptor type 3 (Sstr3) and melanin-concentrating hormone receptor 1 (Mchr1) in the brain, indicating that BBS proteins are essential for targeting of specific receptors to the ciliary membrane (Berbari et al., 2008). Consistent with these results, the ciliary TRPV channel OSM-9 was found to be mislocalized in *C. elegans bbs* mutants (Tan et al., 2007). However, BBS proteins have also been suggested to participate in the regulation of anterograde IFT in certain subsets of *C. elegans* sensory cilia via coordination of heterotrimeric and homodimeric kinesin-2 motors (Ou et al., 2005a), and several BBS proteins have been implicated in ubiquitination (Chiang et al., 2006) and proteasome-mediated degradation of β-catenin (Gerdes et al., 2007), suggesting that BBS proteins have multiple functions related to ciliary signaling, spanning from ciliary membrane vesicle/receptor transport and coordination of kinesin-2 motors to proteasome-mediated degradation of Wnt pathway components. This is in line with the pleiotropic clinical manifestations of BBS patients (Blacque and Leroux, 2006; Tobin and Beales, 2007).

As mentioned previously, the Golgi-derived vesicles destined for the ciliary compartment are probably exocytosed at or near the site where transition fibers contact the ciliary membrane (Rosenbaum and Witman, 2002), which is the site where IFT proteins accumulate (Deane et al., 2001). The transition fibers appear to be derived from the distal appendages of the mother centriole (Paintrand et al., 1992), and hence the formation of these appendages is likely to be essential for ciliogenesis. Indeed, studies in mammalian cells suggested that the Odf2 and Cep164 proteins are involved in forming the distal appendages on the mother centriole and depletion of these proteins resulted in failure to assemble primary cilia (Graser et al., 2007; Ishikawa et al., 2005). Interestingly, Odf2 was found to interact directly with Rab8 in yeast-2-hybrid and pull-down assays, suggesting a direct link between the Odf2-containing distal appendages and the post-Golgi vesicle docking site at the ciliary base (Yoshimura et al., 2007). Additional components of these appendages/transition fibers may include importin β-1, an essential component of the nuclear pore complex that interacts with Ran, regulates nuclear import and export and is required for ciliogenesis in MDCK cells (Fan et al., 2007), as well as centrosomal proteins such as pericentrin (Jurczyk et al., 2004; for a review on centrosomal proteins involved in ciliogenesis, see Pedersen et al., 2008). Furthermore, emerging evidence suggests that several proteins associated with nephronophthisis, MKS, and related disorders such as Joubert and Senior-Løken syndromes also play a critical role in ciliogenesis at the ciliary base (for recent reviews, see Lehman et al., 2008; Salomon et al., 2008; von Schnakenburg et al., 2007). In mammals several of the nephrocystins (NPHPs) such as NPHP-4, -5, -6, and -8 as well as some of their binding partners localize to connecting cilia of photoreceptors (Arts et al., 2007; Khanna et al., 2005; Salomon et al., 2008), and for some of the NPHPs a requirement in ciliogenesis has been established. For example, siRNA-mediated depletion of NPHP-6/Cep290 impairs ciliogenesis in cultured retinal pigment epithelial cells (Graser et al., 2007). Further, in C. elegans NPHP-1 and NPHP-4 localize to basal bodies and transition zones of sensory cilia, and mutant animals defective in the genes encoding these proteins display cilia in which certain GFP-tagged IFT proteins display abnormal and reduced ciliary localization patterns, suggesting a role for NPHP-1 and NPHP-4 in regulating ciliary access of the IFT machinery (Jauregui et al., 2008). How might NPHPs regulate such selective entry of proteins into the ciliary compartment? At the nuclear pore access of proteins harboring nuclear localization signals depends on a G protein that, when in its GDP state, binds the protein to be transferred and releases it again in the nucleus when a nuclear guanine nucleotide exchange factor (GEF) converts the G protein to its GTP state (Chook and Blobel, 1999). Given that GEFs and small G proteins have been localized to cilia and ciliary transition zones in a variety of cell types and organisms (Hong et al., 2003; Leroux, 2007; Nair et al., 1999), and that some of the mammalian NPHPs (e.g., NPHP-4, NPHP-5, and NPHP-8) interact with the putative GEF RPGR-ORF15 that is also localized near the ciliary transition zone (Arts et al., 2007; Khanna et al., 2005), one can speculate that NPHPs and associated GEFs may similarly regulate access of ciliary proteins at the base of the cilium via small G proteins.

While the available evidence suggests a role for NPHPs in regulating ciliary access of proteins at the ciliary base, the molecular mechanisms involved are likely to be complex and involve a number of additional proteins that are only now beginning to be uncovered. Thus a recent study in C. elegans (Williams et al., 2008) showed that NPHP-1 and NPHP-4 function redundantly with a complex of three B9 domain-containing proteins (XBX-7, TZA-1, and TZA-2), which are homologous to human MKS1, B9D2/Stumpy, and B9D1 proteins, respectively. Williams and colleagues found that *C. elegans* XBX-7, TZA-1, and TZA-2 localize to the base of sensory cilia, but animals harboring mutations in any one of the three genes encoding these proteins do not have any overt ciliary structural or functional defects. Although the lack of ciliary defects in the mutant animals could be explained by partial retention of B9 protein function (the mutants were not null mutants), when mutations in the genes encoding XBX-7, TZA-1, or TZA-2 were combined with mutations in *NPHP-1* or *NPHP-4*, severe ciliary structural defects were observed (Williams *et al.*, 2008). Consistent with a role for B9 domain proteins in ciliogenesis, loss of the mammalian homologs of XBX-7 (MKS1) or TZA-1 (Stumpy) impairs formation of primary cilia (Dawe *et al.*, 2007b; Town *et al.*, 2008). In addition, for MKS1 as well as for its binding partner MKS3 it was shown that depletion of the proteins impaired migration of centrioles to the ciliary assembly site at the apical plasma membrane, suggesting a role for these MKS proteins in centriole migration and/or docking (Dawe *et al.*, 2007b). Whether MKS proteins contribute to centriole migration/docking in all cell types is uncertain and may depend on cell morphology, because primary fibroblasts from patients with mutations in the newly identified *MKS6/CC2D2A* gene lack primary cilia but display normal localization of centrioles at the apical plasma membrane (Tallila *et al.*, 2008). As more and more ciliary disease genes are being identified and characterized, it is also becoming evident that the phenotype/clinical feature of a specific disease-causing mutation varies greatly depending on the nature of the mutation (Lehman *et al.*, 2008; Marshall, 2008). Consequently, determining the molecular mechanism(s) by which individual ciliary disease proteins function will require careful examination of these proteins at the genetic, biochemical, and cellular levels.

12. IFT AND CILIA-MEDIATED SIGNALING

The ciliary membrane is a repository for receptors and ion channels involved in a number of signaling pathways that control cell growth, behavior, and development, including the vertebrate Shh, PDGFRα, polycystin, and Wnt pathways (Christensen *et al.*, 2007; Eggenschwiler and Anderson, 2007; Pazour and Witman, 2003; Singla and Reiter, 2006). Given the requirement for IFT in building the ciliary axoneme it is perhaps not surprising that mutations in genes encoding IFT particle or motor subunits adversely affect signaling mediated via these pathways (Corbit *et al.*, 2008; Huangfu *et al.*, 2003; Pazour *et al.*, 2000; Schneider *et al.*, 2005). Emerging evidence suggests, however, that IFT also plays a more direct role in cilia-mediated signaling. For example, during mating in

Chlamydomonas, targeting of a cGMP-dependent protein kinase to a discrete flagellar compartment was shown to depend directly on IFT (Wang *et al.*, 2006), and studies of the Shh pathway in vertebrates also strongly support a direct role for IFT in signaling (reviewed in Scholey and Anderson, 2006). One way that IFT could participate directly in signaling is by transporting specific signaling proteins (e.g., receptors and ion channels) into the ciliary compartment in response to environmental cues. As discussed above and summarized in Fig. 2.3, there is ample evidence suggesting that IFT, in conjunction with a number of vesicle transport and adapter proteins such as Rab8 and BBS proteins, is directly involved in transporting specific membrane proteins into the ciliary compartment. Other lines of evidence suggest that such IFT-mediated transport could be modulated in response to environmental cues. For example, recent work in *C. elegans* indicated that sensory signaling input from specific ciliary chemosensory signaling molecules is required to maintain the structure and architecture of AWB olfactory neuron cilia, and that this signaling-mediated remodeling of cilia architecture depends on kinesin-II as well as on BBS proteins and Rab8 (Mukhopadhyay *et al.*, 2008). Further support for the idea that IFT-mediated ciliary targeting of specific signaling molecules may be modulated by environmental signals are provided by observations that IFT mediates vectorial movement of TRPV channels within the ciliary membrane of *C. elegans* sensory neurons (Qin *et al.*, 2005) and that the 7TM receptor protein Smoothened shuttles into and out of the cilia in response to Shh ligand (Corbit *et al.*, 2005; Rohatgi *et al.*, 2007). The mechanisms by which environmental signals affect IFT-mediated transport of specific signaling molecules are still elusive, however, but appear not to involve alterations of IFT motor activity (Mukhopadhyay *et al.*, 2008). Rather, it is likely that environmental cues affect the loading/unloading of specific IFT cargoes at the ciliary base and tip. Indeed, there is evidence from *Chlamydomonas* indicating that signals that trigger disassembly of the flagellum can regulate the association of axonemal proteins with IFT particles at the flagellar base and tip (Pan and Snell, 2005). How such environmental signals are translated to regulate IFT–cargo protein interactions is, however, unclear.

While the role for IFT in targeting specific signaling proteins to the ciliary compartment appears to be important for cilia-mediated signal transduction, the downregulation, processing and/or removal of ciliary signaling proteins is likely to be equally important. Indeed, partial or transient inhibition of IFT in *Chlamydomonas*, *C. elegans*, or vertebrate cells was shown to result in accumulation of polycystin-2 in the cilium, indicating that IFT plays a role in removing polycystin-2 from the ciliary compartment (Bae and Barr, 2008; Huang *et al.*, 2007; Pazour *et al.*, 2002). Moreover, evidence from vertebrates indicated a requirement for IFT in the processing of Gli transcription factors of the Shh pathway, which directly impinges on transcription of Shh-responsive genes (Haycraft *et al.*, 2005). The notion that IFT-mediated processing of ciliary signaling components affects transcription of downstream target genes is also supported by studies indicating that the ciliary polycystin-1 receptor becomes processed following mechanical stimulation, and subsequently translocates to the nucleus to directly regulate transcription via the transcription factor STAT6 and its coactivator P100 (Chauvet et al., 2004; Low et al., 2006). The extent to which IFT is involved in this processing, however, and the exact mechanisms by which such IFT-mediated processing might occur are not clear.

13. CONCLUSIONS AND PERSPECTIVES

Since its discovery in 1993 (Kozminski et al., 1993), significant progress has been made toward understanding the molecular mechanisms and functions of IFT. Virtually all the IFT particle polypeptides and motor subunits have been identified, and most of them characterized at the functional level in multiple organisms. However, we still know relatively little about how IFT is regulated; what is clear, however, is that IFT regulation is complex involving many different molecules and mechanisms that may affect motor activity and/or the association between IFT particles and their cargo at the ciliary base and tip. What is also quite clear is that there is an intimate relationship between IFT regulation, environmental signals, and the cell cycle, which is only now beginning to be uncovered.

One important aspect of IFT regulation involves the association of ciliary precursors and IFT particles at the ciliary base, which likely is controlled, at least in part, by sensory input via the ciliary membrane. Combined efforts from many different groups have revealed a growing list of disease-related proteins, including BBS, NPHPs, and MKS-associated proteins, which may be critical for regulating access of proteins into the ciliary compartment and potentially the association of such proteins with IFT particles. A future challenge will be to define the exact interactions between the large number of proteins that constitute or function in the vicinity of the ciliary transition fibers, and to determine how these interactions are regulated in response to environmental and cellular signals. To this end, it will be necessary to employ biochemical methods such as fractionation and sucrose density gradient analysis of isolated ciliary preparations, for which organisms like *Chlamydomonas* are ideally suited. Finally, it is becoming apparent that while IFT plays a critical role in bringing precursors and signaling proteins into the ciliary compartment, it may have an equally important role in their subsequent processing and removal. There is good evidence that processed ciliary proteins (e.g., receptors and transcription factors) can affect transcription of specific target genes. This crosstalk between the cilium and nucleus will be an interesting avenue for future research. Among candidate proteins involved in regulating transcription in response to ciliary signals are proteins associated with the development of ciliopathies, but which do not localize to the cilium, such as the recently identified Seahorse protein from zebrafish (Kishimoto et al., 2008).

ACKNOWLEDGMENTS

Research in the authors' laboratories is supported by grants from the Danish Natural Science Research Council (nr. 272-05-0411 and 272-07-0503), The Novo Nordisk Foundation, and The Lundbeck Foundation (LBP) as well as the National Institutes of Health (GMO14642; JLR). We are grateful to Drs. Karl F. Lechtreck, George B. Witman, David R. Mitchell, Cynthia G. Jensen, Randy H. Cardon, and Stefan Geimer for EM images of cilia and flagella and to Jacob M. Schrøder for Figs. 2.1 and 2.3.

REFERENCES

Absalon, S., Blisnick, T., Kohl, L., Toutirais, G., Dore, G., Julkowska, D., Tavenet, A., and Bastin, P. (2008). Intraflagellar transport and functional analysis of genes required for flagellum formation in trypanosomes. *Mol. Biol. Cell* **19,** 929–944.

Adams, G. M. W., Huang, B., and Luck, D. J. L. (1982). Temperature-sensitive, assembly-defective flagella mutants of *Chlamydomonas reinhardtii*. *Genetics* **100,** 579–586.

Arts, H. H., Doherty, D., van Beersum, S. E., Parisi, M. A., Letteboer, S. J., Gorden, N. T., Peters, T. A., Marker, T., Voesenek, K., Kartono, A., Ozyurek, H., Farin, F. M., et al. (2007). Mutations in the gene encoding the basal body protein RPGRIP1L, a nephro-cystin-4 interactor, cause Joubert syndrome. *Nat. Genet.* **39,** 882–888.

Avidor-Reiss, T., Maer, A. M., Koundakjian, E., Polyanovsky, A., Keil, T., Subramaniam, S., and Zuker, C. S. (2004). Decoding cilia function: Defining specialized genes required for compartmentalized cilia biogenesis. *Cell* **117,** 527–539.

Azimzadeh, J., and Bornens, M. (2007). Structure and duplication of the centrosome. *J. Cell Sci.* **120,** 2139–2142.

Badano, J. L., Mitsuma, N., Beales, P. L., and Katsanis, N. (2006). The ciliopathies: An emerging class of human genetic disorders. *Annu. Rev. Genomics Hum. Genet.* **7,** 125–148.

Bae, Y. K., and Barr, M. M. (2008). Sensory roles of neuronal cilia: Cilia development, morphogenesis, and function in *C. elegans*. *Front. Biosci.* **13,** 5959–5974.

Bae, Y. K., Qin, H., Knobel, K. M., Hu, J., Rosenbaum, J. L., and Barr, M. M. (2006). General and cell-type specific mechanisms target TRPP2/PKD-2 to cilia. *Development* **133,** 3859–3870.

Baron, D. M., Ralston, K. S., Kabututu, Z. P., and Hill, K. L. (2007). Functional genomics in *Trypanosoma brucei* identifies evolutionarily conserved components of motile flagella. *J. Cell Sci.* **120,** 478–491.

Barr, M. M., and Sternberg, P. W. (1999). A polycystic kidney-disease gene homologue required for male mating behaviour in *C. elegans*. *Nature* **401,** 386–389.

Beales, P. L., Bland, E., Tobin, J. L., Bacchelli, C., Tuysuz, B., Hill, J., Rix, S., Pearson, C. G., Kai, M., Hartley, J., Johnson, C., Irving, M., et al. (2007). IFT80, which encodes a conserved intraflagellar transport protein, is mutated in Jeune asphyxiating thoracic dystrophy. *Nat. Genet.* **39,** 727–729.

Bengs, F., Scholz, A., Kuhn, D., and Wiese, M. (2005). LmxMPK9, a mitogen-activated protein kinase homologue affects flagellar length in *Leishmania mexicana*. *Mol. Microbiol.* **55,** 1606–1615.

Berbari, N. F., Lewis, J. S., Bishop, G. A., Askwith, C. C., and Mykytyn, K. (2008). Bardet-Biedl syndrome proteins are required for the localization of G protein-coupled receptors to primary cilia. *Proc. Natl Acad. Sci. USA* **105,** 4242–4246.

Berman, S. A., Wilson, N. F., Haas, N. A., and Lefebvre, P. A. (2003). A novel MAP kinase regulates flagellar length in *Chlamydomonas*. *Curr. Biol.* **13,** 1145–1149.

Blacque, O. E., and Leroux, M. R. (2006). Bardet-Biedl syndrome: An emerging pathomechanism of intracellular transport. *Cell. Mol. Life Sci.* **63,** 2145–2161.

Blacque, O. E., Reardon, M. J., Li, C., McCarthy, J., Mahjoub, M. R., Ansley, S. J., Badano, J. L., Mah, A. K., Beales, P. L., Davidson, W. S., Johnsen, R. C., Audeh, M., et al. (2004). Loss of *C. elegans* BBS-7 and BBS-8 protein function results in cilia defects and compromised intraflagellar transport. *Genes Dev.* **18,** 1630–1642.

Blacque, O. E., Perens, E. A., Boroevich, K. A., Inglis, P. N., Li, C., Warner, A., Khattra, J., Holt, R. A., Ou, G., Mah, A. K., McKay, S. J., Huang, P., et al. (2005). Functional genomics of the cilium, a sensory organelle. *Curr. Biol.* **15,** 935–941.

Blacque, O. E., Li, C., Inglis, P. N., Esmail, M. A., Ou, G., Mah, A. K., Baillie, D. L., Scholey, J. M., and Leroux, M. R. (2006). The WD repeat-containing protein IFTA-1 is required for retrograde intraflagellar transport. *Mol. Biol. Cell* **17,** 5053–5062.

Blacque, O. E., Cevik, S., and Kaplan, O. I. (2008). Intraflagellar transport: From molecular characterisation to mechanism. *Front. Biosci.* **13,** 2633–2652.

Bouck, G. B. (1971). The structure, origin, isolation, and composition of the tubular mastigonemes of the *Ochromonas* flagellum. *J. Cell Biol.* **50,** 362–384.

Brazelton, W. J., Amundsen, C. D., Silflow, C. D., and Lefebvre, P. A. (2001). The *bld1* mutation identifies the *Chlamydomonas osm-6* homolog as a gene required for flagellar assembly. *Curr. Biol.* **11,** 1591–1594.

Brown, J. M., Marsala, C., Kosoy, R., and Gaertig, J. (1999). Kinesin-II is preferentially targeted to assembling cilia and is required for ciliogenesis and normal cytokinesis in *Tetrahymena*. *Mol. Biol. Cell* **10,** 3081–3096.

Brown, J. M., Fine, N. A., Pandiyan, G., Thazhath, R., and Gaertig, J. (2003). Hypoxia regulates assembly of cilia in suppressors of *Tetrahymena* lacking an intraflagellar transport subunit gene. *Mol. Biol. Cell* **14,** 3192–3207.

Burghoorn, J., Dekkers, M. P., Rademakers, S., de Jong, T., Willemsen, R., and Jansen, G. (2007). Mutation of the MAP kinase DYF-5 affects docking and undocking of kinesin-2 motors and reduces their speed in the cilia of *Caenorhabditis elegans*. *Proc. Natl Acad. Sci. USA* **104,** 7157–7162.

Chang, B., Khanna, H., Hawes, N., Jimeno, D., He, S., Lillo, C., Parapuram, S. K., Cheng, H., Scott, A., Hurd, R. E., Sayer, J. A., Otto, E. A., et al. (2006). In-frame deletion in a novel centrosomal/ciliary protein CEP290/NPHP6 perturbs its interaction with RPGR and results in early-onset retinal degeneration in the rd16 mouse. *Hum. Mol. Genet.* **15,** 1847–1857.

Chauvet, V., Tian, X., Husson, H., Grimm, D. H., Wang, T., Hiesberger, T., Igarashi, P., Bennett, A. M., Ibraghimov-Beskrovnaya, O., Somlo, S., and Caplan, M. J. (2004). Mechanical stimuli induce cleavage and nuclear translocation of the polycystin-1 C terminus. *J. Clin. Invest.* **114,** 1433–1443.

Chiang, A. P., Beck, J. S., Yen, H. J., Tayeh, M. K., Scheetz, T. E., Swiderski, R. E., Nishimura, D. Y., Braun, T. A., Kim, K. Y., Huang, J., Elbedour, K., Carmi, R., et al. (2006). Homozygosity mapping with SNP arrays identifies TRIM32, an E3 ubiquitin ligase, as a Bardet-Biedl syndrome gene (BBS11). *Proc. Natl Acad. Sci. USA* **103,** 6287–6292.

Chook, Y. M., and Blobel, G. (1999). Structure of the nuclear transport complex karyopherin-beta2-Ran × GppNHp. *Nature* **399,** 230–237.

Christensen, S. T., Pedersen, L. B., Schneider, L., and Satir, P. (2007). Sensory cilia and integration of signal transduction in human health and disease. *Traffic* **8,** 97–109.

Cole, D. G. (2003). The intraflagellar transport machinery of *Chlamydomonas reinhardtii*. *Traffic* **4,** 1–8.

Cole, D. G., Cande, W. Z., Baskin, R. J., Skoufias, D. A., Hogan, C. J., and Scholey, J. M. (1992). Isolation of a sea urchin egg kinesin-related protein using peptide antibodies. *J. Cell Sci.* **101,** 291–301.

Cole, D. G., Chinn, S. W., Wedaman, K. P., Hall, K., Vuong, T., and Scholey, J. M. (1993). Novel heterotrimeric kinesin-related protein purified from sea urchin eggs. *Nature* **366,** 268–270.

Cole, D. G., Diener, D. R., Himelblau, A. L., Beech, P. L., Fuster, J. C., and Rosenbaum, J. L. (1998). *Chlamydomonas* kinesin-II-dependent intraflagellar transport (IFT): IFT particles contain proteins required for ciliary assembly in *Caenorhabditis elegans* sensory neurons. *J. Cell Biol.* **141,** 993–1008.

Collet, J., Spike, C. A., Lundquist, E. A., Shaw, J. E., and Herman, R. K. (1998). Analysis of *osm-6*, a gene that affects sensory cilium structure and sensory neuron function in *C. elegans*. *Genetics* **148,** 187–200.

Corbit, K. C., Aanstad, P., Singla, V., Norman, A. R., Stainier, D. Y., and Reiter, J. F. (2005). Vertebrate Smoothened functions at the primary cilium. *Nature* **437,** 1018–1021.

Corbit, K. C., Shyer, A. E., Dowdle, W. E., Gaulden, J., Singla, V., Chen, M. H., Chuang, P. T., and Reiter, J. F. (2008). Kif3a constrains beta-catenin-dependent Wnt signalling through dual ciliary and non-ciliary mechanisms. *Nat. Cell Biol.* **10,** 70–76.

Dabdoub, A., and Kelley, M. W. (2005). Planar cell polarity and a potential role for a Wnt morphogen gradient in stereociliary bundle orientation in the mammalian inner ear. *J. Neurobiol.* **64,** 446–457.

Davenport, J. R., and Yoder, B. K. (2005). An incredible decade for the primary cilium: A look at a once-forgotten organelle. *Am. J. Physiol. Renal Physiol.* **289,** F1159–F1169.

Dawe, H. R., Farr, H., and Gull, K. (2007a). Centriole/basal body morphogenesis and migration during ciliogenesis in animal cells. *J. Cell Sci.* **120,** 7–15.

Dawe, H. R., Smith, U. M., Cullinane, A. R., Gerrelli, D., Cox, P., Badano, J. L., Blair-Reid, S., Sriram, N., Katsanis, N., Attie-Bitach, T., Afford, S. C., Copp, A. J., *et al.* (2007b). The Meckel-Gruber syndrome proteins MKS1 and meckelin interact and are required for primary cilium formation. *Hum. Mol. Genet.* **16,** 173–186.

Deane, J. A., Cole, D. G., Seeley, E. S., Diener, D. R., and Rosenbaum, J. L. (2001). Localization of intraflagellar transport protein IFT52 identifies basal body transitional fibers as the docking site for IFT particles. *Curr. Biol.* **11,** 1586–1590.

Deretic, D., and Papermaster, D. S. (1991). Polarized sorting of rhodopsin on post-Golgi membranes in frog retinal photoreceptor cells. *J. Cell Biol.* **113,** 1281–1293.

Dirksen, E. R. (1991). Centriole and basal body formation during ciliogenesis revisited. *Biol. Cell* **72,** 31–38.

Dwyer, N. D., Adler, C. E., Crump, J. G., L'Etoile, N. D., and Bargmann, C. I. (2001). Polarized dendritic transport and the AP-1 mu1 clathrin adaptor UNC-101 localize odorant receptors to olfactory cilia. *Neuron* **31,** 277–287.

Efimenko, E., Bubb, K., Mak, H. Y., Holzman, T., Leroux, M. R., Ruvkun, G., Thomas, J. H., and Swoboda, P. (2005). Analysis of *xbx* genes in *C. elegans*. *Development* **132,** 1923–1934.

Eggenschwiler, J. T., and Anderson, K. V. (2007). Cilia and developmental signaling. *Annu. Rev. Cell Dev. Biol.* **23,** 345–373.

Fan, S., Fogg, V., Wang, Q., Chen, X. W., Liu, C. J., and Margolis, B. (2007). A novel Crumbs3 isoform regulates cell division and ciliogenesis via importin beta interactions. *J. Cell Biol.* **178,** 387–398.

Feistel, K., and Blum, M. (2006). Three types of cilia including a novel 9 + 4 axoneme on the notochordal plate of the rabbit embryo. *Dev. Dyn.* **235,** 3348–3358.

Flannery, R. J., French, D. A., and Kleene, S. J. (2006). Clustering of cyclic-nucleotide-gated channels in olfactory cilia. *Biophys. J.* **91,** 179–188.

Fliegauf, M., Benzing, T., and Omran, H. (2007). When cilia go bad: Cilia defects and ciliopathies. *Nat. Rev. Mol. Cell Biol.* **8,** 880–893.

Follit, J. A., Tuft, R. A., Fogarty, K. E., and Pazour, G. J. (2006). The intraflagellar transport protein IFT20 is associated with the Golgi complex and is required for cilia assembly. *Mol. Biol. Cell* **17,** 3781–3792.

Fujiwara, M., Ishihara, T., and Katsura, I. (1999). A novel WD40 protein, CHE-2, acts cell-autonomously in the formation of *C. elegans* sensory cilia. *Development* **126,** 4839–4848.

Gerdes, J. M., Liu, Y., Zaghloul, N. A., Leitch, C. C., Lawson, S. S., Kato, M., Beachy, P. A., Beales, P. L., DeMartino, G. N., Fisher, S., Badano, J. L., and Katsanis, N. (2007). Disruption of the basal body compromises proteasomal function and perturbs intracellular Wnt response. *Nat. Genet.* **39,** 1350–1360.

Gibbons, B. H., Asai, D. J., Tang, W. J., Hays, T. S., and Gibbons, I. R. (1994). Phylogeny and expression of axonemal and cytoplasmic dynein genes in sea urchins. *Mol. Biol. Cell* **5,** 57–70.

Gilula, N. B., and Satir, P. (1972). The ciliary necklace. A ciliary membrane specialization. *J. Cell Biol.* **53,** 494–509.

Graser, S., Stierhof, Y. D., Lavoie, S. B., Gassner, O. S., Lamla, S., Le Clech, M., and Nigg, E. A. (2007). Cep164, a novel centriole appendage protein required for primary cilium formation. *J. Cell Biol.* **179,** 321–330.

Grissom, P. M., Vaisberg, E. A., and McIntosh, J. R. (2002). Identification of a novel light intermediate chain (D2LIC) for mammalian cytoplasmic dynein 2. *Mol. Biol. Cell* **13,** 817–829.

Han, Y. G., Kwok, B. H., and Kernan, M. J. (2003). Intraflagellar transport is required in *Drosophila* to differentiate sensory cilia but not sperm. *Curr. Biol.* **13,** 1679–1686.

Haycraft, C. J., Swoboda, P., Taulman, P. D., Thomas, J. H., and Yoder, B. K. (2001). The *C. elegans* homolog of the murine cystic kidney disease gene *Tg737* functions in a ciliogenic pathway and is disrupted in *osm-5* mutant worms. *Development* **128,** 1493–1505.

Haycraft, C. J., Schafer, J. C., Zhang, Q., Taulman, P. D., and Yoder, B. K. (2003). Identification of CHE-13, a novel intraflagellar transport protein required for cilia formation. *Exp. Cell Res.* **284,** 251–263.

Haycraft, C. J., Banizs, B., Aydin-Son, Y., Zhang, Q., Michaud, E. J., and Yoder, B. K. (2005). Gli2 and Gli3 localize to cilia and require the intraflagellar transport protein polaris for processing and function. *PLoS Genet.* **1,** e53.

Hirokawa, N., Tanaka, Y., Okada, Y., and Takeda, S. (2006). Nodal flow and the generation of left–right asymmetry. *Cell* **125,** 33–45.

Hong, D. H., Pawlyk, B., Sokolov, M., Strissel, K. J., Yang, J., Tulloch, B., Wright, A. F., Arshavsky, V. Y., and Li, T. (2003). RPGR isoforms in photoreceptor connecting cilia and the transitional zone of motile cilia. *Invest. Ophthalmol. Vis. Sci.* **44,** 2413–2421.

Hou, Y., Pazour, G. J., and Witman, G. B. (2004). A dynein light intermediate chain, D1bLIC, is required for retrograde intraflagellar transport. *Mol. Biol. Cell* **15,** 4382–4394.

Hou, Y., Qin, H., Follit, J. A., Pazour, G. J., Rosenbaum, J. L., and Witman, G. B. (2007). Functional analysis of an individual IFT protein: IFT46 is required for transport of outer dynein arms into flagella. *J. Cell Biol.* **176,** 653–665.

Houde, C., Dickinson, R. J., Houtzager, V. M., Cullum, R., Montpetit, R., Metzler, M., Simpson, E. M., Roy, S., Hayden, M. R., Hoodless, P. A., and Nicholson, D. W. (2006). Hippi is essential for node cilia assembly and Sonic hedgehog signaling. *Dev. Biol.* **300,** 523–533.

Huang, B., Rifkin, M. R., Luck, D. J. L., and Kozler, V. (1977). Temperature-sensitive mutations affecting flagellar assembly and function in *Chlamydomonas reinhardtii*. *J. Cell Biol.* **72,** 67–85.

Huang, K., Diener, D. R., Mitchell, A., Pazour, G. J., Witman, G. B., and Rosenbaum, J. L. (2007). Function and dynamics of PKD2 in *Chlamydomonas reinhardtii* flagella. *J. Cell Biol.* **179,** 501–514.

Huangfu, D., and Anderson, K. V. (2005). Cilia and Hedgehog responsiveness in the mouse. *Proc. Natl Acad. Sci. USA* **102,** 11325–11330.

Huangfu, D., Liu, A., Rakeman, A. S., Murcia, N. S., Niswander, L., and Anderson, K. V. (2003). Hedgehog signalling in the mouse requires intraflagellar transport proteins. *Nature* **426,** 83–87.

Ibanez-Tallon, I., Heintz, N., and Omran, H. (2003). To beat or not to beat: Roles of cilia in development and disease. *Hum. Mol. Genet.* **12,** R27–R35.

Insinna, C., and Besharse, J. C. (2008). Intraflagellar transport and the sensory outer segment of vertebrate photoreceptors. *Dev. Dyn.* **237,** 1982–1992.

Insinna, C., Pathak, N., Perkins, B., Drummond, I., and Besharse, J. C. (2008). The homodimeric kinesin, Kif17, is essential for vertebrate photoreceptor sensory outer segment development. *Dev. Biol.* **316,** 160–170.

Iomini, C., Babaev-Khaimov, V., Sassaroli, M., and Piperno, G. (2001). Protein particles in *Chlamydomonas* flagella undergo a transport cycle consisting of four phases. *J. Cell Biol.* **153,** 13–14.

Ishikawa, H., Kubo, A., and Tsukita, S. (2005). Odf2-deficient mother centrioles lack distal/subdistal appendages and the ability to generate primary cilia. *Nat. Cell Biol.* **7,** 517–524.

Jauregui, A. R., Nguyen, K. C., Hall, D. H., and Barr, M. M. (2008). The *Caenorhabditis elegans* nephrocystins act as global modifiers of cilium structure. *J. Cell Biol.* **180,** 973–988.

Jekely, G., and Arendt, D. (2006). Evolution of intraflagellar transport from coated vesicles and autogenous origin of the eukaryotic cilium. *BioEssays* **28,** 191–198.

Jenkins, P. M., Hurd, T. W., Zhang, L., McEwen, D. P., Brown, R. L., Margolis, B., Verhey, K. J., and Martens, J. R. (2006). Ciliary targeting of olfactory CNG channels requires the CNGB1b subunit and the kinesin-2 motor protein, KIF17. *Curr. Biol.* **16,** 1211–1216.

Jensen, C. G., Poole, C. A., McGlashan, S. R., Marko, M., Issa, Z. I., Vujcich, K. V., and Bowser, S. S. (2004). Ultrastructural, tomographic and confocal imaging of the chondrocyte primary cilium in situ. *Cell Biol. Int.* **28,** 101–110.

Johnson, K. A., and Rosenbaum, J. L. (1992). Polarity of flagellar assembly in *Chlamydomonas*. *J. Cell Biol.* **119,** 1605–1611.

Jurczyk, A., Gromley, A., Redick, S., San Agustin, J., Witman, G., Pazour, G. J., Peters, D. J., and Doxsey, S. (2004). Pericentrin forms a complex with intraflagellar transport proteins and polycystin-2 and is required for primary cilia assembly. *J. Cell Biol.* **166,** 637–643.

Kessel, R., and Kardon, R. (1979). "Tissues and Organs: A Text-Atlas of Scanning Electron Microscopy." WH Freeman and Company, San Francisco, CA.

Khanna, H., Hurd, T. W., Lillo, C., Shu, X., Parapuram, S. K., He, S., Akimoto, M., Wright, A. F., Margolis, B., Williams, D. S., and Swaroop, A. (2005). RPGR-ORF15, which is mutated in retinitis pigmentosa, associates with SMC1, SMC3, and microtubule transport proteins. *J. Biol. Chem.* **280,** 33580–33587.

Kim, J. C., Badano, J. L., Sibold, S., Esmail, M. A., Hill, J., Hoskins, B. E., Leitch, C. C., Venner, K., Ansley, S. J., Ross, A. J., Leroux, M. R., Katsanis, N., *et al.* (2004). The Bardet-Biedl protein BBS4 targets cargo to the pericentriolar region and is required for microtubule anchoring and cell cycle progression. *Nat. Genet.* **36,** 462–470.

King, S. M. (2003). Organization and regulation of the dynein microtubule motor. *Cell Biol. Int.* **27,** 213–215.

Kishimoto, N., Cao, Y., Park, A., and Sun, Z. (2008). Cystic kidney gene seahorse regulates cilia-mediated processes and Wnt pathways. *Dev. Cell* **14,** 954–961.

Kobayashi, T., Gengyo-Ando, K., Ishihara, T., Katsura, I., and Mitani, S. (2007). IFT-81 and IFT-74 are required for intraflagellar transport in C. elegans. *Genes Cells* **12,** 593–602.

Kovacs, J. J., Whalen, E. J., Liu, R., Xiao, K., Kim, J., Chen, M., Wang, J., Chen, W., and Lefkowitz, R. J. (2008). Beta-arrestin-mediated localization of smoothened to the primary cilium. *Science* **320,** 1777–1781.

Kozminski, K. G., Johnson, K. A., Forscher, P., and Rosenbaum, J. L. (1993). A motility in the eukaryotic flagellum unrelated to flagellar beating. *Proc. Natl Acad. Sci. USA* **90,** 5519–5523.

Kozminski, K. G., Beech, P. L., and Rosenbaum, J. L. (1995). The *Chlamydomonas* kinesin-like protein FLA10 is involved in motility associated with the flagellar membrane. *J. Cell Biol.* **131,** 1517–1527.

Kunitomo, H., and Iino, Y. (2008). *Caenorhabditis elegans* DYF-11, an orthologue of mammalian Traf3ip1/MIP-T3, is required for sensory cilia formation. *Genes Cells* **13,** 13–25.

Lansbergen, G., and Akhmanova, A. (2006). Microtubule plus end: A hub of cellular activities. *Traffic* **7,** 499–507.

Lehman, J. M., Michaud, E. J., Schoeb, T. R., Aydin-Son, Y., Miller, M., and Yoder, B. K. (2008). The Oak Ridge Polycystic Kidney mouse: Modeling ciliopathies of mice and men. *Dev. Dyn.* **237,** 1960–1971.

Leroux, M. R. (2007). Taking vesicular transport to the cilium. *Cell* **129,** 1041–1043.

Li, J. B., Gerdes, J. M., Haycraft, C. J., Fan, Y., Teslovich, T. M., May-Simera, H., Li, H., Blacque, O. E., Li, L., Leitch, C. C., Lewis, R. A., Green, J. S., *et al.* (2004). Comparative genomics identifies a flagellar and basal body proteome that includes the BBS5 human disease gene. *Cell* **117,** 541–552.

Li, C., Inglis, P. N., Leitch, C. C., Efimenko, E., Zaghloul, N. A., Mok, C. A., Davis, E. E., Bialas, N. J., Healey, M. P., Heon, E., Zhen, M., Swoboda, P., *et al.* (2008). An essential role for DYF-11/MIP-T3 in assembling functional intraflagellar transport complexes. *PLoS Genet.* **4,** e1000044.

Low, S. H., Vasanth, S., Larson, C. H., Mukherjee, S., Sharma, N., Kinter, M. T., Kane, M. E., Obara, T., and Weimbs, T. (2006). Polycystin-1, STAT6, and P100 function in a pathway that transduces ciliary mechanosensation and is activated in polycystic kidney disease. *Dev. Cell* **10,** 57–69.

Lucker, B. F., Behal, R. H., Qin, H., Siron, L. C., Taggart, W. D., Rosenbaum, J. L., and Cole, D. G. (2005). Characterization of the intraflagellar transport complex B core: Direct interaction of the IFT81 and IFT74/72 subunits. *J. Biol. Chem.* **280,** 27688–27696.

Marshall, W. F. (2008). The cell biological basis of ciliary disease. *J. Cell Biol.* **180,** 17–21.

Marshall, W. F., and Rosenbaum, J. L. (2001). Intraflagellar transport balances continuous turnover of outer doublet microtubules: Implications for flagellar length control. *J. Cell Biol.* **155,** 405–414.

Marszalek, J. R., Ruiz-Lozano, P., Roberts, E., Chien, K. R., and Goldstein, L. S. (1999). Situs inversus and embryonic ciliary morphogenesis defects in mouse mutants lacking the KIF3A subunit of kinesin-II. *Proc. Natl Acad. Sci. USA* **96,** 5043–5048.

May, S. R., Ashique, A. M., Karlen, M., Wang, B., Shen, Y., Zarbalis, K., Reiter, J., Ericson, J., and Peterson, A. S. (2005). Loss of the retrograde motor for IFT disrupts localization of Smo to cilia and prevents the expression of both activator and repressor functions of Gli. *Dev. Biol.* **287,** 378–389.

Merchant, S. S., Prochnik, S. E., Vallon, O., Harris, E. H., Karpowicz, S. J., Witman, G. B., Terry, A., Salamov, A., Fritz-Laylin, L. K., Marechal-Drouard, L., Marshall, W. F., Qu, L. H., et al. (2007). The *Chlamydomonas* genome reveals the evolution of key animal and plant functions. *Science* **318**, 245–250.

Mikami, A., Tynan, S. H., Hama, T., Luby-Phelps, K., Saito, T., Crandall, J. E., Besharse, J. C., and Vallee, R. B. (2002). Molecular structure of cytoplasmic dynein 2 and its distribution in neuronal and ciliated cells. *J. Cell Sci.* **115**, 4801–4808.

Miller, M. S., Esparza, J. M., Lippa, A. M., Lux, F. G., III, Cole, D. G., and Dutcher, S. K. (2005). Mutant kinesin-2 motor subunits increase chromosome loss. *Mol. Biol. Cell* **16**, 3810–3820.

Mitchell, B. F., Pedersen, L. B., Feely, M., Rosenbaum, J. L., and Mitchell, D. R. (2005). ATP production in *Chlamydomonas reinhardtii* flagella by glycolytic enzymes. *Mol. Biol. Cell* **16**, 4509–4518.

Moritz, O. L., Tam, B. M., Hurd, L. L., Peranen, J., Deretic, D., and Papermaster, D. S. (2001). Mutant rab8 Impairs docking and fusion of rhodopsin-bearing post-Golgi membranes and causes cell death of transgenic Xenopus rods. *Mol. Biol. Cell* **12**, 2341–2351.

Morris, R. L., and Scholey, J. M. (1997). Heterotrimeric kinesin-II is required for the assembly of motile 9 + 2 ciliary axonemes on sea urchin embryos. *J. Cell Biol.* **138**, 1009–1022.

Mrug, M., Li, R., Cui, X., Schoeb, T. R., Churchill, G. A., and Guay-Woodford, L. M. (2005). Kinesin family member 12 is a candidate polycystic kidney disease modifier in the cpk mouse. *J. Am. Soc. Nephrol.* **16**, 905–916.

Mueller, J., Perrone, C. A., Bower, R., Cole, D. G., and Porter, M. E. (2005). The *FLA3* KAP subunit is required for localization of kinesin-2 to the site of flagellar assembly and processive anterograde intraflagellar transport. *Mol. Biol. Cell* **16**, 1341–1354.

Mukhopadhyay, S., Lu, Y., Shaham, S., and Sengupta, P. (2008). Sensory signaling-dependent remodeling of olfactory cilia architecture in *C. elegans*. *Dev. Cell* **14**, 762–774.

Murayama, T., Toh, Y., Ohshima, Y., and Koga, M. (2005). The *dyf-3* gene encodes a novel protein required for sensory cilium formation in *Caenorhabditis elegans*. *J. Mol. Biol.* **346**, 677–687.

Nachury, M. V., Loktev, A. V., Zhang, Q., Westlake, C. J., Peranen, J., Merdes, A., Slusarski, D. C., Scheller, R. H., Bazan, J. F., Sheffield, V. C., and Jackson, P. K. (2007). A core complex of BBS proteins cooperates with the GTPase Rab8 to promote ciliary membrane biogenesis. *Cell* **129**, 1201–1213.

Nair, S., Guerra, C., and Satir, P. (1999). A Sec7-related protein in *Paramecium*. *FASEB J.* **13**, 1249–1257.

Nakagawa, T., Setou, M., Seog, D., Ogasawara, K., Dohmae, N., Takio, K., and Hirokawa, N. (2000). A novel motor, KIF13A, transports mannose-6-phosphate receptor to plasma membrane through direct interaction with AP-1 complex. *Cell* **103**, 569–581.

Nonaka, S., Tanaka, Y., Okada, Y., Takeda, S., Harada, A., Kanai, Y., Kido, M., and Hirokawa, N. (1998). Randomization of left–right asymmetry due to loss of nodal cilia generating leftward flow of extraembryonic fluid in mice lacking KIF3B motor protein. *Cell* **95**, 829–837.

Omori, Y., Zhao, C., Saras, A., Mukhopadhyay, S., Kim, W., Furukawa, T., Sengupta, P., Veraksa, A., and Malicki, J. (2008). Elipsa is an early determinant of ciliogenesis that links the IFT particle to membrane-associated small GTPase Rab8. *Nat. Cell Biol.* **10**, 437–444.

Orozco, J. T., Wedaman, K. P., Signor, D., Brown, M., Rose, L., and Scholey, J. M. (1999). Movement of motor and cargo along cilia. *Nature* **398**, 674.

Ou, G., Blacque, O. E., Snow, J. J., Leroux, M. R., and Scholey, J. M. (2005a). Functional coordination of intraflagellar transport motors. *Nature* **436**, 583–587.

Ou, G., Qin, H., Rosenbaum, J. L., and Scholey, J. M. (2005b). The PKD protein qilin undergoes intraflagellar transport. *Curr. Biol.* **15,** R410–R411.

Paintrand, M., Moudjou, M., Delacroix, H., and Bornens, M. (1992). Centrosome organization and centriole architecture: Their sensitivity to divalent cations. *J. Struct. Biol.* **108,** 107–128.

Pan, J., and Snell, W. J. (2005). *Chlamydomonas* shortens its flagella by activating axonemal disassembly, stimulating IFT particle trafficking, and blocking anterograde cargo loading. *Dev. Cell* **9,** 431–438.

Pan, J., and Snell, W. (2007). The primary cilium: Keeper of the key to cell division. *Cell* **129,** 1255–1257.

Pan, J., Wang, Q., and Snell, W. J. (2005). Cilium-generated signaling and cilia-related disorders. *Lab. Invest.* **85,** 452–463.

Pan, X., Ou, G., Civelekoglu-Scholey, G., Blacque, O. E., Endres, N. F., Tao, L., Mogilner, A., Leroux, M. R., Vale, R. D., and Scholey, J. M. (2006). Mechanism of transport of IFT particles in *C. elegans* cilia by the concerted action of kinesin-II and OSM-3 motors. *J. Cell Biol.* **174,** 1035–1045.

Park, T. J., Mitchell, B. J., Abitua, P. B., Kintner, C., and Wallingford, J. B. (2008). Dishevelled controls apical docking and planar polarization of basal bodies in ciliated epithelial cells. *Nat. Genet.* **40,** 871–879.

Pathak, N., Obara, T., Mangos, S., Liu, Y., and Drummond, I. A. (2007). The zebrafish *fleer* gene encodes an essential regulator of cilia tubulin polyglutamylation. *Mol. Biol. Cell* **18,** 4353–4364.

Pazour, G. J., and Rosenbaum, J. L. (2002). Intraflagellar transport and cilia-dependent diseases. *Trends Cell Biol.* **12,** 551–555.

Pazour, G. J., and Witman, G. B. (2003). The vertebrate primary cilium is a sensory organelle. *Curr. Opin. Cell Biol.* **15,** 105–110.

Pazour, G. J., Wilkerson, C. G., and Witman, G. B. (1998). A dynein light chain is essential for retrograde particle movement in intraflagellar transport (IFT). *J. Cell Biol.* **141,** 979–992.

Pazour, G. J., Dickert, B. L., and Witman, G. B. (1999). The DHC1b (DHC2) isoform of cytoplasmic dynein is required for flagellar assembly. *J. Cell Biol.* **144,** 473–481.

Pazour, G. J., Dickert, B. L., Vucica, Y., Seeley, E. S., Rosenbaum, J. L., Witman, G. B., and Cole, D. G. (2000). *Chlamydomonas* IFT88 and its mouse homologue, polycystic kidney disease gene *tg737*, are required for assembly of cilia and flagella. *J. Cell Biol.* **151,** 709–718.

Pazour, G. J., San Agustin, J. T., Follit, J. A., Rosenbaum, J. L., and Witman, G. B. (2002). Polycystin-2 localizes to kidney cilia and the ciliary level is elevated in orpk mice with polycystic kidney disease. *Curr. Biol.* **12,** R378–R380.

Pazour, G. J., Agrin, N., Leszyk, J., and Witman, G. B. (2005). Proteomic analysis of a eukaryotic cilium. *J. Cell Biol.* **170,** 103–113.

Peden, E. M., and Barr, M. M. (2005). The KLP-6 kinesin is required for male mating behaviors and polycystin localization in *Caenorhabditis elegans*. *Curr. Biol.* **15,** 394–404.

Pedersen, L. B., Geimer, S., Sloboda, R. D., and Rosenbaum, J. L. (2003). The microtubule plus end-tracking protein EB1 is localized to the flagellar tip and basal bodies in *Chlamydomonas reinhardtii*. *Curr. Biol.* **13,** 1969–1974.

Pedersen, L. B., Miller, M. S., Geimer, S., Leitch, J. M., Rosenbaum, J. L., and Cole, D. G. (2005). *Chlamydomonas* IFT172 is encoded by *FLA11*, interacts with CrEB1, and regulates IFT at the flagellar tip. *Curr. Biol.* **15,** 262–266.

Pedersen, L. B., Geimer, S., and Rosenbaum, J. L. (2006). Dissecting the molecular mechanisms of intraflagellar transport in *Chlamydomonas*. *Curr. Biol.* **16,** 450–459.

Pedersen, L. B., Veland, I. R., Schroder, J. M., and Christensen, S. T. (2008). Assembly of primary cilia. *Dev. Dyn.* **237,** 1993–2006.

Perkins, L. A., Hedgecock, E. M., Thomson, J. N., and Culotti, J. G. (1986). Mutant sensory cilia in the nematode *Caenorhabditis elegans*. *Dev. Biol.* **117,** 456–487.

Perrone, C. A., Tritschler, D., Taulman, P., Bower, R., Yoder, B. K., and Porter, M. E. (2003). A novel dynein light intermediate chain colocalizes with the retrograde motor for intraflagellar transport at sites of axoneme assembly in *Chlamydomonas* and mammalian cells. *Mol. Biol. Cell* **14,** 2041–2056.

Pfister, K. K., Fisher, E. M., Gibbons, I. R., Hays, T. S., Holzbaur, E. L., McIntosh, J. R., Porter, M. E., Schroer, T. A., Vaughan, K. T., Witman, G. B., King, S. M., and Vallee, R. B. (2005). Cytoplasmic dynein nomenclature. *J. Cell Biol.* **171,** 411–413.

Piperno, G., and Mead, K. (1997). Transport of a novel complex in the cytoplasmic matrix of *Chlamydomonas* flagella. *Proc. Natl Acad. Sci. USA* **94,** 4457–4462.

Piperno, G., Siuda, E., Henderson, S., Segil, M., Vaananen, H., and Sassaroli, M. (1998). Distinct mutants of retrograde intraflagellar transport (IFT) share similar morphological and molecular defects. *J. Cell Biol.* **143,** 1591–1601.

Plotnikova, O. V., Golemis, E. A., and Pugacheva, E. N. (2008). Cell cycle-dependent ciliogenesis and cancer. *Cancer Res.* **68,** 2058–2061.

Porter, M. E., Bower, R., Knott, J. A., Byrd, P., and Dentler, W. (1999). Cytoplasmic dynein heavy chain 1b is required for flagellar assembly in *Chlamydomonas*. *Mol. Biol. Cell* **10,** 693–712.

Qin, H., Rosenbaum, J. L., and Barr, M. M. (2001). An autosomal recessive polycystic kidney disease gene homolog is involved in intraflagellar transport in *C. elegans* ciliated sensory neurons. *Curr. Biol.* **11,** 457–461.

Qin, H., Diener, D. R., Geimer, S., Cole, D. G., and Rosenbaum, J. L. (2004). Intraflagellar transport (IFT) cargo: IFT transports flagellar precursors to the tip and turnover products to the cell body. *J. Cell Biol.* **164,** 255–266.

Qin, H., Burnette, D., Bae, Y. K., Forscher, P., Barr, M. M., and Rosenbaum, J. L. (2005). Intraflagellar transport is required for the vectorial movement of TRPV channels in the ciliary membrane. *Curr. Biol.* **15,** 1695–1699.

Qin, H., Wang, Z., Diener, D., and Rosenbaum, J. (2007). Intraflagellar transport protein 27 is a small G protein involved in cell-cycle control. *Curr. Biol.* **17,** 193–202.

Quarmby, L. M., and Parker, J. D. (2005). Cilia and the cell cycle? *J. Cell Biol.* **169,** 707–710.

Rana, A. A., Barbera, J. P., Rodriguez, T. A., Lynch, D., Hirst, E., Smith, J. C., and Beddington, R. S. (2004). Targeted deletion of the novel cytoplasmic dynein mD2LIC disrupts the embryonic organiser, formation of the body axes and specification of ventral cell fates. *Development* **131,** 4999–5007.

Reese, T. S. (1965). Olfactory cilia in the frog. *J. Cell Biol.* **25,** 209–230.

Reiter, J. F., and Mostov, K. (2006). Vesicle transport, cilium formation, and membrane specialization: The origins of a sensory organelle. *Proc. Natl Acad. Sci. USA* **103,** 18383–18384.

Rohatgi, R., Milenkovic, L., and Scott, M. P. (2007). Patched1 regulates hedgehog signaling at the primary cilium. *Science* **317,** 372–376.

Rompolas, P., Pedersen, L. B., Patel-King, R. S., and King, S. M. (2007). *Chlamydomonas* FAP133 is a dynein intermediate chain associated with the retrograde intraflagellar transport motor. *J. Cell Sci.* **120,** 3653–3665.

Rosenbaum, J. L., and Witman, G. B. (2002). Intraflagellar transport. *Nat. Rev. Mol. Cell Biol.* **3,** 813–825.

Salomon, R., Saunier, S., and Niaudet, P. (2008). Nephronophthisis. *Pediatr. Nephrol.* [Epub ahead of print].

Santos, N., and Reiter, J. F. (2008). Building it up and taking it down: The regulation of vertebrate ciliogenesis. *Dev. Dyn.* **237,** 1972–1981.

Sarpal, R., Todi, S. V., Sivan-Loukianova, E., Shirolikar, S., Subramanian, N., Raff, E. C., Erickson, J. W., Ray, K., and Eberl, D. F. (2003). *Drosophila* KAP interacts with the kinesin II motor subunit KLP64D to assemble chordotonal sensory cilia, but not sperm tails. *Curr. Biol.* **13,** 1687–1696.

Satir, P., and Christensen, S. T. (2006). Overview of structure and function of mammalian cilia. *Annu. Rev. Physiol.* **69,** 377–400.

Schafer, J. C., Haycraft, C. J., Thomas, J. H., Yoder, B. K., and Swoboda, P. (2003). *XBX-1* encodes a dynein light intermediate chain required for retrograde intraflagellar transport and cilia assembly in *Caenorhabditis elegans*. *Mol. Biol. Cell* **14,** 2057–2070.

Schmitt, A., and Wolfrum, U. (2001). Identification of novel molecular components of the photoreceptor connecting cilium by immunoscreens. *Exp. Eye Res.* **73,** 837–849.

Schneider, L., Clement, C. A., Teilmann, S. C., Pazour, G. P., Hoffmann, E. K., Satir, P., and Christensen, S. T. (2005). PDGFR$\alpha\alpha$ signaling is regulated through the primary cilium in fibroblasts. *Curr. Biol.* **15,** 1861–1866.

Scholey, J. M. (2003). Intraflagellar transport. *Annu. Rev. Cell Dev. Biol.* **19,** 423–443.

Scholey, J. M. (2008). Intraflagellar transport motors in cilia: Moving along the cell's antenna. *J. Cell Biol.* **180,** 23–29.

Scholey, J. M., and Anderson, K. V. (2006). Intraflagellar transport and cilium-based signaling. *Cell* **125,** 439–442.

Schrøder, J. M., Schneider, L., Christensen, S. T., and Pedersen, L. B. (2007). EB1 is required for primary cilia assembly in fibroblasts. *Curr. Biol.* **17,** 1134–1139.

Shakir, M. A., Fukushige, T., Yasuda, H., Miwa, J., and Siddiqui, S. S. (1993). *C. elegans osm-3* gene mediating osmotic avoidance behaviour encodes a kinesin-like protein. *NeuroReport* **4,** 891–894.

Shalom, O., Shalva, N., Altschuler, Y., and Motro, B. (2008). The mammalian Nek1 kinase is involved in primary cilium formation. *FEBS Lett.* **582,** 1465–1470.

Signor, D., Wedaman, K. P., Orozco, J. T., Dwyer, N. D., Bargmann, C. I., Rose, L. S., and Scholey, J. M. (1999a). Role of a class DHC1b dynein in retrograde transport of IFT motors and IFT raft particles along cilia, but not dendrites, in chemosensory neurons of living *Caenorhabditis elegans*. *J. Cell Biol.* **147,** 519–530.

Signor, D., Wedaman, K. P., Rose, L. S., and Scholey, J. M. (1999b). Two heteromeric kinesin complexes in chemosensory neurons and sensory cilia of *Caenorhabditis elegans*. *Mol. Biol. Cell* **10,** 345–360.

Singla, V., and Reiter, J. F. (2006). The primary cilium as the cell's antenna: Signaling at a sensory organelle. *Science* **313,** 629–633.

Sloboda, R. D. (2002). A healthy understanding of intraflagellar transport. *Cell Motil. Cytoskeleton* **52,** 1–8.

Sloboda, R. D., and Howard, L. (2007). Localization of EB1, IFT polypeptides, and kinesin-2 in *Chlamydomonas* flagellar axonemes via immunogold scanning electron microscopy. *Cell Motil. Cytoskeleton* **64,** 446–460.

Snow, J. J., Ou, G., Gunnarson, A. L., Walker, M. R. S., Zhou, H. M., Brust-Mascher, I., and Scholey, J. M. (2004). Two anterograde intraflagellar transport motors cooperate to build sensory cilia on *C. elegans* neurons. *Nat. Cell Biol.* **6,** 1109–1113.

Sorokin, S. (1962). Centrioles and the formation of rudimentary cilia by fibroblasts and smooth muscle cells. *J. Cell Biol.* **15,** 363–377.

Stolc, V., Samata, M. P., Tongprasit, W., and Marshall, W. F. (2005). Genome-wide transcriptional analysis of flagellar regeneration in *Chlamydomonas reinhardtii* identifies orthologs of ciliary disease genes. *Proc. Natl Acad. Sci. USA* **102,** 3703–3707.

Sun, Z., Amsterdam, A., Pazour, G. J., Cole, D. G., Miller, M. S., and Hopkins, N. (2004). A genetic screen identifies cilia genes as a principal cause of cystic kidney. *Development* **131,** 4085–4093.

Surpili, M. J., Delben, T. M., and Kobarg, J. (2003). Identification of proteins that interact with the central coiled-coil region of the human protein kinase NEK1. *Biochemistry* **42,** 15369–15376.

Takeda, S., Yonekawa, Y., Tanaka, Y., Okada, Y., Nonaka, S., and Hirokawa, N. (1999). Left–right asymmetry and kinesin superfamily protein KIF3A: New insights in determination of laterality and mesoderm induction by kif3A−/− mice analysis. *J. Cell Biol.* **145,** 825–836.

Tallila, J., Jakkula, E., Peltonen, L., Salonen, R., and Kestila, M. (2008). Identification of CC2D2A as a Meckel syndrome gene adds an important piece to the ciliopathy puzzle. *Am. J. Hum. Genet.* **82,** 1361–1367.

Tan, P. L., Barr, T., Inglis, P. N., Mitsuma, N., Huang, S. M., Garcia-Gonzalez, M. A., Bradley, B. A., Coforio, S., Albrecht, P. J., Watnick, T., Germino, G. G., Beales, P. L., et al. (2007). Loss of Bardet-Biedl syndrome proteins causes defects in peripheral sensory innervation and function. *Proc. Natl Acad. Sci. USA* **104,** 17524–17529.

Tobin, J. L., and Beales, P. L. (2007). Bardet-Biedl syndrome: Beyond the cilium. *Pediatr. Nephrol.* **22,** 926–936.

Town, T., Breunig, J. J., Sarkisian, M. R., Spilianakis, C., Ayoub, A. E., Liu, X., Ferrandino, A. F., Gallagher, A. R., Li, M. O., Rakic, P., and Flavell, R. A. (2008). The stumpy gene is required for mammalian ciliogenesis. *Proc. Natl Acad. Sci. USA* **105,** 2853–2858.

Tran, P. V., Haycraft, C. J., Besschetnova, T. Y., Turbe-Doan, A., Stottmann, R. W., Herron, B. J., Chesebro, A. L., Qiu, H., Scherz, P. J., Shah, J. V., Yoder, B. K., and Beier, D. R. (2008). THM1 negatively modulates mouse sonic hedgehog signal transduction and affects retrograde intraflagellar transport in cilia. *Nat. Genet.* **40,** 403–410.

Tsao, C. C., and Gorovsky, M. A. (2008a). Different effects of *Tetrahymena* IFT172 domains on anterograde and retrograde intraflagellar transport. *Mol. Biol. Cell* **19,** 1450–1461.

Tsao, C. C., and Gorovsky, M. A. (2008b). *Tetrahymena* IFT122A is not essential for cilia assembly but plays a role in returning IFT proteins from the ciliary tip to the cell body. *J. Cell Sci.* **121,** 428–436.

Tsujikawa, M., and Malicki, J. (2004). Intraflagellar transport genes are essential for differentiation and survival of vertebrate sensory neurons. *Neuron* **42,** 703–716.

Vieira, O. V., Gaus, K., Verkade, P., Fullekrug, J., Vaz, W. L., and Simons, K. (2006). FAPP2, cilium formation, and compartmentalization of the apical membrane in polarized Madin-Darby canine kidney (MDCK) cells. *Proc. Natl Acad. Sci. USA* **103,** 18556–18561.

von Schnakenburg, C., Fliegauf, M., and Omran, H. (2007). Nephrocystin and ciliary defects not only in the kidney? *Pediatr. Nephrol.* **22,** 765–769.

Walther, Z., Vashishtha, M., and Hall, J. L. (1994). The *Chlamydomonas FLA10* gene encodes a novel kinesin-homologous protein. *J. Cell Biol.* **126,** 175–188.

Wang, Q., Pan, J., and Snell, W. J. (2006). Intraflagellar transport particles participate directly in cilium-generated signaling in *Chlamydomonas*. *Cell* **125,** 549–562.

White, M. C., and Quarmby, L. M. (2008). The NIMA-family kinase, Nek1 affects the stability of centrosomes and ciliogenesis. *BMC Cell Biol.* **9,** 29.

Wickstead, B., and Gull, K. (2006). A "holistic" kinesin phylogeny reveals new kinesin families and predicts protein functions. *Mol. Biol. Cell* **17,** 1734–1743.

Williams, C. L., Winkelbauer, M. E., Schafer, J. C., Michaud, E. J., and Yoder, B. K. (2008). Functional redundancy of the B9 proteins and nephrocystins in *Caenorhabditis elegans* ciliogenesis. *Mol. Biol. Cell* **19,** 2154–2168.

Wilson, N. F., and Lefebvre, P. A. (2004). Regulation of flagellar assembly by glycogen synthase kinase 3 in *Chlamydomonas reinhardtii*. *Eukaryot. Cell* **3,** 1307–1319.

Witman, G. B., Carlson, L., Berliner, J., and Rosenbaum, J. L. (1972). *Chlamydomonas* flagella. I. Isolation and electrophoretic analysis of microtubules, matrix, membranes, and mastigonemes. *J. Cell Biol.* **54,** 507–539.

Yoder, B. K., Hou, X., and Guay-Woodford, L. M. (2002). The polycystic kidney disease proteins, polycystin-1, polycystin-2, polaris, and cystin, are co-localized in renal cilia. *J. Am. Soc. Nephrol.* **13,** 2508–2516.

Yoshimura, S., Egerer, J., Fuchs, E., Haas, A. K., and Barr, F. A. (2007). Functional dissection of Rab GTPases involved in primary cilium formation. *J. Cell Biol.* **178,** 363–369.

CHAPTER THREE

How Did the Cilium Evolve?

Peter Satir,* David R. Mitchell,[†] and Gáspár Jékely[‡]

Contents

1. Introduction 64
2. The Link Between Ciliary Evolution and Eukaryotic Divergence 66
3. Hypotheses of the Origin of Cilia 67
4. The Viral Hypothesis of Ciliary Origin 68
   4.1. Statement of the hypothesis 68
   4.2. What the viral hypothesis can explain 69
   4.3. Unusual predictions as a consequence of the viral hypothesis 70
   4.4. Objections to the viral hypothesis 70
5. The Autogenous Model for the Origin of Cilia 71
   5.1. Autogenous origin of microtubule-organizing centers 71
   5.2. Autogenous origin of ninefold symmetric basal bodies and the axoneme 74
6. Origin of Intraflagellar Transport and the Sensory Function of Cilia 76
7. The Evolution of Ciliary Motility 78
Acknowledgments 79
References 79

Abstract

The cilium is a characteristic organelle of eukaryotes constructed from over 600 proteins. Bacterial flagella are entirely different. 9 + 2 motile cilia evolved before the divergence of the last eukaryotic common ancestor (LECA). This chapter explores, compares, and contrasts two potential pathways of evolution: (1) via invasion of a centriolar-like virus and (2) via autogenous formation from a pre-existing microtubule-organizing center (MTOC). In either case, the intraflagellar transport (IFT) machinery that is nearly universally required for the assembly and maintenance of cilia derived from the evolving intracellular vesicular transport system. The sensory function of cilia evolved first and the ciliary axoneme evolved gradually with ciliary motility, an important selection mechanism, as one of the driving forces.

* Department of Anatomy and Structural Biology, Albert Einstein College of Medicine, Bronx, New York
[†] Department of Cell and Developmental Biology, SUNY Upstate Medical University, Syracuse, New York
[‡] Max Planck Institute for Developmental Biology, Spemannstrasse 35, 72076 Tübingen, Germany

1. INTRODUCTION

In this chapter, we explore the evolution of one of the most characteristic organelles of the eukaryotic cell, the cilium. Cilia, sometimes called flagella in the literature, are found exclusively on eukaryotic cells, throughout the protista, in many plant phyla on gametes, and on somatic cells and/or gametes in virtually every metazoan phylum. (Bacterial flagella are entirely different organelles.) In humans, motile cilia are present on nasal and tracheal lining cells, ependymal cells of the brain, and cells lining the oviduct. The sperm tail is a cilium with accessory elements. The sensory hair cells of the ear and semicircular canals possess a single "kinocilium" (the "stereocilia" of the ear and epididymis are microvilli, not cilia). Olfactory sense cells each possess multiple cilia, which may be immotile. During embryogenesis, motile cilia present at the node during gastrulation evidently initiate the determination of left–right body asymmetry. In addition, many cells in the body possess a single nonmotile cilium called a primary cilium; a primary cilium with extensive membrane modifications forms the outer segment of the ciliary photoreceptors of the vertebrate eye. A current review of the structure and function of motile and primary mammalian cilia is found in Satir and Christensen (2007).

A series of combined genomic–proteomic studies on protistan (*Chlamydomonas*, *Tetrahymena*) motile cilia, invertebrate (primarily *Caenorhabditis elegans*) sensory cilia, and mammalian and human cilia reveal that cilia are comprised of upwards of 600 proteins (Avidor Reiss *et al.*, 2004; Li *et al.*, 2004; Ostrowski *et al.*, 2002; Pazour *et al.*, 2005; Scholey, 2003; Smith *et al.*, 2005). Many or most of the genes related to cilia are missing in genomes of nonciliated organisms, such as *Arabidopsis*. When cilia are present, a remarkable portion of the nuclear genome, perhaps 3–5% of the typical genome, is devoted to this single organelle. The proteins produced by these genes are highly conserved. Three well-studied illustrative examples of proteins equally important in the single-celled alga *Chlamydomonas reinhardtii* and man are (1) axonemal dyneins, for example, OADα (the α-subunit of outer arm dynein) (Wilkes *et al.*, 2008); (2) hydin (Lechtreck *et al.*, 2008), a central pair projection; and (3) polaris or IFT88, a component of the intraflagellar transport (IFT) system (Pazour *et al.*, 2000; Yoder *et al.*, 2002).

Proteins such as these form defined parts of complex structures in the cilium. As has been understood for over half a century, almost all motile cilia are built on a similar plan, a 200 nm diameter cytoskeletal bundle of nine peripheral doublet microtubules surrounding a central singlet pair, the so-called 9 + 2 axoneme, which grows from a basal body and is enclosed in an extension of the cell membrane called the ciliary membrane (Fig. 3.1). The ciliary membrane incorporates specialized receptors and channels and confers on all cilia a sensory function, the ability to respond to external molecular signals (Christensen *et al.*, 2007). The basal body itself is almost universally an

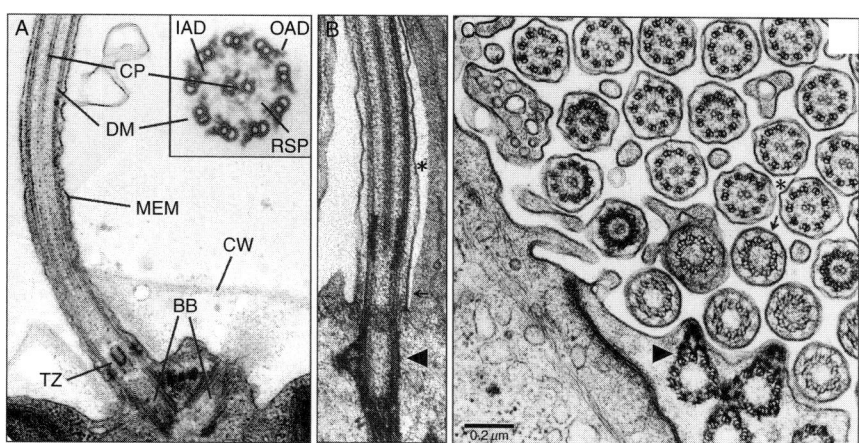

Figure 3.1 Architecture of the motile 9 + 2 cilium and the basal body. (A) Longitudinal section through one *Chlamydomonas reinhardtii* flagellum (a second flagellum is out of the plane of section, although its basal body is visible). *Abbreviations*: CP, central pair; DM, doublet microtubule; MEM, plasma membrane; CW, cell wall; BB, basal bodies; TZ, transition zone. Inset, cross section through an isolated axoneme shows the location of radial spokes (RSP), outer arm dyneins (OAD), and inner arm dyneins (IAD). (B) and (C) Longitudinal and cross sections of a field of oviduct cilia (courtesy Dirksen and Satir, 1972, unpublished micrograph). In these cilia, the basal foot (arrowhead) points in the direction of the effective stroke. A multirowed ciliary necklace (arrow) is present in the transition zone near the IFT loading area, above which the cilium projects from the cell and the central pair of the 9 + 2 axoneme originates. Each axoneme is surrounded by the ciliary membrane, in which special receptors and channels become concentrated. The diameter of the cilium (0.25 μm) provides a scale for the micrographs.

arrangement of nine triplet microtubules, essentially identical to a centriole that has become attached to the cell membrane and formed a specialized structure called the ciliary necklace (Gilula and Satir, 1972). The cilium grows beyond the necklace with each of the nine triplet microtubules reduced to an axonemal doublet. In the transition zone, the central pair arises. Growth to lengths often greater than 10 μm occurs, again almost universally, by a process discovered more recently that has been named intraflagellar transport (Blacque et al., 2008; Rosenbaum and Witman, 2002; Scholey, 2003).

The motile cilium is a nanomachine powered by molecular motors, the axonemal dyneins attached to one edge of each axonemal doublet. The ATPase activity of the OADs and IADs (inner arm dyneins) produces sliding between the doublets, powers the axoneme, and causes ciliary beat. The form and frequency of beat are modulated in a highly complex manner by second messengers generated via ciliary membrane channels or coupled receptors and by actions of the radial spoke–central pair proteins. The radial spokes (Fig. 3.1) extend from each doublet microtubule to specific protein projections surrounding the central singlet microtubules (Smith and Yang, 2004). Orthologous proteins of different organisms form the identical structures and perform the same functions in corresponding axonemes. Moreover, mutations in these proteins lead to structural and motility defects in *Chlamydomonas* and, because of the corresponding defects in human cilia, to ciliopathies in man (Badano *et al.*, 2006; Li *et al.*, 2004; Satir and Christensen, 2008). As examples of human ciliopathies, defects in OADs lead to primary ciliary dyskinesia (PCD), defects in hydin lead to hydrocephalus and defects in IFT88 (polaris) lead to autosomal recessive polycystic kidney disease (PKD).

Specialized sensory cilia like the photoreceptor outer segment and primary cilia also possess an axoneme, a ciliary membrane, a ciliary necklace and they grow above a basal body by IFT. However, although axonemal structure is largely preserved, these cilia are often nonmotile because the dynein arms, the radial spokes, and proteins involved in dynein regulation are missing. In most cases, the central microtubules are also missing and the axonemes are referred to as 9 − 0. Examples abound in vertebrates, as such 9 + 0 "primary" cilia are found on most cell types. In contrast to motile function, the presence or absence of sensory function, regulated adhesion, and surface motility are less easily detected. These functions may be retained by most motile as well as nonmotile cilia.

2. The Link Between Ciliary Evolution and Eukaryotic Divergence

Conservation of primary sequence, protein composition, structure, and function throughout eukaryotic history as recorded in current phyletic divergence suggests a unitary origin of cilia at or very near the beginning of eukaryotic cell evolution. Certainly, the presence of cilia can be traced back to the last eukaryotic common ancestor (LECA) through analysis of gene/protein sequences and studies of ultrastructure in existing organisms. These determinations have been made possible only by the extensive recent advances in our overall understanding of eukaryotic phylogeny (e.g., Brinkmann and Philippe, 2007; Rodriguez-Ezpeleta *et al.*, 2007; Yoon *et al.*, 2008). Using such phylogenetic approaches, it has been possible to confidently state, regardless of specific branch points and groupings, that all of the major identified branches of existing eukaryotes include organisms with nearly indistinguishable 9 + 2 motile cilia, and therefore such cilia evolved prior to the divergence from the LECA (Cavalier-Smith, 2002; Mitchell, 2004). Although the possibility remains that an organism with a more primitive motile apparatus will be definitively identified as a member of a new phyletic clade, not closely related to any of the presently well-defined clades, and thus displaying traits of eukaryotes that existed before the great divergence of most present day eukaryotes, this possibility is becoming increasingly remote as sequence coverage expands. Thus singlet, doublet, and triplet microtubules, and α-, β-, γ-, δ-, and ε-tubulin, a central pair complex, radial spokes, at least five axonemal dynein complexes, and IFT complexes with their associated kinesin and dynein motors, all appear to have evolved prior to eukaryotic divergence.

Further evolution in these organelles during the *ca.* 1 billion years that have passed since the LECA (Brinkmann and Philippe, 2007) has generated variations that are more apparent at the level of gene duplication than at the level of overall structure, a common example being the diversification of single-headed IAD heavy chains (Wickstead and Gull, 2007; Wilkes *et al.*, 2008). However, diversification in the use of cilia appears to be largely the result of refinements of traits already common to these organelles prior to divergence from the LECA. These common uses include planar beating to generate fluid flow, ciliary membrane surface motility to produce particle transport (or gliding of single cells), sensory receptor display, and regulated adhesion, including adhesion-based signaling (Mitchell, 2007). Some of these traits have been lost when no longer needed, such as the loss of beating motility in organelles specialized for sensory functions.

3. Hypotheses of the Origin of Cilia

Three hypotheses of the origin and subsequent evolution of the cilium have been proposed. The first, the symbiotic model, derives cilia from a spirochete bacterium (Margulis, 1981). However, no comparative genomic evidence has been found to support a spirochete ancestry of the ciliary proteome. Given this and the cell biological difficulties with such a symbiotic cell fusion model we and others think that this explanation is highly unlikely.

The second hypothesis is autogenous origin of the centriole and cilium within the evolving eukaryotic cytoplasm (Cavalier-Smith, 1978; Jékely and Arendt, 2006; Mitchell, 2004). There is little controversy that some aspects of the evolution of cilia are autogenous, in part because axonemes are built of microtubules mostly made of the same proteins as cytoplasmic microtubules, in part because of the commonality of molecular motors, and in part because of the relationship of IFT proteins to endocytic scaffold proteins and perhaps to nuclear pore proteins. This chapter will explore what these relationships are in detail and whether autogenous formation of the cilium can account for the whole of evolutionary history of the organelle.

The third hypothesis, relatively new (Satir *et al.*, 2007), assigns a supporting role to the developing eukaryotic cytoplasm. In this hypothesis, the cytoplasm plays host to an invading virus which becomes the primitive centriole. We begin by restating the main features of this sequence of ciliary evolution and the questions it poses. We then consider the pros and cons of this hypothesis, present an alternative autogenous scenario for centriole evolution, and discuss those features of ciliary evolution which are independent of the specific model of centriole origins.

4. THE VIRAL HYPOTHESIS OF CILIARY ORIGIN

4.1. Statement of the hypothesis

The viral hypothesis (Fig. 3.2) suggests that the cilium evolves at a time after the invention of membrane trafficking in the protoeukaryotic cytoplasm. The sequence of events postulated is as follows.

An RNA enveloped virus with ninefold symmetry invades the cell. The basal structure of the virus is the cartwheel, each spoke of which initiates a

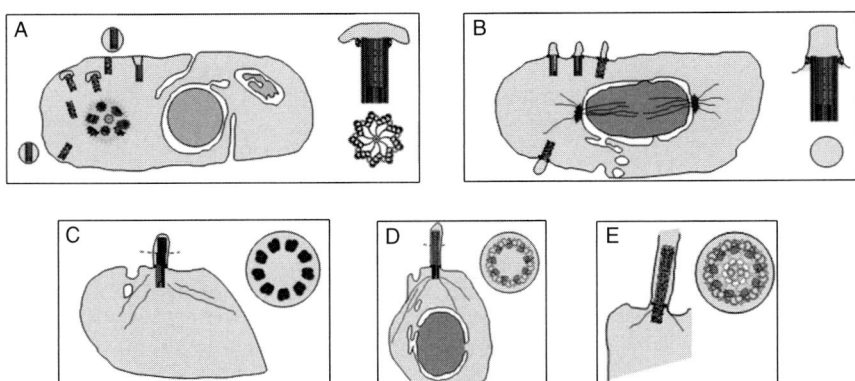

Figure 3.2 The viral hypothesis of ciliary evolution. (A) Membrane trafficking in the protoeukaryotic bacterial cytoplasm leads to endocytosis of cell organelles. An RNA enveloped virus with ninefold symmetry invades this cytoplasm. Cross section shows ninefold symmetry at the cartwheel. (B) The centriolar virus remains attached to the cell membrane and a sensory bulge capable of positional signaling develops. GPS function leads to restriction in organelle number and development of cell polarity. Cross section shows area of attachment at the ciliary necklace. (C) Elongation of the centriolar capsid, utilizing α- and β-tubulin, IFT and motor proteins of the host cell produces the 9 + 0 axoneme (cross section) in a sensory organelle. (D) Cytoplasmic dynein diversifies to give axonemal isoforms—arms between doublets in cross section—motility begins. (E) Misplaced cytoplasmic MTs give rise to the central pair leading to an increase in efficient motility in a stabilized 9 + 2 axoneme (cross section). Efficient signaling and motility in the 9 + 2 axoneme are advantageous for protistan survival and outcompete earlier forms. Images from Satir *et al.* (2007). Courtesy *Cell Motil. Cytoskeleton*. (See Color Insert.)

triplet microtubule. The virus disassembles with the nucleic acid integrating into the evolving eukaryotic cell nucleus.

The virus reassembles in a cytoplasmic viral factory. Many capsids (protocentrioles) form at once. Originally RNA is reincorporated into each protocentriole, but eventually some capsids do not incorporate all or even part of the viral genome. The maturing virus then attaches to a specialized exocytic vesicle, which attachment site eventually gives rise to the ciliary necklace. The virus then becomes encapsulated by the cell membrane when it exits the cell.

A mutation in the encapsulation process occurs so that the centriolar capsid remains attached to the cell membrane with an expanded membrane bulge above it. As IFT proteins and motors evolve within the cell, specific receptors become incorporated into the membrane in this bulge, which then becomes capable of positional signaling. This signaling is used by the cell for spatial identification of cellular axes, which in many, but not all, cells leads to restriction in centriolar number and/or to development of cell polarity.

A second mutation permitting elongation of the centriolar capsid, utilizing α- and β-tubulin, probably manufactured by the host cell, together with IFT motors and proteins, produces a 9 + 0 sensory cilium, where the original protocentriole is now a true basal body. Elongation mutants are common in some viruses (e.g., bacteriophage—Kellenberger, 1965).

4.2. What the viral hypothesis can explain

The viral hypothesis offers relatively easy explanation of several remarkable features of ciliary morphogenesis:

1. In many cells, basal body (centriole) formation is not templated as a mother–daughter relationship, but is explosive and *de novo*, with many basal bodies, sometimes hundreds, arising at once in a cytoplasmic factory (reminiscent of a viral factory). Some examples are anarchic oral field formation in ciliates (Bell *et al.*, 2008), gametogenesis in *Gingko* or *Marsilia* (Mizukami and Gall, 1966), and ciliogenesis in mammalian respiratory epithelia. In the ameboflagellate *Naegleria* and similar genera, normally only two to four cilia form, but ciliogenesis is *de novo* (Fulton and Walsh, 1980).
2. Centriole assembly is stereotypic self-assembly with microtubules polymerizing on a basal plate (the cartwheel) (Salisbury, 2008), length regulation, defined ninefold symmetry and a single enantiomorphic form, all reminiscent of viral assembly.
3. Basal bodies contain unique tubulins, tektins, and ribbon proteins not present in cytoplasmic microtubules, permitting the formation of stable triplet and doublet microtubules. Similarly, all direct attachment of

centrioles to membranes or membrane vesicles that fuse with the cell membrane involves a ciliary necklace (Satir et al., 1976).
4. There may be specific centriolar or basal body RNAs with sequences reminiscent of retroviruses sometimes found within the organelle (Alliegro et al., 2006). Although the centriolar genome is now completely or largely nuclear and dispersed, during morphogenesis of the organelle centriolar genes must be turned on as cassettes in specific sequences.
5. Ciliary axonemal microtubules of the nine doublets are the only microtubules that grow directly from basal body (centriolar) microtubules. Other cytoplasmic microtubules including spindle microtubules arise from γ-tubulin-containing foci within pericentriolar centrosomal material (Salisbury, 2008). These observations imply that only the ciliary doublet microtubules are templated by the centriole, and not vice versa.
6. Without a centriole, there are no cilia, but there are microtubules, microtubule-containing cell extensions such as axopodia, and mitosis. In a particularly informative experiment, Basto et al. (2006) ablated centrioles in *Drosophila* cells. These observations uncouple centriole and cilia evolution from cell division, suggesting that centriolar duplication connected to the cell cycle is a secondary event.

4.3. Unusual predictions as a consequence of the viral hypothesis

Two specific predictions of the viral hypothesis as stated are that:
1. The sensory cilium originates before the motile cilium.
2. The first ciliated cells were neither unikonts nor bikonts, but rather multi-ciliated, as might be expected from an abortive viral budding process.

These predictions are quite different from those made from the autogenous hypothesis and might be testable via extensive genomic analysis of primitive protists.

4.4. Objections to the viral hypothesis

Several objections that may be raised to the viral hypothesis are now discussed:
1. The basal body would be an unusually large and complicated virus.

This is correct, and it might imply that the host cell which the virus invaded was unusually large for its time also. The initial invader need not be exactly like the modern centriole, but if the hypothesis is correct, all the essential features, especially ninefold symmetry, and probably the unique proteins involved in genesis of the centriole, would be present.

2. There are no viruses with ninefold symmetry.

This also appears to be largely true, although an artificial virus with ninefold symmetry has been constructed (Martin-Benito *et al.*, 2001), which suggests that viruses with such symmetry could potentially exist. Of course, finding such a modern day virus, especially if some of its proteins were centriole-related, would provide a strong demonstration in favor of the viral origin hypothesis.

3. There are organisms whose gametes have large numbers of doublet microtubules that do not originate from a centriole (Phillips, 1974); alternatively there are gametes with less than nine doublets and minimal centrioles (Goldstein and Schrevel, 1982).

These organisms are not primitive and have no general significance for early evolution, but rather show that the selection pressure on gametes results in the fixation or divergence from the ancestral state.

4. It should be possible to find more evidence of the ancient retroviral sequences in centriolar genes and more viral RNA in extant centrioles than has been reported.

Maybe so, but retroviral sequences are found throughout the eukaryotic genome and tracing lineages in this way may be futile. It seems likely that RNA could be found in other centrioles, but its presence could be confined to specific times in development of the organism or to specific stages of centriologenesis.

5. There is no sequence similarity between centriolar proteins and viral capsid proteins (G. Jékely, unpublished data).

This is true, but our sampling of viral diversity is still scarce, and such viruses may one day be found.

5. THE AUTOGENOUS MODEL FOR THE ORIGIN OF CILIA

5.1. Autogenous origin of microtubule-organizing centers

An alternative to the viral origin of basal bodies and cilia is an autogenous model. In this model, the basal body structure evolved from pre-existing cellular components via the process of gene duplication and divergence and the final ninefold symmetry is the result of a coevolutionary process to optimize cilium-based motility. According to this model, basal body structure and axonemal structure tightly coevolved from the very beginning, and therefore neither structure was templated on the other one. There is strong indication that the symmetry and structure of the basal body and the axoneme are tightly linked and if the axoneme is lost, the centriole is lost completely (e.g., flowering plants). The starting point according to this model was a cell already possessing a microtubule cytoskeleton, which regulated cell shape and served as a scaffold for the motility of intracellular membranes. (In the viral hypothesis, this is the host cell that is invaded by the centriolar virus.) The microtubule cytoskeleton could have been unpolarized initially. Basal body and cilium evolution started when the cell first acquired polarity by evolving a primitive microtubule-organizing center (MTOC) that served as the major site to nucleate microtubules (Fig. 3.3).

Such MTOCs that also form the ends of the mitotic spindle must have evolved very early as structures needed for microtubule organization and cell polarity determination. In the autogenous model, the evolution of a cilium is assumed to have occurred through modification of a single MTOC to support assembly of a specialized set of microtubules that became the axoneme (Fig. 3.3). According to this model the need to anchor the axoneme drove evolution directly from a simpler MTOC that supported a simpler protoaxoneme into (eventually) a centriole–basal body system that serves dual purposes in axoneme assembly and in microtubule nucleation during mitosis. This dual-purpose MTOC evolved into the centrosome during animal evolution. This association of centrioles with a mitotic organizing function has been secondarily lost in multiciliated cells, including both protists such as *Tetrahymena* and vertebrate cells such as those of ciliated airway epithelia, where tens or hundreds of centrioles dock with the cell membrane to become the basal bodies of cilia, separated from sites of spindle assembly. Plausible scenarios exist for evolution of an axoneme from a unique organizing center that only later developed an association with the mitotic MTOC, but analysis of the evolution of centrin, a protein essential for spindle pole body duplication in yeasts and also important for basal body/centriole replication (Loncarek *et al.*, 2008; Salisbury, 2007), provides support to the concept that basal bodies evolved linearly through modification of an ancestral mitotic MTOC (Bornens and Azimzadeh, 2007).

A MTOC could have originated due to the self-organizational capacity of the microtubule system and associated molecular motors. One can conceive a scenario where the minus end of microtubules became focused by the action of dimeric minus end-directed motors (Surrey *et al.*, 2001). Such foci could have nucleated further microtubules either on the locally concentrated microtubules (self-nucleation) or on an emerging MTOC material that could have accumulated at the foci as cargo carried by molecular motors. Self-nucleation and transport could have allowed the emergence of a robust MTOC via positive feedback loops (i.e., more localized microtubule nucleation leading to more localized transport of MTOC material). During further evolution, these MTOC became independent and were maintained also in the absence of microtubules. MTOCs evolved

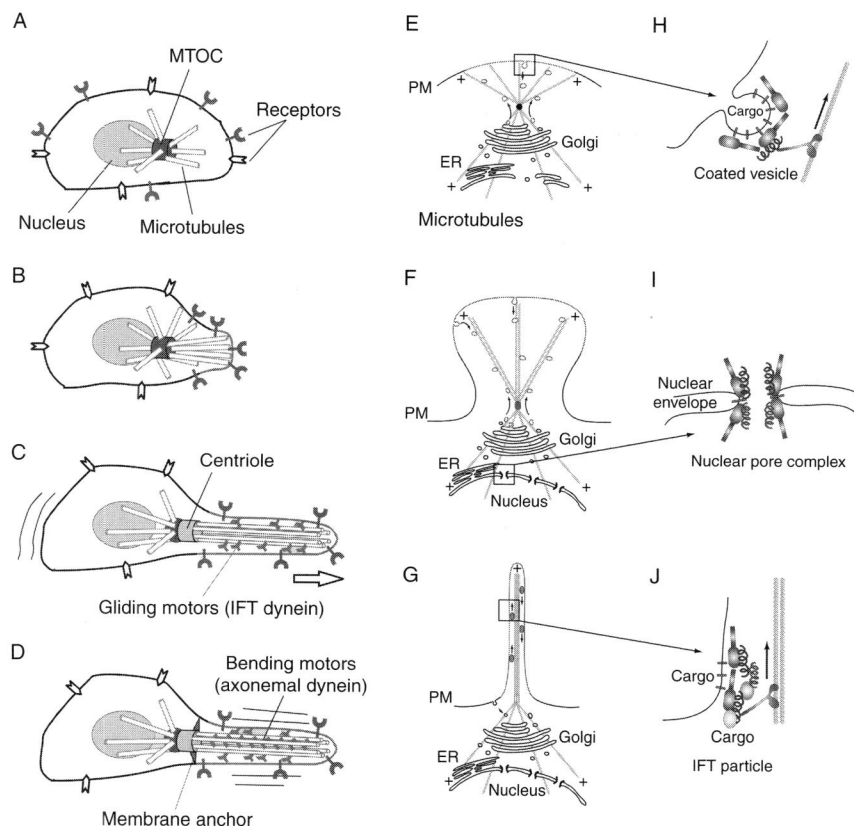

Figure 3.3 The autogenous hypothesis of the evolution of motile ciliary functions and IFT from coated vesicle transport. (A) The evolving eukaryotic cell possesses microtubules emanating from an organizing center (MTOC) and receptors randomly distributed on the cell surface. (B) Sensory receptor capping and directional transport lead to MT elongation and bundling. A specialized membrane patch (red) appears at the ends of elongated MT. (C) The bundled MTs form the primitive basal body (centriole) and a primitive axoneme. Vesicle transport motors evolve into IFT kinesin and dynein (green). The axoneme uses IFT motors for gliding motility, promoting selection of the polarized MT bundle/protoaxoneme. (D) Axonemal dynein isoforms evolve (red) that permit bending movement of the axoneme, which drives selection of doublet microtubules and of a centriole with membrane anchors. The 9 + 2 pattern became fixed at a later time based on its superior regulation of bending parameters. (E) The proto-IFT complex in a protoeukaryote mediated transport of transmembrane proteins (endocytosis and/or recycling) at a specialized region of the plasma membrane (shown in red). (F) Microtubule growth exposed this membrane domain to the environment. Transmembrane proteins were transported in vesicles, proteins of the developing axoneme arrived by diffusion. Nuclear pore complexes also diverged from a protocoatomer complex. (G) Bidirectional IFT in modern cilia with kinesin-II as the anterograde and dynein 1b/2 as the retrograde motor. The motors are thought to have evolved from the ones that moved coated vesicles. IFT cargo can be either transmembrane or cytoplasmic. During the subsequent stages of cilia evolution basal body structure coevolved with axonemal structure. (H), (I), and (J) Organization of coated vesicle complexes, nuclear pore complexes, and the IFT complex. Adapted from Mitchell (2004) and Jékely and Arendt (2006). (See Color Insert.)

the ability to self-assemble either in a templated manner on a mother MTOC or independently. Mechanisms to coordinate MTOC duplication and cell division also evolved. At this stage γ-tubulin diverged from cytoplasmic tubulin (following gene duplication) and specialized to nucleate microtubules in the MTOC. By regulating the microtubule skeleton and cell polarity, these MTOCs also became indispensable for the positioning of the nucleus and for cell division. The MTOC had no ninefold symmetry at this stage, and was similar to the pericentriolar material, being able to nucleate microtubules in all directions. During subsequent stages the MTOC acquired polarity, triplet microtubules, and the ability to nucleate the developing axonemal microtubules from one end only (Fig. 3.3C). This origin of centriolar complexity from a simpler MTOC is of course not necessary in the viral hypothesis. However, other examples of microtubules forming complex symmetrical structures are seen in the suctorian tentacle, where a sevenfold microtubule array develops (Tucker, 1974) and the heliozoan axopodium, where large 12-fold symmetric double-spiral helical microtubule arrays form (Tilney and Porter, 1965).

5.2. Autogenous origin of ninefold symmetric basal bodies and the axoneme

Formation of triplet microtubules, from which axonemal doublets arise, requires both δ- and ε-tubulin (Dutcher, 2003), which are present in all organisms with motile cilia and flagella. In contrast to the viral hypothesis which assumes that these proteins were brought into the protoeukaryotic cell with the centriolar virus, the autogenous hypothesis assumes that these tubulin forms arose via gene duplication and divergence of pre-existing tubulin forms, most likely γ-tubulin, within the nuclear genome (Dutcher, 2001). The forces that led to the evolution of these more complex microtubule structures are likely linked to the asymmetry provided by a doublet, which provides differentiation into a cargo-binding surface for permanent dynein attachment (the A-tubule), and a track for dynein-based movement (the B-tubule). Further development into a triplet may have more to do with the anchoring of basal bodies to withstand the forces of axonemal beating, since basal bodies in *C. elegans*, which lacks motile cilia, have only doublet microtubules (Dutcher, 2003), as do the basal bodies of the 9 + 0 sperm axonemes of a termite (Baccetti and Dallai, 1978), whose sperm are described as being "feebly motile."

Recent study of mutations that alter basal body structure and function in *Chlamydomonas* have revealed for the first time at least part of the basis of the ninefold symmetry of basal bodies. Two conserved gene products, BLD10 and SAS-6 in *Chlamydomonas* (homologs of previously identified metazoan centriolar proteins CEP135 and SAS-6) are both located in the cartwheel found at the proximal end of basal bodies and centrioles. BLD10 immunolocalizes to the outer ends of the cartwheel spokes, at the point where they interact with the inner-most microtubule (A-tubule) of each basal body triplet (Hiraki et al., 2007). Particularly interesting is the observation that C-terminal truncations of BLD10 can partially rescue the null mutant, but 25% of such cells have basal bodies with a perfect cylinder of only eight instead of the normal nine triplets. While the hub of the cartwheel often appeared normal in these BLD10 ΔC basal bodies, and cartwheels retained nine spokes, the spokes were shorter than normal and failed to interact properly with the triplet A-tubules. In *bld12* mutant cells (lacking SAS-6), cartwheel spokes are present and retain their association with the A-tubule of each triplet, but the central hub of the cartwheel is missing and the number of triplets varies from 7 to 12, with only 70% retaining the normal number of 9 (Nakazawa et al., 2007). Importantly, the basal bodies with abnormal numbers of triplets retain their cylindrical shape. Immunolocalization of SAS-6 indicates that it is near the center of the cartwheel at about the spoke-hub junction, consistent with work on its homolog in *C. elegans* (Pelletier et al., 2006). Thus basal bodies have a built in propensity for radial symmetry, and the properties of only one or two proteins appear to stabilize that symmetry at ninefold.

In the autogenous model, what then has driven the evolution of ninefold symmetry? One hint may come from examination of the flagellar axonemes that assemble onto the abnormal basal bodies of *bld12* mutant cells. Axonemes with fewer than nine doublets retain dynein arms and radial spokes, but have a central lumen that lacks any central pair apparatus, presumably because the central pair complex could not fit in the remaining space. Those with 10 or 11 doublets retain the central pair, but radial spokes from only about half of the doublets can make contact with this central apparatus. Clearly the current geometry of radial spokes and central apparatus, once established, acted to fix the number of doublets in the axoneme. Some nonmotile insect spermatozoa that naturally lack radial spokes, dynein arms and a central apparatus have 12 or 14 doublets, which are assembled from basal bodies of 12 or 14 triplet microtubules (Baccetti et al., 1973). Naturally occurring nonmotile axonemes with 10 doublets have also been reported in sensory cilia of a parasitic nematode (Ross, 1967). Thus there is nothing inherent in the architecture of a basal body that requires ninefold symmetry.

Is ninefold symmetry linked to the evolution of motility? Some of the more divergent insect spermatozoa that retain weak motility have large numbers of doublet microtubules arranged in spiral rows (Baccetti et al., 1974b). These lack radial spokes or a central apparatus, and have a single row of dyneins. A few organisms have evolved simplified axonemes with only six (Kuriyama et al., 2005) or three (Prensier et al., 1980) doublets, and also lack radial spokes and a central apparatus. Therefore, many geometries can support motility, and cylindrical basal bodies can exist that have 10-, 12-, or 14-fold symmetry, rather than ninefold symmetry.

There is, however, a unique aspect of the motility of axonemes with ninefold symmetric basal bodies. Such motile axonemes come in three forms, the most common is the 9 + 2 variety with an asymmetric central apparatus built on a platform of two singlet microtubules. The 9 + 2 pattern is present on virtually all somatic cilia of eukaryotes and on most sperm tails as well. The other varieties are radially symmetric 9 + 0 axonemes, that completely lack a central apparatus and may or may not possess radial spokes (Gibbons *et al.*, 1983), and in rare cases, axonemes that retain radial spokes but have a radially symmetric central apparatus built from three microtubules (Baccetti *et al.*, 1970, 1974a) or from nonmicrotubular structures (Baccetti, 1986). The sperm tail flagella of such organisms with radially symmetric axonemes have invariably shown a simplified helical motility pattern, rather than the more complex planar or three-dimensional beat patterns typical of organelles that retain a 9 + 2 architecture. The planar motility supported by an asymmetric central apparatus may have provided an evolutionary advantage that drove fixation of this structural geometry (Mitchell, 2004). An interesting possibility of why fixation of a 9 + 2 pattern might be advantageous is suggested by an early speculation by Astbury and Beighton (1955), consistent with the discovery of central pair twisting or rotation in certain protists. They proposed that just "nine circles can be arranged in a ring to enclose two others, . . . to allow the latter to 'roll around' inside" touching each of the peripheral circles (the spoke heads?) in turn. Clearly an axoneme with other geometries could serve as well for the other functions common to these organelles (surface gliding motility and sensory receptor display), but only those with a 9 + 2 geometry survived to become the ancestor of all extant eukaryotes. Based on these arguments, and consistent with the autogenous origin of the centriole and cilium, ninefold symmetry was fixed late during the evolution of eukaryotes, but still prior to the divergence from our LECA. These considerations support the notion that ninefold symmetry is not the historical consequence of the invasion of a host cell with a virus with ninefold symmetry, but is the result of a long evolutionary process to optimize motility and the regulation and pattern of axonemal beating.

6. Origin of Intraflagellar Transport and the Sensory Function of Cilia

The evolutionary history of centriolar and axonemal components suggests that protoeukaryotes must have already possessed a dynamic cytoskeleton when cilium evolution started (Cavalier-Smith, 2002). In addition, the history of IFT suggests that this protoeukaryotic cell also contained a secretory endomembrane system that was shaped by vesicle coat complexes and molecular motors (Jékely and Arendt, 2006) (Fig. 3.3). Based on these we can infer that the protoeukaryotic cell that started to evolve a cilium was already rather developed, with dynamic membranes and cytoskeleton, and thus the acquisition of cilia is a relatively late event in the history of eukaryote evolution. Below we will discuss what the evolutionary history of IFT has revealed about cilium evolution.

IFT machinery is nearly universally required for the assembly and maintenance of cilia (Rosenbaum and Witman, 2002). IFT is characterized by a bidirectional, microtubule motor-driven transport process between the ciliary axoneme and the ciliary membrane (Kozminski et al., 1993). A large multiprotein complex, the IFT complex, serves as a platform or raft that carries both membrane and axonemal proteins along the length of cilia, using kinesin and dynein motor proteins. Transport is bidirectional, ensuring a constant turnover of the ciliary proteome, including the IFT particles and the associated motor proteins as well.

The core proteins of IFT show structural similarity to components of vesicle coats (COPI and clathrin) (Avidor-Reiss et al., 2004; Jékely and Arendt, 2006). These proteins all have unique domain architecture, with an N-terminal β-propeller and a C-terminal α-solenoid. Besides, distant sequence similarity can also be detected between these proteins using sensitive sequence similarity searches. The vesicle coat proteins and the IFT complex all evolved from a primordial membrane curving complex, the protocoatomer, which also gave rise to the nuclear pore complex in the protoeukaryotic cell as illustrated in Fig. 3.3D (Devos et al., 2004).

This deep homology between vesicle coats and IFT core proteins suggests that IFT evolved from an intracellular bidirectional coated vesicle transport process (Fig. 3.3E–J), most likely one operating between the plasma membrane and the evolving Golgi complex (Jékely and Arendt, 2006). This evolutionary scenario is also supported by the similar mechanism of membrane recruitment of vesicle coats and the IFT complex. In both cases members of the Ras-like small GTPase family play a critical role, by binding and recruiting the complexes in their active, GTP-bound form (Omori et al., 2008).

This scenario provides interesting insights into early cilium evolution. It is conceivable that the transport process that was the evolutionary precursor of IFT delivered membrane proteins to a specialized region of the plasma membrane, perhaps delineated by the ciliary necklace. This transport could have occurred along the polarized microtubule cytoskeleton that was organized by the primitive MTOC (see above). Such localized transport could have created a membrane domain with a distinct composition and could have allowed the cell to localize its membrane receptors in one membrane domain (Fig. 3.3). This could have functioned as a "sensory membrane patch" that later protruded when microtubules polymerized from the MTOC and pushed the membrane out. Alternatively, the membrane domain produced by the precursor of IFT might have been a preferred site for centriolar virus attachment and budding. According to these models, the sensory function of cilia evolved first and motility evolved later to increase exposure of sensory receptors to the environment.

7. THE EVOLUTION OF CILIARY MOTILITY

Regardless of whether centrioles and basal bodies evolved from a virus or via an autogenous mechanism, it is widely accepted that the ciliary axoneme evolved gradually and that ciliary motility was one of the driving forces. Motility provides an important selection mechanism, since a motile single-celled organism is far better able to find nutrients and to avoid predators and unfavorable conditions. Since primitive eukaryotes likely had actin and myosin and a well-developed amoeboid motility, the motility provided by a microtubule-based axoneme must have had specific advantages that drove its continued evolution. Gliding motility may have evolved first, through simple regulated coupling of microtubule motor-dependent transport to membrane adhesion sites (Fig. 3.3C). Many single-celled organisms move by flagellar surface gliding (Bloodgood, 1988; Chantangsi et al., 2008; Tamm, 1967), which can transport cells at up to 30 μm/s (Saito et al., 2003). Such flagellar surface gliding might have had several selective advantages over amoeboid motility, including inherent polarity of the microtubule track, close association of the motile machinery with polarized surface receptors to aid chemotactic signaling, and a higher transport rate than could be achieved by amoeboid movement. Thus development of gliding motility, based on a polarized microtubule bundle, cytoplasmic dynein as a retrograde motor, and a vesicular trafficking kinesin as an anterograde motor to return dynein to the tip, might have been a selective force that drove formation of a simple axoneme.

The addition of bending motility to the repertoire of this protocilium might first be driven by the ability of bending to create currents in the fluid environment, increasing both nutrient exchange and the rate of capture of suspended particulates, as is commonly seen among benthic protists today. To produce bending motility within the cilium, cytoplasmic dynein diversifies into axonemal dynein, becomes stabilized as an arm capable of producing sliding between adjacent axonemal microtubules (Fig. 3.3D) and ultimately drives selection for doublet microtubules. In the alternative viral hypothesis, axonemal doublets arose by lengthening of the viral capsid (Fig. 3.2C). The axonemal dyneins then diversify through selection for greater motile efficiency. In a further step, microtubules polymerize independently of the basal body within the axoneme, giving rise first to a symmetric central apparatus, and finally to the asymmetric central pair, which permits planar bending and provides a further increase in motility regulation. Efficient signaling and regulated motility provided by the 9 + 2 axoneme are so advantageous for protistan survival that other early eukaryote forms are outcompeted and disappear.

ACKNOWLEDGMENTS

We thank Michael Cammer and Ann Holland for help in preparing this manuscript.

REFERENCES

Alliegro, M. C., Alliegro, M. A., and Palazzo, R. E. (2006). Centrosome-associated RNA in surf clam oocytes. *Proc. Natl. Acad. Sci. USA* **103**, 9034–9038.
Astbury, W. T., and Beighton, E. (1955). "The Structure of Bacterial Flagella. Symposia of the Society for Experimental Biology," pp. 299–301. Academic Press, New York.
Avidor-Reiss, T., Maer, A. M., Koundakjian, E., Polyanovsky, A., Keil, T., Subramaniam, S., and Zuker, C. S. (2004). Decoding cilia function: Defining specialized genes required for compartmentalized cilia biogenesis. *Cell* **117**, 527–539.
Baccetti, B. (1986). Evolutionary trends in sperm structure. *Comp. Biochem. Physiol. A* **85**(1), 29–36.
Baccetti, B., and Dallai, R. (1978). The spermatozoon of arthropoda. XXX. The multiflagellate spermatozoon in the termite *Mastotermes darwiniensis*. *J. Cell Biol.* **76**, 569–576.
Baccetti, B., Dallai, R., and Rosati, F. (1970). The spermatozoon of arthropoda. 8. The 9 + 3 flagellum of spider sperm cells. *J. Cell Biol.* **44**, 681–682.
Baccetti, B., Dallai, R., and Fratello, B. (1973). The spermatozoon of arthropoda. XXII. The '12 + 0', '14 + 0' or aflagellate sperm of protura. *J. Cell Sci.* **13**, 321–335.
Baccetti, B., Dallai, R., Giusti, F., and Bernini, F. (1974a). The spermatozoon of arthropoda. 23. The "9 plus 9 plus 3" spermatozoon of simuliid Diptera. *J. Ultrastruct. Res.* **46**(3), 427–440.
Baccetti, B., Dallai, R., Giusti, F., and Bernini, F. (1974b). The spermatozoon of arthropoda. XXV. A new model of tail having up to 170 doublets: *Monarthropalpus buxi*. *Tissue Cell* **6**(2), 269–278.
Badano, J. L., Mitsuma, N., Beales, P. L., and Katsanis, N. (2006). The ciliopathies: An emerging class of human genetic disorders. *Annu. Rev. Genomics Hum. Genet.* **7**, 125–148.
Basto, R., Lau, J., Vinogradova, T., Gardiol, A., Woods, C. G., Khodjakov, A., and Raff, J. W. (2006). Flies without centrioles. *Cell* **125**, 1375–1386.
Bell, A. J., Satir, P., and Grimes, G. W. (2008). Mirror-imaged doublets of *Tetmemena pustulata*: Implications for the development of left–right asymmetry. *Dev. Biol.* **314**, 150–160.
Blacque, O. E., Cevik, S., and Kaplan, O. I. (2008). Intraflagellar transport: From molecular characterisation to mechanism. *Front. Biosci.* **13**, 2633–2652.
Bloodgood, R. A. (1988). Gliding motility and the dynamics of flagellar membrane glycoproteins in *Chlamydomonas reinhardtii*. *J. Protozool.* **35**(4), 552–558.
Bornens, M., and Azimzadeh, J. (2007). Origin and evolution of the centrosome. *Adv. Exp. Med. Biol.* **607**, 119–129.
Brinkmann, H., and Philippe, H. (2007). The diversity of eukaryotes and the root of the eukaryotic tree. *Adv. Exp. Med. Biol.* **607**, 20–37.
Cavalier-Smith, T. (1978). The evolutionary origin and phylogeny of microtubules, mitotic spindles and eukaryote flagella. *BioSystems* **10**, 93–114.

Cavalier-Smith, T. (2002). The phagotrophic origin of eukaryotes and phylogenetic classification of Protozoa. *Int. J. Syst. Evol. Microbiol.* **52**(Pt. 2), 297–354.
Chantangsi, C., Esson, H. J., and Leander, B. S. (2008). Morphology and molecular phylogeny of a marine interstitial tetraflagellate with putative endosymbionts: *Aurantocordis quadriverberis* n. gen. et sp. (Cercozoa). *BMC Microbiol.* **8**, 123.
Christensen, S. T., Pedersen, L., Schneider, L., and Satir, P. (2007). Sensory cilia and integration of signal transduction in human health and disease. *Traffic* **8**, 97–109.
Devos, D., Dokudovskaya, S., Alber, F., Williams, R., Chait, B. T., Sali, A., and Rout, M. P. (2004). Components of coated vesicles and nuclear pore complexes share a common molecular architecture. *PLoS Biol.* **2**, e380.
Dutcher, S. K. (2001). The tubulin fraternity: Alpha to eta. *Curr. Opin. Cell Biol.* **13**, 49–54.
Dutcher, S. K. (2003). Long-lost relatives reappear: Identification of new members of the tubulin superfamily. *Curr. Opin. Microbiol.* **6**, 634–640.
Fulton, C., and Walsh, C. (1980). Cell differentiation and flagellar elongation in *Naegleria gruberi*: Dependence on transcription and translation. *J. Cell Biol.* **85**, 346–360.
Gibbons, B. H., Gibbons, I. R., and Baccetti, B. (1983). Structure and motility of the 9 + 0 flagellum of eel spermatozoa. *J. Submicrosc. Cytol.* **15**, 15–20.
Gilula, N. B., and Satir, P. (1972). The ciliary necklace. A ciliary membrane specialization. *J. Cell Biol.* **53**, 494–509.
Goldstein, S. F., and Schrevel, J. (1982). Motility of the 6 + 0 flagellum of *Lecudina tuzetae*. *Cell Motil.* **2**, 369–384.
Hiraki, M., Nakazawa, Y., Kamiya, R., and Hirono, M. (2007). Bld10p constitutes the cartwheel-spoke tip and stabilizes the 9-fold symmetry of the centriole. *Curr. Biol.* **17**(20), 1778–1783.
Jékely, G., and Arendt, D. (2006). Evolution of intraflagellar transport from coated vesicles and autogenous origin of the eukaryotic cilium. *BioEssays* **28**, 191–198.
Kellenburger, E. (1965). Control mechanisms in bacteriophage morphogenesis. *In* "CIBA Foundation Symposium Principles of Biomolecular Organization" (G. E. W. Wolstenholme and M. O'Conner, eds.), pp. 192–226. Little Brown, Boston.
Kozminski, K. G., Johnson, K. A., Forscher, P., and Rosenbaum, J. L. (1993). A motility in the eukaryotic flagellum unrelated to flagellar beating. *Proc. Natl Acad. Sci. USA* **90**, 5519–5523.
Kuriyama, R., Besse, C., Gèze, M., Omoto, C. K., and Schrével, J. (2005). Dynamic organization of microtubules and microtubule-organizing centers during the sexual phase of a parasitic protozoan, *Lecudina tuzetae* (Gregarine, Apicomplexa). *Cell Motil. Cytoskeleton* **62**, 195–209.
Lechtreck, K. F., Delmotte, P., Robinson, M. L., Sanderson, M. J., and Witman, G. B. (2008). Mutations in Hydin impair ciliary motility in mice. *J. Cell Biol.* **180**, 633–643.
Li, J. B., Gerdes, J. M., Haycraft, C. J., Fan, Y., Leitch, C. C., Lweis, R. A., Green, J. S., Parfrey, P. S., Leroux, M. R., Davidson, W. S., Beales, P. L., Guay-Woodford, L. M., et al. (2004). Comparative genomics identifies a flagellar and basal body proteome that includes the *BBS5* human disease gene. *Cell* **117**, 541–552.
Loncarek, J., Hergert, P., Magidson, V., and Khodjakov, A. (2008). Control of daughter centriole formation by the pericentriolar material. *Nat. Cell Biol.* **10**(3), 322–328.
Margulis, L. (1981). Symbiosis in Cell Evolution. Life and Its Environment on the Early Earth. San Francisco: W. H. Freeman, San Francisco.
Martin-Benito, J., Area, E., Ortega, J., Llorca, O., Valpuesta, J. M., Carrascosa, J. L., and Ortin, J. (2001). Three-dimensional reconstruction of a recombinant influenza virus ribonucleoprotein particle. *EMBO Rep.* **2**(4), 313–317.
Mitchell, D. R. (2004). Speculations on the evolution of 9 + 2 organelles and the role of central pair microtubules. *Biol. Cell* **96**, 691–696.

Mitchell, D. R. (2007). The evolution of eukaryotic cilia and flagella as motile and sensory organelles. *Adv. Exp. Med. Biol.* **607**, 130–140.

Mizukami, I., and Gall, J. (1966). Centriole replication. II. Sperm formation in the fern, *Marsilea*, and the cycad, *Zamia*. *J. Cell Biol.* **29**, 97–111.

Nakazawa, Y., Hiraki, M., Kamiya, R., and Hirono, M. (2007). SAS-6 is a cartwheel protein that establishes the 9-fold symmetry of the centriole. *Curr. Biol.* **17**, 2169–2174.

Omori, Y., Zhao, C., Saras, A., Mukhopadhyay, S., Kim, W., Furukawa, T., Sengupta, P., Veraksa, A., and Malicki, J. (2008). Elipsa is an early determinant of ciliogenesis that links the IFT particle to membrane-associated small GTPase Rab8. *Nat. Cell Biol.* **10**, 437–444.

Ostrowski, L. E., Blackburn, K., Radde, K. M., Moyer, M. B., Schlatzer, D. M., Moseley, A., and Boucher, R. C. (2002). A proteomic analysis of human cilia: Identification of novel components. *Mol. Cell. Proteomics* **1**, 451–465.

Pazour, G. J., Dickert, B. L., Vucica, Y., Seeley, E. S., Rosenbaum, J. L., Witman, G. B., and Cole, D. G. (2000). Chlamydomonas IFT88 and its mouse homologue, polycystic kidney disease gene *Tg737*, are required for assembly of cilia and flagella. *J. Cell Biol.* **151**, 709–718.

Pazour, G. J., Agrin, N., Leszyk, J., and Witman, G. B. (2005). Proteomic analysis of a eukaryotic cilium. *J. Cell Biol.* **170**, 103–113.

Pelletier, L., O'Toole, E., Schwager, A., Hyman, A. A., and Muller-Reichert, T. (2006). Centriole assembly in *Caenorhabditis elegans*. *Nature* **444**, 619–623.

Phillips, D. M. (1974). Structural variants in invertebrate sperm flagella and their relationship to motility. In "Cilia and Flagella" (M. A. Sleigh, ed.), pp. 379–402. Academic Press, New York.

Prensier, G., Vivier, E., Goldstein, S., and Schrevel, J. (1980). Motile flagellum with a "3 + 0" ultrastructure. *Science* **207**, 1493–1494.

Rodriguez-Ezpeleta, N., Brinkmann, H., Burger, G., Roger, A. J., Gray, M. W., Philippe, H., and Lang, B. F. (2007). Toward resolving the eukaryotic tree: The phylogenetic positions of jakobids and cercozoans. *Curr. Biol.* **17**, 1420–1425.

Rosenbaum, J. L., and Witman, G. B. (2002). Intraflagellar transport. *Nat. Rev. Mol. Cell Biol.* **3**, 813–825.

Ross, M. M. (1967). Modified cilia in sensory organs of juvenile stages of a parasitic nematode. *Science* **156**, 1494–1495.

Saito, A., Suetomo, Y., Arikawa, M., Omura, G., Khan, S. M., Kakuta, S., Suzaki, E., Kataoka, K., and Suzaki, T. (2003). Gliding movement in *Peranema trichophorum* is powered by flagellar surface motility. *Cell Motil. Cytoskeleton* **55**, 244–253.

Salisbury, J. L. (2007). A mechanistic view on the evolutionary origin for centrin-based control of centriole duplication. *J. Cell. Physiol.* **213**(2), 420–428.

Salisbury, J. L. (2008). Breaking the ties that bind centriole numbers. *Nat. Cell Biol.* **10**, 255–257.

Satir, P., and Christensen, S. T. (2007). Overview of structure and function of mammalian cilia. *Annu. Rev. Physiol.* **69**, 377–400.

Satir, P., and Christensen, S. T. (2008). Structure and function of mammalian cilia. *Histochem. Cell Biol.* **129**(6), 687–693.

Satir, B., Sale, W. S., and Satir, P. (1976). Membrane renewal after dibucaine deciliation of *Tetrahymena*. *Exp. Cell Res.* **97**, 83–91.

Satir, P., Guerra, C., and Bell, A. J. (2007). Evolution and persistence of the cilium. *Cell Motil. Cytoskeleton* **64**, 906–913.

Scholey, J. M. (2003). Intraflagellar transport. *Annu. Rev. Cell Dev. Biol.* **19**, 423–443.

Smith, E. F., and Yang, P. (2004). The radial spokes and central apparatus: Mechanochemical transducers that regulate flagellar motility. *Cell Motil. Cytoskeleton* **57**, 8–17.

Smith, J. C., Northey, J. G., Garg, J., Pearlman, R. E., and Siu, K. W. (2005). Robust method for proteome analysis by MS/MS using an entire translated genome: Demonstration on the ciliome of *Tetrahymena thermophila*. *J. Proteome. Res.* **4,** 909–919.

Surrey, T., Nedelec, F., Leibler, S., and Karsenti, E. (2001). Physical properties determining self-organization of motors and microtubules. *Science* **292,** 1167–1171.

Tamm, S. L. (1967). Flagellar development in the protozoan *Peranema trichophorum*. *J. Exp. Zool.* **164,** 163–186.

Tilney, L. G., and Porter, K. R. (1965). Studies on microtubules in Heliozoa. I. The fine structure of *Actinosphaerium nucleofilum* (Barrett), with particular reference to the axial rod structure. *Protoplasma* **60,** 317–344.

Tucker, J. B. (1974). Microtubule arms and cytoplasmic streaming and microtubule bending and stretching of intertubule links in the feeding tentacle of the suctorian ciliate Tokophrya. *J. Cell Biol.* **62,** 424–437.

Wickstead, B., and Gull, K. (2007). Dyneins across eukaryotes: A comparative genomic analysis. *Traffic* **8,** 1708–1721.

Wilkes, D. E., Watson, H. E., Mitchell, D. R., and Asai, D. J. (2008). Twenty-five dyneins in *Tetrahymena*: A re-examination of the multidynein hypothesis. *Cell Motil. Cytoskeleton* **65,** 342–351.

Yoder, B. K., Tousson, A., Millican, L., Wu, J. H., Bugg, C. E., Schafer, J. A., and Balkovetz, D. F. (2002). Polaris, a protein disrupted in orpk mutant mice, is required for assembly of renal cilium. *Am. J. Physiol. Renal Physiol.* **282,** F541–F552.

Yoon, H. S., Grant, J., Tekle, Y. I., Wu, M., Chaon, B. C., Cole, J. C., Logsdon, J. M., Jr., Patterson, D. J., Bhattacharya, D., and Katz, L. A. (2008). Broadly sampled multigene trees of eukaryotes. *BMC Evol. Biol.* **8,** 14.

CHAPTER FOUR

Ciliary Tubulin and Its Post-Translational Modifications

Jacek Gaertig *and* Dorota Wloga

Contents

1. Introduction — 83
2. Production and Turnover of Ciliary Tubulin — 84
3. Post-Translational Modifications of Tubulin — 89
   3.1. Acetylation of α-tubulin — 90
   3.2. Tail domains of tubulins and their post-translational modifications — 93
4. Conclusions — 103
Acknowledgments — 103
References — 104

Abstract

Tubulin, the most abundant axonemal protein, is extensively modified by several highly conserved post-translational mechanisms including acetylation, detyrosination, glutamylation, and glycylation. We discuss the pathways that contribute to the assembly and maintenance of axonemal microtubules, with emphasis on the potential functions of post-translational modifications that affect tubulin. The recent identification of a number of tubulin modifying enzymes and mutational studies of modification sites on tubulin have allowed for significant functional insights. Polymeric modifications of tubulin (glutamylation and glycylation) have emerged as important determinants of the 9 + 2 axoneme assembly and motility.

1. Introduction

Microtubules are made by polymerization of tubulin (dimers of α- and β-tubulin subunits). In the axoneme, microtubules provide structural support, create binding sites for protein complexes (including dynein arms and

Department of Cellular Biology, University of Georgia, Athens, Georgia radial spokes), and function as tracks for intraflagellar transport (IFT). Generally, axonemal microtubules contain only one isotype of both α- and β-tubulin, even in organisms that express multiple isotypes, possibly because the axoneme requires isotypic homogeneity of tubulin (Silflow, 1991). The α- and β-tubulin isotypes that are used to build axonemal microtubules tend to be of a highly conserved kind, possibly because the participation in numerous protein–protein interactions within the axoneme constrains tubulin evolution (Gaertig et al., 1993; Nielsen and Raff, 2002; Raff, 1994). On the other hand, the axonemal tubulin isotypes are multifunctional, as in the same cell type axonemal tubulin is often used for assembly of nonaxonemal microtubules such as spindles (Hoyle and Raff, 1990; Thazhath et al., 2004). Despite the lack of variation in the primary sequence, axonemal tubulin is likely the most biochemically heterogeneous tubulin type known, due to a series of extensive post-translational modifications that occur shortly after axoneme assembly.

2. Production and Turnover of Ciliary Tubulin

Tubulin destined for cilia is derived from the pool of α/β-tubulin dimers that are produced in the cell body. The pathway that generates the cytosolic pool of α/β-tubulin dimers has been studied extensively in vertebrate cells (Fig. 4.1). After translation, the nascent α- and β-tubulin polypeptides bind to the prefoldin chaperone complex (Vainberg et al., 1998). Prefoldin facilitates the insertion of monomers into the cavity of CCT, a cytosolic type II chaperonin complex that folds tubulin, actin, and some other proteins (Dekker et al., 2008; Frydman et al., 1992). Inside the lumen of CCT, α- and β-tubulin undergo partial folding. The monomers released from CCT associate with tubulin-specific chaperones or cofactors (TBCA, TBCB, TBCC, TBCD, and TBCE), which promote the formation of α/β-dimers (Tian et al., 1996). When the cilium assembles, incorporation of new tubulin takes places at the tip of the axoneme (Johnson and Rosenbaum, 1992). In most ciliated cell types, the precursor tubulin is probably carried to the tips of growing axonemes by the IFT motors that move along the sides of outer doublet microtubules (Rosenbaum and Witman, 2002). It is not known in what form tubulin is transported inside cilia. Interestingly, inside nerve projections, tubulin is transported along microtubules in the polymerized form (oligomers or short microtubules) (Baas et al., 2005; Terada et al., 2000). Thus, by analogy, cilia-destined tubulin could also be transported in larger polymeric complexes.

Unexpectedly, multiple components known to be involved in the cytosolic folding reactions on tubulin have been found inside cilia. A prefoldin subunit was detected in the detergent-soluble fraction of

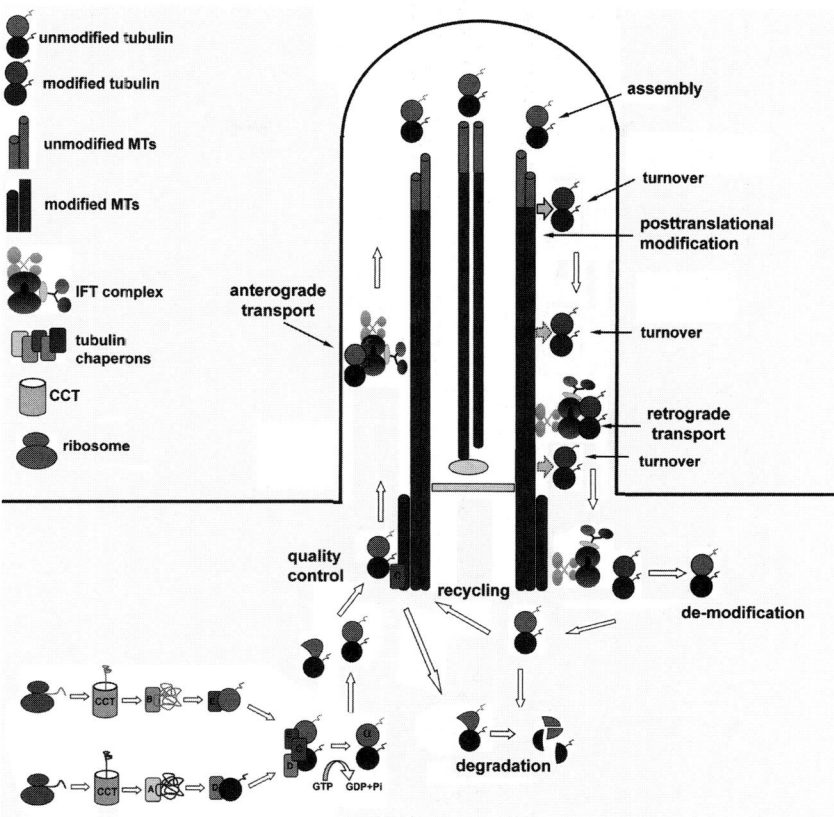

Figure 4.1 The major steps involved in folding, transport, and turnover of tubulin in a ciliated cell. (See Color Insert.)

Chlamydomonas flagella (Pazour et al., 2005). Prefoldin is important to ciliary functions. Mice lacking Pfd1, one of the six subunits of prefoldin, have defects in the brain and the immune system and display phenotypes consistent with ciliary dyskinesia: hydrocephaly and accumulation of mucus in the lungs (Cao et al., 2008). Four CCT subunits (CCTα, CCTη, CCTε, and CCTδ) were found inside cilia of *Tetrahymena* (Seixas et al., 2003) and CCTε was detected in *Chlamydomonas* flagella (Pazour et al., 2005). Prefoldin and CCT subunits are not specific to tubulin and their presence inside cilia could reflect activities on nontubulin targets. However, TBCB, a chaperone that appears to be dedicated to folding of α-tubulin (Radcliffe and Toda, 2000; Tian et al., 1997), is enriched in ependymal and respiratory cilia in mice (Lopez-Fanarraga et al., 2007). The presence of the CCT subunits and tubulin chaperons in cilia could simply reflect the fact that folding of tubulin continues inside ciliary matrix, implying that at least some cilia-destined tubulin is delivered in an unfolded (monomeric) state.

However, it is more likely that ciliary tubulin folding factors have cilia-specific function, for example, in the transport or turnover of tubulin. There is already evidence that some tubulin chaperones have cytosolic functions distinct from tubulin folding. For example, TBCE promotes degradation of tubulin dimers (Keller and Lauring, 2005), while TBCB regulates the dynamics of cytosolic microtubules (Vadlamudi et al., 2005).

Studies that address the cilia-specific functions of tubulin folding factors are complicated by the fact that these proteins are essential for generation of the cytosolic pool of tubulin and some of them (CCT and prefoldin) also fold nontubulin targets such as actin. Not surprisingly, in *Tetrahymena*, knockouts of CCTα, CCTδ, and TBCB genes cause lethality, in association with depletion of tubulin in the cell body (C. Seixas, J. Gaertig, and H. Soares, unpublished data; D. Dave and J. Gaertig, unpublished data). However, some evolutionary lineages have proteins that are closely related to tubulin folding factors that have evolved cilia-related functions. In humans, a mutation in RP2, a protein related to TBCC, causes X-linked retinitis pigmentosa, a hereditary retinal dystrophy (Schwahn et al., 1998). A knockdown of RP2 expression in *Trypanosoma* did not disturb the mitotic or pellicular microtubules, but disorganized the axonemal microtubules. RP2 was detected around the old and new basal body and near foci of tyrosinated α-tubulin, a likely precursor tubulin destined to cilia (see below). Thus, RP2 could have a role in the "quality control" of cilia-destined tubulin (Stephan et al., 2007). During the folding reactions on cytosolic tubulin, TBCC stimulates GTPase activity of tubulin monomers, which is required for dimerization (Tian et al., 1997). The human RP2 functionally complements a null mutation of TBCC in yeast (Bartolini et al., 2002). This argues that RP2 is a cilia-specialized variant that has not lost the ability to perform general functions of TBCC. RP2 could bind to tubulin destined to cilia, probe its conformation, and possibly redirect improperly folded tubulin back to the cell body for degradation (Stephan et al., 2007) (Fig. 4.1). This mechanism could prevent incorporation of defective tubulin subunits, a function that might be particularly important inside the axonemal microtubule lattice, as its tubulin turns over relatively slowly (see below).

Two additional proteins have cilia-specific functions and their sequences show homology to subunits of CTT. BBS6 and BBS10, two animal-specific proteins whose mutations cause a ciliopathic disease, the Bardet-Biedl syndrome (BBS), are related to CCTα and CCTζ, respectively (Kim et al., 2005; Stoetzel et al., 2006). It is likely that the ciliary function of BBS6 is nonidentical with the cytosolic function of its CCT subunit counterpart, as BBS6 does not belong to a large complex of the CCT size range (Kim et al., 2005). In organisms such as *Tetrahymena* that do not have BBS6 and BBS10 homologs, the functions of these proteins in cilia are probably fulfilled by the general CCT subunits that could shuttle between the cell body and cilia (Seixas et al., 2003).

Once tubulins are incorporated into axonemes, the axonemal microtubule lattice undergoes a relatively slow subunit exchange. The turnover rate is between 1% and 5% per hour (Gorovsky et al., 1970; Nelsen, 1975; Rosenbaum and Child, 1967; Rosenbaum et al., 1969; Song and Dentler, 2001; Thazhath et al., 2004), but a much higher exchange rate was observed in sea urchin embryonic cilia (Stephens, 1994). In addition, a large number of other axonemal proteins turnover, including dynein heavy chains and radial spoke proteins (Song and Dentler, 2001).

In preformed flagella of *Chlamydomonas*, the newly added tubulin subunits are mainly incorporated into the distal portions of axonemes, likely by cycles of end depolymerization/repolymerization (Marshall and Rosenbaum, 2001). Experiments in *Tetrahymena* suggest that the entire axoneme turns over. In *Tetrahymena*, the locomotory cilia never resorb, and new cilia are inserted during each cell cycle (Allen, 1969). Thus, the cortex of *Tetrahymena* contains a mosaic of multiple generations of cilia. In preformed axonemes, the newly synthesized epitope-tagged tubulin is initially incorporated at the distal ends of axonemes, but with time new tubulin gradually spreads over the entire axoneme. A complete exchange of all pre-existing tubulin in axonemes required about 13 cell cycles (Thazhath et al., 2004). In another experiment, Thazhath et al. (2004) followed the replacement of wild-type tubulin by a β-tubulin carrying a recessive mutation that supports assembly of short cilia. The mutant tubulin slowly replaced the pre-existing wild-type tubulin along the entire length of the axoneme and caused gradual shortening of the axoneme to the size of organelle that assembles *de novo* from mutant tubulin. These data suggest that in *Tetrahymena*, the axoneme turns over slowly but completely. At the tips, the turnover could be based primarily on the cycles of distal end depolymerization/polymerization based on studies in *Chlamydomonas* (Marshall and Rosenbaum, 2001). However, it appears that a complete turnover of the axoneme requires a lattice displacement mechanism similar to treadmilling as proposed by Stephens for sea urchin cilia (Stephens, 2000) or could be based on extraction and replacement of individual dimers within the lattice.

It appears that turnover of tubulin is critical to the length regulation of cilia. When the IFT pathway is inhibited, cilia gradually shorten, suggesting that IFT-dependent delivery of tubulin balances its loss due to turnover (Brown et al., 1999; Marshall and Rosenbaum, 2001). However, the relationship between axonemal turnover and length regulation is more complex. For example, incorporation of new tubulin continues under experimental conditions that induce flagellar shortening, and the turnover rate does not decrease in a *Chlamydomonas* mutant that has excessively long flagella (Song and Dentler, 2001).

Kinesin-13 is among the proteins that directly interact with tubulin and could mediate tubulin turnover specifically at the tips of axonemes. Members of the kinesin-13 subfamily act as depolymerizers of the ends of microtubules (Howard and Hyman, 2007). In *Leishmania*, overexpression of kinesin-13 shortened flagella, while an RNAi knockdown made flagella longer (Blaineau *et al.*, 2007). In *Giardia*, kinesin-13 localized to the tips of flagella, and expression of its rigor-type mutant form caused elongation of flagella (Dawson *et al.*, 2007). Axonemal kinesin-13 could be associated with the cap complexes that are located at the ends of axonemes and have been proposed to regulate the dynamics of axonemal microtubules (Dentler, 1980).

Experiments in *Chlamydomonas* and *Tetrahymena* have identified additional proteins that could mediate axoneme depolymerization during turnover, possibly also at sites that are distinct from the axoneme tips. The IFT mutant of *Chlamydomonas*, *fla10*, has a temperature-sensitive mutation in the subunit of the anterograde motor of IFT, kinesin-2 (Walther *et al.*, 1994). When the *fla10* cells are transferred to restrictive temperature, they lose flagella either by gradual resorption or by deflagellation (Parker and Quarmby, 2003; Walther *et al.*, 1994). Resorption can be seen as an outcome of turnover that is not balanced by addition of new axonemal components via IFT. Deflagellation is a rapid loss of the axoneme that is caused by severing of doublet microtubules within the transitional zone (Quarmby, 2004). Deflagellation requires the function of Fa1p and Fa2p proteins (Finst *et al.*, 2000; Mahjoub *et al.*, 2002). Double mutants *fla10–fa1* or *fla10–fa2* are unable to undergo deflagellation (as expected), but remarkably, also have a reduction in the rate of resorption of flagella (Parker and Quarmby, 2003). These experiments indicate that slow resorption and deflagellation pathways share components. Thus, the deflagellation factors could also mediate loss of tubulin during flagellar resorption by stimulating axonemal turnover. Fa2p is a Nima A-related kinase (NRK) that localizes to the transitional region where the axoneme breaks during deflagellation (Mahjoub *et al.*, 2002, 2004). Thus, it is possible that during the slow turnover, tubulin subunits are removed from the transitional zone with assistance of Fa2p and other deflagellation proteins.

It appears that additional NRKs controls turnover of axonemal microtubules at multiple locations by stimulating microtubule depolymerization. Another NRK of *Chlamydomonas*, Cnk2p, localizes along the length of the axoneme. A knockdown of Cnk2p produced longer flagella and its overexpression rapidly shortened flagella (Bradley and Quarmby, 2005). Nrk2p of *Tetrahymena*, a protein phylogenetically related to Cnk2p, when overexpressed, accumulated at the tips of cilia and caused their rapid shortening, by gradual depolymerization of the distal end of the axoneme. Expression of a dominant-negative variant of Nrk2p made cilia longer, suggesting that Nrk2p is active in axoneme depolymerization during steady-state turnover. Intriguingly, an epitope-tagged Nrk2p that was expressed in the endogenous locus, accumulated preferentially at the tips of assembling cilia, suggesting that the turnover occurs even during axoneme assembly. Nrk30p and Nrk17p, when overexpressed, also caused rapid depolymerization of axonemes but the sites of preferential depolymerization were distinct: remarkably the central pair depolymerized faster compared to the rest of the axoneme. In the case of overproduced Nrk17p, the depolymerization occurred first at the proximal end of the central microtubules, above the transitional zone (Wloga et al., 2006). These observations are consistent with a model in which tubulin subunits are removed from the axoneme at multiple sites including the proximal and distal ends of microtubules and the transitional zone.

The fate of tubulin that is removed from the axoneme during steady-state turnover or ciliary resorption is poorly understood but it appears that this tubulin can be reutilized. When one of the two flagella of *Chlamydomonas* was removed by mechanical stress, the remaining flagellum underwent partial shortening, concomitant with assembly of the new flagellum (Rosenbaum et al., 1969). This observation suggests, tubulin that originates from the turnover is reutilized for axoneme assembly. However, the recycled tubulin may need to be demodified before its transport and incorporation into the axoneme (see below).

3. Post-Translational Modifications of Tubulin

Axonemal tubulin undergoes several highly conserved post-translational modifications, including acetylation, detyrosination, glutamylation, and glycylation (Verhey and Gaertig, 2007; Westermann and Weber, 2003). As some of these modifications result in variable length side chain polymers, and individual tubulin molecules can contain multiple types of modifications or the same type of modifications at multiple sites, the mature axonemes are composed of hundreds of distinct tubulin isoforms (Redeker et al., 2005) (Fig. 4.2). Tubulin PTMs are not restricted to axonemes. However, tubulin PTMs are especially abundant on axonemes, and the polymeric PTMs (glutamylation and glycylation) tend to have the longest side chains on microtubules of axonemes within the cell (Wloga et al., 2008). Organisms that lack cilia, such as fungi and higher plants, lack most of tubulin PTMs (Alfa and Hyams, 1991; Janke et al., 2005), suggesting that these mechanisms are particularly important inside cilia. Modified tubulins have been used as cytological markers of cilia and basal bodies.

The soluble tubulin destined to cilia is largely unmodified and the extent of modifications increases dramatically in the course of axoneme assembly (Adoutte et al., 1991; Brown et al., 1999; Iftode et al., 2000; Johnson, 1998; Lechtreck and Geimer, 2000; Sasse and Gull, 1988; Sharma et al., 2007; Sherwin et al., 1987; Stephens, 1992). Thus, the levels of tubulin PTMs distinguish between the assembling and mature cilia.

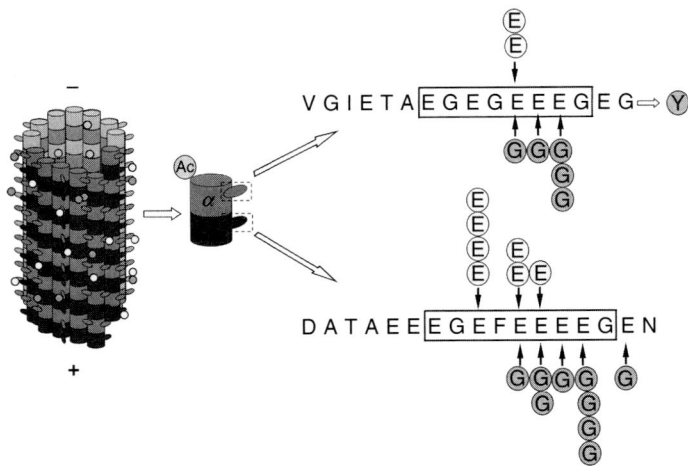

Figure 4.2 The post-translational modifications that affect ciliary tubulin. The sequences of CTTs of tubulin subunits of *Tetrahymena thermophila* are shown. Boxes mark the location of the "axoneme motif." (See Color Insert.)

The tubulin that is recovered from the mature axoneme by resorption or steady-state turnover is initially modified post-translationally. For example, in the *fla10* mutant of *Chlamydomonas* that resorbed flagella at restrictive temperature, there was an increase in the content of acetylated form of α-tubulin in the detergent-soluble compartment (Qin *et al.*, 2004). Thus, tubulin that is recovered from the axoneme could undergo demodification reactions (including deacetylation) inside the cell body before its reutilization (Fig. 4.1). Indeed a high level of tubulin deacetylase enzyme activity is present in the cell body of *Chlamydomonas* (Maruta *et al.*, 1986).

3.1. Acetylation of α-tubulin

Acetylation of K40 on α-tubulin was discovered in the flagella of *Chlamydomonas* (Greer *et al.*, 1985; L'Hernault and Rosenbaum, 1985). Acetylation is highly enriched on axonemal microtubules (LeDizet and Piperno, 1991), and can be very abundant; in *Tetrahymena* most of the axonemal α-tubulin is acetylated (Gaertig *et al.*, 1995). The tubulin acetyltransferase was partially purified from *Chlamydomonas* flagella (Maruta *et al.*, 1986), but the catalytic subunit has not yet been identified. Two conserved enzymes, HDAC6 and SIRT2, were identified as tubulin deacetylases in mammals (Hubbert *et al.*, 2002; Matsuyama *et al.*, 2002; North *et al.*, 2003; Zhang *et al.*, 2003). K40 is located on an N-terminal domain of α-tubulin that is exposed to the lumen of the microtubules (Nogales *et al.*, 1999). This indicates that acetylation could be important in the interactions with proteins that reside inside the microtubule lumen (Garvalov et al., 2006), and specifically inside the outer doublets (Nicastro et al., 2006; Sui and Downing, 2006).

K40 acetylation on α-tubulin has been documented as a marker of the age of microtubules. Thus, older cytoplasmic microtubules are enriched in acetylation and the levels of acetylation increase upon treatment with a microtubule stabilizer, paclitaxel (Piperno et al., 1987; Webster and Borisy, 1989). Acetylation did not detectably influence the dynamics of microtubules assembled *in vitro* (Maruta et al., 1986), suggesting that accumulation of acetylation on microtubules is the result of their stabilization by other factors. On the other side, experimentally induced changes in the levels of HDAC6 tubulin deacetylase mildly stabilize microtubules *in vivo* (Matsuyama et al., 2002; Tran et al., 2007), but it is not certain that these effects are mediated by the activity on α-tubulin, as HDAC6 has additional substrates (see below).

Ectopic expression of a nonacetylatable K40R α-tubulin in *Chlamydomonas* did not affect the gross phenotype (Kozminski et al., 1993). A complete replacement of the single conventional α-tubulin gene of *Tetrahymena* with a K40R-encoding variant also did not affect cell multiplication, motility, and sensitivity to antimicrotubule drugs (Gaertig et al., 1995). It appears unlikely that in *Tetrahymena*, another acetylation site on α-tubulin can be used in the absence of K40. On an isoelectric focusing gel, the entire set of axonemal R40 α-tubulin isoforms shifted in mobility toward the more basic region (Gaertig et al., 1995), and R40 α-tubulin is not recognized by a pan acetyl-K antibody (S. Akella, D. Wloga, and J. Gaertig, unpublished data), indicating that there is no compensatory acetylation in the absence of K40.

In *Caenorhabditis elegans*, only one of several α-tubulin isotypes, MEC-12, has K40. The MEC-12 α-tubulin, together with MEC-7 β-tubulin is expressed in the touch receptor neurons that have unusual 15 protofilament microtubules. Because MEC-12 is not expressed in cilia, in *C. elegans* axonemes lack tubulin acetylation. Deletion of either MEC-7 or MEC-12 leads to a loss of touch receptor function and a corresponding loss of the 15 protofilament microtubules (Fukushige et al., 1999; Savage et al., 1989). Surprisingly, the MEC-12 loss-of-function mutants are rescued by introduction of a gene encoding a MEC-12 α-tubulin with a K40Q acetylation-mimicking substitution (Fukushige et al., 1999), suggesting that deregulation of this PTM is well tolerated.

While in protists and invertebrates K40 acetylation is not required for axoneme assembly and the function of this modification on ciliary and nonciliary microtubules is unclear, recent studies suggest that vertebrate cell types are more sensitive to the loss of K40 α-tubulin acetylation. Kinesin-1 is the major motor responsible for transport of cargoes along microtubules inside nerve projections. *In vitro*, kinesin-1 has higher binding affinity and produced faster gliding of acetylated microtubules as compared to nonacetylated (K40R α-tubulin) microtubules. Moreover, treating neural cells with inhibitors of HDAC6 increased the levels of K40 acetylation in the neurites and stimulated transport of kinesin-1 cargo protein, JIP1, to neurite tips (Reed et al., 2006). Dompierre et al. (2007) showed that in cultured cortical neurons, the rate of transport of vesicles containing the brain-derived neurotrophic factor (BDNF) is stimulated by chemical inhibition of HDAC6 and inhibited by overexpression of a nonacetylatable (but not wild-type) α-tubulin. It is surprising that a luminal microtubule PTM can affect motor proteins that move on the outside surface of microtubules. However, luminal features can affect the structure of the microtubule surface. For example, a mutation of G56 on β-tubulin, which is exposed to the lumen of the microtubule, inhibits the assembly of outer dynein arms in sperm axonemes of *Drosophila* (Raff et al., 2008).

A recent study hints at a potential role of K40 α-tubulin acetylation in vertebrate primary cilia, and specifically during cilia resorption. Activation of the Aurora kinase is required for resorption of flagella in *Chlamydomonas* (Pan et al., 2004). Pugacheva et al. (2007) showed that a related mammalian kinase, AurA, promotes resorption of primary cilia in cultured cells. AurA copurified with HDAC6 and phosphorylated HDAC6 *in vitro*, and chemical inhibitors of HDAC6 blocked resorption of primary cilia. These data opened the possibility that HDAC6 promotes ciliary resorption by deacetylating α-tubulin. However, HDAC6 has at least two additional deacetylation substrates, HSP90 (Kovacs et al., 2005) and cortactin (Zhang et al., 2007). A protein related to HSP90 was found inside cilia of *Tetrahymena* (Williams and Nelsen, 1997). HSP90 and HSP70 are among the proteins that interact with PACRG (Imai et al., 2003), a component of ciliary doublet microtubules (Ikeda et al., 2007) that is required for their stability (Dawe et al., 2005). A further complication is the fact that HDAC6 can act independently of its deacetylase activity, by binding to cytoplasmic dynein/dynactin and polyubiquitinated proteins (Kawaguchi et al., 2003). Thus, it is not clear whether the effects of HDAC6 on cilia are mediated by acetylation of α-tubulin.

Given that HDAC6 has multiple substrates, it is surprising that mice with a deletion of HDAC6 develop normally and are fertile (Zhang et al., 2008). The levels of α-tubulin acetylation at K40 have increased several fold in multiple tissues of the HDAC6-null mice. In the wild-type testes, the levels of K40 acetylation are low at the early stage of sperm differentiation and increase greatly during the formation of axonemes. In the HDAC6-null mice, the levels of K40 acetylation are high during the entire spermatogenesis but no abnormalities were detected during sperm development, indicating that severe deregulation of the acetylation/deacetylation cycle on α-tubulin has no detectable consequences on the axoneme assembly. However, the HDAC6-null mice have mild defects in the bone structure and a weaker immune response (Zhang et al., 2008). It will be important to establish whether any of the deficiencies in the HDAC6-null mice are mediated by defects in the ciliary functions and specifically in resorption of cilia.

3.2. Tail domains of tubulins and their post-translational modifications

Most of the microtubule PTMs occur on the C-terminal tail domains (CTTs) of either α- or β-tubulin. CTTs are flexible domains that are located on the surface of microtubules (Luchko et al., 2008; Nogales et al., 1998, 1999). CTTs are highly negatively charged due to the presence of multiple glutamic acids (Fig. 4.2). CTTs are the sites of PTMs of diverse nature, including detyrosination, glutamylation, and glycylation. CTTs are known to strongly interact with motors and microtubule lattice-destabilizing proteins (katanin, spastin, and end-depolymerizing kinesin-13) (Lakamper and Meyhofer, 2006; Lu et al., 2004; Ovechkina et al., 2002; Roll-Mecak and Vale, 2008; Skiniotis et al., 2004; Wang and Sheetz, 2000). CTTs also affect unpolymerized tubulin; a CCT-less β-tubulin was outcompeted by a coexpressed intact β-tubulin for incorporation into α/β-dimers (Hoyle et al., 2001).

In vitro, apparently normal microtubules can polymerize from tubulin in which CTTs were removed proteolytically (Serrano et al., 1984). However, CTTs are important *in vivo*. Deletion of a CTT from either from α- or β-tubulin is lethal in *Tetrahymena* (Duan and Gorovsky, 2002). Moreover, the CTT of β-tubulin plays a critical role in axoneme assembly. *Drosophila* expresses several isotypes of β-tubulin but only β_2 participates in axoneme assembly, and it cannot be functionally substituted by other isotypes (Hoyle and Raff, 1990). Most of the amino acid differences between β_2 and axoneme-incompetent β-tubulins of *Drosophila* are contained within the CTT and in the "internal variable region" (amino acids 55–57) in the microtubule lumen (Nielsen and Raff, 2002). Flies that express solely a CTT-less β_2 in the postmitotic male germline, do assemble microtubules including doublet microtubules, but these microtubules fail to organize into a 9 + 2 axoneme (Fackenthal et al., 1993). The CTT of β_2 contains the so-called "axoneme motif" EGEF(E/D)$_3$ (Fig. 4.2), an amino acid sequence that is well conserved on β-tubulins in cell types that have motile 9 + 2 cilia (Raff et al., 1997). In mammals only the B_{IV}-isotype has this motif, and this protein is highly enriched in cilia (Jensen-Smith et al., 2003). Duan and Gorovsky (2002) noticed that a similar sequence exists on the axoneme-competent α-tubulin, and that the consensus motif for both subunits is EGE – F/G(E/D)$_{2-5}$ – G/A (Fig. 4.2). Indeed, the CTT of α-tubulin interacts extensively with its counterpart on the β-tubulin subunit and under certain conditions can supply its function (see below).

While *Tetrahymena* requires the presence of a CTT on both α- and β-tubulin for survival, cells that are carrying a CTT sequence of α-tubulin identity on both α- and β-tubulin or carrying a CTT of β-tubulin identity on both subunits are viable and have motile cilia (Duan and Gorovsky, 2002). While the tail switch experiments reveal functional equivalency of the CTT sequences of α- and β-tubulin identity, other experiments show that the CTT of β-tubulin and not its counterpart on α-tubulin, carries sequence features that are critical for survival and axoneme assembly. However, even these essential features of the CTT of β-tubulin can be artificially transferred to the CTT of α-tubulin (see below).

Both CTTs are highly acidic. The CTT of α-tubulin of *Tetrahymena* has a total of six E residues while the CTT of β-tubulin has nine Es (Fig. 4.2). Some if not all of these residues are subjected to PTMs known as polymodifications (glycylation and glutamylation, see below). All six Es on α-tubulin can be replaced with Ds and the resulting *Tetrahymena* cells have motile cilia albeit these cells grow and move more slowly (Wloga et al., 2008). However, a replacement of three or more specific Es on CTT of β-tubulin is lethal (Xia et al., 2000), severely shortens the axoneme and inhibits central pair assembly (Thazhath et al., 2002, 2004). Remarkably, lethality of the triple E/D mutation of the CTT domain of β-tubulin ($\beta DDDE_{440}$) can be rescued if the sequence of the CTT of β-tubulin that spans the mutated region replaces the corresponding sequence of the CTT of α-tubulin, and the double mutants have motile cilia (Xia et al., 2000). Moreover, when all nine Es of the CTT of β-tubulin are changed to Ds, the mutation is lethal on its own, but can be rescued by changing the entire CTT sequence of α-tubulin to that of β-tubulin identity (Duan and Gorovsky, 2002). These data show that in the wild-type tubulin dimer, the tail of β-tubulin has sequence features that are essential for viability and axoneme assembly, but that these features can be transplanted over to the CTT of α-tubulin. The simplest explanation is that critical factors that interact with axonemal microtubules initially bind to CTTs and that the precise tail location is not important as long as a sufficient number of CTT-binding sites are present on the surface of microtubules. Moreover, the tail mutational studies not only undermine the importance of the acidity of the tail domain, but also uncover a specific requirement for glutamic acids, many of which are known to be subjected to polymodifications (see below).

3.2.1. Detyrosination of α-tubulin

The genomically encoded C-terminal Y is removed proteolytically by the process called detyrosination. This modification was discovered in the rat brain tubulin (Arce et al., 1975; Argarana et al., 1978). The Nna1/CCP1 zinc metallocarboxypeptidase could be the enzyme responsible for detyrosination. Mice deficient in Nna1/CCP1 have greatly reduced levels of detyrosinated tubulin and increased levels of unmodified tyrosinated tubulin in the mitral cells of olfactory bulb (Fernandez-Gonzalez et al., 2002; Kalinina et al., 2007). An Nna1/CCP mutant mouse develops the Purkinje cell degeneration (pcd) syndrome characterized by altered gait, loss of Purkinje cells, loss of the mitral cells, degeneration of photoreceptor cells, and male sterility (Greer and Shepherd, 1982; Landis and Mullen, 1978; LaVail et al., 1982; Wang and Morgan, 2007). These defects are consistent with aberrant functions of diverse microtubules, possibly including those forming the sperm axonemes.

The detyrosination reaction is reversible at least in some organisms. The detyrosination reaction occurs mainly on microtubules while the retyrosination reaction occurs on the soluble tubulin dimer (Gundersen et al., 1987). In vertebrates and some nonvertebrates (e.g., trypanosomatids), the enzyme tubulin-tyrosine ligase (TTL) is responsible for readdition of Y (Schroder et al., 1985). Mice lacking TTL function have decreased levels of tyrosinated microtubules and die during embryonic development, mainly due to defects in the formation of the central nervous system (Erck et al., 2005). CLIP-170 and p150 Glued, plus-end-tracking proteins that have a CAP-GLY domain, associate preferentially with tyrosinated microtubule segments in wild-type cells, and are mislocalized in fibroblasts taken from TTL-null mice (Peris et al., 2006). Thus, detyrosination could play a role in negative regulation of binding of subsets of plus-end-tracking proteins.

In sea urchin embryonic cilia, α-tubulin present in the soluble fraction is primarily unmodified (tyrosinated), likely because this fraction represent the precursor pool (Stephens, 1992). Moreover, *Trypanosoma* has foci of detergent-insoluble tyrosinated α-tubulin near the basal bodies, that could represent the sites of storage of cilia-destined tubulin (Stephan et al., 2007). As mentioned above, tyrosinated α-tubulin is a preferred substrate for proteins with a CAP-GLY domain (Peris et al., 2006; Weisbrich et al., 2007). Intriguingly, some tubulin chaperones, namely TBCB and TBCE, have such a domain (Grynberg et al., 2003). TBCB is a known component of cilia (Lopez-Fanarraga et al., 2007). TBCE could also be associated more directly with cilia, since the mRNA for this protein and not several other tubulin chaperones, is strongly upregulated during cilia regeneration in *Tetrahymena* (M. Coelho and H. Soares, personal communication). It is tempting to speculate that CAP-GLY containing chaperones interact with the tyrosinated ciliary tubulin during its transport or specifically associate with segments of axonemal microtubules that remain tyrosinated after assembly.

The levels of detyrosination greatly increase during the assembly of axonemes (Johnson, 1998; Redeker et al., 2005; Sasse and Gull, 1988; Sherwin et al., 1987). In *Trypanosoma*, the newly assembled flagellum can be distinguished from the old flagellum based on the higher content of tyrosinated tubulin (Sasse and Gull, 1988; Sherwin et al., 1987), revealing a significant gap between the end of axoneme assembly and the acquisition of detyrosination. Intriguingly, in the mature axoneme, detyrosinated tubulin is unevenly distributed. In *Chlamydomonas*, the central microtubules contain mainly tyrosinated isoforms, while the outer doublets have a mixture of tyrosinated and detyrosinated tubulin. Within the doublets, the detyrosinated modified isoforms are highly enriched on the B-subfiber. In assembling axonemes, the modification accumulates at the time when flagella acquire vigorous beating (Johnson, 1998). As the B-tubule is the site for binding of the motor domain of axonemal dynein, detyrosinated α-tubulin could be important in regulating the force production by ciliary dynein.

3.2.2. Tubulin glutamylation

Glutamylation is an unusual modification generated by post-translational synthesis of a polypeptide side chain made of a variable number of glutamic acids, using a γ-carboxyl group of a primary sequence glutamic acid as an attachment site (Eddé *et al.*, 1990). Glutamylation occurs on both α- and β-tubulin and often on multiple glutamic acids within the CTTs (Redeker *et al.*, 1991, 1992). Glutamylation is enriched on axonemes, basal bodies/centrioles, the mitotic spindle, and a subset of cytoplasmic microtubules inside nerve projections (Audebert *et al.*, 1993; Bobinnec *et al.*, 1998b; Bré *et al.*, 1994; Kann *et al.*, 2003; Lechtreck and Geimer, 2000; Plessmann *et al.*, 2004; Wolff *et al.*, 1992). Glutamylation appears to accumulate on axonemes very early in the course of assembly, with a maximum level that peaks prior to the maximal level of detyrosination, acetylation, and glycylation (Lechtreck and Geimer, 2000; Sharma *et al.*, 2007). Thus, tubulin glutamylation could play a role in the assembly of axonemes.

The functional studies on the role of this PTM have been greatly facilitated by the recent identification of glutamic acid ligase enzymes (glutamylases). Janke *et al.* (2005) purified the enzyme from developing murine brains. The complex contained several proteins, but only one of them, TTLL1 (TTL-like 1), a protein related to TTL, is conserved in diverse organisms that have tubulin glutamylation. TTLL1 and several other related and conserved proteins, including TTLL4, TTLL5, TTLL6 (and mammalian paralogs TTLL7, TTLL11, and TTLL13), and TTLL9 function as catalytic subunits of tubulin glutamylation (Ikegami *et al.*, 2006; Janke *et al.*, 2005; van Dijk *et al.*, 2007; Wloga *et al.*, 2008). All these proteins have a domain homologous to the catalytic domain of TTL, which reflects the structural relatedness of glutamylation and tyrosination reactions. Some of the glutamylating TTLLs have a strong chain-initiating activity (e.g., Ttll1p of *Tetrahymena*, murine TTLL4, TTLL5, and TTLL7) while others act primarily as chain elongases (e.g., Ttll6Ap of *Tetrahymena*, murine TTLL11 and TTLL13) (van Dijk *et al.*, 2007; Wloga *et al.*, 2008). Some TTLLs prefer α-tubulin (e.g., TTLL1, TTLL5) while others prefer β-tubulin as a substrate (e.g., Ttll6Ap of *Tetrahymena* and murine TTLL7) (Ikegami *et al.*, 2006; van Dijk *et al.*, 2007; Wloga *et al.*, 2008).

Most of the subtypes of known tubulin glutamic acid ligases are conserved across diverse eukaryotes, but all these enzymes are missing in fungi and higher plants, suggesting that they coevolved with cilia and centrioles/ basal bodies (Janke et al., 2005). The murine TTLL1 is associated specifically with glutamylation on α-tubulin (Janke et al., 2005). A *Tetrahymena* ortholog, Ttll1p, is primarily associated with basal bodies, and its deletion led to a strong decrease in the levels of tubulin glutamylation in basal bodies and defects in the maturation of basal bodies (Wloga et al., 2008). Injection of GT335 antibodies that specifically recognize glutamylated proteins into HeLa cells caused transient disassembly of centrioles that was associated with dispersal of pericentriolar proteins and apparent loss of the centrosome (Bobinnec et al., 1998a). Thus, glutamylation could be important for assembly and maintenance of centrioles and basal bodies.

Overproduction of *Tetrahymena* Ttll6Ap glutamic acid ligase stimulates the levels of tubulin glutamylation on axonemes and leads to ciliary paralysis, showing that proper levels of tubulin glutamylation are critical for ciliary motility (Janke et al., 2005). Inside *Tetrahymena* cilia, glutamylation is primarily located on the B-tubule of doublet microtubules (S. Suryavanshi and J. Gaertig, unpublished data), which agrees with observations in other species (Fouquet et al., 1996; Lechtreck and Geimer, 2000; Multigner et al., 1996). Antiglutamylation-specific antibodies inhibited the motility of *in vitro* reactivated sea urchin sperm axonemes (Gagnon et al., 1996). Thus, tubulin glutamylation could regulate the activity of ciliary dynein.

Some observations indicate a role for tubulin glutamylation in the assembly of axonemes. In the mouse, TTLL1 is in a complex with a noncatalytic vertebrate-specific subunit PGs1 (Regnard et al., 2003). The mutation of PGs1 in the mouse leads to pleiotropic defects, including decreased male aggression and male sterility (Campbell et al., 2002). The PGs1-null mice have greatly decreased levels of glutamylation on α-tubulin (Ikegami et al., 2007). The axonemes in PGs1-null males are truncated and lack a subset of either doublet or central microtubules (Campbell et al., 2002). Depletion of TTLL6 tubulin glutamic acid ligase in zebrafish led to shortening of olfactory cilia, possibly due to impaired assembly (Pathak et al., 2007).

How tubulin glutamylation affects axoneme assembly or stability is not clear. The levels of glutamylation appear to be high in the assembling cilia. As the cilium matures, the apparent levels of glutamylation decrease, and the levels of glycylation increase suggesting that polymodifications undergo remodeling and that high levels of glutamylation are important for IFT (Sharma et al., 2007). While effects on IFT are possible, TTLL1 and TTLL6 orthologs are present in *Plasmodium* species that lack IFT (Briggs et al., 2004). Moreover, glutamylation appears to affect only a fraction of tubulin subunits within the axoneme (Hoyle et al., 2008; Redeker et al., 2005), and the functions of polymodified amino acids sites are not sensitive to severe dilution (e.g., mutations of essential polymodification sites on β-tubulin are recessive; Thazhath et al., 2002), which argues against a direct involvement with processive motor proteins such as kinesin-2. Moreover, in several species, glutamylation appears to be primarily on the B-tubule (Fouquet et al., 1996; Lechtreck and Geimer, 2000; Multigner et al., 1996), while the IFT motors may use primarily the A-tubule of outer doublets as a track (Kozminski et al., 1995). It is also puzzling that there are significant variations in the patterns of glutamylation among different species. While *Chlamydomonas* has more glutamylation on the doublets (and the B-tubule) and less on the central pair (Lechtreck and Geimer, 2000), there is approximately equal partition of glutamylated isoforms between the outer and central microtubules in the sea urchin cilia (Hoyle et al., 2008). In the *Drosophila* sperm axoneme, most of the glutamylation was detected in a fraction that contained the central and insect-specific accessory singlet microtubules and not in the doublets (Hoyle et al., 2008). Thus, despite conservation of this PTM, the pattern of glutamylation appears variable. Hoyle et al. (2008) proposed that tubulin glutamylation, while highly conserved, could be adapted to multiple roles within the cilium that are regulatory and species specific. However, there are currently significant methodological limits, when the patterns of glutamylation are compared across different species. There is currently only one antibody available (GT335) that can recognize a side chain made of a single E, and the epitope of this antibody includes the primary sequence of tubulin (Wolff et al., 1992). As glutamylation may affect multiple Es of the primary sequence of CTT, GT335 may not be able to detect all possible glutamylation sites. Detection of glutamylation by mass spectrometry has also been problematic, as glutamylated tubulin peptides are underrepresented in the mixtures of tubulin peptides, possibly due to less efficient ionization (Redeker et al., 2005). Thus, it is possible that the levels of glutamyl side chains are underestimated in some species.

3.2.3. Tubulin glycylation

The first indication of the existence of glycylation came from studies done by Adoutte et al. (1985) who produced an antibody against axonemal tubulin of *Paramecium*, and showed that this antibody detects axonemal tubulin but fails to react with nonciliary cytosolic tubulin in diverse species. A subsequent study showed that the tubulin epitope recognized by this antibody appears on axonemes late in ciliogenesis (Adoutte et al., 1991). A mass spectrometry study of axonemal tubulin of *Paramecium* detected the presence of multiple post-translationally added glycines to CTTs (Redeker et al., 1994). Glycylation is caused by addition of side chains made of one or more glycines to glutamic acids of the CTT of both α- and β-tubulin. Glycylation is relatively conserved but not as conserved as glutamylation. Immunological studies showed that this PTM occurs mainly if not exclusively in cell types that have cilia, such as sperm and ciliated epithelia in vertebrates, as well as in ciliated protozoans (Bré et al., 1996). However, some organisms have cilia and apparently lack tubulin glycylation, including *Plasmodium* (Fennell et al., 2008) and *Trypanosoma* (Schneider et al., 1997). The side chain length can be probed using monoclonal antibodies that recognize either a single G (TAP952) or three or more Gs (AXO49) (Bré et al., 1996; Callen et al., 1994; Levilliers et al., 1995). Interestingly, based on these antibodies, murine sperm tubulin is polyglycylated while human sperm is only monoglycylated (Bré et al., 1996). *Chlamydomonas* flagella also appear to have only monoglycylated tubulin (Bré et al., 1996). The internal patterns of glycylation also appear species specific. In sea urchin and mammalian sperm, the outer doublets are more glycylated then the central pair (Kann et al., 1998; Multigner et al., 1996; Péchart et al., 1999). However, in *Paramecium*, the central and outer microtubules appear to be equally glycylated (Iftode et al., 2000; Péchart et al., 1999). However, it is not certain that the available antibody probes can detect glycyl side chains of all length and at all modifications sites, especially when these antibodies are used in diverse species. Thus, as is certainly the case of tubulin glutamylation, the levels of tubulin glycylation may be underestimated and the observed patterns may not represent the entire population of glycyl side chains that exists in axonemes.

The assembling axoneme has a high content of monoglycylated tubulin recognized by the TAP952 antibody (Brown et al., 1999; Iftode et al., 2000). As the cilium matures, the signal of TAP952 decreases, and the signal of AXO49, an antibody that recognizes glycyl side chains that have three or more Gs per chain, increases, indicating that the maturation of cilia coincides with elongation of glycyl chains (Adoutte et al., 1991; Sharma et al., 2007). In the mature cilium, glycylation could regulate ciliary motility, based on the observation that glycylation-specific antibodies inhibit motility of ATP reactivated sea urchin sperm axonemes (Bré et al., 1996). However, nonmotile cilia, such as primary cilia of mammalian cells, also have tubulin glycylation (Davenport et al., 2007; Ong and Wheatley, 2003), suggesting a more general role in axoneme assembly or maintenance.

Mass spectrometry analyses have established that several consecutive glutamic acids within or near the "axoneme motif" on both α- and β-tubulin undergo glycylation in *Paramecium* (Vinh et al., 1999). The *in vivo* role of the homologous amino acid sites has been addressed in *Tetrahymena*. On α-tubulin, substitution of three adjacent Es by Ds abolished most glycylation (Xia et al., 2000). A substitution of all six Es of the CTT of α-tubulin abolished all glycylation and glutamylation. Surprisingly, the resulting strains lacking ability to polymodify α-tubulin, are viable and motile. However, these mutants grow more slowly and have a reduced rate of phagocytosis, a function that is cilia dependent in *Tetrahymena* (Wloga et al., 2008). Thus, both polymodification types (glutamylation and glycylation) on α-tubulin are not essential in *Tetrahymena*. Since the tail domain on α-tubulin is essential (Duan and Gorovsky, 2002), it is clear that the tail has additional functions that are not mediated by polymodifications.

On the β-tubulin CTT of *Paramecium*, the sites of glycylation were mapped to a cluster of four Es downstream of the GEF sequence within the axoneme motif (Vinh et al., 1999). On β-tubulin of *Tetrahymena* there are consecutive five glutamic acids located downstream of the GEF part of the axoneme motif (Fig. 4.2). Single and double aspartate substitutions did not affect cell viability. However, triple mutations led to slow growth, slow motility, and defects in cytokinesis, or were lethal (Xia et al., 2000). Viable triple mutant cells (βEDDD$_{440}$) grow and move more slowly. The outer doublets showed frequent lack of closure of the B-tubule on the surface of the A-tubule. Electron-dense aggregates were detected between the doublets and the membrane, possibly due to defects in IFT (Redeker et al., 2005).

Surprisingly, studies on the viable glycylation site mutants reveal that there are extensive interactions among distinct polymodifications types and between CTTs of partner tubulin subunits of the dimer (trans-tail interactions). The βEDDD$_{440}$ *Tetrahymena* mutant has greatly reduced levels of polyglycylation on β-tubulin. However, in the same strain, there are dramatic changes in the levels of other PTMs. Strikingly, major changes occur on α-tubulin, the subunit that was not mutated. While wild-type α-tubulin is both glycylated and extensively glutamylated, the mass spectrum of α-tubulin in the βEDDD$_{440}$ mutant revealed mostly glycylated isoforms, suggesting that there was a strong reduction in α-tubulin glutamylation. Moreover, the α-tubulin of the βEDDD$_{440}$ mutant was hyperglycylated (Redeker et al., 2005). Thus, there is crosstalk among PTMs that affect CTTs of the partner subunits. Several mechanisms could mediate such interactions: (1) competition for the modifying enzymes: α- and β-tubulin are structurally similar and could be modified by the same enzyme, (2) participation of the nonmodified subunit as an enzyme-binding site, and (3) indirect effects mediated by the polymer dynamics.

The finding that some of the mutations within the glycylation cluster on β-tubulin are lethal (Xia et al., 2000) was initially surprising, as tubulin glycylation has been considered a marker of cilia and these organelles are not essential for survival in *Tetrahymena*. To reconstruct the events that lead to lethality of substitutions of polymodification sites, Thazhath et al. (2002) placed a lethal triple mutation (βDDDE$_{440}$), into the germline micronucleus of *Tetrahymena* and brought the mutation to expression by mating. The progeny cells that exclusively expressed zygotic mutant tubulin, βDDDE$_{440}$, showed a novel phenotype of "cell chains." The mutant cells grew for several cell cycles and expanded the cell cortex, while duplicating basal bodies within the ciliary rows. The βDDDE$_{440}$ cells initiated cytokinesis, but the contractile ring was arrested half way. The fidelity of duplication of the basal bodies seemed unaffected. However, the mutation had a strong effect on the assembly and maintenance of cilia. The newly assembled cilia were excessively short, generally lacked the central microtubules and occasionally lacked the B-tubule on outer doublets.

An earlier study in *Drosophila* showed the CTT of axoneme-competent β-tubulin carries sequence elements that are required for central pair formation (Nielsen et al., 2001). The β_1-isotype is not competent for assembly of the 9 + 2 axoneme but it is used for assembly of basal bodies. When β_1 was expressed in place of β_2 in the postmitotic germ cells of *Drosophila*, this isoform supported the assembly of short 9 + 0 axonemes (Raff et al., 2000). Remarkably a chimera of β_1 and the CTT of β_2 supported the 9 + 2 axoneme assembly. Moreover, introduction of the full "axoneme motif" sequence into the CTT of β_1 by a mutation of only two amino acids, provided β_1 with the ability to support assembly of a 9 + 2 axoneme (Nielsen et al., 2001). These two substitutions were located immediately upstream of the cluster of glycylation sites within the axoneme motif (Nielsen et al., 2001; Vinh et al., 1999). The *Tetrahymena* and *Drosophila* studies taken together show that the axoneme motif is needed for the central pair assembly and that, within this motif, there is a requirement for a minimal number of glutamic acids that could be polymodifiable. Thus, polymodifications may be specifically required for central pair assembly and axoneme elongation. It should be considered however, that the lethality of the substitutions at the glycylation sites could result from elimination of sites of tubulin glutamylation. The glutamylation sites have not been mapped in *Paramecium* or *Tetrahymena* but the known sites of glutamylation mapped in other organisms are adjacent or overlap with the sites of glycylation (Redeker et al., 1996, 2004; Vinh et al., 1999).

The lethality in the $\beta DDDE_{440}$ *Tetrahymena* mutant is caused by a defect in cytokinesis. An analysis of the cell body phenotype of this mutant has provides further clues to the function of the polymodification sites and linked polymodifications to the activity of microtubule-severing factors, specifically to katanin. In *Tetrahymena*, the cortical microtubules are polyglycylated and monoglutamylated (Thazhath et al., 2002; Wloga et al., 2008). A proper rearrangement of the longitudinal microtubule bundles (LMs) is required for completion of cytokinesis. These are bundles of partly overlapping microtubules that are positioned on one side of the ciliary rows (Pitelka, 1961). In interphase cells, LMs are continuous across the entire cell. As the contractile ring is formed, coincident with the initial phase of constriction, a zone develops that lacks basal bodies. As the contractile ring ingresses, the segments of LMs within the equatorial area bend and break. Strikingly, the $\beta DDDE_{440}$ mutants have an excessive number of microtubules in LM bundles and some of LMs fail to break during cytokinesis (Thazhath et al., 2002). Thus, excessive polymerization or stability of LM bundles could block the cleavage furrow ingression. Moreover, it appears that the arrest in cytokinesis in the $\beta DDDE_{440}$ mutants is caused by insufficient microtubule-severing activity (see below).

Sharma et al. (2007) showed that a mutation in katanin, a microtubule-severing factor (McNally and Vale, 1993), precisely phenocopies the $\beta DDDE_{440}$ glycylation region mutation in *Tetrahymena*. Katanin is a protein complex containing a p60 catalytic subunit and a p80 noncatalytic subunit (Hartmann et al., 1998; McNally and Vale, 1993). A knockout of either p60- or p80-encoding genes in *Tetrahymena* has produced cell chains with short, paralyzed 9 + 0 cilia that were remarkably similar to the $\beta DDDE_{440}$ cell chains (Sharma et al., 2007). The katanin knockouts accumulate high levels of glutamylation and glycylation on microtubules in the cell cortex and cell body. However, these microtubules also accumulate acetylation of K40 on α-tubulin (Sharma et al., 2007). Thus, the simplest explanation is that microtubules in the cell body of katanin mutants are more long lived. Indeed, the cell body microtubules in the katanin knockouts are more stable because most did not depolymerize in the presence of nocodazole, while the same microtubules in wild-type cells did (Sharma et al., 2007).

The question is whether the potential katanin interactions with poly-modified microtubules affect the assembly of cilia directly or indirectly via katanin functions in the cell body. The simplest model is that katanin works outside of cilia in the management of cell body (cortical and cytoplasmic) microtubules, and that its role in the assembly of cilia is indirect (e.g., katanin could be important in the generation of tubulin destined for cilia). However, when katanin-null mutants were deciliated, they regenerated short 9 + 0 cilia (Sharma et al., 2007). Thus, the katanin knockout strains have a specific defect in the assembly of the central pair and elongation of outer doublet microtubules and maintain a pool of cilia-destined tubulin. Moreover, a GFP-tagged katanin was targeted to both basal bodies and axonemes. Within cilia, katanin associated with the outer and not with the central microtubules, a surprising result given that katanin deficiency had a major influence on the central microtubules (Sharma et al., 2007). Importantly, an earlier study in *Chlamydomonas* provides strong support to the direct ciliary role of katanin (Dymek et al., 2004). Dymek has mapped the mutation in the *pf15* mutant that has paralyzed flagella and lacks the central microtubules, to p80 katanin. An epitope-tagged p80 gene rescues the flagellar paralysis phenotype of the *pf15* mutant. Fractionation of flagella in the rescued strain showed that p80-HA was present in the isolated axonemes, and it was present at the normal levels in the pf20 axonemes that lack central microtubules, suggesting that p80 katanin is associated primarily with the outer doublets (Dymek et al., 2004). The *Tetrahymena* and *Chlamydomonas* studies taken together indicate that katanin is an important factor in the assembly of axonemes, regulates PTM levels on microtubules, and could interact preferentially with polymodified microtubules.

Sorting out the respective contributions of glutamylation and glycylation to the functions of CTTs during axoneme assembly is an important future goal. Moreover, it is important to determine whether katanin interacts preferentially with either glutamylated or glycylated microtubules. Answers to both questions would be facilitated by identification of enzymes that catalyze tubulin glycylation. A recent study showed that TTLL10, a member of the TTLL superfamily, catalyzes glycylation of NAP1, a chromatin assembly factor (Ikegami *et al.*, 2008). It remains to be determined whether TTLL10 or another uncharacterized member of the TTLL superfamily acts as a tubulin glycine ligase.

4. Conclusions

Precursor tubulin is largely unmodified and axonemal tubulin acquires diverse PTMs as the cilium assembles. Subtypes of axonemal microtubules may differ in the levels of specific PTMs and segments of the same microtubules may be modified to an extent that is dependent on their position within the axoneme. Distinct PTMs, especially those on the tubulin CTTs, can influence each other within the same tubulin subunit or across the dimer. Based on the loss-of-function studies of modifying enzymes, tubulin glutamylation appears to be important in either axoneme assembly or stability, but the exact mechanism is unclear. Studies that have addressed the function of α-tubulin acetylation and detyrosination on nonaxonemal microtubules indicate that tubulin PTMs can influence interactions with motors and plus-end-tracking proteins. Thus, PTMs that are important for the axonemal functions, such as tubulin glutamylation, could also regulate the binding and activity of microtubule interactor proteins. The sites of tubulin glycylation are important for axoneme assembly, and are required for the central microtubule assembly, but the same sites can potentially be used by tubulin glutamylation. Thus, the role of tubulin glycylation needs to be reassessed once the enzymes that generate this modification are identified. The proper levels of PTMs on ciliary and nonciliary microtubules require the activity of katanin and loss of katanin phenocopies a lethal mutation of the polymodification sites on β-tubulin. Thus, katanin is a major candidate for an effector of polymodified microtubules in the context of ciliogenesis.

ACKNOWLEDGMENTS

This work was supported by National Science Foundation grant MBC-033965 to JG. The authors are grateful to Drashti Dave and Scott Dougan (University of Georgia), Joseph Frankel (University of Iowa) Martin A. Gorovsky (University of Rochester), Lynne Quarmby (Simon Fraser University), Elena Pugacheva (West Virginia University), Elizabeth Raff (Indiana University), Cecilia Seixas (CRBM, Montpellier), and Helena Soares (Instituto Gulbenkian de Ciência) for helpful comments.

REFERENCES

Adoutte, A., Claisse, M., Maunoury, R., and Beisson, J. (1985). Tubulin evolution: Ciliate-specific epitopes are conserved in the ciliary tubulin in metazoa. *J. Mol. Evol.* **22,** 220–229.

Adoutte, A., Delgado, P., Fleury, A., Levilliers, N., Lainé, M.-C., Marty, M.-C., Boisvieux-Ulrich, E., and Sandoz, D. (1991). Microtubule diversity in ciliated cells: Evidence for its generation by post-translational modification in the axonemes of *Paramecium* and quail oviduct cells. *Biol. Cell* **71,** 227–245.

Alfa, C. E., and Hyams, J. S. (1991). Microtubules in the fission yeast *Schizosaccharomyces pombe* contain only the tyrosinated form of alpha-tubulin. *Cell Motil. Cytoskeleton* **18,** 86–93.

Allen, R. D. (1969). The morphogenesis of basal bodies and accessory structures of the cortex of the ciliate protozoan *Tetrahymena pyriformis*. *J. Cell Biol.* **40,** 716–733.

Arce, C. A., Rodriguez, J. A., Barra, H. S., and Caputo, R. (1975). Incorporation of L-tyrosine, L-phenylalanine and L-3,4-dihydroxyphenylalanine as single units into rat brain tubulin. *Eur. J. Biochem.* **59,** 145–149.

Argarana, C. E., Barra, H. S., and Caputo, R. (1978). Release of [^{14}C]-tyrosine from tubulinyl-[^{14}C]-tyrosine by brain extract. Separation of a carboxypeptidase from tubulin tyrosine ligase. *Mol. Cell. Biochem.* **19,** 17–22.

Audebert, S., Desbruyères, E., Gruszczynski, C., Koulakoff, A., Gros, F., Denoulet, P., and Eddé, B. (1993). Reversible polyglutamylation of α- and β-tubulin and microtubule dynamics in mouse brain neurons. *Mol. Biol. Cell* **4,** 615–626.

Baas, P. W., Karabay, A., and Qiang, L. (2005). Microtubules cut and run. *Trends Cell Biol.* **15,** 518–24.

Bartolini, F., Bhamidipati, A., Thomas, S., Schwahn, U., Lewis, S. A., and Cowan, N. J. (2002). Functional overlap between retinitis pigmentosa 2 protein and the tubulin-specific chaperone cofactor C. *J. Biol. Chem.* **277,** 14629–14634.

Blaineau, C., Tessier, M., Dubessay, P., Tasse, L., Crobu, L., Pages, M., and Bastien, P. (2007). A novel microtubule-depolymerizing kinesin involved in length control of a eukaryotic flagellum. *Curr. Biol.* **17,** 778–82.

Bobinnec, Y., Khodjakov, A., Mir, L. M., Rieder, C. L., B., E., and Bornens, M. (1998a). Centriole disassembly *in vivo* and its effect on centrosome structure and function in vertebrate cells. *J. Cell Biol.* **143,** 1575–1589.

Bobinnec, Y., Moudjou, M., Fouquet, J. P., Desbruyères, E., Eddé, B., and Bornens, M. (1998b). Glutamylation of centriole and cytoplasmic tubulin in proliferating non-neuronal cells. *Cell Motil. Cytoskeleton* **39,** 223–232.

Bradley, B. A., and Quarmby, L. M. (2005). A NIMA-related kinase, Cnk2p, regulates both flagellar length and cell size in *Chlamydomonas*. *J. Cell Sci.* **118,** 3317–3326.

Bré, M.-H., Redeker, V., Quibell, M., Darmanaden-Delome, J., Bressac, C., Cosson, J., Huitore, P., Schmitte, J.-M., Rossier, J., Johnson, T., Adoutte, A., and Levilliers, N. (1996). Axonemal tubulin polyglycylation probed with two monoclonal antibodies: Widespread evolutionary distribution, appearance during spermatozoan maturation and possible function in motility. *J. Cell Sci.* **109,** 727–738.

Bré, M. H., de Néchaud, B., Wolff, A., and Fleury, A. (1994). Glutamylated tubulin probed in ciliates with the monoclonal antibody GT335. *Cell Motil. Cytoskeleton* **27,** 337–349.

Briggs, L. J., Davidge, J. A., Wickstead, B., Ginger, M. L., and Gull, K. (2004). More than one way to build a flagellum: Comparative genomics of parasitic protozoa. *Curr. Biol.* **14,** R611–R612.

Brown, J. M., Marsala, C., Kosoy, R., and Gaertig, J. (1999). Kinesin-II is preferentially targeted to assembling cilia and is required for ciliogenesis and normal cytokinesis in *Tetrahymena*. *Mol. Biol. Cell* **10,** 3081–3096.

Callen, A.-M., Adoutte, A., Andrew, J. M., Baroin-Tourancheau, A., Bré, M.-H., Ruiz, P. C., Clérot, J.-C., Delgado, P., Fleury, A., Jeanmaire-Wolf, R., Viklicky, V., Villalobo, E., and Levilliers, N. (1994). Isolation and characterization of libraries of monoclonal antibodies directed against various forms of tubulin in *Paramecium*. *Biol. Cell* **81**, 95–119.

Campbell, P. K., Waymire, K. G., Heier, R. L., Sharer, C., Day, D. E., Reimann, H., Jaje, J. M., Friedrich, G. A., Burmeister, M., Bartness, T. J., Russell, L. D., Young, L. J., *et al.* (2002). Mutation of a novel gene results in abnormal development of spermatid flagella, loss of intermale aggression and reduced body fat in mice. *Genetics* **162**, 307–320.

Cao, S., Carlesso, G., Osipovich, A. B., Llanes, J., Lin, Q., Hoek, K. L., Khan, W. N., and Ruley, H. E. (2008). Subunit 1 of the prefoldin chaperone complex is required for lymphocyte development and function. *J. Immunol.* **181**, 476–484.

Davenport, J. R., Watts, A. J., Roper, V. C., Croyle, M. J., van Groen, T., Wyss, J. M., Nagy, T. R., Kesterson, R. A., and Yoder, B. K. (2007). Disruption of intraflagellar transport in adult mice leads to obesity and slow-onset cystic kidney disease. *Curr. Biol.* **17**, 1586–1594.

Dawe, H. R., Farr, H., Portman, N., Shaw, M. K., and Gull, K. (2005). The Parkin co-regulated gene product, PACRG, is an evolutionarily conserved axonemal protein that functions in outer-doublet microtubule morphogenesis. *J. Cell Sci.* **118**, 5421–5430.

Dawson, S. C., Sagolla, M. S., Mancuso, J. J., Woessner, D. J., House, S. A., Fritz-Laylin, L., and Cande, W. Z. (2007). Kinesin-13 regulates flagellar, interphase, and mitotic microtubule dynamics in *Giardia intestinalis*. *Eukaryot. Cell* **6**, 2354–2364.

Dekker, C., Stirling, P. C., McCormack, E. A., Filmore, H., Paul, A., Brost, R. L., Costanzo, M., Boone, C., Leroux, M. R., and Willison, K. R. The interaction network of the chaperonin CCT. *EMBO J.* **27**(13), 1827–1839.

Dentler, W. L. (1980). Structures linking the tips of ciliary and flagellar microtubules to the membrane. *J. Cell Sci.* **42**, 207–220.

Dompierre, J. P., Godin, J. D., Charrin, B. C., Cordelieres, F. P., King, S. J., Humbert, S., and Saudou, F. (2007). Histone deacetylase 6 inhibition compensates for the transport deficit in Huntington's disease by increasing tubulin acetylation. *J. Neurosci.* **27**, 3571–3583.

Duan, J., and Gorovsky, M. A. (2002). Both carboxy terminal tails of alpha and beta tubulin are essential, but either one will suffice. *Curr. Biol.* **12**, 313–316.

Dymek, E. E., Lefebvre, P. A., and Smith, E. F. (2004). PF15p is the *Chlamydomonas* homologue of the katanin p80 subunit and is required for assembly of flagellar central microtubules. *Eukaryot. Cell* **3**, 870–879.

Eddé, B., Rossier, J., Le Caer, J.-P., Desbruyères, E., Gros, F., and Denoulet, P. (1990). Posttranslational glutamylation of α-tubulin. *Science* **247**, 83–85.

Erck, C., Peris, L., Andrieux, A., Meissirel, C., Gruber, A. D., Vernet, M., Schweitzer, A., Saoudi, Y., Pointu, H., Bosc, C., Salin, P. A., Job, D., *et al.* (2005). A vital role of tubulin-tyrosine-ligase for neuronal organization. *Proc. Natl. Acad. Sci. USA* **102**, 7853–8.

Fackenthal, J. D., Turner, F. R., and Raff, E. C. (1993). Tissue-specific microtubule functions in *Drosophila* spermatogenesis require the $\beta 2$ isotype-specific carboxy terminus. *Dev. Biol.* **158**, 213–227.

Fennell, B. J., Al-Shatr, Z. A., and Bell, A. (2008). Isotype expression, post-translational modification and stage-dependent production of tubulins in erythrocytic *Plasmodium falciparum*. *Int. J. Parasitol.* **38**, 527–539.

Fernandez-Gonzalez, A., La Spada, A. R., Treadaway, J., Higdon, J. C., Harris, B. S., Sidman, R. L., Morgan, J. I., and Zuo, J. (2002). Purkinje cell degeneration (pcd) phenotypes caused by mutations in the axotomy-induced gene, Nna1. *Science* **295**, 1904–1906.

Finst, R. J., Kim, P. J., Griffis, E. R., and Quarmby, L. M. (2000). Fa1p is a 171 kDa protein essential for axonemal microtubule severing in *Chlamydomonas*. *J. Cell Sci.* **113**(Pt 11), 1963–1971.

Fouquet, J. P., Prigent, Y., and Kann, M. L. (1996). Comparative immunogold analysis of tubulin isoforms in the mouse sperm flagellum: unique distribution of glutamylated tubulin. *Mol. Reprod. Dev.* **43**, 358–365.

Frydman, J., Nimmesgern, E., Erdjument-Bromage, H., Wall, J. S., Tempst, P., and Hartl, F. U. (1992). Function in protein folding of TRiC, a cytosolic ring complex containing TCP-1 and structurally related subunits. *EMBO J.* **11**, 4767–4778.

Fukushige, T., Siddiqui, Z. K., Chou, M., Culotti, J. G., Gogonea, C. B., Siddiqui, S. S., and Hamelin, M. (1999). MEC-12, an alpha-tubulin required for touch sensitivity in *C. elegans*. *J. Cell Sci.* **112**, 395–403.

Gaertig, J., Cruz, M. A., Bowen, J., Gu, L., Pennock, D. G., and Gorovsky, M. A. (1995). Acetylation of lysine 40 in alpha-tubulin is not essential in *Tetrahymena thermophila*. *J. Cell Biol.* **129**, 1301–1310.

Gaertig, J., Thatcher, T. H., McGrath, K. E., Callahan, R. C., and Gorovsky, M. A. (1993). Perspectives on tubulin isotype function and evolution based on the observations that *Tetrahymena thermophila* microtubules contain a single α- and β-tubulin. *Cell Motil. Cytoskeleton* **25**, 243–253.

Gagnon, C., White, D., Cosson, J., Huitorel, P., Eddé, B., Desbruyères, E., Paturle-Lafanèchere, L., Multigner, L., Job, D., and Cibert, C. (1996). The polyglutamylated lateral chain of alpha-tubulin plays a key role in flagellar motility. *J. Cell Sci.* **109**, 1545–1553.

Garvalov, B. K., Zuber, B., Bouchet-Marquis, C., Kudryashev, M., Gruska, M., Beck, M., Leis, A., Frischknecht, F., Bradke, F., Baumeister, W., Dubochet, J., and Cyrklaff, M. (2006). Luminal particles within cellular microtubules. *J. Cell Biol.* **174**, 759–765.

Gorovsky, M. A., Carlson, K., and Rosenbaum, J. L. (1970). Simple method for quantitative densitometry of polyacrylamide gels using fast green. *Anal. Biochem.* **35**, 359–370.

Greer, C. A., and Shepherd, G. M. (1982). Mitral cell degeneration and sensory function in the neurological mutant mouse Purkinje cell degeneration (PCD). *Brain Res.* **235**, 156–161.

Greer, K., Maruta, H., L'Hernault, S. W., and Rosenbaum, J. L. (1985). Alpha-tubulin acetylase activity in isolated *Chlamydomonas* flagella. *J. Cell Biol.* **10**, 2081–2084.

Grynberg, M., Jaroszewski, L., and Godzik, A. (2003). Domain analysis of the tubulin cofactor system: A model for tubulin folding and dimerization. *BMC Bioinformatics* **4**, 46.

Gundersen, G. G., Khawaja, S., and Bulinski, J. C. (1987). Postpolymerization detyrosination of α-tubulin: A mechanism for subcellular differentiation of microtubules. *J. Cell Biol.* **105**, 251–264.

Hartmann, J. J., Mahr, J., McNally, K. P., Okawa, K., Iwamatsu, A., Thomas, S., Cheesman, S., Heuser, J., Vale, R. D., and McNally, F. J. (1998). Katanin, a microtubule-severing protein, is a novel AAA ATPase that targets to the centrosome using a WD40 domain. *Cell* **93**, 277–287.

Howard, J., and Hyman, A. A. (2007). Microtubule polymerases and depolymerases. *Curr. Opin. Cell Biol.* **19**, 31–35.

Hoyle, H. D., and Raff, E. C. (1990). Two *Drosophila* beta tubulin isoforms are not functionally equivalent. *J. Cell Biol.* **111**, 1009–1026.

Hoyle, H. D., Turner, F. R., Brunick, L., and Raff, E. C. (2001). Tubulin sorting during dimerization *in vivo*. *Mol. Biol. Cell* **12**, 2185–2194.

Hoyle, H. D., Turner, F. R., and Raff, E. C. (2008). Axoneme-dependent tubulin modifications in singlet microtubules of the *Drosophila* sperm tail. *Cell Motil. Cytoskeleton* **65**, 295–313.

Hubbert, C., Guardiola, A., Shao, R., Kawaguchi, Y., Ito, A., Nixon, A., Yoshida, M., Wang, X. F., and Yao, T. P. (2002). HDAC6 is a microtubule-associated deacetylase. *Nature* **417,** 455–458.

Iftode, F., Clerot, J. C., Levilliers, N., and Bré, M. H. (2000). Tubulin polyglycylation: A morphogenetic marker in ciliates. *Biol. Cell* **92,** 615–628.

Ikeda, K., Ikeda, T., Morikawa, K., and Kamiya, R. (2007). Axonemal localization of *Chlamydomonas* PACRG, a homologue of the human Parkin-coregulated gene product. *Cell Motil Cytoskeleton* **64,** 814–821.

Ikegami, K., Heier, R. L., Taruishi, M., Takagi, H., Mukai, M., Shimma, S., Taira, S., Hatanaka, K., Morone, N., Yao, I., Campbell, P. K., Yuasa, S., *et al.* (2007). Loss of alpha-tubulin polyglutamylation in ROSA22 mice is associated with abnormal targeting of KIF1A and modulated synaptic function. *Proc. Natl. Acad. Sci. USA* **104,** 3213–3218.

Ikegami, K., Horigome, D., Mukai, M., Livnat, I., Macgregor, G. R., and Setou, M. (2008). TTLL10 is a protein polyglycylase that can modify nucleosome assembly protein 1. *FEBS Lett.* **582,** 1129–1134.

Ikegami, K., Mukai, M., Tsuchida, J., Heier, R. L., Macgregor, G. R., and Setou, M. (2006). TTLL7 Is a Mammalian beta-Tubulin Polyglutamylase Required for Growth of MAP2-positive Neurites. *J. Biol Chem.* **281,** 30707–30716.

Imai, Y., Soda, M., Murakami, T., Shoji, M., Abe, K., and Takahashi, R. (2003). A product of the human gene adjacent to parkin is a component of Lewy bodies and suppresses Pael receptor-induced cell death. *J. Biol. Chem.* **278,** 51901–51910.

Janke, C., Rogowski, K., Wloga, D., Regnard, C., Kajava, A. V., Strub, J.-M., Temurak, N., van Dijk, J., Boucher, D., van Dorsselaer, A., Suryavanshi, S., Gaertig, J., *et al.* (2005). Tubulin polyglutamylase enzymes are members of the TTL domain protein family. *Science* **308,** 1758–1762.

Jensen-Smith, H. C., Luduena, R. F., and Hallworth, R. (2003). Requirement for the betaI and betaIV tubulin isotypes in mammalian cilia. *Cell Motil. Cytoskeleton* **55,** 213–220.

Johnson, K. A. (1998). The axonemal microtubules of the *Chlamydomonas* flagellum differ in tubulin isoform content. *J. Cell Sci.* **111,** 313–320.

Johnson, K. A., and Rosenbaum, J. L. (1992). Polarity of flagellar assembly in *Chlamydomonas*.

Kalinina, E., Biswas, R., Berezniuk, I., Hermoso, A., Aviles, F. X., and Fricker, L. D. (2007). A novel subfamily of mouse cytosolic carboxypeptidases. *FASEB J.* **21,** 836–50.

Kann, M. L., Prigent, Y., Levilliers, N., Bré, M. H., and Fouquet, J. P. (1998). Expression of glycylated tubulin during the differentiation of spermatozoa in mammals. *Cell Motil. Cytoskeleton* **41,** 341–352.

Kann, M. L., Soues, S., Levilliers, N., and Fouquet, J. P. (2003). Glutamylated tubulin: Diversity of expression and distribution of isoforms. *Cell Motil. Cytoskeleton.* **55,** 14–25.

Kawaguchi, Y., Kovacs, J. J., McLaurin, A., Vance, J. M., Ito, A., and Yao, T. P. (2003). The deacetylase HDAC6 regulates aggresome formation and cell viability in response to misfolded protein stress. *Cell* **115,** 727–738.

Keller, C. E., and Lauring, B. P. (2005). Possible regulation of microtubules through destabilization of tubulin. *Trends Cell Biol.* **15,** 571–573.

Kim, J. C., Ou, Y. Y., Badano, J. L., Esmail, M. A., Leitch, C. C., Fiedrich, E., Beales, P. L., Archibald, J. M., Katsanis, N., Rattner, J. B., and Leroux, M. R. (2005). MKKS/BBS6, a divergent chaperonin-like protein linked to the obesity disorder Bardet-Biedl syndrome, is a novel centrosomal component required for cytokinesis. *J. Cell Sci.* **118,** 1007–1020.

Kovacs, J. J., Murphy, P. J., Gaillard, S., Zhao, X., Wu, J. T., Nicchitta, C. V., Yoshida, M., Toft, D. O., Pratt, W. B., and Yao, T. P. (2005). HDAC6 regulates Hsp90 acetylation and chaperone-dependent activation of glucocorticoid receptor. *Mol. Cell* **18,** 601–607.

Kozminski, K. G., Beech, P. L., and Rosenbaum, J. L. (1995). The *Chlamydomonas* kinesin-like protein FLA10 is involved in motility associated with the flagellar membrane. *J. Cell Biol.* **131,** 1517–1527.

Kozminski, K. G., Diener, D. R., and Rosenbaum, J. L. (1993). High level expression of nonacetylatable α-tubulin in *Chlamydomonas reinhardtii*. *Cell Motil. Cytoskeleton* **25,** 158–170.

L'Hernault, S. W., and Rosenbaum, J. L. (1985). *Chlamydomonas* α–tubulin is posttranslationally modified by acetylation on the e-amino group of a lysine. *Biochem.* **24,** 473–478.

Lakamper, S., and Meyhofer, E. (2006). Back on track - on the role of the microtubule for kinesin motility and cellular function. *J. Muscle Res. Cell Motil.* **27,** 161–171.

Landis, S. C., and Mullen, R. J. (1978). The development and degeneration of Purkinje cells in pcd mutant mice. *J. Comp. Neurol.* **177,** 125–143.

LaVail, M. M., Blanks, J. C., and Mullen, R. J. (1982). Retinal degeneration in the pcd cerebellar mutant mouse. I. Light microscopic and autoradiographic analysis. *J. Comp. Neurol.* **212,** 217–230.

Lechtreck, K.-F., and Geimer, S. (2000). Distributon of polyglutamylated tubulin in the flagellar apparatus of green flagellates. *Cell Motil. Cytoskeleton* **47,** 219–235.

LeDizet, M., and Piperno, G. (1991). Detection of acetylated α-tubulin by specific antibodies. *Methods of Enzymology* **196,** 264–274.

Levilliers, N., Fleury, A., and Hill, A. M. (1995). Monoclonal and polyclonal antibodies detect a new type of post-translational modification of axonemal tubulin. *J.Cell Sci.* **108,** 3013–3028.

Lopez-Fanarraga, M., Carranza, G., Bellido, J., Kortazar, D., Villegas, J. C., and Zabala, J. C. (2007). Tubulin cofactor B plays a role in the neuronal growth cone. *J. Neurochem.* **100,** 1680–1687.

Lu, C., Srayko, M., and Mains, P. E. (2004). The *Caenorhabditis elegans* Microtubule-severing Complex MEI-1/MEI-2 Katanin Interacts Differently with Two Superficially Redundant β-Tubulin Isotypes. *Mol. Biol. Cell* **15,** 142–150.

Luchko, T., Huzil, J. T., Stepanova, M., and Tuszynski, J. (2008). Conformational analysis of the carboxy-terminal tails of human beta-tubulin isotypes. *Biophys. J.* **94,** 1971–1982.

Mahjoub, M. R., Montpetit, B., Zhao, L., Finst, R. J., Goh, B., Kim, A. C., and Quarmby, L. M. (2002). The FA2 gene of *Chlamydomonas* encodes a NIMA family kinase with roles in cell cycle progression and microtubule severing during deflagellation. *J. Cell Sci.* **115,** 1759–1768.

Mahjoub, M. R., Qasim Rasi, M., and Quarmby, L. M. (2004). A NIMA-related kinase, Fa2p, localizes to a novel site in the proximal cilia of *Chlamydomonas* and mouse kidney cells. *Mol. Biol. Cell* **15,** 5172–5186.

Marshall, W. F., and Rosenbaum, J. L. (2001). Intraflagellar transport balances continuous turnover of outer doublet microtubules: Implications for flagellar length control. *J. Cell Biol.* **155,** 405–414.

Maruta, H., Greer, K., and Rosenbaum, J. L. (1986). The acetylation of alpha-tubulin and its relationship to the assembly and disassembly of microtubules. *J. Cell Biol.* **103,** 571–579.

Matsuyama, A., Shimazu, T., Sumida, Y., Saito, A., Yoshimatsu, Y., Seigneurin-Berny, D., Osada, H., Komatsu, Y., Nishino, N., Khochbin, S., Horinouchi, S., and Yoshida, M. (2002). In vivo destabilization of dynamic microtubules by HDAC6-mediated deacetylation. *EMBO J.* **21,** 6820–6831.

McNally, F. J., and Vale, R. D. (1993). Identification of katanin, an ATPase that severs and disassembles stable microtubules. *Cell* **75,** 419–429.

Multigner, L., Pignot-Paintrand, I., Saoudi, Y., Job, D., Plessmann, U., Rüdiger, M., and Weber, K. (1996). The A and B tubules of the outer doublets of sea urchin sperm axonemes are composed of different tubulin variants. *Biochemistry* **35,** 10862–10871.

Nelsen, E. M. (1975). Regulation of tubulin during ciliary regeneration of nongrowing *Tetrahymena*. *Exp. Cell Res.* **94,** 152–158.

Nicastro, D., Schwartz, C., Pierson, J., Gaudette, R., Porter, M. E., and McIntosh, J. R. (2006). The molecular architecture of axonemes revealed by cryoelectron tomography. *Science* **313**, 944–948.

Nielsen, M. G., and Raff, E. C. (2002). The best of all worlds or the best possible world? Developmental constraint in the evolution of beta-tubulin and the sperm tail axoneme. *Evol. Dev.* **4**, 303–315.

Nielsen, M. G., Turner, F. R., Hutchens, J. A., and Raff, E. C. (2001). Axoneme-specific beta-tubulin specialization: A conserved C-terminal motif specifies the central pair. *Curr. Biol.* **11**, 529–533.

Nogales, E., Whittaker, M., Milligan, R. A., and Downing, K. H. (1999). High-resolution model of the microtubule. *Cell* **96**, 79–88.

Nogales, E., Wolff, S. G., and Downing, K. H. (1998). Structure of the $\alpha\beta$ tubulin dimer by electron crytallography. *Nature* **391**, 199–202.

North, B. J., Marshall, B. L., Borra, M. T., Denu, J. M., and Verdin, E. (2003). The human Sir2 ortholog, SIRT2, is an NAD+-dependent tubulin deacetylase. *Mol. Cell* **11**, 437–444.

Ong, A. C., and Wheatley, D. N. (2003). Polycystic kidney disease–the ciliary connection. *Lancet* **361**, 774–776.

Ovechkina, Y., Wagenbach, M., and Wordeman, L. (2002). K-loop insertion restores microtubule depolymerizing activity of a "neckless" MCAK mutant. *J. Cell Biol.* **159**, 557–562.

Pan, J., Wang, Q., and Snell, W. J. (2004). An aurora kinase is essential for flagellar disassembly in Chlamydomonas. *Dev. Cell* **6**, 445–451.

Parker, J. D., and Quarmby, L. M. (2003). Chlamydomonas fla mutants reveal a link between deflagellation and intraflagellar transport. *BMC Cell Biol.* **4**, 11.

Pathak, N., Obara, T., Mangos, S., Liu, Y., and Drummond, I. A. (2007). The Zebrafish fleer gene encodes an essential regulator of cilia tubulin polyglutamylation. *Mol. Biol. Cell* **18**, 4353–4364.

Pazour, G. J., Agrin, N., Leszyk, J., and Witman, G. B. (2005). Proteomic analysis of a eukaryotic cilium. *J. Cell Biol.* **170**, 103–113.

Péchart, I., Kann, M. L., Levilliers, N., Bre, M. H., and Fouquet, J. P. (1999). Composition and organization of tubulin isoforms reveals a variety of axonemal models. *Biol. Cell* **91**, 685–697.

Peris, L., Thery, M., Faure, J., Saoudi, Y., Lafanechere, L., Chilton, J. K., Gordon-Weeks, P., Galjart, N., Bornens, M., Wordeman, L., Wehland, J., Andrieux, A., *et al.* (2006). Tubulin tyrosination is a major factor affecting the recruitment of CAP-Gly proteins at microtubule plus ends. *J. Cell Biol.* **174**, 839–849.

Piperno, G., Ledizet, M., and Chang, X. J. (1987). Microtubules containing acetylated α-tubulin in mammalian cells in culture. *J. Cell Biol.* **104**, 289–302.

Pitelka, D. R. (1961). Fine structure and silver line and fibrillar systems of three tetrahyminid species. *J. Protozool.* **8**, 75–89.

Plessmann, U., Reiter-Owona, I., and Lechtreck, K. F. (2004). Posttranslational modifications of alpha-tubulin of *Toxoplasma gondii*. *Parasitol. Res.* **94**, 386–389.

Pugacheva, E. N., Jablonski, S. A., Hartman, T. R., Henske, E. P., and Golemis, E. A. (2007). HEF1-dependent Aurora A activation induces disassembly of the primary cilium. *Cell* **129**, 1351–1363.

Qin, H., Diener, D. R., Geimer, S., Cole, D. G., and Rosenbaum, J. L. (2004). Intraflagellar transport (IFT) cargo: IFT transports flagellar precursors to the tip and turnover products to the cell body. *J. Cell Biol.* **164**, 255–266.

Quarmby, L. M. (2004). Cellular deflagellation. *Int. Rev. Cytol.* **233**, 47–91.

Radcliffe, P. A., and Toda, T. (2000). Characterisation of fission yeast alp11 mutants defines three functional domains within tubulin-folding cofactor B. *Mol. Gen. Genet.* **263**, 752–760.

Raff, E. C. (1994). The Role of Multiple Tubulin Isoforms in Cellular Microtubule Function. *In* "Microtubules" (J. S. Hyams and C. W. Lloyd, eds.), pp. 85–109. Wiley-Liss, Inc.

Raff, E. C., Fackenthal, J. D., Hutchens, J. A., Hoyle, H. D., and Turner, H. F. (1997). Microtubule architecture specified by a β-tubulin isoform. *Science* **275**, 70–73.

Raff, E. C., Hoyle, H. D., Popodi, E. M., and Turner, F. R. (2008). Axoneme beta-tubulin sequence determines attachment of outer dynein arms. *Curr. Biol* **18**, 911–914.

Raff, E. C., Hutchens, J. A., Hoyle, H. D., Nielsen, M. G., and Turner, F. R. (2000). Conserved axoneme symmetry altered by a component beta-tubulin. *Curr. Biol.* **10**, 1391–1394.

Redeker, V., Frankfurter, A., Parker, S. K., Rossier, J., and Detrich, H. W., 3rd. (2004). Posttranslational modification of brain tubulins from the Antarctic fish *Notothenia coriiceps*: Reduced C-terminal glutamylation correlates with efficient microtubule assembly at low temperature. *Biochemistry* **43**, 12265–12274.

Redeker, V., Le Caer, J. P., Rossier, J., and Promé, J. C. (1991). Structure of the polyglutamyl side chain posttranslationally added to alpha-tubulin. *J. Biol. Chem.* **266**, 23461–23466.

Redeker, V., Levilliers, N., Schmitter, J.-M., Le Caer, J.-P., Rossier, J., Adoutte, A., and Bré, M.-H. (1994). Polyglycylation of tubulin: A post-translational modification in axonemal microtubules. *Science* **266**, 1688–1691.

Redeker, V., Levilliers, N., Vinolo, E., Rossier, J., Jaillard, D., Burnette, D., Gaertig, J., and Bré, M. H. (2005). Mutations of tubulin glycylation sites reveal cross-talk between the C termini of alpha- and beta-tubulin and affect the ciliary matrix in *Tetrahymena*. *J. Biol. Chem.* **280**, 596–606.

Redeker, V., Melki, R., Promé, D., Le Caer, J.-P., and Rossier, J. (1992). Structure of tubulin C-terminal domain obtained by subtilisin treatment. The major α- and β- tubulin isotypes from pig brain are glutamylated. *FEBS Lett.* **313**, 185–192.

Redeker, V., Rusconi, F., Mary, J., Prome, D., and Rossier, J. (1996). Structure of the C-terminal tail of alpha-tubulin: Increase of heterogeneity from newborn to adult. *J. Neurochem* **67**, 2104–2114.

Reed, N. A., Cai, D., Blasius, L., Jih, G. T., Meyhofer, E., Gaertig, J., and Verhey, K. J. (2006). Microtubule acetylation promotes kinesin-1 binding and transport. *Curr. Biol* **16**, 2166–2172.

Regnard, C., Fesquet, D., Janke, C., Boucher, D., Desbruyères, E., Koulakoff, A., Insina, C., Travo, P., and Edde, B. (2003). Characterization of PGs1, a subunit of a protein complex co-purifying with tubulin polyglutamylase. *J. Cell Sci.* **116**, 4181–4190.

Roll-Mecak, A., and Vale, R. D. (2008). Structural basis of microtubule severing by the hereditary spastic paraplegia protein spastin. *Nature* **451**, 363–367.

Rosenbaum, J. L., and Child, F. M. (1967). Flagellar regeneration in protozoan flagellates. *J. Cell Biol.* **34**, 345–364.

Rosenbaum, J. L., Moulder, J. E., and Ringo, D. L. (1969). Flagellar elongation and shortening in *Chlamydomonas*. The use of cycloheximide and colchicine to study the sythesis and assembly of flagellar proteins. *J. Cell Biol.* **41**, 600–619.

Rosenbaum, J. L., and Witman, G. B. (2002). Intraflagellar transport. *Nat. Rev. Mol. Cell. Biol.* **3**, 813–825.

Sasse, R., and Gull, K. (1988). Tubulin post-translational modifications and the construction of microtubular organelles in *Trypanosoma brucei*. *J. Cell Sci.* **90**, 577–589.

Savage, C., Hamelin, M., Culotti, J. G., Coulson, A., Albertson, D. G., and Chalfie, M. (1989). *mec-7* is a β-tubulin gene required for the production of 15-protofilament microtubules in *Caenorhabditis elegans*. *Genes Dev.* **3**, 870–881.

Schneider, A., Plessmann, U., and Weber, K. (1997). Subpellicular and flagellar microtubules of *Trypanosoma brucei* are extensively glutamylated. *J. Cell Sci.* **110**, 431–437.

Schroder, H. C., Wehland, J., and Weber, K. (1985). Purification of brain tubulin-tyrosine ligase by biochemical and immunological methods. *J. Cell Biol.* **100**, 276–281.

Schwahn, U., Lenzner, S., Dong, J., Feil, S., Hinzmann, B., van Duijnhoven, G., Kirschner, R., Hemberger, M., Bergen, A. A., Rosenberg, T., Pinckers, A. J., Fundele, R., *et al.* (1998). Positional cloning of the gene for X-linked retinitis pigmentosa 2. *Nat. Genet.* **19**, 327–332.

Seixas, C., Casalou, C., Melo, L. V., Nolasco, S., Brogueira, P., and Soares, H. (2003). Subunits of the chaperonin CCT are associated with *Tetrahymena* microtubule structures and are involved in cilia biogenesis. *Exp. Cell Res.* **290**, 303–321.

Serrano, L., De La Torre, J., Maccioni, R. B., and Avila, J. (1984). Involvement of the carboxy-terminal domain of tubulin in the regulation of its assembly. *Proc. Natl. Acad. Sci. USA* **81**, 5989–5993.

Sharma, N., Bryant, J., Wloga, D., Donaldson, R., Davis, R. C., Jerka-Dziadosz, M., and Gaertig, J. (2007). Katanin regulates dynamics of microtubules and biogenesis of motile cilia. *J. Cell Biol.* **178**, 1065–1079.

Sherwin, T., Schneider, A., Sasse, R., Seebeck, T., and Gull, K. (1987). Distinct localization and cell cycle dependence of COOH terminally tyrosinolated alpha-tubulin in the microtubules of *Trypanosoma brucei*. *J. Cell Biol.* **104**, 439–446.

Silflow, C. D. (1991). Why do tubulin gene families lack diversity in flagellate/ciliate protists? *Protoplasma* **164**, 9–11.

Skiniotis, G., Cochran, J. C., Muller, J., Mandelkow, E., Gilbert, S. P., and Hoenger, A. (2004). Modulation of kinesin binding by the C-termini of tubulin. *EMBO J.* **23**, 989–999.

Song, L., and Dentler, W. (2001). Flagellar protein dynamics in *Chlamydomonas*. *J. Biol. Chem.* **276**, 29754–29763.

Stephan, A., Vaughan, S., Shaw, M. K., Gull, K., and McKean, P. G. (2007). An essential quality control mechanism at the eukaryotic basal body prior to intraflagellar transport. *Traffic* **8**, 1323–1330.

Stephens, R. E. (1992). Tubulin in sea urchin embryonic cilia: Post-translational modifications during regeneration. *J. Cell Sci.* **101**, 837–845.

Stephens, R. E. (1994). Tubulin and tektin in sea urchin embryonic cilia: Pathways of protein incorporation during turnover and regeneration. *J. Cell Sci.* **107**, 683–692.

Stephens, R. E. (2000). Preferential incorporation of tubulin into the junctional region of ciliary outer doublet microtubules: A model for treadmilling by lattice dislocation. *Cell Motil. Cytoskeleton* **47**, 130–140.

Stoetzel, C., Laurier, V., Davis, E. E., Muller, J., Rix, S., Badano, J. L., Leitch, C. C., Salem, N., Chouery, E., Corbani, S., Jalk, N., and Vicaire, S. (2006). BBS10 encodes a vertebrate-specific chaperonin-like protein and is a major BBS locus. *Nat. Genet.* **38**, 521–524.

Sui, H., and Downing, K. H. (2006). Molecular architecture of axonemal microtubule doublets revealed by cryo-electron tomography. *Nature* **442**, 475–478.

Terada, S., Kinjo, M., and Hirokawa, N. (2000). Oligomeric tubulin in large transporting complex is transported via kinesin in squid giant axons. *Cell* **103**, 141–155.

Thazhath, R., Jerka-Dziadosz, M., Duan, J., Wloga, D., Gorovsky, M. A., Frankel, J., and Gaertig, J. (2004). Cell context-specific effects of the beta-tubulin glycylation domain on assembly and size of microtubular organelles. *Mol. Biol. Cell* **15**, 4136–4147.

Thazhath, R., Liu, C., and Gaertig, J. (2002). Polyglycylation domain of beta-tubulin maintains axonemal architecture and affects cytokinesis in *Tetrahymena*. *Nature Cell Biol.* **4,** 256–259.

Tian, G., Huang, Y., Rommelaere, H., Vandekerckhove, J., Ampe, C., and Cowan, N. J. (1996). Pathway leading to correctly folded beta-tubulin. *Cell* **86,** 287–296.

Tian, G., Lewis, S. A., Feierbach, B., Stearns, T., and Rommelaere, H. Tubulin subunits exist in an activated conformational state generated and maintained by protein cofactors. *J. Cell Biol.* **138**(4), 821–832.

Tran, A. D., Marmo, T. P., Salam, A. A., Che, S., Finkelstein, E., Kabarriti, R., Xenias, H. S., Mazitschek, R., Hubbert, C., Kawaguchi, Y., Sheetz, M. P., Yao, T. P., et al. (2007). HDAC6 deacetylation of tubulin modulates dynamics of cellular adhesions. *J. Cell. Sci.* **120,** 1469–1479.

Vadlamudi, R. K., Barnes, C. J., Rayala, S., Li, F., Balasenthil, S., Marcus, S., Goodson, H. V., Sahin, A. A., and Kumar, R. (2005). p21-activated kinase 1 regulates microtubule dynamics by phosphorylating tubulin cofactor B. *Mol. Cell Biol.* **25,** 3726–3736.

Vainberg, I. E., Lewis, S. A., Rommelaere, H., Ampe, C., Vandekerckhove, J., Klein, H. L., and Cowan, N. J. (1998). Prefoldin, a chaperone that delivers unfolded proteins to cytosolic chaperonin. *Cell* **93,** 863–873.

van Dijk, J., Rogowski, K., Miro, J., Lacroix, B., Eddé, B., and Janke, C. (2007). A targeted multienzyme mechanism for selective microtubule polyglutamylation. *Mol. Cell* **26,** 437–448.

Verhey, K. J., and Gaertig, J. (2007). The tubulin code. *Cell Cycle* **6,** 2152–2160.

Vinh, J., Langridge, J. I., Bré, M.-H., Levilliers, N., Redeker, V., Loyaux, D., and Rossier, J. (1999). Structural characterization by tandem spectroscopy of the posttranslational modifications of tubulin. *Biochemistry* **38,** 3133–3139.

Walther, Z., Vashistha, M., and Hall, J. L. (1994). The *Chlamydomonas FLA10* gene encodes a novel kinesin-homologous protein. *J. Cell Biol.* **126,** 175–188.

Wang, T., and Morgan, J. I. (2007). The Purkinje cell degeneration (pcd) mouse: An unexpected molecular link between neuronal degeneration and regeneration. *Brain Res* **1140,** 26–40.

Wang, Z., and Sheetz, M. P. (2000). The C-terminus of tubulin increases cytoplasmic dynein and kinesin processivity. *Biophys. J.* **78,** 1955–1964.

Webster, D. R., and Borisy, G. G. (1989). Microtubules are acetylated in domains that turn over slowly. *J. Cell Sci.* **92,** 57–65.

Weisbrich, A., Honnappa, S., Jaussi, R., Okhrimenko, O., Frey, D., Jelesarov, I., Akhmanova, A., and Steinmetz, M. O. (2007). Structure-function relationship of CAP-Gly domains. *Nat. Struct. Mol. Biol.* **14,** 959–967.

Westermann, S., and Weber, K. (2003). Post-translational modifications regulate microtubule function. *Nat. Rev. Mol. Cell. Biol.* **4,** 938–947.

Williams, N. E., and Nelsen, E. M. (1997). HSP70 and HSP90 homologs are associated with tubulin in hetero-oligomeric complexes, cilia and the cortex of Tetrahymena. *J. Cell Sci.* **110**(Pt 14), 1665–1672.

Wloga, D., Camba, A., Rogowski, K., Manning, G., Jerka-Dziadosz, M., and Gaertig, J. (2006). Members of the Nima-related kinase family prmote disassembly of cilia by multiple mechanisms. *Mol. Biol. Cell* **17,** 2799–2810.

Wloga, D., Rogowski, K., Sharma, N., van Dijk, J., Janke, C., Edde, B., Bré, M. H., Levilliers, N., Redeker, V., Duan, J., Gorovsky, M. A., Jerka-Dziadosz, M., et al. (2008). Glutamylation on α-tubulin is not essential but affects the assembly and functions of a subset of microtubules in *Tetrahymena*. *Eukaryot. Cell* **7,** 1362–1372.

Wolff, A., De Néchaud, B., Chillet, D., Mazarguil, H., Desbruyères, E., Audebert, S., Eddé, B., Gros, F., and Denoulet, P. (1992). Distribution of glutamylated α- and β-tubulin in mouse tissues using a specific monoclonal antibody, GT335. *Eur. J. Cell Biol.* **59,** 425–432.

Xia, L., Hai, B., Gao, Y., Burnette, D., Thazhath, R., Duan, J., Bré, M.-H., Levilliers, N., Gorovsky, M. A., and Gaertig, J. (2000). Polyglycylation of tubulin is essential and affects cell motility and division in *Tetrahymena thermophila. J. Cell Biol.* **149,** 1097–1106.

Zhang, X., Yuan, Z., Zhang, Y., Yong, S., Salas-Burgos, A., Koomen, J., Olashaw, N., Parsons, J. T., Yang, X. J., Dent, S. R., Yao, T. P., Lane, W. S., *et al.* (2007). HDAC6 modulates cell motility by altering the acetylation level of cortactin. *Mol. Cell* **27,** 197–213.

Zhang, Y., Kwon, S., Yamaguchi, T., Cubizolles, F., Rousseaux, S., Kneissel, M., Cao, C., Li, N., Cheng, H. L., Chua, K., Lombard, D., Mizeracki, A., Matthias, G., Alt, F. W., Khochbin, S., and Matthias, P. (2008). Mice lacking histone deacetylase 6 have hyperacetylated tubulin but are viable and develop normally. *Mol. Cell. Biol.* **28,** 1688–1701.

Zhang, Y., Li, N., Caron, C., Matthias, G., Hess, D., Khochbin, S., *et al.* (2003). HDAC-6 interacts with and deacetylates tubulin and microtubules *in vivo. EMBO J.* **22,** 1168–1179.

CHAPTER FIVE

Targeting Proteins to the Ciliary Membrane

Gregory J. Pazour* and Robert A. Bloodgood[†]

Contents

1. Introduction	116
2. Structure of the Ciliary Membrane	118
3. Functions of the Ciliary Membrane	121
3.1. Regulation of ciliary beat pattern	121
3.2. Adhesion	122
3.3. Sensory functions	123
3.4. Cell signaling	127
4. Protein Machinery Involved in Trafficking to the Ciliary Membrane	127
4.1. Intraflagellar transport	128
4.2. Bardet-Biedl syndrome proteins	129
4.3. Polarity proteins	131
4.4. Small GTPases	132
5. Ciliary Targeting Sequences	134
References	139

Abstract

Most vertebrate cell types display solitary nonmotile cilia on their surface that serve as cellular antennae to sense the extracellular environment. These organelles play key roles in the development of mammals by coordinating the actions of a single cell with events occurring around them. Severe defects in cilia lead to midgestational lethality in mice while more subtle defects lead to pathology in most organs of the body. These pathologies range from cystic diseases of the kidney, liver, and pancreas, to retinal degeneration, to bone and skeletal defects, hydrocephaly, and obesity. The sensory functions of cilia rely on proteins localized specifically to the ciliary membrane. Even though the ciliary membrane is a subdomain of the plasma membrane and is continuous with the plasma membrane, cells have the ability to specifically localize

* Program in Molecular Medicine, University of Massachusetts Medical School, Biotech II, Worcester, Massachusetts
[†] Department of Cell Biology, University of Virginia School of Medicine, Charlottesville, Virginia

proteins to this domain. In this chapter, we will review what is currently known about the structure and function of the ciliary membrane. We will further discuss ongoing work to understand how the ciliary membrane is assembled and maintained, and discuss protein machinery that is thought to play a role in sorting or trafficking proteins to the ciliary membrane.

ABBREVIATIONS

BBS	Bardet-Biedl syndrome
CTS	ciliary targeting sequence
GPCR	G protein-coupled receptor
IFT	intraflagellar transport
SHH	Sonic Hedgehog
PDGF	platelet-derived growth factors

1. INTRODUCTION

Eukaryotic cells possess many functionally unique compartments defined by biological membranes. Evolution has provided unique mechanisms for delivery of proteins and lipids from their sites of synthesis to these compartments. In this regard, cilia and flagella are one of the least understood compartments of the cell. While most of the surface of the cilium is covered by a membrane, there is a small opening at the base of the cilium where materials can enter the organelle. Thus cilia are more akin to the nucleus than to truly sealed membranous organelles like mitochondria and lysosomes. The entry point into the cilium may be analogous to the nuclear pore and has been called the "ciliary pore" (Bloodgood, 2000; Jekely and Arendt, 2006; Rosenbaum and Witman, 2002). The functional ciliary pore probably lies at the level of the transition zone between the basal body and the ciliary axoneme (Rosenbaum and Witman, 2002). This structurally complex and sterically congested region prevents entry of membrane vesicles and ribosomes into the cilium. In addition, this region is presumed to prevent the diffusion of most proteins between the cytosolic and ciliary compartments allowing the ciliary matrix to have a composition distinct from that of the cytosol. Similarly, there are membrane barriers that maintain the composition of the ciliary membrane distinct from that of the other plasma membrane domains (Fig. 5.1).

Because cilia have no machinery for protein synthesis, all ciliary and flagellar proteins must be synthesized elsewhere in the cell and imported into the cilium. Presumably, axonemal and soluble matrix proteins are

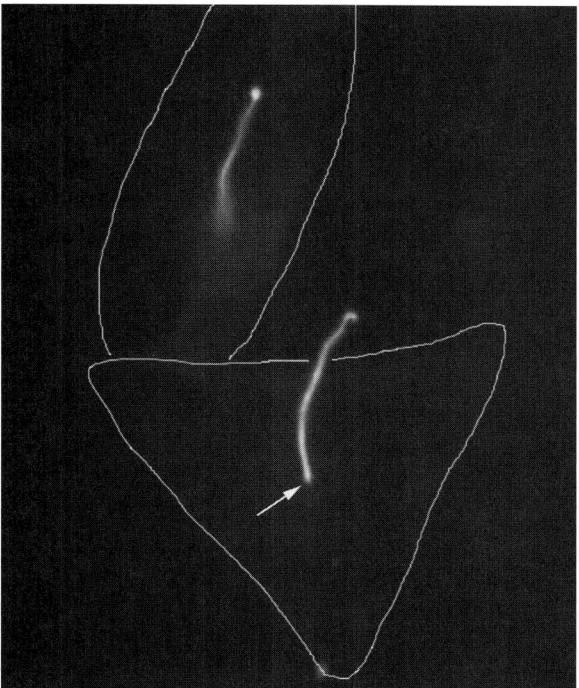

Figure 5.1 Image of live IMCD3 mouse kidney epithelial cells expressing GFP-tagged somatostatin receptor 3. Note the high degree to which SSTR3 is concentrated to the cilia and the sharp boundary (arrow) between the cilium and cell body (outlined in white). (See Color Insert.)

synthesized in the cytosol on free polyribosomes and post-translationally imported into cilia through the ciliary pore while integral membrane proteins are synthesized in the rough endoplasmic reticulum, processed in the Golgi and transported by a vesicular pathway from the Golgi to a specialized plasma membrane domain at the base of the cilium. It is not known how ciliary membrane and ciliary axoneme assembly are coordinated. While axonemal elongation necessitates formation of the ciliary membrane, it is less clear whether the converse is true. Interference with Golgi function partially inhibits ciliary assembly in algae and sea urchin embryos (Dentler, 2006; Haller and Fabry, 1998; Stephens, 2001). Data obtained using trypanosomes (Absalon et al., 2008; Davidge et al., 2006) suggest that the flagellar membrane can assemble in the absence of axonemal assembly.

The process of delivering proteins synthesized on free polyribosomes to the cilium is likely to involve intraflagellar transport (IFT), which is the subject of Chapter 2 in this issue. The process of delivering newly synthesized ciliary membrane proteins to the cilium is the focus of this chapter. We will discuss the structure and functions of the ciliary membrane and then review mechanisms and machinery utilized for the transport and targeting of integral membrane proteins to the ciliary membrane.

2. Structure of the Ciliary Membrane

The ciliary membrane is a domain of the plasma membrane that definitely possesses a unique protein composition and is likely to also have a unique lipid composition. Difficulties in the purification of mammalian primary cilia (Huang *et al.*, 2006; Mitchell *et al.*, 2004) have impeded our knowledge about its lipid composition, but studies in protistans indicate that the lipid composition of cilia in these organisms differs from that of the bulk plasma membrane. In general, ciliary membranes have high concentrations of sterols relative to phospholipids (Chailley and Boisvieux-Ulrich, 1985; Kaneshiro, 1990). Interestingly, lipid rafts or lipid microdomains are also enriched in sterols and these structures appear to be platforms for organizing signal transduction machinery (Simons and Toomre, 2000). With the important role that cilia play in sensory perception, it will be interesting to understand how similar the ciliary membrane is to known lipid microdomains. There is emerging evidence for lipid rafts or raft-like properties in cilia. For example, purification of detergent resistant membranes from *Chlamydomonas* flagella identified a group of proteins involved in behavioral responses to light (Iomini *et al.*, 2006) and glycosphingolipids, lipids found in rafts, were found to be enriched in mammalian cilia (Janich and Corbeil, 2007).

In most cases, the ciliary membrane is closely adherent to the axoneme and hence is cylindrical in shape. However, in some specialized situations, the ciliary membrane is expanded to carry out sensory or adhesive functions of cilia. The most dramatic example occurs in the vertebrate photoreceptors, where the ciliary membrane is formed into elaborate stacks of membrane for the detection of light (Besharse and Horst, 1990). Other examples include the AWB and AWC classes of *Caenorhabditis elegans* neurons where the distal end of the cilia exhibit expanded fork-like and fan-like specializations of the ciliary membrane, respectively (Ward *et al.*, 1975). Sensory input can modify the structure of the ciliary membrane in the AWB neurons converting from the fork-like to the fan-like morphology (Mukhopadhyay *et al.*, 2008). Adhesion of parasitic protozoa to the gut of host insects by cilia is aided by expansion of the ciliary membrane at the sites of attachment (Vickerman and Tetley, 1990). Another example of ciliary adhesion-induced modification of the ciliary membrane can be found in *Chlamydomonas* mating. Adhesion of flagella on plus and minus gametes (via specific sexual agglutinin glycoproteins) results in activation of a signaling cascade that involves a rise in cAMP and the swelling of the distal tips of the gametic flagella (Mesland *et al.*, 1980).

Studies on the localization of receptors to cilia indicate that cells have the ability to specifically localize membrane proteins to this domain (Fig. 5.1).

This could be accomplished by anchoring the proteins to ciliary axonemes to prevent diffusion or by creating diffusional barriers that isolate the ciliary membrane from the plasma membrane. It appears that many integral membrane proteins are directly anchored to the cytoskeleton as they are not extracted from the cytoskeleton by detergent (Pazour et al., 2005). This is supported by both transmission electron microscopic studies that show morphological links between the ciliary membrane and the axoneme (Dentler, 1990) and by freeze fracture studies that show linear arrays of intramembrane particles that may represent membrane proteins anchored on outer doublet microtubules (Bergman et al., 1975). Diffusional barriers may also play an important role is maintaining a unique composition of ciliary membranes. Direct evidence for such a diffusion barrier has been found in *Chlamydomonas* (Hunnicutt et al., 1990; Musgrave et al., 1986). Furthermore, it appears that this diffusion barrier can be regulated during mating such that integral and peripheral membrane proteins are recruited from the plasma membrane to the flagellar membrane in response to flagellar adhesion between the gametes. The structural and biochemical basis for the diffusional barrier is not well understood. It may involve lipid structure itself, as Veira et al. (2006) discovered a region of unusually high lipid ordering in the transition zone at the basal end of the membrane of kidney primary cilia. All ciliary membranes have a unique structural feature called the ciliary necklace located at the level of the transition zone (Gilula and Satir, 1972). The ciliary necklace is composed of extensive cytoskeleton to membrane linkages. These can be observed as multiple, circular rows of particles in the ciliary membrane using freeze-etch microscopy, or as structural linkages between the ciliary membrane and the underlying axoneme by transmission electron microscopy (Besharse and Horst, 1990). This domain may play a role in the formation of the barrier or may be part of the specialized zone of plasma membrane for insertion of ciliary membrane components (Deane et al., 2001) discussed below. The fact that vesicles are sterically prevented from entering cilia requires that fusion of vesicles containing ciliary membrane proteins occur on the plasma membrane outside of the cilium. Accumulating evidence in many organisms suggests that this docking and fusion occurs in a specialized plasma membrane domain. This is most clearly seen in Trypanosomatids. These flagellate protozoa possess three plasma membrane domains: the flagellar membrane, the flagellar pocket membrane, and the cell body plasma membrane (Fig. 5.2B). All endocytosis and exocytosis occurs at the flagellar pocket from which proteins are then sorted to flagellar or cell body membranes (Overath et al., 1997). Similarly, *Euglena* possesses a flagellar "reservoir" plasma membrane domain where flagellar membrane proteins initially appear on the cell surface (Rogalski and Bouck, 1982). This process also appears to occur in the alga *Pyramimonas* (McFadden and Wetherbee, 1985), which possesses a periciliary plasma membrane domain called a "scale

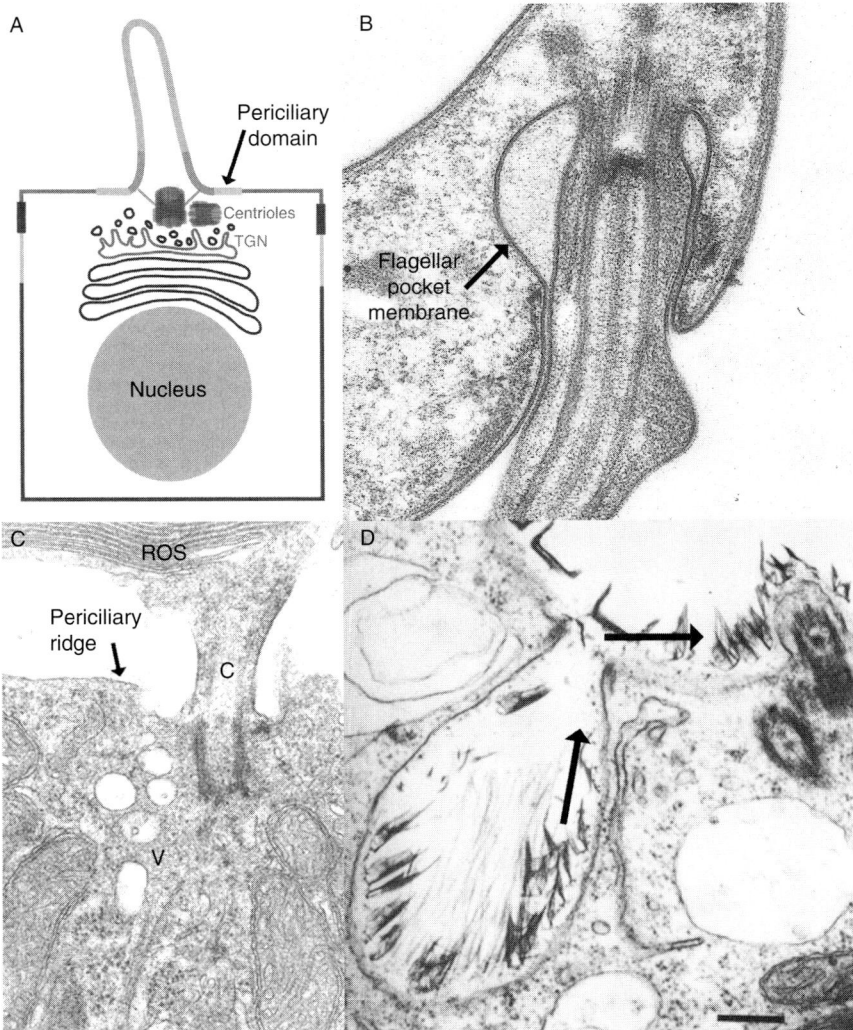

Figure 5.2 Four examples of putative periciliary sorting domains. (A) Diagram from Reiter and Mostov (2006) showing location of periciliary domains in canine kidney cells. (B) Transmission electron micrograph (courtesy of Y.-D. Stierhof) showing the location of the flagellar membrane pocket in *Trypanosoma brucei*. (C) Transmission electron micrograph from Papermaster *et al.* (1985) showing the apical end of a frog rod photoreceptor inner segment and connecting cilium, with the periciliary ridge domain indicated by the arrow. (D) Transmission electron micrograph from McFadden and Wetherbee (1985) showing a portion of the cell surface of the algal cell *Pyramimonas* showing the scale reservoir located near the base of a flagellum. It is presumed that the flagellar membrane scales migrate from the scale reservoir membrane onto the flagellar membrane. (See Color Insert.)

reservoir" (Fig. 5.2D) from which scales move onto the flagellar membrane. Similar domains for vesicle fusion appear to exist in vertebrates. The best evidence comes from studies of opsin transport in which it was observed that opsin carriers dock with ninefold symmetry around the base of the cilium in a structure called the periciliary ridge complex (Peters et al., 1983) (Fig. 5.2C). This structure may be highly elaborated in photoreceptors because of the large amount of transport into these cilia but is probably a general feature of all vertebrate cilia. Veira et al. (2006) described a periciliary domain in cultured mammalian cells that may play an equivalent role (Fig. 5.2A). This domain is characterized by the presence of the protein galectin-3 and by the absence of two GPI-linked proteins which are found in the rest of the apical plasma membrane.

3. Functions of the Ciliary Membrane

The functions of the ciliary membrane are a reflection of the many functions of cilia in general (Bloodgood, 1990). These ciliary membrane-associated functions vary from controlling the activity of molecular motors that power cells, to adhesive properties, to receiving and amplifying sensory input before sending signals on to the cell body.

3.1. Regulation of ciliary beat pattern

The ionic environment in the interior of the cilium is regulated, at least in part, by ion channels and pumps located within the ciliary membrane. Calcium is a key ionic regulator of axonemal bending in motile cilia (Salathe, 2007) and calcium channels and calcium pumps are found in cilia (Preston and Saimi, 1990). The role of calcium in cilia has been extensively studied in *Chlamydomonas* where an influx of calcium ions converts the flagellar waveform from a ciliary beat to a flagellar beat causing the cell to reverse its direction of movement (Hyams and Borisy, 1978; Witman, 1993). By coupling sensory inputs at the ciliary surface to the regulation of waveform, motile cilia can both sense and respond to external stimuli. The avoidance response in *Paramecium* is a classic example of this (Jennings, 1906). When a swimming *Paramecium* encounters a solid object, mechanical stimulation of the cilia by contact with the object opens calcium channels in the ciliary membrane. The resulting influx of calcium ions causes the axonemes to alter their waveform and the organism swims backward away from the object (Dunlap, 1977). Another example of ciliary motility responding to external stimuli occurs in the mammalian respiratory tract and oviduct. Cilia in these locations sense viscosity using TRPV channels in the ciliary membrane (Andrade et al., 2005; Lorenzo et al., 2008; Teilmann et al., 2005) and modulate beat frequency as viscous load changes.

3.2. Adhesion

In certain specialized situations, the ciliary membrane displays adhesive properties that can be utilized by organisms for a variety of interactions with their environment. The term haptocilia has been coined for adhesive cilia (Tyler, 1973). The uses of adhesive cilia are quite diverse. For example, adhesive properties of cilia allow the turbellarian worm *Paratomella* to adhere to structures in its environment (Tyler, 1973). Other organisms, like certain protistans, couple adhesive properties with motility for the capture of prey (Wetherbee and Andersen, 1992). Similarly, *Chlamydomonas* cells use adhesive properties of glycoproteins in cilia coupled with molecular motors to move over solid surfaces by gliding motility (Bloodgood, 1990, 2009). The initial step in mating of *Chlamydomonas* (van den Ende *et al.*, 1990) and a number of ciliate protozoa including *Paramecium*, *Tetrahymena*, and *Euplotes* (Hiwatashi, 1988) involves the specific adhesion of cilia from individuals of two different mating types. In *Chlamydomonas*, this interaction is mediated by a pair of mating type-specific glycoproteins associated with the flagellar membranes of the respective gametes (Ferris *et al.*, 2005). The interaction of these proteins leads to a cascade of downstream signaling events necessary for fertilization (Pan and Snell, 2000). Adhesive properties also play important roles in mammalian fertilization including the pickup of newly ovulated oocyte–cumulus complexes by oviduct cilia (Norwood *et al.*, 1978; Talbot *et al.*, 2003) and in some species, sperm bind to the ciliary membrane of the ciliated epithelial cells of the oviduct (Lefebvre *et al.*, 1995).

Pathogens take advantage of adhesive properties of cilia to anchor themselves to their hosts. For example, eukaryotic parasites (including *Leishmania*, *Trypanosoma*, and *Crithidia*) utilize hemidesmosome-like attachments between their flagellar membranes and the gut wall of the host insects (Brooker, 1971; Killick-Kendrick *et al.*, 1974; Vickerman and Tetley, 1990). Bacterial pathogens like *Mycoplasma pneumoniae* and *Bordetella pertussis* adhere to the ciliary membrane of epithelial cells of the human respiratory tract (Tuomanen, 1990; Tuomanen and Hendley, 1983). The binding sites on the cilia are not well understood, but the bacterial proteins that mediate the interaction are known (Edwards *et al.*, 2005; Hsu and Minion, 1998) and are being used for vaccine development (Chen *et al.*, 2006). Pathogen interaction with the ciliary membrane can disrupt the ciliary necklace as well as the axoneme (Ballenger, 1988; Carson *et al.*, 1979). Binding of pathogens to the ciliary membrane can also activate signaling cascades. For example, binding of *Mycoplasma hyopneumoniae* to the ciliary membrane activates a G protein-coupled receptor (GPCR), which in turn activates a phospholipase C pathway resulting in a rise in intracellular calcium in the ciliated epithelial cells of the respiratory epithelium (Park *et al.*, 2002).

3.3. Sensory functions

3.3.1. Introduction

In recent years, it has become clear that sensory reception is one of the major functions of cilia. In particular, the primary cilium of mammals serves as an antenna for the cell, reaching out into the extracellular space to project various types of receptors to monitor the environment (Pazour and Witman, 2003). It is likely that all cilia, including motile and nonmotile primary cilia, perform sensory functions. While primary cilia were described as early as 1898 (Zimmerman, 1898), it was only with the advent of transmission electron microscopy that it became clear that most types of vertebrate cells possess primary cilia and that major sensory receptors like the retinal photoreceptor cell, the olfactory cells of the nose, and insect mechanoreceptors utilized modified cilia for sensory perception.

To be useful, sensory functions of cilia must be coupled to downstream signaling functions so one cannot separate a discussion of the sensory functions of cilia from their signaling functions. Cilium-associated signaling pathways involved in mammalian development (such as SHH, Wnt, and PDGF) will be discussed extensively elsewhere in this volume.

3.3.2. Mechanoreception

It has long been known that many invertebrate sensory receptors are built around modified cilia. One of the best studied examples of this is the Type I class of sensory organs in insects (bristles, campaniform sensilla, and chordotonal organs). These organs contain sensory neurons each containing modified cilia that are mechanoreceptors for a wide variety of stimuli including sound and proprioception. Mutational analysis has implicated TRP channels, primarily TRPV, but also TRPN, as being key to the function of these cilia (Kernan, 2007; Kim et al., 2003; Liedtke and Kim, 2005). In *Drosophila*, TRPP channels (polycystins) are found only in the sperm flagellar membrane, where they appear to be used for sensory feedback related to sperm migration (Gao et al., 2003; Watnick et al., 2003).

The nematode *Caenorhabditis* contains cilium-based sensory receptors for mechanoreception, chemoreception, osmolarity reception, and nose touch. These cilia contain TRPV channels (OSM-9, OCR-1, OCR-2, OCR-3, OCR-4) essential for these functions (Kahn-Kirby and Bargmann, 2006; Liedtke and Kim, 2005). In addition, cilia play important roles in male mating behavior in these worms. This behavior is driven by TRPP channels (polycystins) localized on male-specific sensory neurons in the tail of the worm (Barr and Sternberg, 1999).

Statocysts are a particularly interesting category of cilium-based sensory receptors found in invertebrates such as mollusks, crustaceans, ctenophores, and coelenterates. These balance organs consist of a fluid-filled sac-containing statoliths or crystals that are displaced when the organism moves.

The movement of the statoliths mechanically stimulates sensory cells, which cause a depolarization of the membrane on motile "balancer" cilia resulting in a transient increase in the beat frequency of these cilia (Lowe, 1997).

In recent years, enormous attention has been paid to the mechanoreceptor properties of primary cilia in mammals. Mammals measure fluid flow through blood vessels and kidney tubules in order to effect feedback regulation and it appears that cilia are key to flow detection. Schwartz et al. (1997) demonstrated that primary cilia on kidney epithelial cells bend in response to fluid shear. Praetorius and Spring (2001) showed that bending these primary cilia with a micropipette or through increased flow rate of a perfusate resulted in an increase in cytoplasmic free calcium. This response was dependent on extracellular calcium and the calcium influx resulted in calcium-induced calcium release from intracellular stores mediated by inositol 1,4,5-trisphosphate (IP_3). Loss of the cilium resulted in a loss of flow sensing (Praetorius and Spring, 2002). The TRPP polycystins (products of the PKD1 and PKD2 genes) localize to the primary cilium membrane (Pazour et al., 2002b; Yoder et al., 2002). Defects in either polycystin-1 or polycystin-2 abolish flow sensing in the kidney tubule (Nauli et al., 2003) and result in autosomal dominant polycystic kidney disease (ADPKD) (Kottgen, 2007). A similar flow-sensing role for polycystins in primary cilia has been shown for other epithelial duct systems, such as in the cells that line bile ducts in the liver (cholangiocytes). Flow-induced bending of primary cilia on cholangiocytes induces polycystin-dependent influx of calcium that activates both an IP_3-dependent release of calcium from intracellular stores and inhibits cAMP-dependent signaling (Masyuk et al., 2006). Primary cilia on these same cells also appear to use a TRPV receptor (and calcium influx) for osmosensation (Gradilone et al., 2007).

Endothelial cells (another type of epithelial cell) perform flow-sensing functions for the circulatory system (Poelmann et al., 2008). The situation in the circulatory system appears to be more complicated than in kidney tubules. Primary cilium formation on endothelial cells is stimulated in areas of low and disturbed/oscillating flow while cilium formation is inhibited in areas of continuous high flow (Iomini et al., 2004; Van der Heiden et al., 2006, 2008). Nauli et al. (2008) showed that flow sensing in cultured endothelial cells requires primary cilia and polycystin-1 (PKD-1) and triggers a rise in cytosolic calcium and stimulation of nitric oxide signaling.

Although calcium and membrane channel proteins that allow calcium entry into the cilium have been repeated themes when discussing the sensory functions of the cilium, it is important to note that not all mechanosensory roles of primary cilia are mediated directly through calcium; this can be illustrated using another example of flow sensing by primary cilia (Malone et al., 2007). Osteoblasts and osteocytes in bone respond to fluid flow by increasing the expression of genes encoding osteopontin and cyclooxygenase-2 as well as increasing release of cytokines such as prostaglandin E_2. These responses are dependent on the presence of primary cilia. Although flow induces an increase in intracellular calcium, there is no correlation between the presence or absence of a cilium and the flow-induced rise in intracellular calcium (Malone et al., 2007). In contrast to the case with kidney epithelial cells, removal of extracellular calcium or addition of gadolinium (a calcium channel blocker) did not affect flow-induced signaling events dependent on primary cilia in bone, suggesting that cilium-based flow sensing in bone cells is unlikely to involve calcium influx or the polycystin-2 calcium channel. A unique mechanosensory role for primary cilia occurs in the node of developing embryos and is involved in the breaking of symmetry that leads to the establishment of left–right asymmetry in mammals. During a brief period during early development, nodal cells contain primary cilia, at least some of which exhibit a unique form of motility that generates a left–right flow in the extraembryonic fluid over the node (Nonaka et al., 1998; Supp et al., 1999). The presence of nodal cilia, the motility of the nodal cilia, and the leftward flow of the extracellular fluid are all necessary for proper establishment of left–right asymmetry in the embryo. The exact function of the nodal flow is still controversial with one hypothesis being that the leftward flow creates an extracellular gradient of secreted signaling molecules whose increased concentration on the left side of the embryo establishes asymmetric gene expression (Tanaka et al., 2005). An alternative hypothesis (the two cilia model) is that flow generated by motile nodal cilia bends nonmotile primary cilia at the edge of the node, which then stimulates calcium ion influxes through the action of PKD2 channels (McGrath et al., 2003).

3.3.3. Photoreception

The most dramatic example of evolutionary adaptation of a mammalian cilium for a sensory function occurs in the photoreceptor rod and cone cells of the retina (Besharse and Horst, 1990). The outer segments of rods and cones are developmentally derived from primary cilia. During development, the cilium becomes highly modified such that only a small region at the base, called the connecting cilium, is recognizable as being derived from a cilium. However, the entire outer segment is of ciliary origin including the disks of the outer segments, which are derived from the ciliary membrane. The connecting cilium is the only connection between the inner and outer segments of the photoreceptor cells. As is the case in all cilia, there is no protein synthesis in the outer segments and all components of the outer segments are synthesized in the inner segments and trafficked through the connecting cilium. Photoreceptor outer segments turn over about 10% of their disks each day by shedding of the oldest disks off of the distal end of the outer segment (Besharse, 1986). Thus, enormous numbers of opsin molecules (at least 10^7 molecules per day per photoreceptor) are trafficked through the connecting cilium from the inner to the outer segments (Besharse and Horst, 1990). Anything that interferes with this transport could be expected to lead to degeneration of the rod and cone outer segments and vision problems. Transport vesicles containing lipids and membrane proteins (primarily rhodopsin) destined for the outer segment are transported in the inner segment from the *trans*-Golgi to the periciliary ridge complex located near the basal body at the base of the connecting cilium. These transport vesicles presumably fuse with the apical plasma membrane of the inner segment at the periciliary ridge (Fig. 5.2C) and subsequently move into the membrane domain of the connecting cilium. Mutational analysis in mice has shown that microtubule motor proteins are necessary for the maintenance of the outer segment and have been proposed to be involved in the transport of rhodopsin and other proteins through the connecting cilium (Jimeno *et al.*, 2006; Marszalek *et al.*, 2000).

3.3.4. Chemoreception

Invertebrate sperm exhibit chemotaxis toward eggs in response to small peptides released by the eggs. This is best studied in sea urchins, where a species-specific chemotactic peptide binds to and activates guanylyl cyclase in the sperm flagellar membrane (Kaupp *et al.*, 2003; Kirkman-Brown *et al.*, 2003). The resulting rise in cGMP within the sperm tail opens potassium channels in the flagellar membrane leading to a transient hyperpolarization of the cell (Strunker *et al.*, 2006). The recovery from hyperpolarization coincides with an influx of calcium through calcium channels in the flagellar membrane. This produces spikes of calcium in the sperm tail (Bohmer *et al.*, 2005) that regulates beating of the axoneme to direct the sperm toward the egg that had released the chemotactic peptide.

The roundworm, *Caenorhabditis*, possesses a number of classes of specialized chemosensory neurons that can respond to a variety of water-soluble and volatile chemicals. In all of these neurons the sensory transduction molecules are localized in cilia (Inglis *et al.*, 2006). The ciliary membrane contains GPCRs. In contrast to mammals, a single sensory neuron can contain GPCRs for a number of different chemicals. In one of the major classes of chemosensory neurons (AWC cells), the sensory transduction pathway in the cilia appears to involve the GPCRs activating a G protein (odr-3) which activates a membrane-bound guanylate cyclase (or perhaps inactivates a cGMP phosphodiesterase) raising cGMP levels and opening a cGMP-gated cation channel thereby depolarizing the cell (Bargmann, 2006).

In mammals, the olfactory epithelium in the nose contains specialized bipolar sensory neurons whose dendritic bulb is decorated with many olfactory cilia whose membranes contain G protein-coupled odorant receptors (Mombaerts, 1999). Binding of an odorant molecule to its specific receptor results in activation of a G protein ($G\alpha_{olf}$) whose effector protein is adenylyl cyclase, another olfactory ciliary membrane protein. This results in a transient increase in concentration of cAMP, which opens specific ligand-gated cation channels, resulting in increased influx of calcium and sodium into the olfactory cilium (Menashe and Lancet, 2006). This results in depolarization of the sensory neuron (Reisert et al., 2005; Restrepo, 2005).

3.4. Cell signaling

As has already been made clear in this chapter, cilia (both motile and primary cilia) and ciliary membranes contain many proteins that are associated with cell signaling and the generation of second messages such as cyclic nucleotides or the regulation of cytosolic levels of calcium. As is well described in other chapters of this volume, recent work has shown that some of the major signaling pathways associated with development, including the hedgehog (Huangfu et al., 2003; Oro, 2007), Wnt (Ross et al., 2005), and PDGF (Schneider et al., 2005) signaling pathways, are associated with primary cilia and the ciliary membrane can contain a number of important signaling molecules, including somatostatin receptors (Handel et al., 1999), Smoothened (Corbit et al., 2005), and Tie (Teilmann and Christensen, 2005).

The many functions of the ciliary membrane (regulation of beat, adhesion, sensory reception, and signaling) are dependent upon numerous ciliary membrane proteins, which must be properly targeted to the ciliary membrane from their sites of synthesis (RER) and processing (Golgi). The remainder of this chapter will address the mechanisms utilized for targeting of ciliary membrane proteins to the cilium and the targeting information contained within ciliary proteins that serve as zip codes for proper delivery to the cilium.

4. PROTEIN MACHINERY INVOLVED IN TRAFFICKING TO THE CILIARY MEMBRANE

As discussed above, the best evidence suggests that vesicles containing proteins destined for the ciliary membrane fuse with the plasma membrane at the base of the cilium before the proteins are transported into the cilium. The proteins responsible for sorting, transporting, and fusion of vesicles at the base of the cilium are far from understood but a number of recent studies have started to provide information on the central players in this well-choreographed production. The IFT system (as discussed in Chapter 2) is clearly involved in transporting materials needed to build the axoneme both into the cilium and all along the length of the cilium. It is likely that IFT plays a role in transport of at least some ciliary membrane proteins, but its generalized use in delivery of membrane proteins to the cilium remains to be demonstrated. Likewise, the Bardet-Biedl syndrome (BBS) proteins clearly play important roles in ciliary function and have been proposed to facilitate transport to the ciliary membrane. Additionally, polarity proteins function in the formation of the ciliary membrane compartment and in the delivery of at least some membrane proteins to the ciliary membrane and a number of small GTPases have been implicated in ciliary formation and function probably by regulating the transport and fusion of carrier vesicles. Each of these systems will be discussed below as they relate to the ciliary membrane.

4.1. Intraflagellar transport

The IFT system transports materials for building ciliary axonemes (see Chapter 2) and it is logical that it also is involved in the movement of membrane proteins into the cilium but the evidence for this is only starting to accumulate. In mammalian photoreceptors, mutations in either kinesin-2 (Marszalek et al., 2000) or IFT88 (Pazour et al., 2002a) disrupt assembly of photoreceptor outer segments. This also causes an accumulation of rhodopsin in the inner segments suggesting that IFT plays a role in the transport of rhodopsin, but the mechanism could be indirect. In *C. elegans*, the TRPV channels OSM-9 and OCR-2 move along the ciliary membrane at rates comparable to IFT suggesting that they may be carried by IFT along the axoneme (Qin et al., 2005). Similarly, in *Chlamydomonas*, a small fraction of the transmembrane protein PKD2 moves in the anterograde direction at rates similar to IFT (Huang et al., 2007). However, in *C. elegans*, PKD-2 does not appear to be transported along the cilium by IFT (Qin et al., 2005).

In mammalian cells, all of the IFT particle proteins, with the exception of IFT20, are predominantly localized to the cilium and the peri-basal body or centrosomal region at the base of the cilium. IFT20 is unusual in that it is localized to the Golgi complex in addition to the centrosome and cilium (Follit et al., 2006). In cultured mammalian cells, the cilium is usually localized in the center of the Golgi complex; however, occasionally the cilium will be located some distance from the Golgi apparatus. In these instances, a thread of IFT20 can be observed between the Golgi complex and the base of the cilium. GFP-tagged IFT20 is highly dynamic in both the cilium and the Golgi complex and can be observed moving between the two compartments. It is likely that the Golgi pool of IFT20 is required for ciliary assembly as an RNAi knockdown that depleted the Golgi pool but not the centrosome pool, prevented ciliary assembly. Less efficient knockdown of IFT20 left ciliary assembly largely intact but reduced the amount of the membrane protein polycystin-2 found within the ciliary membrane. This suggests that IFT20 functions at the Golgi complex in the sorting of membrane proteins destined for the ciliary membrane into specific transport carriers or in the actual targeting of these transport vesicles, once formed, to the base of the cilium (Follit et al., 2006).

How IFT20 functions in this process is unknown and the identification of IFT20-binding partners will provide information on this process.

Recently, it was shown that the *Elipsa* gene product in zebrafish is an IFT20-binding protein (Omori *et al.*, 2008). The *Elipsa* gene is required for ciliary assembly in *Danio* as is the *Elipsa* ortholog, *Dyf-11*, in *Caenorhabditis* (Bacaj *et al.*, 2008; Kunitomo and Iino, 2008; Li *et al.*, 2008). Affinity purification of Elipsa copurified a complex containing IFT20 and many known complex B subunits (Omori *et al.*, 2008), suggesting that Elipsa and IFT20 interact as part of complex B and not as part of the Golgi-associated IFT20 pool. Furthermore in *Caenorhabditis*, Dyf-11 is transported along cilia with properties that suggest it is a complex B subunit (Li *et al.*, 2008). The yeast two-hybrid screen with the Elipsa that initially identified IFT20 also identified Rabaptin5 as another Elipsa-binding partner. Prior work had established Rabaptin5 as Rab4 and Rab5 effectors that regulated aspects of the endocytic pathway (Vitale *et al.*, 1998). Since Rabaptin5 interacted with Rab4 and Rab5, Omori *et al.* (2008) reasoned that it may also interact with Rab8 and showed this to be the case by yeast two-hybrid and *in vitro* GST pull-down assays. Morpholino-induced reduction of rabaptin5 had no effect on ciliary length and reduction of Rab8 reduced cilia length by 10% but the combined reduction of Rabaptin5 and Rab8 reduced cilia length by 20%. Thus, the authors propose that Elipsa links Rabaptin5/Rab8 to IFT20 and the IFT particle to facilitate movement of membrane proteins via IFT.

It is also possible that the IFT motors function in the movement of membrane proteins independent of the IFT particle. Recent data indicates that targeting of the smoothened seven-transmembrane receptor to cilia requires β-arrestin (Kovacs *et al.*, 2008). β-arrestins are well known for their ability to mediate the desensitization of seven-transmembrane receptors. For example, in photoreceptors, the inactivation of photoactivated rhodopsin initiates with phosphorylation of residues in the C-terminal tail of the protein. This phosphorylation event then allows arrestin to bind to sequences in internal loops and terminate the signaling cascade. In the case of smoothened, β-arrestin binds both smoothened and the Kif3A subunit of the anterograde IFT motor. RNAi reduction of β-arrestin-1 or -2 strongly inhibited the ability of smoothened to localize to cilia and to interact with Kif3A. This suggests that β-arrestin functions as a bridge between the smoothened and kinesin-2 to move the transmembrane protein into the cilium (Kovacs *et al.*, 2008). However, at this point, it is unclear why knockout of either β-arrestin-1 or -2 has little effect on mouse development (Bohn *et al.*, 1999; Conner *et al.*, 1997) if these proteins play such a key role in trafficking of smoothened, which is critical for development (Zhang *et al.*, 2001).

4.2. Bardet-Biedl syndrome proteins

BBS is a pleiotropic disorder of man that causes obesity, mental retardation, retinal degeneration, and cystic kidney disease among other phenotypes. Currently mutations in twelve genes are known to cause the disorder. Even though the function of the BBS proteins clearly involves cilia, the BBS proteins are not absolutely required for ciliary assembly. In *C. elegans,* mutations in the genes encoding BBS7 and BBS8 result in cilia that are shorter than normal but longer than cilia found on IFT mutant worms (Blacque et al., 2004). In mouse a number of BBS genes have been knocked out and the effects on ciliary assembly have varied by cell type. For example, sperm flagella are absent but respiratory and kidney tubule cilia appear normal (Blacque and Leroux, 2006). In the retina, the BBS proteins are not required for formation of the outer segments as photoreceptors are structurally normal in young animals lacking BBS2 or BBS4. Furthermore, there did not appear to be large defects in opsin transport as opsin was largely confined to the outer segments as normal. However, with time, the cells apoptosed resulting in a complete loss of photoreceptors by 10 months (Mykytyn et al., 2004; Nishimura et al., 2004a). The function of the BBS proteins is not completely understood. In *C. elegans,* the BBS proteins are transported with the IFT particle and are required for coupling the A and B IFT subcomplexes together (Ou et al., 2005). In mammalian cells, it is not clear if the BBS proteins enter the cilium as one study did not observe BBS4 in cilia using an antibody against the native protein (Kim et al., 2004) whereas another study detected GFP-tagged BBS4 in cilia (Nachury et al., 2007). Likewise, proteomic analysis of *Chlamydomonas* cilia suggests that if the proteins are in cilia, they are less abundant than the IFT proteins (Pazour et al., 2005). Work in zebrafish indicates that BBS gene products are required for noncanonical Wnt signaling and regulation of the planar cell polarity pathway (Ross et al., 2005). It has been suggested that the BBS proteins alter noncanonical Wnt by regulating the degradation of β-catenin by the proteasome (Gerdes et al., 2007). Nachury et al. (2007) purified a complex containing BBS1, 2, 4, 5, 7, 8, and 9 from cultured human cells that he termed the BBSome. Substoichiometric levels of Rabin8, which is a guanosyl exchange factor for Rab8 was associated with the purified BBSome. Guanosyl exchange factors promote the exchange of GDP for GTP to activate GTPases Finding Rabin8 associated with the BBS complex is important as Rab8 is implicated in regulating the fusion of opsin carrier vesicles at the base of the connecting cilium in frog photoreceptors (Moritz et al., 2001). Nachury further went on to show that Rabin8 was required for the entry of Rab8 into cilia and that a dominant negative mutation of Rab8 blocked the formation of cilia. The implication of this work is that the BBSome, through the actions of Rabin8 and Rab8, regulates the transport of membrane proteins to the cilium. This idea is supported by studies showing that BBS mutant mice fail to localize the SSTR3 somatostatin receptor and the Mchr1 melanin-concentrating hormone receptors on neuronal cilia (Berbari et al., 2008b).

4.3. Polarity proteins

The polarity proteins were initially identified as gene products required for the proper development of polarity in *C. elegans* embryos. These proteins are intimately involved in the formation and maintenance of epithelial polarity in both vertebrates and invertebrates by participating in and regulating the formation of junctional complexes that separate the apical and basal–lateral domains of the cell. Recent work has implicated a complex of polarity proteins containing Crumbs3, Par3, Par6, and aPKC in mammalian cilia assembly in addition to the more traditional role at tight junctions. Crumbs3 is a small transmembrane protein; Par3 and Par6 are PDZ domain proteins; and aPKC is an isoform of protein kinase C. Work from the Margolis group (Fan *et al.*, 2007; Gilchrist *et al.*, 2006) has shown that the CLPI isoform of the polarity protein Crumbs3 localizes to cilia of cultured MDCK cells while the ERLI splice variant does not. In these cells, knockdown of the Crumbs3–CLPI isoform blocks cilia assembly but does not affect the junctional complex while the knockdown of Crumbs3–ERLI disrupts the junctional complex without affecting cilia assembly suggesting that these two isoforms have distinct nonoverlapping functions (Fan *et al.*, 2007). The other members of Crumbs3/Par3/Par6/aPKC complex also localize to cilia and Par3 is required for formation of cilia (Gilchrist *et al.*, 2006; Sfakianos *et al.*, 2007). Interestingly, *Par3* mutant mice die at midgestation with phenotypes that share similarity to the phenotypes of ciliary mutants (Hirose *et al.*, 2006) raising the question of how much of the phenotype is due to defects in junctional complexes and how much is due to the loss of cilia. In neuronal cells, Par3 is transported along the axon by kinesin-2 (Nishimura *et al.*, 2004b) and kinesin-2 coprecipitates with the Crumbs3/Par3/Par6/aPKC complex in epithelial cells suggesting that kinesin-2 could transport this complex along the cilium (Gilchrist *et al.*, 2006). Since Crumbs3–CLPI contains a transmembrane domain, this would connect movement of membrane to the motor. This could play roles either in the delivery to the cilium or movement along the length of the cilium. Delivery into the cilium is an intriguing possibility as Crumbs3–CLPI binds to importin-β (Fan *et al.*, 2007). Importin-β is best known for its role in nuclear transport where it binds to cargo for transport through the nuclear pore. Upon entry into the nucleus, importin-β is released from the cargo by interaction with RAN GTPase. Both RAN and importin-β were found by proteomic analysis of *Chlamydomonas* cilia (Pazour *et al.*, 2005). The interaction between Crumbs3–CLPI and importin-β is particularly intriguing in light of the comparisons that have been made between nuclear transport and transport to the ciliary compartment (Bloodgood, 2000; Deane *et al.*, 2001; Rosenbaum and Witman, 2002).

In zebrafish, the crumbs proteins are important for ciliary assembly and morphogenesis of the apical compartment of epithelial cells. Morpholino-induced reduction of crumbs2b reduced the length of cilia in the pronephric duct by about 50% while knockdown of crumbs3a reduced auditory kinocilium length to about 70% of normal. In zebrafish, the crumbs proteins are found at the base of the cilia but do not appear to be in the cilia themselves. Thus, it is not clear if the Crumbs proteins in zebrafish function in ciliary assembly or if the effect is an indirect result of defects in the development of the apical compartment (Omori and Malicki, 2006). In MDCK cells, RNAi depletions of many proteins needed for morphogenesis of the apical compartment also disrupts ciliary assembly. The most severe effects on ciliary assembly were seen with annexin-13, syntaxin-3, and the phosphatidylinositol-4-phosphate adaptor protein FAPP2 but effects were also seen with caveolin-1, crumbs-3, galectin-3, and VIP17 (Torkko et al., 2008; Veira et al., 2006). With the exception of crumbs3, none of these proteins have been found in cilia and so more work will be needed to understand whether these proteins directly influence the formation of the ciliary compartment or if the ciliary defect is secondary to the defective apical domain. In the case of FAPP2, this protein is required for the formation of a condensed apical membrane domain that surrounds the base of the cilium and is probably important for maintenance of the ciliary membrane as a unique domain of the apical membrane (Veira et al., 2006).

4.4. Small GTPases

Transport through the endomembrane system is a highly regulated process with small GTPases in the Rab, Arf, and Arl subfamilies of the Ras superfamily controlling many steps along the secretory pathway (Gillingham and Munro, 2007). A number of different proteomic and genetic studies have identified small GTPases from these families in the cilium or implicated them in ciliary processes. Work of Deretic and collaborators have identified a cascade of small GTPases that regulates the budding of rhodopsin transport carriers from the Golgi complex and subsequent fusion at the base of the connecting cilium (Deretic, 2006). In this cascade, it is proposed that Arf4, Rab6, and Rab11 regulate the budding of the carriers from the Golgi complex. Arf4 is thought to directly bind to the ciliary targeting sequence in rhodopsin and play a key role in formation of the transport carriers (Deretic et al., 2005). Once the carriers are released from the Golgi, Arf4 is released while Rab8 and Rac1 become associated. Rab8 is thought to regulate the fusion of the carriers with the plasma membrane at the base of the cilium as dominant negative mutations of Rab8 cause an accumulation of rhodopsin carrier vesicles at the base of the connecting cilium (Moritz et al., 2001). Rac1 is proposed to play a role in tethering the carriers at the base of the connecting cilium by regulating the actin cytoskeleton (Deretic et al., 2004).

To characterize the role of Rab family members in primary cilia formation, Yoshimura et al. (2007) expressed 39 predicted RabGAPs in cultured human cells and measured how this affected formation of primary cilia.

RabGAPs promote the hydrolysis of GTP to GDP and tend to inactivate their cognate Rab partners. This analysis identified four RabGAPs that reduced primary cilia formation; three of which (TBC1D7, EVI5like, and XM_037557) did so without global effects on the cell and so are likely to be specifically involved in cilia functions. They then determined which of the 46 Rabs these three RabGAPs were active on. EVI5like was active on Rab23, TBC1D7 was active on Rab17, and XM_037557 was strongly active on Rab8a and showed some activity on the closely related isoform Rab8b. As discussed above, Rab8 had been previously implicated in cilia function (Moritz et al., 2001). Rab23 is the gene product of the mouse open brain mutant which has a hedgehog phenotype (Eggenschwiler et al., 2001). The mammalian hedgehog pathway is regulated by the cilium (Huangfu et al., 2003) and so this suggests that the open brain phenotype is caused by a ciliary defect. Rab17 had not been previously implicated in cilia function. Interestingly, of the 46 human Rabs, only Rab8a could be detected in the primary cilium when the proteins were expressed as GFP fusions. However, one must interpret this cautiously as the GFP tag could interfere with localization.

Proteomic analysis of cilia from various organisms has identified a number of small GTPases in cilia. In photoreceptor rod outer segments one analysis identified 62 different proteins with Rab GTPase domains (Liu et al., 2007) while an independent study identified 29 Rab family members and verified the presence of Rab1b, Rab11b, and Rab18 in outer segments by immunofluorescence (Kwok et al., 2008). Proteomic analysis of *Trypanosome* and *Tetrahymena* axonemes did not detect Rabs or other small G proteins but these studies used detergent in the isolation procedure that would have removed this class of proteins during the purification (Broadhead et al., 2006; Smith et al., 2005). Analysis of purified *Chlamydomonas* cilia identified four Rab family members (Rab1, 2, 5, 7), two Arl family members (Arl3, 6), and an isoform of Ran. In addition, this analysis identified two Rab-like proteins that are part of the IFT complex (IFT27/FAP156/Rabl4 and IFTA-2/FAP9/Rabl5) (Pazour et al., 2005). These two proteins were also found in human axonemes (Ostrowski et al., 2002).

Genomic studies that predict ciliary proteins have also identified a number of small G proteins. One type of analysis compared genomes of ciliated versus nonciliated organisms to predict proteins with ciliary-specific functions. This approach identified Rab4, Rab23, Rab27 (Avidor-Reiss et al., 2004), and IFTA-2/FAP9/Rabl5 (Li et al., 2004; Schafer et al., 2006) as potential cilia-specific G proteins. Another type of analysis takes advantage of the observation that in *C. elegans* many cilia genes are regulated by Daf-19, which binds to X-box sequences upstream of the genes. Examination of genes containing X-box sequences has been very predictive of ciliary genes and this analysis identified Rab10, Rab39, Arl3, Arl6 (Blacque et al., 2005), Rab21, IFTA-2/FAP9/Rabl5, and Arl3 (Efimenko et al., 2005) as potential ciliary GTPases. In *Chlamydomonas*, ciliary genes are often upregulated by deflagellation. A global analysis of genes upregulated by deflagellation identified IFT27/Rabl4/FAP156, IFTA-2/FAP9/Rabl5, and Arl6 as ciliary GTPases (Stolc et al., 2005).

Genetic analysis also has implicated a number of small GTPases in ciliary processes. Most mutations affecting cilia and ciliary processes are in genes encoding members of the Arl subfamily. This includes Arl6 which is the gene product of the BBS3 gene and, when mutated, causes Bardet-Biedl syndrome (Fan et al., 2004) as discussed above. Another Arl family member, Arl3 is found in *Leishmania* cilia and overexpression of the constitutively active form of the protein results in shortened cilia (Cuvillier et al., 2000). Arl3 mutations in mice cause cystic kidney disease and retinal degeneration similar to what is caused by mutations in known ciliary genes (Schrick et al., 2006). Arl13b has been mutated in both zebrafish, where it causes cystic kidney disease (Sun et al., 2004), and mouse, where is causes defects in ciliary structure and results in embryonic lethality (Caspary et al., 2007). In addition, Arl2 localizes to the centrosome (Zhou et al., 2006) and may participate in cilia–centrosome regulated processes.

In the Rab subfamily, the cilia related isoforms of Rab23 and Rab8a have been mutated. Rab23 is the gene product of the *open brain* mutation in mouse. This mutation caused a hedgehog phenotype (Eggenschwiler et al., 2001) similar to what is caused by mutations in IFT genes (Huangfu et al., 2003). This suggests that Rab23 may be involved in cilia function but the role remains to be determined. As described previously, a number of studies have implicated Rab8 involvement in ciliary assembly. In mice there are two isoforms of Rab8 called Rab8a and Rab8b. Rab8a, but not Rab8b, was found in cilia (Yoshimura et al., 2007) suggesting that even though the two isoforms are 83% identical, they have different functions. Recently Rab8a was knocked out in mice. The mutant mice developed intestinal disease due to a failure to correctly localize peptidases and transporters to the apical membrane but the authors did not report the presence of the classic ciliary phenotype of embryonic lethality at midgestation that is typically seen when cilia are missing or the cystic kidney disease, retinal degeneration, and left–right laterality defects seen with more subtle ciliary defects (Sato et al., 2007). It will be interesting to more closely examine these mice for ciliary defects and to determine if Rab8b can substitute for Rab8a when it is missing.

5. Ciliary Targeting Sequences

Proteins that are localized to organelles often contain *cis*-acting sequences that serve as a type of "cellular zip code" to direct the proteins to the particular organelle. For example, membrane proteins with KKxx sequences at their C-terminal ends are packaged into COPI vesicles at the

Golgi complex for transport to the endoplasmic reticulum (Kreis et al., 1995). Proteins destined for the lysosome are marked with a mannose 6-phosphate group that binds to the mannose 6-phosphate receptor for specific delivery to this organelle (Hille-Rehfeld, 1995). It is likely that sorting to the ciliary membrane utilizes an analogous mechanism of cis acting motifs within proteins to direct them to the ciliary compartment. As discussed above, the ciliary membrane is a specialized domain of the plasma membrane. Like sorting to organelles, sorting of proteins to the apical and basal–lateral domains of the plasma membrane also appears to be driven by motifs within the targeted proteins (Rodriguez-Boulan et al., 2005). Basolateral membrane targeting motifs consist of tyrosine, leucine, and dileucine motifs that interact with adaptor protein complexes. These adaptor protein complexes are then thought to bind clathrin coats as part of the sorting mechanism. Apical targeting motifs are much more diverse and do not share significant sequence homology with each other. They have been found in extracellular, transmembrane, and intracellular portions of proteins and in some cases involve lipid or sugar moieties. van Meer and Simons (1988) proposed that all of these diverse motifs function to direct proteins to lipid rafts in the Golgi complex, which are in turn preferentially sorted to the apical surface.

Experiments to identify ciliary membrane targeting sequences have been carried out on a number of different ciliary membrane proteins in a variety of organisms. If a consensus ciliary membrane targeting sequence can be identified, it will allow for the prediction of new ciliary membrane proteins and will also provide a means for identification of the machinery responsible for targeting proteins to the ciliary membrane. The most extensive analysis of a ciliary targeting sequence has been carried out on rhodopsin (Tam et al., 2000). Rhodopsin is a seven-transmembrane receptor that is highly concentrated in the outer segments of photoreceptors. Since the outer segment is developmentally derived from a cilium, the outer segment targeting sequence of opsin can be thought of as a ciliary targeting sequence. In transgenic frogs, the C-terminal 44 residues of rhodopsin efficiently targeted GFP to the outer segment. Further dissection of this 44 residue region, showed that the ciliary targeting sequence is composed of an eight residue motif at the very C-terminal end of the protein and a pair of palmitoylated cysteine residues located near the last transmembrane domain. The post-translational addition of palmitic acid to proteins often serves to anchor them to a lipid bilayer. This appears to be the role in opsin as the palmitoylated cysteines could be functionally replaced by a myristoylated glycine residue. The eight residues at the C-terminal end are likely to contain the motif that is responsible for the active sorting or transport of rhodopsin to the outer segment. Previously, this region had been proposed to be important for rhodopsin trafficking because mutations in humans that cause the most severe forms of autosomal dominant retinal degeneration cluster in this area. Comparison of the C-terminal tails of rhodopsin and S-opsin, which can be transported to rod outer segments (Shi et al., 2007), suggests that the essential targeting motif is VxPx-COOH. Work from Deretic indicates that the C-terminal tail of rhodopsin is required for budding of rhodopsin transport carrier vesicles in the *trans*-Golgi network possibly through interactions with the small GTPase Arf4 (Infante et al., 1999). This region of rhodopsin has also been reported to bind to a light chain of cytoplasmic dynein, which may play a role in transport to the base of the cilium (Tai et al., 1999).

Recently, ciliary targeting sequences were identified in another pair of ciliary-targeted seven-transmembrane receptors (Berbari et al., 2008a). In contrast to rhodopsin, the ciliary targeting sequences in the Sstr3 somatostatin and Htr6 serotonin receptors were not found at the C-terminal end but in the third intracellular loops. The authors then compared the sequences within these loops to identify a consensus sequence Ax[S/A]xQ that they used to search this loop in all of the seven-transmembrane receptors of humans. This analysis identified 11 receptors including rhodopsin and the opsins that are localized to photoreceptor outer segments and three olfactory receptors that are localized to olfactory cilia. Four other seven-transmembrane proteins, not previously known to be localized to cilia, were identified; the authors demonstrated that one of these, the melanin-concentrating hormone receptor was localized to cilia.

Work also is underway to identify the targeting sequences in other types of membrane proteins such as the polycystins. The polycystins are multispan transmembrane proteins that localize to cilia (Pazour et al., 2002b; Yoder et al., 2002). Dissection of the ciliary targeting sequences from these proteins showed that the polycystin-1 targeting sequence is located in the C-terminal 112 residues (Xu et al., 2007) whereas the targeting sequence of polycystin-2 is located within the first 15 amino acids at the N-terminus of the protein (Geng et al., 2006). The boundaries of the polycystin-1 targeting sequence were not further defined and so it is likely that a smaller portion of the C-terminal tail would be sufficient for targeting. Comparison of the 15 amino acid polycystin-2 targeting sequence between orthologues in vertebrates and sea urchins identified a set of conserved residues within the targeting sequence. Mutation of these residues showed that the sequence RVxP is required for the targeting sequence to function. Interestingly, this motif shares similarity with the VxPx motif of the rhodopsin ciliary targeting sequence (Infante et al., 1999). The identification of ciliary targeting sequences has also been examined in parasitic protozoa and *C. elegans*. In *Leishmania*, the iso-1 glucose transporter is targeted to cilia through a bipartite motif found at the N-terminus of the protein (Ignatushchenko et al., 2004). In *Trypanosoma*, the peripheral membrane protein FCaBP is targeted to the cilium by a 24 residue region at the N-terminus that contains palmitoylated cysteine and myristoylated glycine residues (Godsel and Engman, 1999). In *C. elegans*, deletion of the C-terminal tails of STR-1 and ODR-10 prevented the proteins from being specifically targeted to cilia. Comparison of these regions indicated that they were very dissimilar except for a hydrophobic and basic residue pair just after the last transmembrane domain (FR in ODR-10 and YR in STR-1). Mutation of the FR residues prevented ODR-10 from being localized to cilia. This indicates that these residues are critical for ciliary targeting but, without showing that the motif is sufficient for localization of an exogenous protein, one cannot know if the result is direct or indirect. Based on these results, Corbit *et al.* (2005) predicted the ciliary targeting sequences of the seven-transmembrane receptors smoothened, the Sstr3 somatostatin receptor, the Htr6 serotonin receptor, and rhodopsin. Mutation of the predicted residues in smoothened prevented its ciliary localization but, again, without showing that the motif is sufficient to target an exogenous protein to the cilium, one cannot know if the result is direct. In the cases of the somatostatin and serotonin receptors and rhodopsin, the predicted residues are not part of the experimentally determined ciliary targeting sequences (Berbari *et al.*, 2008a; Tam *et al.*, 2000).

One of the main goals of the analysis of ciliary targeting sequences is to be able to devise a search pattern that can be used to find new ciliary membrane proteins. Comparison of the known targeting sequences reveals very little in common (Table 5.1). The only case in which a motif has been successfully used to find a new ciliary protein is in the identification of the melanin concentrating hormone receptor using the Ax[S/A]xQ motif (Berbari *et al.*, 2008a). While this motif was able to predict a new ciliary membrane protein, the motif is not present in other ciliary localized seven-transmembrane receptors, like smoothened or the majority of olfactory receptors, and so either alternative targeting sequences are used in these proteins or the motif is too specific and needs more wobble. Furthermore, even though this sequence is found in rhodopsin, it is not required for the targeting of rhodopsin to the cilium (Tam *et al.*, 2000). As discussed above, the ciliary targeting sequence of polycystin-2 and opsin contain similar motifs of RVxP in polycystin-2 and VxPx in opsin. The low complexity and lack of a specific location within proteins makes it hard to use this sequence for predictive purposes. However, the sequence was found in the ciliary-targeted cyclic nucleotide-gated channel CNGB1b. Mutation of the motif blocked entry into the cilium of CNGB1b into cilia but the sequence was not sufficient to target a nonciliary protein into the cilium (Jenkins *et al.*, 2006) so it is not clear that the motif plays a role in ciliary targeting in this protein or if the effect is indirect.

Lipid modifications are found in two of the ciliary targeting sequences (Table 5.1) and are abundant in ciliary membrane proteins. The opsin ciliary targeting sequence contains two cysteine residues that are palmitoylated and needed for proper targeting to the cilium (Tam *et al.*, 2000) while the

Table 5.1 Comparison of ciliary targeting sequences from integral and peripheral ciliary membrane proteins. Only those targeting sequences that have been characterized to the level of sufficiency by moving them to a heterologous protein are shown

Protein[a]	Ciliary targeting sequence (CTS)	Location within the protein	Reference
Opsin	CC...SSSQ**VS**P**A**	C-terminus	Tam et al. (2000)
SSTR3	VVKVRSTTRRVR<u>APSC**Q**WVQAPAC**Q**RRRSERRVTR</u>	Cytoplasmic loop	Berbari et al. (2008a)
Htr6	TYCRILLAARKQAVQVASLTTGT<u>A</u>TAG**Q**ALETLQV	Cytoplasmic loop	Berbari et al. (2008a)
Polycystin-1	SQLDGLSVSLGRLGTRCEPEPSRLQAVFEA LLTQFDRLNQATEDVYQLEQQLHSLQGRRS SRAPAGSSRGPSPGLRPALPSRLARASRGV DLATGPSRTPLRAKNKVHPSST[b]	C-terminus	Xu et al. (2007)
Polycystin-2	MVNSS**RV**Q**P**QQPGDA	N-terminus	Geng et al. (2006)
Glucose transporter	GMSDRVEVNERRSDSVS...RKDVTDDQE	N-terminus	Ignatushchenko et al. (2004)
Calcium-binding protein	MGACGSKGSTSDKGLASDKDGKNA	N-terminus	Godsel and Engman (1999)

[a] Only proteins in which the CTS was characterized to the level of sufficiency by moving it to a heterologous protein are shown. In a number of other proteins, potential CTSs have been characterized by point mutations but no test of sufficiency was carried out and so are not listed here.
[b] The boundaries of the CTS within this peptide were not defined.

calcium-binding protein targeting sequence contains a palmitoylated cysteine and a myristoylated glycine (Godsel and Engman, 1999). Furthermore, myristoylation motifs are enriched in ciliary proteins from *Chlamydomonas* (Pazour et al., 2005) and also have been observed in other ciliary proteins such as SMP-1 in *Leishmania* (Tull et al., 2004) and cystin in mouse (Hou et al., 2002). In mammalian cells, lipid modifications can play a role in trafficking to the apical membrane. It has been proposed that the lipid increases the association of the targeted protein with lipid rafts, which are preferentially transported to the apical membrane. It is possible that a similar type of process is important in the sorting of proteins to the ciliary membrane.

REFERENCES

Absalon, S., Blisnick, T., Kohl, L., Toutirais, G., Doré, G., Julkowska, D., Tavenet, A., and Bastin, P. (2008). Intraflagellar transport and functional analysis of genes required for flagellum formation in trypanosomes. *Mol. Biol. Cell* **19,** 929–944.

Andrade, Y. N., Fernandes, J., Vazquez, E., Fernandez-Fernandez, J. M., Arniges, M., Sanchez, T. M., Villalon, M., and Valverde, M. A. (2005). TRPV4 channel is involved in the coupling of fluid viscosity changes to epithelial ciliary activity. *J. Cell Biol.* **168,** 869–874.

Avidor-Reiss, T., Maer, A. M., Koundakjian, E., Polyanovsky, A., Keil, T., Subramaniam, S., and Zuker, C. S. (2004). Decoding cilia function: Defining specialized genes required for compartmentalized cilia biogenesis. *Cell* **117,** 527–539.

Bacaj, T., Lu, Y., and Shaham, S. (2008). The conserved proteins CHE-12 and DYF-11 are required for sensory cilium function in *Caenorhabditis elegans*. *Genetics* **178,** 989–1002.

Ballenger, J. J. (1988). Acquired ultrastructural alterations of respiratory cilia and clinical disease. *Ann. Otol. Rhinol. Laryngol.* **97,** 253–258.

Bargmann, C. I. (2006). *In* Chemosensation in *C. elegans*, (WormBook, Ed.). The *C. elegans* Research Community, WormBook. doi/10.1895/wormbook.1.123.1, http://www.wormbook.org.

Barr, M. M., and Sternberg, P. W. (1999). A polycystic kidney-disease gene homologue required for male mating behavior in *C. elegans*. *Nature* **401,** 386–389.

Berbari, N. F., Johnson, A. D., Lewis, J. S., Askwith, C. C., and Mykytyn, K. (2008a). Identification of ciliary localization sequences within the third intracellular loop of G protein-coupled receptors. *Mol. Biol. Cell* **19,** 1540–1547.

Berbari, N. F., Lewis, J. S., Bishop, G. A., Askwith, C. C., and Mykytyn, K. (2008b). Bardet-Biedl syndrome proteins are required for the localization of G protein-coupled receptors to primary cilia. *Proc. Natl Acad. Sci. USA* **105,** 4242–4246.

Bergman, K., Goodenough, U. W., Goodenough, D. A., Jawitz, J., and Martin, H. (1975). Gametic differentiation in *Chlamydomonas reinhardtii*. II. Flagellar membranes and the agglutination reaction. *J. Cell Biol.* **67,** 606–622.

Besharse, J. C. (1986). Photosensitive membrane turnover: Differentiated membrane domains and cell–cell interaction. *In* "The Retina: A Model for Cell Biological Studies" (R. Adler and D. Farber, Eds.), Part I, pp. 297–352. Academic Press, New York.

Besharse, J. C., and Horst, C. J. (1990). The photoreceptor connecting cilium. A model for the transition zone. *In* "Ciliary and Flagellar Membranes" (R. A. Bloodgood, Ed.), pp. 389–417. Plenum Press, New York.

Blacque, O. E., and Leroux, M. R. (2006). Bardet-Biedl syndrome: An emerging pathomechanism of intracellular transport. *Cell. Mol. Life Sci.* **63**, 2145–2161.

Blacque, O. E., Reardon, M. J., Li, C., McCarthy, J., Mahjoub, M. R., Ansley, S. J., Badano, J. L., Mah, A. K., Beales, P. L., Davidson, W. S., Johnsen, R. C., Audeh, M., et al. (2004). Loss of *C. elegans* BBS-7 and BBS-8 protein function results in cilia defects and compromised intraflagellar transport. *Genes Dev.* **18**, 1630–1642.

Blacque, O. E., Perens, E. A., Boroevich, K. A., Inglis, P. N., Li, C., Warner, A., Khattra, J., Holt, R. A., Ou, G., Mah, A. K., McKay, S. J., Huang, P., et al. (2005). Functional genomics of the cilium, a sensory organelle. *Curr. Biol.* **15**, 935–941.

Bloodgood, R. A. (1990). "Ciliary and Flagellar Membranes," 431pp. Plenum Press, New York.

Bloodgood, R. A. (2000). Protein targeting to flagella of trypanosomatid protozoa. *Cell Biol. Int.* **24**, 857–862.

Bloodgood, R. A. (2009). The *Chlamydomonas* flagellar membrane and its dynamic properties. In "The *Chlamydomonas* Source Book" (G. Witman, Ed.). Vol. 3, pp. 305–364. Elsevier, San Diego.

Bohmer, M., Van, Q., Weyand, I., Hagen, V., Beyermann, M., Matsumoto, M., Hoshi, M., Hildebrand, E., and Kaupp, U. B. (2005). Ca^{2+} spikes in the flagellum control chemotactic behavior of sperm. *EMBO J.* **24**, 2741–2752.

Bohn, L. M., Lefkowitz, R. J., Gainetdinov, R. R., Peppel, K., Caron, M. G., and Lin, F. T. (1999). Enhanced morphine analgesia in mice lacking beta-arrestin 2. *Science* **286**, 2495–2498.

Broadhead, R., Dawe, H. R., Farr, H., Griffiths, S., Hart, S. R., Portman, N., Shaw, M. K., Ginger, M. L., Gaskell, S. J., McKean, P. G., and Gull, K. (2006). Flagellar motility is required for the viability of the bloodstream trypanosome. *Nature* **440**, 224–227.

Brooker, B. E. (1971). Flagellar attachment and detachment of *Crithidia fasciculata* to the gut wall of *Anopheles gambiae*. *Protoplasma* **73**, 191–202.

Carson, J. L., Collier, A. M., and Clyde, W. A., Jr. (1979). Ciliary membrane alterations occurring in experimental *Mycoplasma pneumoniae* infection. *Science* **206**, 349–351.

Caspary, T., Larkins, C. E., and Anderson, K. V. (2007). The graded response to Sonic Hedgehog depends on cilia architecture. *Dev. Cell* **12**, 767–778.

Chailley, B., and Boisvieux-Ulrich, E. (1985). Detection of plasma membrane cholesterol by filipin during microvillogenesis and ciliogenesis in quail oviduct. *J. Histochem. Cytochem.* **33**, 1–10.

Chen, A. Y., Fry, S. R., Forbes-Faulkner, J., Daggard, G., and Mukkur, T. K. S. (2006). Evaluation of the immunogenicity of the P97R1 adhesin of *Mycoplasma hyopneumoniae* as a mucosal vaccine in mice. *J. Med. Microbiol.* **55**, 923–929.

Conner, D. A., Mathier, M. A., Mortensen, R. M., Christe, M., Vatner, S. F., Seidman, C. E., and Seidman, J. G. (1997). β-Arrestin1 knockout mice appear normal but demonstrate altered cardiac responses to β-adrenergic stimulation. *Circ. Res.* **81**, 1021–1026.

Corbit, K. C., Aanstad, P., Singla, V., Norman, A. R., Stainier, D. Y., and Reiter, J. F. (2005). Vertebrate Smoothened functions at the primary cilium. *Nature* **437**, 1018–1021.

Cuvillier, A., Redon, F., Antoine, J. C., Chardin, P., DeVos, T., and Merlin, G. (2000). LdARL-3A, a *Leishmania* promastigote-specific ADP-ribosylation factor-like protein, is essential for flagellum integrity. *J. Cell Sci.* **113**, 2065–2074.

Davidge, J. A., Chambers, E., Dickinson, H. A., Towers, K., Ginger, M. L., McKean, P. G., and Gull, K. (2006). *Trypanosome* IFT mutants provide insight into the motor location for mobility of the flagella connector and flagellar membrane formation. *J. Cell Sci.* **119**, 3935–3943.

Deane, J. A., Cole, D. G., Seeley, E. S., Diener, D. R., and Rosenbaum, J. L. (2001). Localization of intraflagellar transport protein IFT52 identifies basal body transitional fibers as the docking site for IFT particles. *Curr. Biol.* **11**, 1586–1590.

Dentler, W. L. (1990). Linkages between microtubules and membranes in cilia and flagella. *In* "Ciliary and Flagellar Membranes" (R. A. Bloodgood, Ed.), pp. 31–64. Plenum Press, New York.

Dentler, W. (2006). Importance of golgi for the regulation of flagellar length in *Chlamydomonas*. *Mol. Biol. Cell* **17**. abstract #1596 (CD ROM).

Deretic, D. (2006). A role for rhodopsin in a signal transduction cascade that regulates membrane trafficking and photoreceptor polarity. *Vis. Res.* **46**, 4427–4433.

Deretic, D., Traverso, V., Parkins, N., Jackson, F., Rodriguez de Turco, E. B., and Ransom, N. (2004). Phosphoinositides, ezrin/moesin, and rac1 regulate fusion of rhodopsin transport carriers in retinal photoreceptors. *Mol. Biol. Cell* **15**, 359–370.

Deretic, D., Williams, A. H., Ransom, N., Morel, V., Hargrave, P. A., and Arendt, A. (2005). Rhodopsin C terminus, the site of mutations causing retinal disease, regulates trafficking by binding to ADP-ribosylation factor 4 (ARF4). *Proc. Natl Acad. Sci. USA* **102**, 3301–3306.

Dunlap, K. (1977). Localization of calcium channels in *Paramecium caudatum*. *J. Physiol.* **271**, 119–133.

Edwards, J. A., Groathouse, N. A., and Boitano, S. (2005). *Bordetella bronchiseptica* adherence to cilia is mediated by multiple adhesin factors and blocked by surfactant protein A. *Infect. Immun.* **73**, 3618–3626.

Efimenko, E., Bubb, K., Mak, H. Y., Holzman, T., Leroux, M. R., Ruvkun, G., Thomas, J. H., and Swoboda, P. (2005). Analysis of *xbx* genes in *C. elegans*. *Development* **132**, 1923–1934.

Eggenschwiler, J. T., Espinoza, E., and Anderson, K. V. (2001). Rab23 is an essential negative regulator of the mouse Sonic hedgehog signalling pathway. *Nature* **412**, 194–198.

Fan, Y., Esmail, M. A., Ansley, S. J., Blacque, O. E., Boroevich, K., Ross, A. J., Moore, S. J., Badano, J. L., May-Simera, H., Compton, D. S., Green, J. S., Lewis, R. A., *et al.* (2004). Mutations in a member of the Ras superfamily of small GTP-binding proteins causes Bardet-Biedl syndrome. *Nat. Genet.* **36**, 989–993.

Fan, S., Fogg, V., Wang, Q., Chen, X. W., Liu, C. J., and Margolis, B. (2007). A novel Crumbs3 isoform regulates cell division and ciliogenesis via importin beta interactions. *J. Cell Biol.* **178**, 387–398.

Ferris, P. J., Waffenschmidt, S., Umen, J. G., Lin, H., Lee, J.-H., Ishida, K., Kubo, T., Lau, J., and Goodenough, U. W. (2005). Plus and minus sexual agglutinins from *Chlamydomonas reinhardtii*. *Plant Cell* **17**, 597–615.

Follit, J. A., Tuft, R. A., Fogarty, K. E., and Pazour, G. J. (2006). The intraflagellar transport protein IFT20 is associated with the Golgi complex and is required for cilia assembly. *Mol. Biol. Cell* **17**, 3781–3792.

Gao, Z., Ruden, D. M., and Lu, X. (2003). PKD2 cation channel is required for directional sperm movement and male fertility. *Curr. Biol.* **13**, 2175–2178.

Geng, L., Okuhara, D., Yu, Z., Tian, X., Cai, Y., Shibazaki, S., and Somlo, S. (2006). Polycystin-2 traffics to cilia independently of polycystin-1 by using an N-terminal RVxP motif. *J. Cell Sci.* **119**, 1383–1395.

Gerdes, J. M., Liu, Y., Zaghloul, N. A., Leitch, C. C., Lawson, S. S., Kato, M., Beachy, P. A., Beales, P. L., DeMartino, G. N., Fisher, S., Badano, J. L., and Katsanis, N. (2007). Disruption of the basal body compromises proteasomal function and perturbs intracellular Wnt response. *Nat. Genet.* **39**, 1350–1360.

Gilchrist, A., Au, C. E., Hiding, J., Bell, A. W., Fernandez-Rodriguez, J., Lesimple, S., Nagaya, H., Roy, L., Gosline, S. J., Hallett, M., Paiement, J., Kearney, R. E., *et al.* (2006). Quantitative proteomics analysis of the secretory pathway. *Cell* **127**, 1265–1281.

Gillingham, A. K., and Munro, S. (2007). The small G proteins of the Arf family and their regulators. *Annu. Rev. Cell Dev. Biol.* **23**, 579–611.

Gilula, N. B., and Satir, P. (1972). The ciliary necklace: A ciliary membrane specialization. *J. Cell Biol.* **53,** 494–509.

Godsel, L. M., and Engman, D. M. (1999). Flagellar protein localization mediated by a calcium-myristoyl/palmitoyl switch mechanism. *EMBO J.* **18,** 2057–2065.

Gradilone, S. A., Masyuk, A. I., Slinter, P. L., Banales, J. M., Huang, B. Q., Tietz, P. S., Masyuk, T. V., and LaRusso, N. F. (2007). Cholangiocyte cilia express TRPV4 and detect changes in luminal tonicity including bicarbonate secretion. *Proc. Natl Acad. Sci. USA* **104,** 19138–19143.

Haller, K., and Fabry, S. (1998). Brefeldin A affects synthesis and integrity of a eukaryotic flagellum. *Biochem. Biophys. Res. Commun.* **242,** 597–601.

Handel, M., Schulz, S., Stanarius, A., Schreff, M., Erdtmann-Vourliotis, M., Schmidt, H., Wolf, G., and Hollt, V. (1999). Selective targeting of somatostatin receptor 3 to neuronal cilia. *Neuroscience* **89,** 909–926.

Hille-Rehfeld, A. (1995). Mannose 6-phosphate receptors in sorting and transport of lysosomal enzymes. *Biochim. Biophys. Acta* **1241,** 177–194.

Hirose, T., Karasawa, M., Sugitani, Y., Fujisawa, M., Akimoto, K., Ohno, S., and Noda, T. (2006). PAR3 is essential for cyst-mediated epicardial development by establishing apical cortical domains. *Development* **133,** 1389–1398.

Hiwatashi, K. (1988). Sexual recognition in *Paramecium*. In Eukaryote Cell Recognition: Concepts and Model Systems, (G. P. Chapman, C. C. Ainsworth, and C. J. Chatham, Eds.), pp. 77–91. Cambridge University Press, Cambridge.

Hou, X., Mrug, M., Yoder, B. K., Lefkowitz, E. J., Kremmidiotis, G., D'Eustachio, P., Beier, D. R., and Guay-Woodford, L. M. (2002). Cystin, a novel cilia-associated protein, is disrupted in the cpk mouse model of polycystic kidney disease. *J. Clin. Invest.* **109,** 533–540.

Hsu, T., and Minion, F. C. (1998). Identification of the cilium binding epitope of the *Mycoplasma hyopneumoniae* P97 adhesion. *Infect. Immun.* **66,** 4762–4766.

Huang, B. Q., Masyuk, T. V., Muff, M. A., Tietz, P. S., Masyuk, A. I., and LaRusso, N. F (2006). Isolation and characterization of cholangiocyte primary cilia. *Am. J. Physiol Gastrointest. Liver Physiol.* **291,** G500–G509.

Huang, K., Diener, D. R., Mitchell, A., Pazour, G. J., Witman, G. B., and Rosenbaum, J. L (2007). Function and dynamics of PKD2 in *Chlamydomonas reinhardtii* flagella. *J. Cell Biol* **179,** 501–514.

Huangfu, D., Liu, A., Rakeman, A. S., Murcia, N. S., Niswander, L., and Anderson, K. V. (2003). Hedgehog signaling in the mouse requires intraflagellar transport proteins. *Nature* **426,** 83–87.

Hunnicutt, G. R., Kosfiszer, M. G., and Snell, W. J. (1990). Cell body and flagella agglutinins in *Chlamydomonas reinhardtii*: The cell body plasma membrane is a reservoir for agglutinins whose migration to the flagella is regulated by a functional barrier. *J. Cell Biol.* **111,** 1605–1616.

Hyams, J. S., and Borisy, G. G. (1978). Isolated flagellar apparatus of *Chlamydomonas*: Characterization of forward swimming and alteration of waveform and reversal of motion by calcium ions *in vitro*. *J. Cell Sci.* **33,** 235–253.

Ignatushchenko, M., Nasser, M. I., and Landfear, S. M. (2004). Sequences required for the flagellar targeting of an integral membrane protein. *Mol. Biochem. Parasitol.* **135,** 89–100.

Infante, C., Ramos-Morales, F., Fedriani, C., Bornens, M., and Rios, R. M. (1999). GMAP-210, A *cis*-Golgi network-associated protein, is a minus end microtubule-binding protein. *J. Cell Biol.* **145,** 83–98.

Inglis, P. N., Ou, G., Leroux, M. R., and Scholey, J. M. (2006). In The Sensory Cilia of *Caenorhabditis elegans*, (WormBook, Ed.). The *C. elegans* Research Community, WormBookdoi/10.1895/wormbook.1.126.1http://www.wormbook.org.

Iomini, C., Tejada, K., Mo, W., Vaananen, H., and Piperno, G. (2004). Primary cilia of human endothelial cells disassemble under laminar flow stress. *J. Cell Biol.* **164,** 811–817.

Iomini, C., Li, L., Mo, W., Dutcher, S. K., and Piperno, G. (2006). Two flagellar genes, AGG2 and AGG3, mediate orientation to light in *Chlamydomonas*. *Curr. Biol.* **16,** 1147–1153.

Janich, P., and Corbeil, D. (2007). GM1 and GM3 gangliosides highlight distinct lipid microdomains within the apical domain of epithelial cells. *FEBS Lett.* **581,** 1783–1787.

Jekely, G., and Arendt, D. (2006). Evolution of intraflagellar transport from coated vesicles and autogenous origin of the eukaryotic cilium. *BioEssays* **28,** 191–198.

Jenkins, P. M., Hurd, T. W., Zhang, L., McEwen, D. P., Brown, R. L., Margolis, B., Verhey, K. J., and Martens, J. R. (2006). Ciliary targeting of olfactory CNG channels requires the CNGB1b subunit and the kinesin-2 motor protein, KIF17. *Curr. Biol.* **16,** 1211–1216.

Jennings, H. S. (1906). *In* "Behavior of Lower Organisms," pp. 47–49. The Columbia University Press, New York.

Jimeno, D., Feiner, L., Lillo, C., Teofilo, K., Goldstein, L. B. S., Pierce, E. A., and Williams, D. S. (2006). Analysis of kinesin-2 function in photoreceptor cells using synchronous Cre–loxP knockout of Kif3a with RHO-Cre. *Invest. Ophthalmol. Vis. Sci.* **47,** 5039–5046.

Kahn-Kirby, A. H., and Bargmann, C. I. (2006). TRP channels in *C. elegans*. *Annu. Rev. Physiol.* **68,** 719–736.

Kaneshiro, E. S. (1990). Lipids of ciliary and flagellar membranes. *In* "Ciliary and Flagellar Membranes" (R. A. Bloodgood, Ed.), pp. 241–265. Plenum Press, New York.

Kaupp, U. B., Solzin, J., Hildebrand, E., Brown, J. E., Helbig, A., Hagen, V., Beyermann, M., Pampaloni, F., and Weyand, I. (2003). The signal flow and motor response controlling chemotaxis of sea urchin sperm. *Nat. New Biol.* **5,** 109–117.

Kernan, M. J. (2007). Mechanotransduction and auditory transduction in *Drosophila*. *Pflugers Arch. Eur. J. Physiol.* **454,** 703–720.

Killick-Kendrick, R., Molyneux, D. H., and Ashford, R. W. (1974). *Leishmania* in phlebotomid sandflies. I. Modifications of the flagellum associated with attachment to the mid-gut and oesophageal valve of the sandfly. *Proc. R. Soc. Lond. B Biol. Sci.* **187,** 409–419.

Kim, J., Chung, Y. D., Park, D.-Y., Choi, S., Shin, D. W., Soh, H., Lee, H. W., Son, W., Yim, J., Park, C.-S., Kernan, M. J., and Kim, C. (2003). A TRPV family ion channel required for hearing in *Drosophila*. *Nature* **424,** 81–84.

Kim, J. C., Badano, J. L., Sibold, S., Esmail, M. A., Hill, J., Hoskins, B. E., Leitch, C. C., Venner, K., Ansley, S. J., Ross, A. J., Leroux, M. R., Katsanis, N., and Beales, P. L. (2004). The Bardet-Biedl protein BBS4 targets cargo to the pericentriolar region and is required for microtubule anchoring and cell cycle progression. *Nat. Genet.* **36,** 462–470.

Kirkman-Brown, J. C., Sutton, K. A., and Florman, H. M. (2003). How to attract a sperm. *Nat. New Biol.* **5,** 93–96.

Kottgen, M. (2007). TRPP2 and autosomal polycystic kidney disease. *Biochim. Biophys. Acta* **1772,** 836–850.

Kovacs, J. J., Whalen, E. J., Liu, R., Xiao, K., Kim, J., Chen, M., Wang, J., Chen, W., and Lefkowitz, R. J. (2008). β-Arrestin-mediated localization of smoothened to the primary cilium. *Science* **320,** 1777–1781.

Kreis, T. E., Lowe, M., and Pepperkok, R. (1995). COPs regulating membrane traffic. *Annu. Rev. Cell Dev. Biol.* **11,** 677–706.

Kunitomo, H., and Iino, Y. (2008). *Caenorhabditis elegans* DYF-11, an orthologue of mammalian Traf3ip1/MIP-T3, is required for sensory cilia formation. *Genes Cells* **13,** 13–25.

Kwok, M. C., Holopainen, J. M., Molday, L. L., Foster, L. J., and Molday, R. S. (2008). Proteomics of photoreceptor outer segments identifies a subset of SNARE and Rab proteins implicated in membrane vesicle trafficking and fusion. *Mol. Cell Proteomics* **7**, 1053–1066.

Lefebvre, R., Chenoweth, P. J., Drost, M., LeClear, C. T., MacCubbin, M., Dutton, J. T., and Suarez, S. S. (1995). Characterization of the oviductal sperm reservoir in cattle. *Biol. Reprod.* **53**, 1066–1074.

Li, J. B., Gerdes, J. M., Haycraft, C. J., Fan, Y., Teslovich, T. M., May-Simera, H., Li, H., Blacque, O. E., Li, L., Leitch, C. C., Lewis, R. A., Green, J. S., et al. (2004). Comparative genomics identifies a flagellar and basal body proteome that includes the BBS5 human disease gene. *Cell* **117**, 541–552.

Li, C., Inglis, P. N., Leitch, C. C., Efimenko, E., Zaghloul, N. A., Mok, C. A., Davis, E. E., Bialas, N. J., Healey, M. P., Heon, E., Zhen, M., Swoboda, P., et al. (2008). An essential role for DYF-11/MIP-T3 in assembling functional intraflagellar transport complexes. *PLoS Genet.* **4**, e1000044.

Liedtke, W., and Kim, C. (2005). Functionality of the TRPV subfamily of TRP ion channels: Add mechano-TRP and osmo-TRP to the lexicon!. *Cell. Mol. Life Sci.* **62**, 2985–3001.

Liu, Q., Tan, G., Levenkova, N., Li, T., Pugh, E. N., Jr., Rux, J. J., Speicher, D. W., and Pierce, E. A. (2007). The proteome of the mouse photoreceptor sensory cilium complex. *Mol. Cell Proteomics* **6**, 1299–1317.

Lorenzo, I. M., Liedtke, W., Sanderson, M. J., and Valverde, M. A. (2008). TRPV4 channel participates in receptor-operated calcium entry and ciliary beat frequency regulation in mouse airway epithelial cells. *Proc. Natl Acad. Sci. USA* **105**, 12611–12616.

Lowe, B. (1997). The role of calcium in deflection-induced excitation of motile, mechanoresponsive balancer cilia in the ctenophore statocyst. *J. Exp. Biol.* **200**, 1593–1606.

Malone, A. M. D., Anderson, C. T., Padmaja, T., Kwon, R. Y., Johnson, T. R., Sterns, T., and Jacobs, C. R. (2007). Primary cilia mediate mechanosensing in bone cells by a calcium-independent mechanism. *Proc. Natl Acad. Sci. USA* **104**, 13325–13330.

Marszalek, J. R., Liu, X., Roberts, E. A., Chui, D., Marth, J. D., Williams, D. S., and Goldstein, L. S. B. (2000). Genetic evidence for selective transport of opsin and arrestin by kinesin-II in mammalian photoreceptors. *Cell* **102**, 175–187.

Masyuk, A. I., Masyuk, T. V., Splinter, P. L., Huang, B. Q., Stroope, A. J., and LaRusso, N. F. (2006). Cholangiocyte cilia detect changes in luminal fluid flow and transmit them into intracellular Ca^{2+} and cAMP signaling. *Gastroenterology* **131**, 911–920.

McFadden, G. I., and Wetherbee, R. (1985). Flagellar regeneration and associated scale deposition in *Pyramimonas gelidicola* (Prasinophyceae, Chlorophyta). *Protoplasma* **128**, 31–37.

McGrath, J., Somlo, S., Makova, S., Tian, X., and Brueckner, M. (2003). Two populations of node monocilia initiate left–right asymmetry in the mouse. *Cell* **114**, 61–73.

Menashe, I., and Lancet, D. (2006). Variations in the human olfactory receptor pathway. *Cell. Mol. Life Sci.* **63**, 1485–1493.

Mesland, D. A. M., Hoffman, J. L., Caligor, E., and Goodenough, U. W. (1980). Flagellar tip activation stimulated by membrane adhesions in *Chlamydomonas* gametes. *J. Cell Biol.* **84**, 599–617.

Mitchell, K. A. P., Gallagher, B. C., Szabo, G., and de S. Otero, A. (2004). NDP kinase moves into developing primary cilia. *Cell Motil. Cytoskeleton* **59**, 62–73.

Mombaerts, P. (1999). Molecular biology of odorant receptors in vertebrates. *Annu. Rev. Neurosci.* **22**, 487–509.

Moritz, O. L., Tam, B. M., Hurd, L. L., Peranen, J., Deretic, D., and Papermaster, D. S. (2001). Mutant rab8 impairs docking and fusion of rhodopsin-bearing post-Golgi membranes and causes cell death of transgenic *Xenopus* rods. *Mol. Biol. Cell* **12**, 2341–2351.

Mukhopadhyay, S., Lu, Y., Shaham, S., and Sengupta, P. (2008). Sensory signaling-dependent remodeling of olfactory cilia architecture in *C. elegans*. *Dev. Cell* **14**, 762–774.

Musgrave, A., DeWildt, P., van Etten, I., Pijst, H., Schholma, C., Kooyman, R., Homan, W., and van den Ende, H. (1986). Evidence for a functional membrane barrier in the transition zone between the flagellum and cell body of *Chlamydomonas eugametos* gametes. *Planta* **167**, 544–553.

Mykytyn, K., Mullins, R. F., Andrews, M., Chiang, A. P., Swiderski, R. E., Yang, B., Braun, T., Casavant, T., Stone, E. M., and Sheffield, V. C. (2004). Bardet-Biedl syndrome type 4 (BBS4)-null mice implicate Bbs4 in flagella formation but not global cilia assembly. *Proc. Natl Acad. Sci. USA* **101**, 8664–8669.

Nachury, M. V., Loktev, A. V., Zhang, Q., Westlake, C. J., Peranen, J., Merdes, A., Slusarski, D. C., Scheller, R. H., Bazan, J. F., Sheffield, V. C., and Jackson, P. K. (2007). A core complex of BBS proteins cooperates with the GTPase Rab8 to promote ciliary membrane biogenesis. *Cell* **129**, 1201–1213.

Nauli, S. M., Alenghat, F. J., Luo, Y., Williams, E., Vassilev, P., Li, X., Elia, A. E., Lu, W., Brown, E. M., Quinn, S. J., Ingber, D. E., and Zhou, J. (2003). Polycystins 1 and 2 mediate mechanosensation in the primary cilium of kidney cells. *Nat. Genet.* **33**, 129–137.

Nauli, S. M., Kawanabe, Y., Kaminski, K. K., Pearce, W. J., Ingber, D. E., and Zhou, J. (2008). Endothelial cilia are fluid shear sensors that regulate calcium signaling and nitric oxide production through polycystin-1. *Circulation* **117**, 1161–1171.

Nishimura, D. Y., Fath, M., Mullins, R. F., Searby, C., Andrews, M., Davis, R., Andorf, J. L., Mykytyn, K., Swiderski, R. E., Yang, B., Carmi, R., Stone, E. M., and Sheffield, V. C. (2004a). Bbs2-null mice have neurosensory deficits, a defect in social dominance, and retinopathy associated with mislocalization of rhodopsin. *Proc. Natl Acad. Sci. USA* **101**, 16588–16593.

Nishimura, T., Kato, K., Yamaguchi, T., Fukata, Y., Ohno, S., and Kaibuchi, K. (2004b). Role of the PAR-3–KIF3 complex in the establishment of neuronal polarity. *Nat. Cell Biol.* **6**, 328–334.

Nonaka, S., Tanaka, Y., Okada, Y., Takeda, S., Harada, A., Kanai, Y., Kido, M., and Hirokawa, N. (1998). Randomization of left–right asymmetry due to loss of nodal cilia generating leftward flow of extraembryonic fluid in mice lacking KIF3B motor protein. *Cell* **95**, 829–837.

Norwood, J. T., Hein, C. E., Halbert, S. A., and Anderson, R. G. W. (1978). Polycationic macromolecules inhibit cilia-mediated ovum transport in the rabbit oviduct. *Proc. Natl Acad. Sci. USA* **75**, 4413–4416.

Omori, Y., and Malicki, J. (2006). oko meduzy and related crumbs genes are determinants of apical cell features in the vertebrate embryo. *Curr. Biol.* **16**, 945–957.

Omori, Y., Zhao, C., Saras, A., Mukhopadhyay, S., Kim, W., Furukawa, T., Sengupta, P., Veraksa, A., and Malicki, J. (2008). Elipsa is an early determinant of ciliogenesis that links the IFT particle to membrane-associated small GTPase Rab8. *Nat. Cell Biol.* **10**, 437–444.

Oro, A. E. (2007). The primary cilia, a "Rab-id" transit system for hedgehog signaling. *Curr. Opin. Cell Biol.* **19**, 691–696.

Ostrowski, L. E., Blackburn, K., Radde, K. M., Moyer, M. B., Schlatzer, D. M., Moseley, A., and Boucher, R. C. (2002). A proteomic analysis of human cilia: Identification of novel components. *Mol. Cell Proteomics* **1**, 451–465.

Ou, G., Blacque, O. E., Snow, J. J., Leroux, M. R., and Scholey, J. M. (2005). Functional coordination of intraflagellar transport motors. *Nature* **436**, 583–587.

Overath, P., Stierhof, Y.-D., and Wiese, M. (1997). Endocytosis and secretion in trypanosomatid parasites—Tumultuous traffic in a pocket. *Trends Cell Biol.* **7**, 27–33.

Pan, J., and Snell, W. J. (2000). Signal transduction during fertilization in the unicellular green alga *Chlamydomonas*. *Curr. Opin. Microbiol.* **3**, 596–602.

Papermaster, D. S., Schneider, B. G., and Besharse, J. C. (1985). Vesicular transport of newly synthesized opsin from the Golgi apparatus toward the rod outer segment. Ultrastructural immunocytochemical and autoradiographic evidence in *Xenopus* retinas. *Invest. Ophthalmol. Vis. Sci.* **26,** 1386–1404.

Park, S. C., Yibchok-Anun, S., Cheng, H., Young, T. F., Thacker, E. L., Minion, F. C., Ross, R. F., and Hsu, W. H. (2002). *Mycoplasma hyopneumoniae* increases intracellular calcium release in porcine ciliated tracheal cells. *Infect. Immun.* **70,** 2502–2506.

Pazour, G. J., and Witman, G. B. (2003). The vertebrate primary cilium is a sensory organelle. *Curr. Opin. Cell Biol.* **15,** 105–110.

Pazour, G. J., Baker, S. A., Deane, J. A., Cole, D. G., Dickert, B. L., Rosenbaum, J. L., Witman, G. B., and Besharse, J. C. (2002a). The intraflagellar transport protein, IFT88, is essential for vertebrate photoreceptor assembly and maintenance. *J. Cell Biol.* **157,** 103–113.

Pazour, G. J., San Agustin, J. T., Follit, J. A., Rosenbaum, J. L., and Witman, G. B. (2002b). Polycystin-2 localizes to kidney cilia and the ciliary level is elevated in orpk mice with polycystic kidney disease. *Curr. Biol.* **12,** R378–R380.

Pazour, G. J., Agrin, N., Leszyk, J., and Witman, G. B. (2005). Proteomic analysis of a eukaryotic cilium. *J. Cell Biol.* **170,** 103–113.

Peters, K. R., Palade, G. E., Schneider, B. S., and Papermaster, D. S. (1983). Fine structure of a periciliary ridge complex of frog retinal rod cells revealed by ultrahigh resolution scanning electron microscopy. *J. Cell Biol.* **96,** 265–276.

Poelmann, R. E., Van den Heiden, K., Gittenberger-de Groot, A., and Hierck, B. P. (2008). Deciphering the endothelial shear stress sensor. *Circulation* **117,** 1124–1126.

Praetorius, H. A., and Spring, K. R. (2001). Bending the MDCK cell primary cilium increases intracellular calcium. *J. Membrane Biol.* **184,** 71–79.

Praetorius, H. A., and Spring, K. R. (2002). Removal of the MDCK cell primary cilium abolishes flow sensing. *J. Membrane Biol.* **191,** 69–76.

Preston, R. R., and Saimi, Y. (1990). Calcium ions and the regulation of motility in *Paramecium*. *In* Ciliary and Flagellar Membranes, (R. A. Bloodgood, Ed.), pp. 173–200. Plenum Press, New York.

Qin, H., Burnette, D. T., Bae, Y. K., Forscher, P., Barr, M. M., and Rosenbaum, J. L. (2005). Intraflagellar transport is required for the vectorial movement of TRPV channels in the ciliary membrane. *Curr. Biol.* **15,** 1695–1699.

Reisert, J., Lai, J., Yau, K. W., and Bradley, J. (2005). Mechanism of the excitatory Cl^-response in mouse olfactory receptor neurons. *Neuron* **45,** 553–561.

Reiter, J. F., and Mostov, K. (2006). Vesicle transport, cilium formation and membrane specialization: The origins of a sensory organelle. *Proc. Natl Acad. Sci. USA* **103,** 18383–18384.

Restrepo, D. (2005). The ins and outs of intracellular chloride in olfactory receptor neurons. *Neuron* **45,** 481–482.

Rodriguez-Boulan, E., Kreitzer, G., and Musch, A. (2005). Organization of vesicular trafficking in epithelia. *Nat. Rev. Mol. Cell Biol.* **6,** 233–247.

Rogalski, A. A., and Bouck, G. B. (1982). Flagellar surface antigens in *Euglena*: Immunological evidence for an external glycoprotein pool and its transfer to the regenerating flagellum. *J. Cell Biol.* **93,** 758–766.

Rosenbaum, J. L., and Witman, G. B. (2002). Intraflagellar transport. *Nat. Rev. Mol. Cell Biol.* **3,** 813–825.

Ross, A. J., May-Simera, H., Eichers, E. R., Kai, M., Hill, J., Jagger, D. J., Leitch, C. C., Chapple, J. P., Munro, P. M., Fisher, S., Tan, P. L., Phillips, H. M., et al. (2005). Disruption of Bardet-Biedl syndrome ciliary proteins perturbs planar cell polarity in vertebrates. *Nat. Genet.* **37,** 1135–1140.

Salathe, M. (2007). Regulation of mammalian ciliary beating. *Annu. Rev. Physiol.* **69,** 401–422.

Sato, T., Mushiake, S., Kato, Y., Sato, K., Sato, M., Takeda, N., Ozono, K., Miki, K., Kubo, Y., Tsuji, A., Harada, R., and Harada, A. (2007). The Rab8 GTPase regulates apical protein localization in intestinal cells. *Nature* **448,** 366–369.

Schafer, J. C., Winkelbauer, M. E., Williams, C. L., Haycraft, C. J., Desmond, R. A., and Yoder, B. K. (2006). IFTA-2 is a conserved cilia protein involved in pathways regulating longevity and dauer formation in *Caenorhabditis elegans*. *J. Cell Sci.* **119,** 4088–4100.

Schneider, L., Clement, C. A., Teilmann, S. C., Pazour, G. J., Hoffmann, E. K., Satir, P., and Christensen, S. T. (2005). PDGFRαα signaling is regulated through the primary cilium in fibroblasts. *Curr. Biol.* **15,** 1861–1866.

Schrick, J. J., Vogel, P., Abuin, A., Hampton, B., and Rice, D. S. (2006). ADP-ribosylation factor-like 3 is involved in kidney and photoreceptor development. *Am. J. Pathol.* **168,** 1288–1298.

Schwartz, E. A., Leonard, M. L., Bizios, R., and Bowser, S. S. (1997). Analysis and modeling of the primary cilium bending response to fluid shear. *Am. J. Physiol.* **272,** F132–F138.

Sfakianos, J., Togawa, A., Maday, S., Hull, M., Pypaert, M., Cantley, L., Toomre, D., and Mellman, I. (2007). Par3 functions in the biogenesis of the primary cilium in polarized epithelial cells. *J. Cell Biol.* **179,** 1133–1140.

Shi, G., Yau, K. W., Chen, J., and Kefalov, V. J. (2007). Signaling properties of a short-wave cone visual pigment and its role in phototransduction. *J. Neurosci.* **27,** 10084–10093.

Simons, K., and Toomre, D. (2000). Lipid rafts and signal transduction. *Nat. Rev. Mol. Cell Biol.* **1,** 31–39.

Smith, J. C., Northey, J. G., Garg, J., Pearlman, R. E., and Siu, K. W. (2005). Robust method for proteome analysis by MS/MS using an entire translated genome: Demonstration on the ciliome of *Tetrahymena thermophila*. *J. Proteome Res.* **4,** 909–919.

Stephens, R. E. (2001). Ciliary protein turnover continues in the presence of inhibitors of golgi function: Evidence for membrane protein pools and unconventional intracellular membrane dynamics. *J. Exp. Zool.* **289,** 335–349.

Stolc, V., Samanta, M. P., Tongprasit, W., and Marshall, W. F. (2005). Genome-wide transcriptional analysis of flagellar regeneration in *Chlamydomonas reinhardtii* identifies orthologs of ciliary disease genes. *Proc. Natl Acad. Sci. USA* **10,** 3703–3707.

Strunker, T., Weyand, I., Bonigk, W., Van, Q., Loogen, A., Brown, J. E., Kashikar, N., Hagen, V., Krause, E., and Kaupp, U. B. (2006). A K^+-selective cGMP-gated ion channel controls chemosensation of sperm. *Nat. Cell Biol.* **8,** 1149–1154.

Sun, Z., Amsterdam, A., Pazour, G. J., Cole, D. G., Miller, M. S., and Hopkins, N. (2004). A genetic screen in zebrafish identifies cilia genes as a principal cause of cystic kidney. *Development* **131,** 4085–4093.

Supp, D. M., Brueckner, M., Kuehn, M. R., Witte, D. P., Lowe, L. A., McGrath, J., Corrales, J., and Potter, S. S. (1999). Targeted deletion of the ATP binding domain of left–right dynein confirms its role in specifying development of left–right asymmetries. *Development* **126,** 5495–5504.

Tai, A. W., Chuang, J.-Z., Bode, C., Wolfrum, U., and Sung, C.-H. (1999). Rhodopsin's carboxy-terminal cytoplasmic tail acts as a membrane receptor for cytoplasmic dynein by binding to the dynein light chain Tctex-1. *Cell* **97,** 877–887.

Talbot, P., Shur, B. D., and Myles, D. G. (2003). Cell adhesion and fertilization: Steps in oocyte transport, sperm–zona pellucida interactions and sperm–egg fusion. *Biol. Reprod.* **68,** 1–9.

Tam, B. M., Moritz, O. L., Hurd, L. B., and Papermaster, D. S. (2000). Identification of an outer segment targeting signal in the COOH terminus of rhodopsin using transgenic *Xenopus laevis*. *J. Cell Biol.* **151,** 1369–1380.

Tanaka, Y., Okada, Y., and Hirokawa, N. (2005). FGF-induced vesicular release of Sonic hedgehog and retinoic acid in leftward nodal flow is critical for left–right determination. *Nature* **435,** 172–177.

Teilmann, S. C., and Christensen, S. T. (2005). Localization of the angiopoietin receptors Tie-1 and Tie-2 on the primary cilia in the female reproductive organs. *Cell Biol. Int.* **29,** 340–346.

Teilmann, S. C., Byskov, A. G., Pedersen, P. A., Wheatley, D. N., Pazour, G. J., and Christensen, S. T. (2005). Localization of transient receptor potential ion channels in primary and motile cilia of the female murine reproductive organs. *Mol. Reprod. Dev.* **71,** 444–452.

Torkko, J. M., Manninen, A., Schuck, S., and Simons, K. (2008). Depletion of apical transport proteins perturbs epithelial cyst formation and ciliogenesis. *J. Cell Sci.* **121,** 1193–1203.

Tull, D., Vince, J. E., Callaghan, J. M., Naderer, T., Spurck, T., Mcfadden, G. I., Currie, G., Ferguson, K., Bacic, A., and McConville, M. J. (2004). SMP-1, a member of a new family of small myristoylated proteins in kinetoplastid parasites, is targeted to the flagellum membrane in *Leishmania*. *Mol. Biol. Cell* **15,** 4775–4786.

Tuomanen, E. (1990). The surface of mammalian respiratory cilia. Interactions between cilia and respiratory pathogens. *In* Ciliary and Flagellar Membranes, (R. A. Bloodgood, Ed.), pp. 363–388. Plenum Press, New York.

Tuomanen, E. I., and Hendley, J. O. (1983). Adherence of *Bordetella pertussis* to human respiratory epithelial cells. *J. Infect. Dis.* **148,** 125–130.

Tyler, S. (1973). An adhesive function for modified cilia in an interstitial turbellarian. *Acta Zool.* **54,** 139–151.

Van den Ende, H., Musgrave, A., and Kis, F. M. (1990). The role of flagella in the sexual reproduction of *Chlamydomonas* gametes. *In* Ciliary and Flagellar Membranes (R. A. Bloodgood, Ed.), pp. 129–147. Plenum Press, New York.

Van der Heiden, K., Groenendijk, B. C., Hierck, B. P., Hogers, B., Koerten, H. K., Mommaas, A. M., Gittenberger-de Groot, A. C., and Poelmann, R. E. (2006). Monocilia on chicken embryonic endocardium in low shear stress areas. *Dev. Dyn.* **235,** 19–28.

Van der Heiden, K., Hierck, B. P., Krams, R., De Crom, R., Cheng, C., Baiker, M., Pourquie, M. J. B. M., Alkemade, F. E., DeRuiter, M. C., Gittenberger-de Groot, A. C., and Poelmann, R. E. (2008). Endothelial primary cilia in areas of disturbed flow are at the base of atherosclerosis. *Atherosclerosis* **196,** 542–550.

van Meer, G., and Simons, K. (1988). Lipid polarity and sorting in epithelial cells. *J. Cell. Biochem.* **36,** 51–58.

Veira, O. V., Gaus, K., Verkade, P., Fullekrug, J., Vaz, W. L. C., and Simons, K. (2006). FAPP2, cilium formation and compartmentalization of the apical membrane in polarized Madin-Darby canine kidney (MDCK) cells. *Proc. Natl Acad. Sci. USA* **103,** 18556–18561.

Vickerman, K., and Tetley, L. (1990). Flagellar surfaces of parasitic protozoa and their role in attachment. *In* Ciliary and Flagellar Membranes, (R. A. Bloodgood, Ed.), pp. 267–304. Plenum Press, New York.

Vitale, G., Rybin, V., Christoforidis, S., Thornqvist, P., McCaffrey, M., Stenmark, H., and Zerial, M. (1998). Distinct Rab-binding domains mediate the interaction of Rabaptin-5 with GTP-bound Rab4 and Rab5. *Embo J.* **17,** 1941–1951.

Ward, S., Thompson, N., White, J. G., and Brenner, S. (1975). Electron microscopic reconstruction of the anterior anatomy of the nematode *Caenorhabditis elegans*. *J. Comp. Neurol.* **160,** 313–337.

Watnick, T. J., Jin, Y., Matunis, E., Kernan, M. J., and Montell, C. (2003). A flagellar polycystin-2 homolog required for male fertility in Drosophila. *Curr. Biol.* **13,** 2179–2184.

Wetherbee, R., and Andersen, R. A. (1992). Flagella of a chrysophycean alga play an active role in prey capture and selection. Direct observations on Epipyxis pulchra using image enhanced video microscopy. *Protoplasma* **166,** 1–7.

Witman, G. B. (1993). *Chlamydomonas* phototaxis. *Trends Cell Biol.* **3,** 403–408.

Xu, C., Rossetti, S., Jiang, L., Harris, P. C., Brown-Glaberman, U., Wandinger-Ness, A., Bacallao, R., and Alper, S. L. (2007). Human ADPKD primary cyst epithelial cells with a novel, single codon deletion in the PKD1 gene exhibit defective ciliary polycystin localization and loss of flow-induced Ca^{2+} signaling. *Am. J. Physiol. Renal Physiol.* **292,** F930–F945.

Yoder, B. K., Hou, X., and Guay-Woodford, L. M. (2002). The polycystic kidney disease proteins, polycystin-1, polycystin-2, polaris, and cystin, are co-localized in renal cilia. *J. Am. Soc. Nephrol.* **13,** 2508–2516.

Yoshimura, S., Egerer, J., Fuchs, E., Haas, A. K., and Barr, F. A. (2007). Functional dissection of Rab GTPases involved in primary cilium formation. *J. Cell Biol.* **178,** 363–369.

Zhang, X. M., Ramalho-Santos, M., and McMahon, A. P. (2001). Smoothened mutants reveal redundant roles for Shh and Ihh signaling including regulation of L/R asymmetry by the mouse node. *Cell* **105,** 781–792.

Zhou, C., Cunningham, L., Marcus, A. I., Li, Y., and Kahn, R. A. (2006). Arl2 and Arl3 regulate different microtubule-dependent processes. *Mol. Biol. Cell* **17,** 2476–2487.

Zimmerman, H. (1898). Beitrage zur Kenntniss einiger Drusen und epithelien. *Arch. Mikrosk. Anat.* **52,** 552–706.

CHAPTER SIX

Cilia: Multifunctional Organelles at the Center of Vertebrate Left–Right Asymmetry

Basudha Basu* and Martina Brueckner*,†

Contents

1. Introduction 152
2. Vertebrate LR Asymmetry Is Chiral Asymmetry 154
3. The Structure and Distribution of Cilia in the Vertebrate Embryo 155
4. Motile Primary Cilia in LR Development 156
5. Mouse and Human Mutations Affecting Cilia Have a Wide-Ranging Effect on Left–Right and Cardiac Development 158
6. A Structure with Prominent Primary Cilia is Essential to the Development of LR Asymmetry in Most Vertebrates 160
7. Cilia at the LR Organizer Generate Directional Fluid Flow 162
8. A Conserved Asymmetric Calcium Signal Is Found at the LR Organizer 164
9. Do Cilia Lie at the Root of All Vertebrate LR Asymmetry, and What Leads from Asymmetric Calcium to Asymmetric Organogenesis? 168
References 169

Abstract

Cilia establish the vertebrate left–right (LR) axis and are integral to the development and function of the kidney, liver, and brain. Left–right asymmetry is established in the ciliated ventral node cells of the mouse. The chiral structure of the cilium provides a reference asymmetry to impose handed LR asymmetric development on the bilaterally symmetric vertebrate embryo. A ciliary mechanism of LR development is evolutionarily conserved, as ciliated organs essential to LR axis formation, called LR organizers, are found in other vertebrates, including rabbit, fish, and *Xenopus*. Mice with mutations affecting ciliary biogenesis, motility, or sensory function have abnormal LR development and

* Department of Pediatrics, Yale University School of Medicine, New Haven, Connecticut
† Department of Genetics, Yale University School of Medicine, New Haven, Connecticut

abnormal development of the heart. The axonemal dynein heavy chain left–right dynein (lrd) localizes to the LR organizer and drives counterclockwise movement of node primary cilia. Node primary cilia are an admixture of 9 + 2 and 9 + 0 cilia. Mutations in lrd result in structurally normal, immotile node monocilia. In the mouse, coordinated, directional beating of motile node monocilia at the neural fold stage generates leftward flow of extraembryonic fluid surrounding the node (nodal flow). Nodal flow triggers a rise in intracellular calcium in cells at the left side of the node. The perinodal asymmetric rise in intracellular calcium generated by nodal flow subsequently leads to asymmetric gene expression and morphogenesis.

1. Introduction

The vertebrate body plan is outwardly symmetrical about the midline, the left side being a mirror image of the right. This outward symmetry belies elaborate internal left–right (LR) asymmetries. For example, in all normal vertebrates, the heart tilts to the left, the liver is on the right, and the stomach and spleen are on the left. Positioning of organs along the left–right axis can be divided into three broad classes: situs solitus (SS), where all organs are positioned normally; situs inversus (SI), where there is mirror image reversal of all organs; and heterotaxy (Htx), where there is any positioning of organs along the LR axis differing from situs solitus and situs inversus (Fig. 6.1A). In humans, pure SI is found in 1:8500 in the general population and is usually not associated with intracardiac defects. In contrast, heterotaxy has a high degree of association with intracardiac defects. It has a reported incidence of 1:10,000 and is associated with at least 3% of cases of human congenital heart disease (Zhu et al., 2006).

Molecular asymmetry is established well before there is visible embryonic asymmetry, and analysis of the hierarchy of molecular asymmetries in mouse, chick, and *Xenopus* has permitted the elucidation of a molecular pathway of left–right development (Levin et al., 1995). In mouse, the first molecular asymmetry arises when the *nodal* gene, initially expressed throughout the node, becomes restricted to the left side of the node (Collignon et al., 1996; Lowe et al., 1996, 2001). In the presence of the *nodal* cofactor *cryptic*, the limited domain of asymmetric *nodal* expression at the left of the node then expands to the left lateral plate mesoderm (Yan et al., 1999), where *lefty-2* is also expressed. *lefty-1* is expressed in the left side of the neural floor plate where it may function as part of a midline barrier (Collignon et al., 1996; Lowe et al., 1996; Meno et al., 1996, 1997, 1998). *Sonic hedgehog* and *Indian hedgehog* acting through *smoothened* at the midline

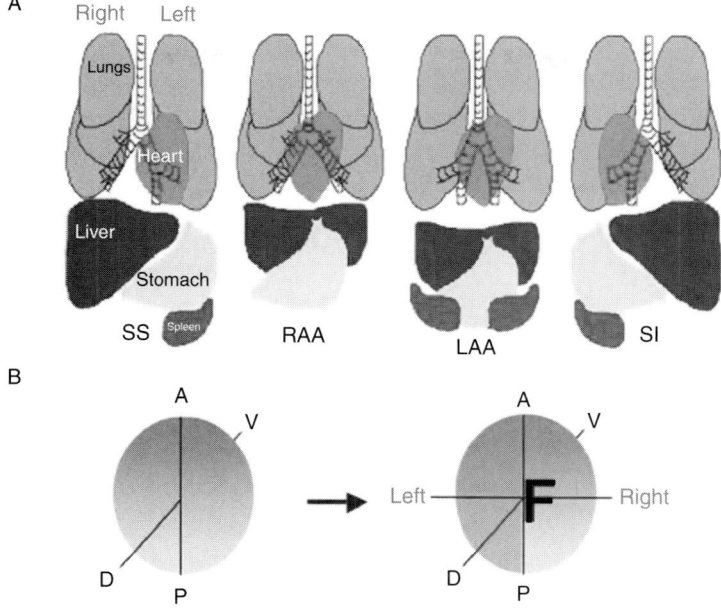

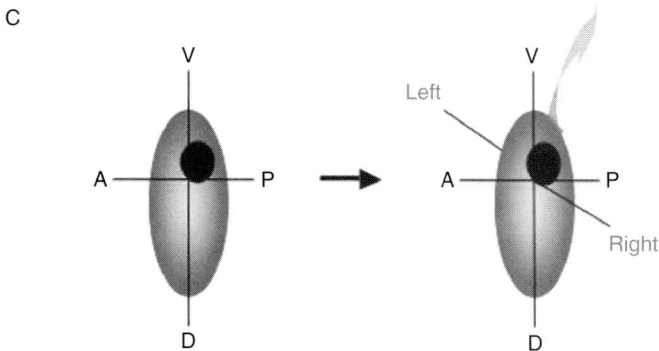

(Adapted from: Brown and Wolpert, development 1990)

Figure 6.1 (A) The anatomic spectrum of organ laterality (SS, situs solitus; RAA, right atrial isomerism). The liver is midline, there are two eparterial bronchi, the position of the stomach and cardiac apex is indeterminate and there is asplenia (LAA, left atrial isomerism). The liver is midline, there are two hyparterial bronchi, the position of the stomach and cardiac apex is indeterminate and there are multiple spleens (SI, situs inversus). (B) Orientation of the hypothetical "F" chiral molecule or macromolecular structure along the AP and DV axes creates obligate chiral asymmetry. (C) The cilium shown as a green arrow is the chiral macromolecular structure that creates chiral asymmetry of the cells at the mouse node. (See Color Insert.)

may be required to maintain these expression patterns. The homeobox genes *PitX2* (Yoshioka *et al.*, 1998) and *NKX3.2* (Schneider *et al.*, 1999) have opposite asymmetric expression patterns in the left and right lateral plate mesoderm, respectively, which may regulate asymmetric expression of downstream effector genes. Overall, asymmetric gene expression is first observed in cells at the left margin of the node, and in wild type mice, the early asymmetry of *nodal* expression at the left edge of the node, is associated with normal LR patterning (Brennan *et al.*, 2002). LR asymmetry in mRNA expression of *nodal* and other lateralized molecular markers is then observed in progressive lateral and caudal domains. It is important to note, however, that asymmetric expression of a signaling molecule like *nodal* cannot arise *de novo*, and requires pre-existing LR positional information.

2. Vertebrate LR Asymmetry Is Chiral Asymmetry

The LR axis is unique in that it is defined with respect to the anteroposterior (AP) and dorsoventral (DV) axes. Therefore, the embryo must have mechanisms to both create asymmetry and to consistently align the asymmetry with the existing AP and DV axes. Wilhelmi (1921) originally proposed that the organism has an underlying mechanism that generates asymmetry. If merely having asymmetry were the full extent of LR development, all organisms within a population would be asymmetric with distinct left and right sides, but the orientation of the LR asymmetry would be random relative to the AP and DV axes: in essence half of the population would manifest situs solitus while the other half would be situs inversus. Since this is not what is observed in vertebrates, a second mechanism must exist which consistently aligns LR to AP and DV. Brown and Wolpert (1990) hypothesized that this asymmetry is biased in a consistent direction by the presence of a handed asymmetric molecule or macromolecular structure, which they represented by the letter "F" (Fig. 6.1B), which can align with the AP and DV axes. The requirement for a pre-existing "reference asymmetry" is a unique feature of LR asymmetry that is not essential for the development of AP and DV asymmetry. Once the reference asymmetry is aligned within a single cell, that cell has handed LR asymmetry. The local asymmetry can then be communicated to the remainder of the developing embryo, and here LR development may utilize the inter- and intracellular signaling pathways already in use by other developmental processes. So the question of what the "F-molecule" or the fundamental molecular asymmetry is lies at the center of understanding the mechanism of LR development.

3. THE STRUCTURE AND DISTRIBUTION OF CILIA IN THE VERTEBRATE EMBRYO

The first hint that cilia are involved in the development of LR asymmetry came from studies of humans with Kartagener's triad. The pathologist Manes Kartagener noted the association between bronchiectasis, sinusitis, and situs inversus in 1933 (Kartagener, 1933). Forty years later, Bjorn Afzelius noticed that the sperm of male patients with Kartagener syndrome were normal appearing but immotile, which lead him to the original discovery that the ciliary axoneme is abnormal in patients with Kartagener syndrome, and that ∼50% of patients with immotile cilia had situs inversus (Afzelius, 1976, 1982). Primary ciliary dyskinesia (PCD) is now used to describe sinopulmonary disease caused by ciliary abnormalities irrespective of situs. Recent phenotypic evaluation of a large number of human patients with PCD identified 47.7% of PCD patients had situs inversus, and 6.5% had heterotaxy and the complex congenital heart disease usually associated with heterotaxy (Kennedy *et al.*, 2007). It was immediately apparent that immotile cilia could lead to sinopulmonary disease and male infertility, however, how ciliary defects lead to situs inversus or heterotaxy was initially more puzzling. Afzelius had proposed that "on the various epithelia of an embryo there are cilia that have determined positions and a fixed beat direction," and that "ciliary beating in normal embryos is assumed to be instrumental in pushing the heart to the left side" (Afzelius, 1976). Although this was an appealing hypothesis, it was long assumed that ciliary function was limited to ciliated epithelia, and that embryos did not have ciliated epithelia before the emergence of the choroid plexus in the developing brain at mouse embryonic day 14.5. With the discovery that the cilium is an essential multifunctional organelle found on virtually every cell type, including most/all cells in the embryo, including embryonic stem cells, it became apparent that cilia are indeed central to many developmental processes including the development of LR asymmetry (Fig. 6.1C).

Cilia have traditionally been classified as 9 + 0 "monocilia" and 9 + 2 "classic cilia" (Fig. 6.2A). This classification is somewhat misleading, as there is significant overlap between the structure and function of 9 + 0 and 9 + 2 cilia. The term "primary cilium" is a better description of the cilia observed in the embryo, specifically those involved in development of LR asymmetry (Poole *et al.*, 1985; reviewed in Davis *et al.*, 2006; Pazour and Witman, 2003). A single primary cilium is found on almost all cell types; primary cilia arise from the centriole of the cell during G1 and are resorbed prior to cell division. The chiral ciliary axoneme, which is composed of 9 outer microtubule doublets consisting of a complete A-tubule fused to an incomplete B-tubule arising from the pinwheel-like template provided by the basal body, is the "skeleton" of all

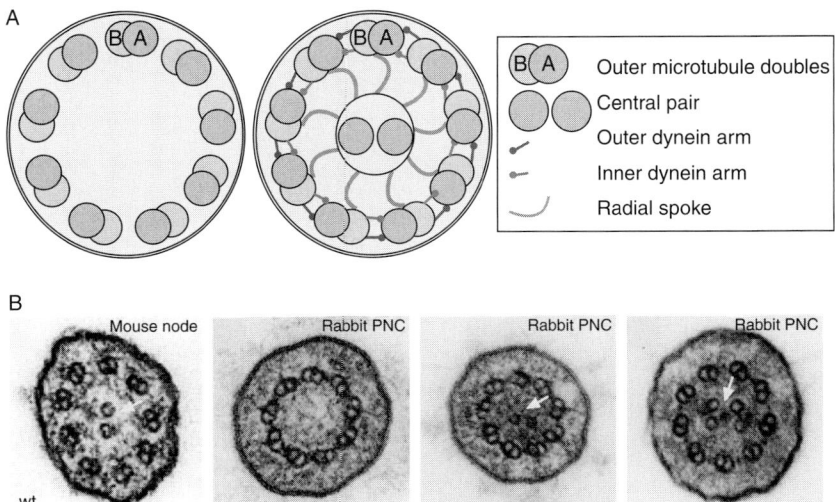

Figure 6.2 (A) Outline of the ciliary axoneme. The axoneme shown at the left can be adapted to a variety of purposes, including the full array of motility components diagrammed on the right. (B) TEM of cilia at the mouse node (Caspary et al., 2007) and rabbit PNC (Feistel and Blum, 2006) showing a wide spectrum of ciliary architecture including 9 + 2, 9 + 0, and 9 + 4 microtubule configurations. Red arrow indicates the location of the central pair(s) of microtubules. (See Color Insert.)

cilia. The contents of primary cilia are extremely varied. By displaying a variety of receptors, these cilia are adapted to function as "cellular antennae," reaching out beyond the cell surface to capture signals ranging from light in the photoreceptor in the retina, to mechanical and chemical stimuli in the ciliated cells found at the anterior tip of *Caenorhabditis elegans*. It is highly likely that all primary cilia can function as sensors. In addition, although most primary cilia are immotile, some carry dynein motors and beat (Pazour and Witman, 2003). The cilia that form the focus of this review, specifically those found at the node and in the heart, are primary cilia and represent a mixture of motile and immotile primary cilia.

4. Motile Primary Cilia in LR Development

Electron microscopy of node cilia initially showed cilia with a 9 + 0 axoneme completely lacking dynein arms, radial spokes, or central pairs (Bellomo et al., 1996; Nonaka et al., 1998). This observation would predict that node cilia be immotile, as dynein motors drive ciliary motility, and radial spokes and central pair microtubules are thought to be essential in regulating ciliary motility. However, at least two axonemal outer arm dyneins, left–right dynein (lrd), and axonemal dynein heavy chain 5 (DnaHC5) had been localized to the node cilia (Olbrich *et al.*, 2002; Supp *et al.*, 1997, 1999). Mutation in lrd (Supp *et al.*, 1997, 1999) or Mdnah5 (Ibanez-Tallon *et al.*, 2002; Olbrich *et al.*, 2002) produces random development of left–right asymmetry and associated congenital heart defects in mice that resemble those observed in humans with abnormal left–right development (Hummel and Chapman, 1959; Layton, 1976). The normal asymmetric expression patterns of genes such as *nodal* and *lefty* become randomized in $lrd^{-/-}$ mutant mice, indicating that the *lrd* functions upstream of these genes (Collignon *et al.*, 1996; Lowe *et al.*, 1996). Dyneins are a family of minus-end-directed microtubule-based motors which are commonly classified as either cytoplasmic or axonemal based on specific sequence characteristics of the component heavy chains (Asai *et al.*, 1994). These proteins function as large multisubunit complexes made of up to three heavy chains in addition to intermediate chains (ICs) and light chains (LCs) (Holzbaur and Vallee, 1994). The dynein heavy chain genes are large proteins (MW $\sim$ 500 kDa) and are encoded by mRNAs ranging from 14 to 18 kb. The heavy chain consists of two major structural domains. The carboxy two thirds of the protein comprises six evenly spaced, highly conserved P-loop consensus sequence elements that correspond to sites of Mg^{2+} ATPase activity, and a single coiled/coil domain that has microtubule-binding ability (Gee *et al.*, 1997). The amino terminal one-third of the protein is more unique to each dynein and contains the region that binds to light and medium chains. To date, at least 15 distinct dynein heavy chain genes have been identified in vertebrates, of which two have sequences identifying them as components of cytoplasmic dyneins, and the remainder are components of multiple outer and inner arm axonemal dynein isoforms (Tanaka *et al.*, 1995; Vaughan *et al.*, 1996). The specialized roles of the many diverse axonemal dynein genes, including lrd, remain unclear. Consistent with the findings of axonemal dyneins in node primary cilia, direct video microscopy of node and KV cilia indeed shows them to be motile. The direction of beating is conserved between species, but there is considerable variability in the beat frequency: overall, the frequencies are slightly lower than those observed in ciliated epithelia, 10 Hz in mouse, 7 Hz in rabbit, and 43 Hz in medaka fish (Nonaka *et al.*, 1998; Okada *et al.*, 2005; Sulik *et al.*, 1994; Supp *et al.*, 1999).

Ciliary motility has been most extensively characterized in the unicellular alga, *Chlamydomonas*. Effective ciliary bending movement results from controlled temporal activation of inner and outer arm dyneins, which slides adjacent outer microtubule doublets against each other. The stereotypical movements observed in *Chlamydomonas* flagella, and the 9 + 2 cilia found on ciliated epithelia such as the airway and choroid plexus consist of an asymmetric effective and recovery stroke. It appears that the central pair microtubules and the radial spokes are important regulators of ciliary bending, and most mutations resulting in abnormal central pair structures result in abnormal or absent ciliary motility (Lechtreck *et al.*, 2008). However, both *Chlamydomonas* flagella in the presence of ATP (Yagi and Kamiya, 2000), and the flagellum of the eel are able to beat without functioning central pair and radial spoke structures (Gibbons *et al.*, 1983), suggesting that CP/RS structures are not absolutely required for any kind of ciliary movement *per se*. More detailed analysis of node primary cilia indicates that there are indeed a variety of ultrastructural arrangements found in node primary cilia (Fig. 6.2B). Recent TEM of mouse node cilia show a mixture of cilia with or without central pairs (Caspary *et al.*, 2007). Analysis of primary cilia in the rabbit notochordal plate showed three types of axonemes, including 9 + 0, 9 + 2, and an unusual axoneme with a 9 + 4 arrangement of microtubules (Feistel and Blum, 2006). Outer dynein arms were found on all three types of axonemes, suggesting that at least some of the stereotypical arrangement of microtubules, motors, and structural proteins found in ciliated epithelial cilia can also be found in primary cilia. It remains unclear whether some of the motile node primary cilia do indeed have a 9 + 0 architecture, or whether the finding of motile primary cilia lacking a central pair is due to the formidable technical difficulties in obtaining optimum fixation of embryo cilia. Initially, node primary cilia were thought to be a rotating stiff cilium, unlike the asymmetric two-phase dynamics observed in traditional 9 + 2 cilia and flagellae (Nonaka *et al.*, 1998). Further videomicroscopic analysis of mouse node cilia demonstrated a two-phase but nonplanar beating pattern, with a leftward power stroke and a rightward, slower recovery stroke (Buceta *et al.*, 2005). Overall, node primary cilia are motile, and the composition and the pattern of ciliary motility may more closely resemble that of classic 9 + 2 cilia than was initially appreciated.

5. Mouse and Human Mutations Affecting Cilia Have a Wide-Ranging Effect on Left–Right and Cardiac Development

Additional evidence pointing to multiple roles for cilia in LR and cardiac development is provided by at least 13 independent mouse mutations affecting ciliary biogenesis and function. These ciliary defects can be classified into three subgroups: first, mutations affecting ciliary motility; second, mutations affecting ciliary sensory function; and third, mutations affecting ciliary biogenesis (Table 6.1). The first class of ciliary mutants (blue in Table 6.1) is exemplified by mutations in the dynein heavy chain genes lrd (Supp *et al.*, 1999) and dynein heavy chain 5 (DNAH5) (Ibanez-Tallon *et al.*, 2002), which are characterized by structurally normal, but immotile cilia. The mice have randomization of LR asymmetry, and some have structural cardiac defects associated with heterotaxy and embryonic

Table 6.1 Classification of constitutive ciliary mutations in mouse

Mutation	Class/ciliary function	Lethality	LR phenotype	Cardiac phenotype
Lrd	1/motility	40% long-term survival	Random loop	40% normal, some AV canal and other CHD
DNAH5	1/motility	2–3 weeks postnatal	Random loop	Normal
mD2LIC, Dnchc2	2/ciliary biogenesis	E10.5	Random loop	90% pericardial effusion, CHF
KIF3A, KIF3B	2/ciliary biogenesis	E10.5	Random loop	100% pericardial effusion, myocardial thinning
IFT88	2/ciliary biogenesis	E11.5	Random loop	100% pericardial effusion, death due to heart failure
IFT172	2/ciliary biogenesis	E12.5 (hypomorphic allele)	Random loop	Pericardial effusion, death due to heart failure
IFT57	2/ciliary biogenesis	E10.5	Random loop	Pericardial effusion
Arl13B	2/ciliary biogenesis	E13.5	Unknown	Unknown
Ofd1	2/ciliary biogenesis	E12.5	Random loop	100% pericardial effusion, 50% myocardial thinning
Pkd2	3/ciliary mechanosensation	E14.5–16.5	Right atrial isomerism	Pericardial effusion, AV canal

lethality. The situs of the surviving mice is either complete situs solitus or complete situs inversus.

The most severely affected mice are those in the second class of ciliary mutations (red in Table 6.1), represented by mutations affecting ciliary biogenesis. Many mutations in this class affect intraflagellar transport (IFT): Heterotrimeric kinesin (Kif3A, Kif3B) (Marszalek et al., 1999; Nonaka et al., 1998), intraflagellar transport components (IFT88, IFT172) and cytoplasmic dynein-2 (Huangfu et al., 2003; Murcia et al., 2000; Rana et al., 2004). Here, cilia are either completely absent, or structurally abnormal. When the heart tube loops, looping is in a random direction. Limited analysis of normally lateralized markers such as *lefty-2* in Kif3B$^{-/-}$ (Nonaka et al., 1998) and IFT88$^{-/-}$ (Murcia et al., 2000) embryos shows bilateral, albeit globally reduced levels of expression. In contrast to mutations affecting ciliary function without affecting ciliary structure, these mutations result in strikingly abnormal hearts with very thin ventricular walls and massive pericardial effusion in addition to random heart looping caused by LR defects. Unlike ciliary function defects, all defects in ciliary structure produce embryonic lethality due to cardiac failure by e13.5 (Slough et al., 2008).

The third class of ciliary mutations (green in Table 6.1) is represented by mice with mutations in polycystin-2 (Boulter et al., 2001; Wu et al., 2000). In Pkd2$^{-/-}$ mice, LR development is abnormal, with 100% of embryos having right atrial isomerism. Expression of *PitX2* is bilateral and posterior, and *nodal* is absent from the LPM (Pennekamp et al., 2002). Unlike mutations affecting ciliary motility, all Pkd2$^{-/-}$ embryos have intracardiac defects with a preponderance of endocardial–cushion defects, and there are no surviving homozygous embryos beyond birth, with most embryos dying by e15.5 with pericardial effusion and edema.

6. A STRUCTURE WITH PROMINENT PRIMARY CILIA IS ESSENTIAL TO THE DEVELOPMENT OF LR ASYMMETRY IN MOST VERTEBRATES

Multiple independent lines of evidence in chick, zebrafish, *Xenopus*, rabbit, and mouse point to a group of cells at Hensen's node (Pagan-Westphal and Tabin, 1998), Kupffer's vesicle (Essner et al., 2005; Kramer-Zucker et al., 2005), the gastrocoel roof plate (GRP) (Schweickert et al., 2007), the posterior notochord (PNC) (Blum et al., 2007), and the mouse node (Davidson et al., 1999), respectively, which are essential to the development of LR asymmetry. These structures all express homologs of the growth factor *nodal*, members of the X-box transcription factor family RFX and genes coding for axonemal motor proteins *Dnah5* and *lrd* (Blum et al., 2007; Essner et al., 2002, 2005). The other striking common feature of these structures is that they have unusually long, motile primary cilia. Thus it appears that most (?all) vertebrates have a "LR organizer."

The best characterized of the putative LR organizers is the mouse node. It is a group of 150-250 cells located at the ventral tip of the mouse embryo, first becoming visible at e7.5 and disappearing by e9.0. It is composed of two distinct cell layers, the dorsal and ventral node, that are in direct contact at the anterior end of the primitive streak, but separated by a basement membrane everywhere else (Bellomo et al., 1996; Sulik et al., 1994). The dorsal node is composed of neuroectodermal, highly proliferative cells facing the amniotic cavity. The mature ventral node consists of closely packed columnar pit cells surrounded by a group of squamous crown cells. The pit cells at the anterior end of the node are contiguous with the PNC, whereas the crown cells are linked to the anterior end of the primitive streak. It is the ventral node cells that are able to induce a secondary axis when transplanted to other regions of the mouse embryo (Beddington, 1994). Node morphogenesis is dependent on the homeobox gene Noto (Beckers et al., 2007), and time-lapse imaging of Noto-GFP-tagged cells shows that node cells arise from underneath the endoderm (Yamanaka et al., 2007). Ventral node cells, in contrast to dorsal node cells, are mitotically relatively silent (Bellomo et al., 1996; Sulik et al., 1994). Node pit cells give rise to the trunk notochord by convergent-extension movements, while node crown cells actively migrate to the posterior of the embryo to give rise to the tail notochord (Yamanaka et al., 2007). Each pit cell carries a long primary cilium extending into the yolk sac cavity, which is covered by Reichert's membrane, creating a small fluid-filled cavity. Observation of the cilia on the node pit cells by video microscopy shows that a significant number of them are motile during a short time window extending from the 0 to 6 somite stage (Hadjantonakis et al., 2008; McGrath et al., 2003; Nonaka et al., 1998; Okada et al., 2005; Supp et al., 1999).

The LR organizers in other species are distinct from the mouse node in their timing and geometry, but share common gene expression patterns, and all except the avian Hensen's node have clearly demonstrated motile primary cilia. In the rabbit the node itself does not have cilia, but the posterior segment of the notochord has many of the features of the mouse node: it consists of densely packed cells with prominent motile cilia, flanked by lateral domains of nodal expression, thus this is the likely "LR organizer" region in the rabbit (Blum et al., 2007). In *Xenopus*, a group of cells derived from mesoderm lies beneath the notochord of the frog gastrula. These cells are arranged in a triangular geometry reminiscent of the rabbit PNC and mouse node, and they too carry prominent motile primary cilia extending into the gastrocoel (Schweickert et al., 2007). The most distinct LR organizer structure is found in fish. Here, a group of ~24 cells called dorsal forerunner cells (DFCs) travel at the leading edge of the embryonic shield (organizer) and migrate into the embryo at the end of gastrulation to form a spherical structure called Kupffer's vesicle located ventral to the notochord in the tailbud (Cooper and D'Amico, 1996). KV cells express genes such as lrd related1 (lrdr1) characteristic of the LR organizer in other organisms, and disruption of lrd1 in KV via antisense morpholinos targeted to KV perturb LR development analogous to what is observed in mice with lrd mutations (Essner et al., 2005). KV cells have long, motile 9 + 2 cilia. They are distributed asymmetrically within the spherical KV, being most abundant in the dorsal anterior portion of KV (Kreiling et al., 2007). Thus, the dorsal surface of KV ends up with a geometry that resembles the mouse node, with the ventral surface creating an enclosed chamber similar to that generated by the covering of the mouse node by Reichert's membrane.

It appears that vertebrates share a unique LR organizer which is temporally and spatially linked to the "true" organizer: the node in mouse, the PNC in rabbit, the GRP in *Xenopus*, and Kupffer's vesicle in fish. All of these structures have prominent motile primary cilia arranged in a triangular geometry, and all are essential to the robust development of handed LR asymmetry. It is remarkable that such a ciliated LR organizer has not yet been clearly identified in avian embryos. Cilia and expression of the chick homolog of lrd has been observed at Hensen's node (Essner et al., 2002), however, no avian LR organizer with motile cilia has thus far been found. It is possible that the avian lineage has diverged, and created an alternate mechanism for the orientation of handed asymmetry; however, it is just as possible that ciliary motility in avian embryos has not been observed due to imaging constraints.

7. CILIA AT THE LR ORGANIZER GENERATE DIRECTIONAL FLUID FLOW

By positioning a cilium, which is inherently asymmetric, on the ventral surface of cell at the LR organizer, each organizer cell has acquired LR positional information. This local cellular asymmetry is conveyed to the embryo via a novel mechanism called "nodal flow." Nodal flow was first demonstrated by elegant experiments studying mice with targeted mutation of heterotrimeric kinesin B (Kif3B) (Nonaka et al., 1998), and subsequent analysis of mice with mutation in Kif3A (Marszalek et al., 1999; Takeda et al., 1999) resulted in similar phenotypes including midgestation lethality and multiple severe developmental abnormalities. Kinesins are a large family of predominantly plus-end-directed microtubule motors that are required for diverse intracellular transport functions. Kif3A and Kif3B proteins form a heterotrimeric complex with an associated protein KAP3 (Kondo et al., 1994; Yamazaki et al., 1995, 1996). Heterotrimeric kinesin homologues are found ubiquitously in many species, including mouse, *C. elegans*

Chlamydomonas, and sea urchin. In sea urchin embryos, disruption of heterotrimeric kinesin function results in a failure to assemble motile ciliary axonemes (Morris and Scholey, 1997). Studies in *Chlamydomonas* and *C. elegans* indicate that this is due to a requirement for heterotrimeric kinesin in intraflagellar transport (IFT); when IFT is defective existing flagella resorb, and no new flagella can be assembled (Cole et al., 1998). Notably, both Kif3A and Kif3B mutants display randomization of cardiac looping and abnormal LR expression patterns of genes such as *lefty-2*, implicating heterotrimeric kinesin in LR development. A striking feature of *kif3A* and *kif3B* homozygous mutants is that they have no cilia, and specifically no node primary cilia. Thus, heterotrimeric kinesin is essential both for assembly of node cilia and normal LR development.

In wild-type mouse embryos, movement of node cilia generates a brisk leftward "nodal flow" of the extraembryonic fluid between the surface of the node and Reichert's membrane. In contrast, mice with absent node cilia due to mutations in Kif3A or Kif3B, or with paralyzed node cilia due to mutations in lrd have no nodal flow (Nonaka et al., 1998; Okada et al., 1999; Supp et al., 1999). Artificially applied nodal flow can rescue LR development in mouse embryos with paralyzed node cilia due to lrd mutation (Nonaka et al., 2002). This proves that nodal flow is essential for LR development, instead of merely being a marker of normal node development. Similar flow of fluid at the LR organizer was observed in *Xenopus* (Schweickert et al., 2007), rabbit (Okada et al., 2005), and zebrafish (Essner et al., 2005; Kramer-Zucker et al., 2005). Elimination of flow at the GRP in *Xenopus* by increasing the viscosity of the fluid in the gastrocoel results in random LR development, as does elimination of directional flow in the zebrafish KV by morpholino antisense knockdown of genes required for ciliary assembly and/or motility.

One of the most puzzling aspects of nodal flow is how an array of relatively few, widely spaced cilia can cooperate to produce vigorous (5–50 $\mu M/s$) laminar flow of the surrounding fluid. The clockwise direction of ciliary movement is determined by the inherent asymmetry of the cilium itself, which drives the order of motor activity and microtubule sliding. However, clockwise movement of rigid cilia positioned at the center of the ventral surface of the node cells would only create a group of clockwise vortices, which when summed up over the entire node do not produce laminar leftward flow. Nodal flow is determined by the fluid properties of the surrounding extraembryonic fluid and the mechanical properties of the node cilia. The fluid properties at the node indicate that nodal flow is happening in an environment of low Reynolds number, where viscosity dominates over inertia (Buceta et al., 2005; Cartwright et al., 2004, 2008). It is postulated that in this environment, differences between active and relaxation phase of the ciliary stroke are unlikely to contribute to the observed directionality of nodal flow (Cartwright et al., 2008). Instead, both mathematical modeling (Buceta *et al.*, 2005) and simulation of node ciliary behavior (Nonaka *et al.*, 2005) predict that giving each cilium a posterior tilt would result in the summation of ciliary movement to produce leftwards flow. Indeed, the cilia of the mouse and rabbit node, Medaka fish Kupffer's vesicle (Okada *et al.*, 2005) and the *Xenopus* GRP (Schweickert *et al.*, 2007) are positioned predominantly in the posterior quadrant of the apical cell surface. Posterior positioning becomes more uniform as development proceeds, and the timing of coordinated posterior ciliary positioning correlates with the timing of nodal flow. The location of the cilium in an oriented quadrant of the node cell across the node epithelium suggests a planar cell polarity mechanism may coordinate cilia position at the LR organizer. One molecule that affects both cilia and planar cell polarity is the ankyrin repeat-containing protein inversin (Simons *et al.*, 2005). Partial deletion of inversin leads to situs inversus (Morgan *et al.*, 1998; Yokoyama *et al.*, 1993, 1995), and analysis of the *inv/inv* mouse embryos showed node cilia that are not uniformly tilted to the posterior (Okada *et al.*, 2005), resulting in nodal flow which is still leftward, but much slower than that observed in wild-type mouse embryos (Okada *et al.*, 1999).

8. A Conserved Asymmetric Calcium Signal Is Found at the LR Organizer

How does extracellular leftward nodal flow signal to drive asymmetric morphogenesis? Asymmetric calcium signals are a striking, conserved element of left–right development downstream of nodal flow. This is not surprising as calcium is a versatile signaling molecule and modulation of its amplitude, frequency, and/or spatial localization can elicit a variety of distinct and specific responses (Berridge *et al.*, 1998). A transient left-sided localization of calcium is observed across the node during left–right patterning in mice (Fig. 6.3A), and at KV in zebrafish. Notably, it is also observed in chick embryos, where thus far no nodal flow equivalent has been identified.

In the mouse node, the left-biased rise in intracellular calcium is observed at embryonic day 8.0 from the 0 to 4 somite stage. This asymmetric calcium signal lies downstream of nodal flow: for example, the calcium signal is not lateralized normally in mouse embryos with paralyzed cilia due to mutations in lrd (McGrath *et al.*, 2003; Tanaka *et al.*, 2005), and it is reduced in intensity in mouse embryos with abnormal ciliary motility due to mutation in the T-box transcription factor Tbx6 (Hadjantonakis *et al.*, 2008). In addition, it depends on the integrity of the cation channel polycystin-2. It is upstream of left-specifying genes like *nodal,lefty* and *PitX2*.

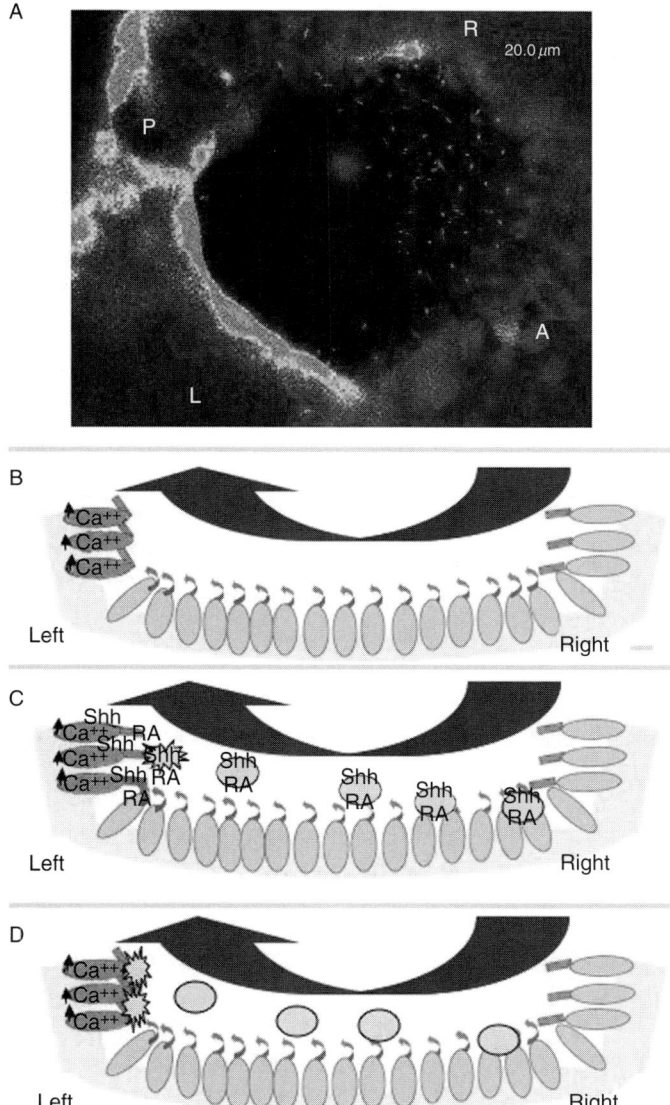

Figure 6.3 (A) Calcium signal at the mouse node. e7.8 embryo showing fluo3 fluorescence at left and posterior margins of the node. Node cilia fluoresce secondary to the GFP-tagged lrd. The cilia are motile, as is indicated by the multiple signals emanating from a single cilium during the acquisition of a single slow scan. (B) The mechanosensory model for left-sided calcium signal. Motile cilia are shown in green, polycystin-containing nonmotile cilia are shown in red. Cilia at the left side of the node are affected differently than on the right to the geometric configuration of the node coupled with mediolateral asymmetry of the cilia themselves. (C) The nodal vesicular parcel model. Motile cilia are shown in green. NVPs loaded with putative morphogens "shatter" preferentially on cilia and cell wall at the left. (D) The "combined" model. Flow-driven interaction of NVPs with mechanosensory cilia leads to an intracellular calcium signal on the left. (See Color Insert.)

The genetic or pharmacological perturbations that cause aberrations in calcium signaling also lead to defects in left–right asymmetry.

In support of evolutionary conservation of a LR asymmetric calcium signal, intracellular calcium is elevated along the left margin of the Kupffer's vesicle in zebrafish. This requires activity of *Ipk*, an IP_5 2-kinase, which when missing results in randomization of laterality (Sarmah et al., 2005). Disrupting the asymmetric calcium signal also affects LR patterning. Inositol phosphates are often found to be involved with calcium signaling. IP_3 causes intracellular calcium release from the endoplasmic reticulum through IP_3 receptors while IP_4 blocks this receptor (Mayrleitner et al., 1995). IP_6 has been reported to enhance voltage-gated L-type calcium channel activity (Yang et al., 2001).

Additional information regarding the involvement of calcium in left–right patterning in zebrafish comes from experiments which reveal a role for Na, K ATPase, which is an enzyme normally involved in maintaining low intracellular calcium levels in resting cells. Downregulation of this gene in the DFCs and the KV results in a failure to reduce intracellular calcium and a consequent randomization of the left–right axis (Shu et al., 2007). Interestingly, this defect can be rescued by inhibiting the calcium/calmodulin-dependent protein kinase II (CaMKII) which implicates this enzyme in patterning.

The formation of the KV itself requires calcium signaling in the KV precursors—the DFCs. This calcium signal antagonizes β-catenin activity. Manipulating the calcium signal affects KV formation and leads to laterality defects (Schneider et al., 2008). A similar phenomenon has also been observed in *Xenopus* embryos (Schneider et al., 2008).

In avian embryos, intracellular as well as extracellular calcium levels are elevated on the left margin of the Hensen's node. The rise in extracellular calcium levels is a result of H^+/K^+ ATPase activity, which generates a membrane potential gradient across Hensen's node causing a calcium ion flux (Raya et al., 2004). The local accumulation of extracellular calcium activates *Notch* signaling specifically on the left side of the node. In zebrafish and mice *Notch* activity has been shown to induce *Nodal* expression in the node, which is a critical player in left–right specification. Intracellular calcium rise on the left side of the Hensen's node is brought about by an interplay of ryanodine receptor signaling and extracellular calcium, which triggers a calcium-induced calcium release (Garic-Stankovic et al., 2008). Perturbation of the calcium signal leads to the randomization of heart laterality and changes *Nodal* expression patterns. How the asymmetric calcium signal is initiated in avian embryos remains unclear, as thus far motile cilia and a nodal flow equivalent have not been demonstrated in avian embryos, although this could at least in part be due to limitations on imaging imposed by the geometry of avian embryos.

Calcium is also involved in left–right axis patterning in invertebrates like the sea urchin (Hibino et al., 2006). In this case, using a calcium ionophore significantly reversed the expression pattern of the left-coelom marker, *HpFoxFQ-like* and caused significant left–right patterning defects.

Therefore, calcium plays a significant role in left–right axis patterning in a large number of species encompassing vertebrates and invertebrates, and evidence in mice and zebrafish place the calcium signal downstream of directional flow of extracellular fluid at the LR organizer. How does flow lead to a calcium signal? Two hypotheses exist regarding the mechanism by which nodal flow translates to elevated calcium levels on the left side of the mouse node. The "chemical" hypothesis is based on the ability of nodal flow to create a gradient of a morphogen across the left–right axis of the node. Such a gradient is theoretically possible under the physiological conditions found at the node, consisting of flow velocity, diffusion and potential rate of degradation of a putative morphogen. However, when this delicate balance is perturbed by changing the velocity of nodal flow, without changing the rate of diffusion or kinetics of the morphogen, LR development still proceeds normally (Nonaka et al., 2002). This observation suggests that it is unlikely that asymmetry is created solely by a gradient of a diffusible morphogen. The finding of small vesicles bearing retinoic acid and sonic hedgehog, which are transported to the left margin of the node by the nodal flow (Tanaka et al., 2005) provides an alternate mechanism for generating a morphogen gradient at the node (Fig. 6.3C). These vesicles, called the nodal vesicular particles (NVPs), have been observed to fragment and release their contents when they strike the rotating nodal monocilia or the left wall of the node. This could initiate *shh* and retinoic acid (RA) signaling and lead to elevated calcium levels. It is interesting to note, however, that the force of nodal flow alone would be inadequate to break the relatively stable membrane encompassing NVPs in the low Reynolds number environment of the node.

On the other hand, the "mechanosensory" hypothesis does not require the involvement of a morphogen (Fig. 6.3B). It is based on the presence of mechanosensory cilia at the node, in addition to the motile cilia responsible for nodal flow. In support of this, node cilia also express *pkd2*, which is a calcium-activated cation channel. The mechanosensory cilia present on the left side of the node bend due to the pressure of the nodal flow and respond by elevating intracellular calcium levels (McGrath et al., 2003). This is similar to the phenomenon observed in the renal epithelial cell line, MDCK, where bending of the primary cilium results in a rise in intracellular calcium (Praetorius and Spring, 2001). The primary problem inherent in the mechanosensory hypothesis lies in the ability of mechanosensory cilia at the left border of the node to perceive a different input from nodal flow than those found at the right border of the node. One possible explanation is that existing mediolateral polarity of the cilia at the rim of the node can interpret directionality of nodal flow to produce a calcium signal on the left (Marshall and Nonaka, 2006). Another interesting possibility that would at least partially reconcile a requirement for nodal vesicular parcels and polycystin-mediated signaling is a "combined NVP and mechanosensation" hypothesis. Here direct contact between a nodal vesicular parcel, or even cellular debris transported by nodal flow, and a mechanosensory cilium initiates the calcium signal (Fig. 6.3D). The calcium signal could result from mechanical interaction of the NVP and the cilium, without a direct requirement for shh or RA; alternatively, interaction of the NVP with the sensory cilium could generate a calcium signal which mediates release of the NVP contents.

9. Do Cilia Lie at the Root of All Vertebrate LR Asymmetry, and What Leads from Asymmetric Calcium to Asymmetric Organogenesis?

A conundrum is that in both chick and *Xenopus* LR asymmetries have been observed prior to the emergence of a ciliated LR organizer (Levin and Mercola, 1999; Levin *et al.*, 2002), while no such asymmetries have been identified in mouse. However, these observations of earlier asymmetries are not mutually exclusive with the presence of an essential ciliated LR organizer in all vertebrates. It is distinctly possible that at least chick and *Xenopus* have parallel early, nonflow-dependent and late, flow-dependent pathways for specifying LR asymmetry. One could postulate that such dual mechanisms are necessitated by different temporal progression of development, and different geometric organization found in different species. However, as long as the chiral reference point(s) of distinct LR pathways are concordant, they could actually reinforce each other and provide additional robustness to LR development.

One of the major remaining questions concerns the mechanism by which the calcium asymmetry at the node is converted to asymmetric gene expression and asymmetric morphogenesis of organs such as the heart that lie at quite a distance from the node. Here, study of the cilia-dependent signaling in polycystic kidney disease may provide insights. Links have been established between cilia and Wnt signaling via *inversin* (Simons *et al.*, 2005) and *seahorse* (Kishimoto *et al.*, 2008). Cilia are intimately associated with the control of cell division (Pugacheva *et al.*, 2007; Qin *et al.*, 2007), and planar cell polarity is disturbed in renal tubules of cystic kidney mutations. These pathways are all capable of influencing LR development, and it is likely that in LR development, as in other cilia-dependent processes, cilia are multifunctional organelles that generate, interpret and integrate signals.

REFERENCES

Afzelius, B. A. (1976). A human syndrome caused by immotile cilia. *Science* **193,** 317–319.
Afzelius, B. A. (1982). Immotile-cilia syndrome: Ultrastructural features. *Eur. J. Respir. Dis. Suppl.* **118,** 117–122.
Asai, D. J., Beckwith, S. M., Kandl, K. A., Keating, H. H., Tjandra, H., and Forney, J. D. (1994). The dynein genes of Paramecium tetraurelia. Sequences adjacent to the catalytic P-loop identify cytoplasmic and axonemal heavy chain isoforms. *J. Cell Sci.* **107,** 839–847.
Beckers, A., Alten, L., Viebahn, C., Andre, P., and Gossler, A. (2007). The mouse homeobox gene Noto regulates node morphogenesis, notochordal ciliogenesis, and left right patterning. *Proc. Natl. Acad. Sci. USA* **104,** 15765–15770.
Beddington, R. S. (1994). Induction of a second neural axis by the mouse node. *Development* **120,** 613–620.
Bellomo, D., Lander, A., Harragan, I., and Brown, N. A. (1996). Cell proliferation in mammalian gastrulation: The ventral node and notochord are relatively quiescent. *Dev. Dyn.* **205,** 471–485.
Berridge, M. J., Bootman, M. D., and Lipp, P. (1998). Calcium–a life and death signal. *Nature* **395,** 645–648.
Blum, M., Andre, P., Muders, K., Schweickert, A., Fischer, A., Bitzer, E., Bogusch, S., Beyer, T., van Straaten, H. W., and Viebahn, C. (2007). Ciliation and gene expression distinguish between node and posterior notochord in the mammalian embryo. *Differentiation* **75,** 133–146.
Boulter, C., Mulroy, S., Webb, S., Fleming, S., Brindle, K., and Sandford, R. (2001). Cardiovascular, skeletal, and renal defects in mice with a targeted disruption of the Pkd1 gene. *Proc. Natl. Acad. Sci. USA* **98,** 12174–12179.
Brennan, J., Norris, D. P., and Robertson, E. J. (2002). Nodal activity in the node governs left-right asymmetry. *Genes Dev.* **16,** 2339–2344.
Brown, N. A., and Wolpert, L. (1990). The development of handedness in left/right asymmetry. *Development* **109,** 1–9.
Buceta, J., Ibanes, M., Rasskin-Gutman, D., Okada, Y., Hirokawa, N., and Izpisua-Belmonte, J. C. (2005). Nodal cilia dynamics and the specification of the left/right axis in early vertebrate embryo development. *Biophys. J.* **89,** 2199–2209.
Cartwright, J. H., Piro, N., Piro, O., and Tuval, I. (2008). Fluid dynamics of establishing left-right patterning in development. *Birth Defects Res. C. Embryo Today* **84,** 95–101.
Cartwright, J. H., Piro, O., and Tuval, I. (2004). Fluid-dynamical basis of the embryonic development of left-right asymmetry in vertebrates. *Proc. Natl. Acad. Sci. USA* **101,** 7234–7239.
Caspary, T., Larkins, C. E., and Anderson, K. V. (2007). The graded response to Sonic Hedgehog depends on cilia architecture. *Dev. Cell* **12,** 767–778.
Cole, D. G., Diener, D. R., Himelblau, A. L., Beech, P. L., Fuster, J. C., and Rosenbaum, J. L. (1998). Chlamydomonas kinesin-II-dependent intraflagellar transport (IFT): IFT particles contain proteins required for ciliary assembly in Caenorhabditis elegans sensory neurons. *J. Cell Biol.* **141,** 993–1008.
Collignon, J., Varlet, I., and Robertson, E. J. (1996). Relationship between asymmetric nodal expression and the direction of embryonic turning [see comments]. *Nature* **381,** 155–158.
Cooper, M. S., and D'Amico, L. A. (1996). A cluster of noninvoluting endocytic cells at the margin of the zebrafish blastoderm marks the site of embryonic shield formation. *Dev. Biol.* **180,** 184–198.

Davidson, B. P., Kinder, S. J., Steiner, K., Schoenwolf, G. C., and Tam, P. P. (1999). Impact of node ablation on the morphogenesis of the body axis and the lateral asymmetry of the mouse embryo during early organogenesis. *Dev. Biol.* **211,** 11–26.

Davis, E. E., Brueckner, M., and Katsanis, N. (2006). The emerging complexity of the vertebrate cilium: New functional roles for an ancient organelle. *Dev. Cell* **11,** 9–19.

Essner, J. J., Amack, J. D., Nyholm, M. K., Harris, E. B., and Yost, H. J. (2005). Kupffer's vesicle is a ciliated organ of asymmetry in the zebrafish embryo that initiates left-right development of the brain, heart and gut. *Development* **132,** 1247–1260.

Essner, J. J., Vogan, K. J., Wagner, M. K., Tabin, C. J., Yost, H. J., and Brueckner, M. (2002). Conserved function for embryonic nodal cilia. *Nature* **418,** 37–38.

Feistel, K., and Blum, M. (2006). Three types of cilia including a novel 9 + 4 axoneme on the notochordal plate of the rabbit embryo. *Dev. Dyn.* **235,** 3348–3358.

Garic-Stankovic, A., Hernandez, M., Flentke, G. R., Zile, M. H., and Smith, S. M. (2008). A ryanodine receptor-dependent $Ca_{i}2+$ asymmetry at Hensen's node mediates avian lateral identity. *Development* **135,** 3271–3280.

Gee, M. A., Heuser, J. E., and Vallee, R. B. (1997). An extended microtubule-binding structure within the dynein motor domain. *Nature* **390,** 636–639.

Gibbons, B. H., Gibbons, I. R., and Baccetti, B. (1983). Structure and motility of the 9 + 0 flagellum of eel spermatozoa. *J. Submicrosc. Cytol.* **15,** 15–20.

Hadjantonakis, A. K., Pisano, E., and Papaioannou, V. E. (2008). Tbx6 regulates left/right patterning in mouse embryos through effects on nodal cilia and perinodal signaling. *PLoS ONE* **3,** e2511.

Hibino, T., Ishii, Y., Levin, M., and Nishino, A. (2006). Ion flow regulates left-right asymmetry in sea urchin development. *Dev. Genes Evol.* **216,** 265–276.

Holzbaur, E. L., and Vallee, R. B. (1994). DYNEINS: Molecular structure and cellular function. *Annu. Rev. Cell Biol.* **10,** 339–372.

Huangfu, D., Liu, A., Rakeman, A. S., Murcia, N. S., Niswander, L., and Anderson, K. V. (2003). Hedgehog signalling in the mouse requires intraflagellar transport proteins. *Nature* **426,** 83–87.

Hummel, K. P., and Chapman, D. B. (1959). Visceral inversion and associated anomalies in the mouse. *Journal of Heredity* **50,** 9–13.

Ibanez-Tallon, I., Gorokhova, S., and Heintz, N. (2002). Loss of function of axonemal dynein Mdnah5 causes primary ciliary dyskinesia and hydrocephalus. *Hum. Mol. Genet.* **11,** 715–721.

Kartagener, M. (1933). Zur Pathogenese der Bronchiektasien bei Situs viscerum inversus. *Beitr. Klin. Tuberk.* **83,** 489–501.

Kennedy, M. P., Omran, H., Leigh, M. W., Dell, S., Morgan, L., Molina, P. L., Robinson, B. V., Minnix, S. L., Olbrich, H., Severin, T., et al. (2007). Congenital heart disease and other heterotaxic defects in a large cohort of patients with primary ciliary dyskinesia. *Circulation* **115,** 2814–2821.

Kishimoto, N., Cao, Y., Park, A., and Sun, Z. (2008). Cystic kidney gene seahorse regulates cilia-mediated processes and Wnt pathways. *Dev. Cell* **14,** 954–961.

Kondo, S., Sato-Yoshitake, R., Noda, Y., Aizawa, H., Nakata, T., Matsuura, Y., and Hirokawa, N. (1994). KIF3A is a new microtubule-based anterograde motor in the nerve axon. *J. Cell Biol.* **125,** 1095–1107.

Kramer-Zucker, A. G., Olale, F., Haycraft, C. J., Yoder, B. K., Schier, A. F., and Drummond, I. A. (2005). Cilia-driven fluid flow in the zebrafish pronephros, brain and Kupffer's vesicle is required for normal organogenesis. *Development* **132,** 1907–1921.

Kreiling, J. A., Williams, G., and Creton, R. (2007). Analysis of Kupffer's vesicle in zebrafish embryos using a cave automated virtual environment. *Dev. Dyn.* **236,** 1963–1969.

Layton, W. M., Jr. (1976). Random determination of a developmental process: Reversal of normal visceral asymmetry in the mouse. *J. Hered.* **67,** 336–338.

Lechtreck, K. F., Delmotte, P., Robinson, M. L., Sanderson, M. J., and Witman, G. B. (2008). Mutations in Hydin impair ciliary motility in mice. *J. Cell Biol.* **180,** 633–643.

Levin, M., Johnson, R. L., Stern, C. D., Kuehn, M., and Tabin, C. (1995). A molecular pathway determining left-right asymmetry in chick embryogenesis. *Cell* **82,** 803–814.

Levin, M., and Mercola, M. (1999). Gap junction-mediated transfer of left-right patterning signals in the early chick blastoderm is upstream of Shh asymmetry in the node. *Development* **126,** 4703–4714.

Levin, M., Thorlin, T., Robinson, K. R., Nogi, T., and Mercola, M. (2002). Asymmetries in H+/K+-ATPase and cell membrane potentials comprise a very early step in left-right patterning. *Cell* **111,** 77–89.

Lowe, L. A., Supp, D. M., Sampath, K., Yokoyama, T., Wright, C. V., Potter, S. S., Overbeek, P., and Kuehn, M. R. (1996). Conserved left-right asymmetry of nodal expression and alterations in murine situs inversus [see comments]. *Nature* **381,** 158–161.

Lowe, L. A., Yamada, S., and Kuehn, M. R. (2001). Genetic dissection of nodal function in patterning the mouse embryo. *Development* **128,** 1831–1843.

Marshall, W. F., and Nonaka, S. (2006). Cilia: Tuning in to the cell's antenna. *Curr. Biol.* **16,** R604–R614.

Marszalek, J. R., Ruiz-Lozano, P., Roberts, E., Chien, K. R., and Goldstein, L. S. (1999). Situs inversus and embryonic ciliary morphogenesis defects in mouse mutants lacking the KIF3A subunit of kinesin-II. *Proc. Natl. Acad. Sci. USA* **96,** 5043–5048.

Mayrleitner, M., Schafer, R., and Fleischer, S. (1995). IP3 receptor purified from liver plasma membrane is an (1,4,5)IP3 activated and (1,3,4,5)IP4 inhibited calcium permeable ion channel. *Cell Calcium* **17,** 141–153.

McGrath, J., Somlo, S., Makova, L., Tian, X., and Brueckner, M. (2003). Two populations of node monocilia initiate left-right asymmetry in the mouse. *Cell* **114,** 61–73.

Meno, C., Ito, Y., Saijoh, Y., Matsuda, Y., Tashiro, K., Kuhara, S., and Hamada, H. (1997). Two closely-related left-right asymmetrically expressed genes, lefty-1 and lefty-2: Their distinct expression domains, chromosomal linkage and direct neuralizing activity in Xenopus embryos. *Genes Cells* **2,** 513–524.

Meno, C., Saijoh, Y., Fujii, H., Ikeda, M., Yokoyama, T., Yokoyama, M., Toyoda, Y., and Hamada, H. (1996). Left-right asymmetric expression of the TGF beta-family member lefty in mouse embryos [see comments]. *Nature* **381,** 151–155.

Meno, C., Shimono, A., Saijoh, Y., Yashiro, K., Mochida, K., Ohishi, S., Noji, S., Kondoh, H., and Hamada, H. (1998). lefty-1 is required for left-right determination as a regulator of lefty-2 and nodal. *Cell* **94,** 287–297.

Morgan, D., Turnpenny, L., Goodship, J., Dai, W., Majumder, K., Matthews, L., Gardner, A., Schuster, G., Vien, L., Harrison, W., *et al.* (1998). Inversin, a novel gene in the vertebrate left-right axis pathway, is partially deleted in the inv mouse. *Nat. Genet.* **20,** 149–156 [published erratum appears in *Nat. Genet.* 1998 Nov; 20(3):312].

Morris, R. L., and Scholey, J. M. (1997). Heterotrimeric kinesin-II is required for the assembly of motile 9 + 2 ciliary axonemes on sea urchin embryos. *J. Cell Biol.* **138,** 1009–1022.

Murcia, N. S., Richards, W. G., Yoder, B. K., Mucenski, M. L., Dunlap, J. R., and Woychik, R. P. (2000). The Oak Ridge Polycystic Kidney (orpk) disease gene is required for left-right axis determination. *Development* **127,** 2347–2355.

Nonaka, S., Shiratori, H., Saijoh, Y., and Hamada, H. (2002). Determination of left right patterning of the mouse embryo by artificial nodal flow. *Nature* **418,** 96–99.

Nonaka, S., Tanaka, Y., Okada, Y., Takeda, S., Harada, A., Kanai, Y., Kido, M., and Hirokawa, N. (1998). Randomization of left-right asymmetry due to loss of nodal cilia generating leftward flow of extraembryonic fluid in mice lacking KIF3B motor protein. *Cell* **95,** 829–837 [published erratum appears in *Cell* 1999 Oct 1;99(1):117].

Nonaka, S., Yoshiba, S., Watanabe, D., Ikeuchi, S., Goto, T., Marshall, W. F., and Hamada, H. (2005). De novo formation of left-right asymmetry by posterior tilt of nodal cilia. *PLoS Biol.* **3,** e268.

Okada, Y., Nonaka, S., Tanaka, Y., Saijoh, Y., Hamada, H., and Hirokawa, N. (1999). Abnormal nodal flow precedes situs inversus in iv and inv mice. *Mol. Cell.* **4,** 459–468.

Okada, Y., Takeda, S., Tanaka, Y., Belmonte, J. C., and Hirokawa, N. (2005). Mechanism of nodal flow: A conserved symmetry breaking event in left-right axis determination. *Cell* **121,** 633–644.

Olbrich, H., Haffner, K., Kispert, A., Volkel, A., Volz, A., Sasmaz, G., Reinhardt, R., Hennig, S., Lehrach, H., Konietzko, N., et al. (2002). Mutations in DNAH5 cause primary ciliary dyskinesia and randomization of left-right asymmetry. *Nat. Genet.* **30,** 143–144.

Pagan-Westphal, S. M., and Tabin, C. J. (1998). The transfer of left-right positional information during chick embryogenesis. *Cell* **93,** 25–35.

Pazour, G. J., and Witman, G. B. (2003). The vertebrate primary cilium is a sensory organelle. *Curr. Opin. Cell. Biol.* **15,** 105–110.

Pennekamp, P., Karcher, C., Fischer, A., Schweickert, A., Skryabin, B., Horst, J., Blum, M., and Dworniczak, B. (2002). The ion channel polycystin-2 is required for left-right axis determination in mice. *Curr. Biol.* **12,** 938–943.

Poole, C. A., Flint, M. H., and Beaumont, B. W. (1985). Analysis of the morphology and function of primary cilia in connective tissues: A cellular cybernetic probe? *Cell Motil.* **5,** 175–193.

Praetorius, H. A., and Spring, K. R. (2001). Bending the MDCK cell primary cilium increases intracellular calcium. *J. Membr. Biol.* **184,** 71–79.

Pugacheva, E. N., Jablonski, S. A., Hartman, T. R., Henske, E. P., and Golemis, E. A. (2007). HEF1-dependent Aurora A activation induces disassembly of the primary cilium. *Cell* **129,** 1351–1363.

Qin, H., Wang, Z., Diener, D., and Rosenbaum, J. (2007). Intraflagellar transport protein 27 is a small G protein involved in cell-cycle control. *Curr. Biol.* **17,** 193–202.

Rana, A. A., Barbera, J. P., Rodriguez, T. A., Lynch, D., Hirst, E., Smith, J. C., and Beddington, R. S. (2004). Targeted deletion of the novel cytoplasmic dynein mD2LIC disrupts the embryonic organiser, formation of the body axes and specification of ventral cell fates. *Development* **131,** 4999–5007.

Raya, A., Kawakami, Y., Rodriguez-Esteban, C., Ibanes, M., Rasskin-Gutman, D., Rodriguez-Leon, J., Buscher, D., Feijo, J. A., and Izpisua Belmonte, J. C. (2004). Notch activity acts as a sensor for extracellular calcium during vertebrate left-right determination. *Nature* **427,** 121–128.

Sarmah, B., Latimer, A. J., Appel, B., and Wente, S. R. (2005). Inositol polyphosphates regulate zebrafish left-right asymmetry. *Dev. Cell* **9,** 133–145.

Schneider, A., Mijalski, T., Schlange, T., Dai, W., Overbeek, P., Arnold, H. H., and Brand, T. (1999). The homeobox gene NKX3.2 is a target of left-right signalling and is expressed on opposite sides in chick and mouse embryos. *Curr. Biol.* **9,** 911–914.

Schneider, I., Houston, D. W., Rebagliati, M. R., and Slusarski, D. C. (2008). Calcium fluxes in dorsal forerunner cells antagonize beta-catenin and alter left-right patterning. *Development* **135,** 75–84.

Schweickert, A., Weber, T., Beyer, T., Vick, P., Bogusch, S., Feistel, K., and Blum, M. (2007). Cilia-driven leftward flow determines laterality in Xenopus. *Curr. Biol.* **17,** 60–66.

Shu, X., Huang, J., Dong, Y., Choi, J., Langenbacher, A., and Chen, J. N. (2007). Na,K-ATPase alpha2 and Ncx4a regulate zebrafish left-right patterning. *Development* **134,** 1921–1930.

Simons, M., Gloy, J., Ganner, A., Bullerkotte, A., Bashkurov, M., Kronig, C., Schermer, B., Benzing, T., Cabello, O. A., Jenny, A., et al. (2005). Inversin, the gene product mutated in nephronophthisis type II, functions as a molecular switch between Wnt signaling pathways. *Nat. Genet.* **37,** 537–543.

Slough, J., Cooney, L., and Brueckner, M. (2008). Monocilia in the embryonic mouse heart suggest a direct role for cilia in cardiac morphogenesis. *Dev. Dyn.* **237,** 2304–2314.

Sulik, K., Dehart, D. B., Iangaki, T., Carson, J. L., Vrablic, T., Gesteland, K., and Schoenwolf, G. C. (1994). Morphogenesis of the murine node and notochordal plate. *Dev. Dyn.* **201,** 260–278.

Supp, D. M., Brueckner, M., Kuehn, M. R., Witte, D. P., Lowe, L. A., McGrath, J., Corrales, J., and Potter, S. S. (1999). Targeted deletion of the ATP binding domain of left-right dynein confirms its role in specifying development of left-right asymmetries. *Development* **126,** 5495–5504.

Supp, D. M., Witte, D. P., Potter, S. S., and Brueckner, M. (1997). Mutation of an axonemal dynein affects left-right asymmetry in inversus viscerum mice. *Nature* **389,** 963–966.

Takeda, S., Yonekawa, Y., Tanaka, Y., Okada, Y., Nonaka, S., and Hirokawa, N. (1999). Left-right asymmetry and kinesin superfamily protein KIF3A: New insights in determination of laterality and mesoderm induction by kif3A- /- mice analysis. *J. Cell Biol.* **145,** 825–836.

Tanaka, Y., Okada, Y., and Hirokawa, N. (2005). FGF-induced vesicular release of Sonic hedgehog and retinoic acid in leftward nodal flow is critical for left-right determination. *Nature* **435,** 172–177.

Tanaka, Y., Zhang, Z., and Hirokawa, N. (1995). Identification and molecular evolution of new dynein-like protein sequences in rat brain. *J. Cell Sci.* **108,** 1883–1893.

Vaughan, K. T., Mikami, A., Paschal, B. M., Holzbaur, E. L. F., Hughes, S. M., Echeverri, C. J., Moore, K. J., Gilbert, D. J., Copeland, N. G., Jenkins, N. A., and Vallee, R. B. (1996). Multiple mouse chromosomal loci for dynein-based motility. *Genomics* **36,** 29–38.

Wilhelmi, H. (1921). Experimentelle Untersuchungen ueber situs inversus viscerum. *Archive der Entwicklungsmechanik* **48,** 517–532.

Wu, G., Markowitz, G. S., Li, L., D'Agati, V. D., Factor, S. M., Geng, L., Tibara, S., Tuchman, J., Cai, Y., Park, J. H., et al. (2000). Cardiac defects and renal failure in mice with targeted mutations in Pkd2. *Nat. Genet.* **24,** 75–78.

Yagi, T., and Kamiya, R. (2000). Vigorous beating of Chlamydomonas axonemes lacking central pair/radial spoke structures in the presence of salts and organic compounds. *Cell Motil. Cytoskeleton* **46,** 190–199.

Yamanaka, Y., Tamplin, O. J., Beckers, A., Gossler, A., and Rossant, J. (2007). Live imaging and genetic analysis of mouse notochord formation reveals regional morphogenetic mechanisms. *Dev. Cell* **13,** 884–896.

Yamazaki, H., Nakata, T., Okada, Y., and Hirokawa, N. (1995). KIF3A/B: A heterodimeric kinesin superfamily protein that works as a microtubule plus end-directed motor for membrane organelle transport. *J. Cell Biol.* **130,** 1387–1399.

Yamazaki, H., Nakata, T., Okada, Y., and Hirokawa, N. (1996). Cloning and characterization of KAP3: A novel kinesin superfamily- associated protein of KIF3A/3B. *Proc. Natl. Acad. Sci. USA* **93,** 8443–8448.

Yan, Y. T., Gritsman, K., Ding, J., Burdine, R. D., Corrales, J. D., Price, S. M., Talbot, W. S., Schier, A. F., and Shen, M. M. (1999). Conserved requirement for EGF-CFC genes in vertebrate left-right axis formation. *Genes Dev.* **13,** 2527–2537.

Yang, S. N., Yu, J., Mayr, G. W., Hofmann, F., Larsson, O., and Berggren, P. O. (2001). Inositol hexakisphosphate increases L-type Ca2+ channel activity by stimulation of adenylyl cyclase. *FASEB J.* **15,** 1753–1763.

Yokoyama, T., Copeland, N. G., Jenkins, N. A., Montgomery, C. A., Elder, F. F., and Overbeek, P. A. (1993). Reversal of left-right asymmetry: A situs inversus mutation [see comments]. *Science* **260,** 679–682.

Yokoyama, T., Harrison, W. R., Elder, F. F. B., and Overbeek, P. A. (eds.) (1995). *In* Molecular Analysis of the inv insertional mutation. Futura Publishing Co. Armonk, NY.

Yoshioka, H., Meno, C., Koshiba, K., Sugihara, M., Itoh, H., Ishimaru, Y., Inoue, T., Ohuchi, H., Semina, E. V., Murray, J. C., *et al.* (1998). Pitx2, a bicoid-type homeobox gene, is involved in a lefty-signaling pathway in determination of left-right asymmetry. *Cell* **94,** 299–305.

Zhu, L., Belmont, J. W., and Ware, S. M. (2006). Genetics of human heterotaxias. *Eur. J. Hum. Genet.* **14,** 17–25.

CHAPTER SEVEN

Ciliary Function and Wnt Signal Modulation

Jantje M. Gerdes *and* Nicholas Katsanis

Contents

1. Introduction	176
2. Overview of Wnt Signaling	176
3. Emerging Roles of the Primary Cilium and Basal Body in Wnt Signaling Modulation	179
3.1. Suppression of canonical Wnt signaling	179
3.2. Transmission of noncanonical Wnt signaling	181
4. Wnt and Ciliogenesis	184
5. Wnt Dysfunction and Ciliopathy Phenotypes	185
6. What Drives the Phenotype: Gain of Canonical Wnt Signaling or Loss of Noncanonical Wnt Signaling?	185
7. Other Phenotypes	187
8. Discussion	189
Acknowledgments	190
References	191

Abstract

With the increase in complexity of morphogenetic signaling cascades over the course of evolution and the emergence of broadly ciliated organisms, the cilium seems to have acquired a role as regulator of paracrine signal transduction. Recently, several lines of evidence have provided a link between basal body and ciliary proteins and Wnt signaling. In this chapter, we will evaluate the evidence linking the basal body and cilium with the regulation of β-catenin-dependent (canonical) and β-catenin-independent (noncanonical) signaling processes as well as which role(s) Wnt signaling might play in ciliogenesis. In addition, we will discuss aberrant Wnt signaling could contribute to phenotypes common to most ciliopathies and why these phenotypes might be driven by loss of noncanonical rather than gain of noncanonical Wnt signaling.

McKusick-Nathans Institute of Genetic Medicine, Johns Hopkins University School of Medicine, Baltimore, Maryland

1. Introduction

Over 30 years ago, the identification of Kartagener syndrome as a defect of motile cilia demonstrated the first causal link between ciliary dysfunction and human disease (Afzelius, 1976; for a recent review, see Marshall and Kintner, 2008). The role of the primary (immotile) cilium, however, has only recently begun to come into the focus of biomedical research, first with the identification of ciliary lesions in mouse models of polycystic kidney disease and left–right axis determination, followed by mutations in genes encoding ciliary and basal body proteins in multisystemic disease (Badano et al., 2006). An emerging paradigm from the studies of both humans and model organisms with structural and/or functional ciliary defects is that the primary cilium participates in a broad range of morphogenetic signaling, a phenomenon particularly prominent in vertebrate cilia and coincident with the expansion of ciliated cell types (Davis et al., 2006).

Presently, basal body and cilia are known to be involved in at least three important signaling pathways in vertebrates: Sonic hedgehog signaling (Shh) (Corbit et al., 2005; Huangfu et al., 2003; Rohatgi et al., 2007; see Chapter 9), platelet-derived growth factor (PDGF) signaling (Schneider et al., 2005; see Chapter 10), and Wnt signaling (Corbit et al., 2008; Gerdes et al., 2007; Ross et al., 2005; Simons et al., 2005). Indeed, defects in some of these pathways potentially provide causal links to some ciliopathy phenotypes in humans. However, there are still many open questions, including what the precise role of the cilium in signal transduction is and which portion of signaling processes are dependent directly on ciliary function.

To answer some of these questions, a combination of *in vitro* and *in vivo* models are commonly utilized: Loss of either *intraflagellar transport 88 homolog* (*Ift88*) or the ciliary-specific kinesin-II subunit *Kif3a* impair ciliary maintenance (Marszalek et al., 1999); therefore, conditional *Ift88* or *Kif3a* knockout mice can be used to elucidate the role of ciliary function in time and tissue-specific contexts. In addition, ciliopathy models such as *oral–facial–digital syndrome* (*Ofd*) and *Bardet-Biedl syndrome* (*Bbs*) mice help to understand the mechanisms underlying the phenotypes observed in ciliopathy patients. In this chapter, we will focus on the connection between ciliary function and Wnt signaling and how perturbed ciliary/basal body function might contribute to the phenotypes common to many ciliary disorders.

2. Overview of Wnt Signaling

The Wnt signaling pathways are remarkably conserved between invertebrates (*Drosophila*) and vertebrates, albeit considerably expanded in mammals: Both in the human and mouse genome, there are 19 genes encoding Wnt ligands, 10 Frizzled (Fz) receptors, and two coreceptors: low density lipoprotein receptor-related protein (LRP) 5 and 6 (refer to the Wnt signaling homepage: http://www.stanford.edu/~rnusse/wntwindow. html). The signaling cascade is initiated by binding of extracellular Wnt ligand to one of the Frizzled receptors and LRP5 or 6, triggering events that play roles both in development and adult homeostasis (for comprehensive reviews, refer to Clevers, 2006; Logan and Nusse, 2004). The signaling pathways converge at Disheveled (DVL), a downstream target of Wnt signaling that is thought to act as a switch between two major branches of the Wnt signaling pathways: canonical (or β-catenin dependent) and noncanonical (or β-catenin independent). In mammals, there are three different DVL proteins, whose individual roles are not sufficiently understood; nevertheless, it has been shown by staining with antibodies that recognize all three members of the protein family that DVLs predominantly localize to the membrane in the context of noncanonical Wnt signaling (Na et al., 2006; Wang et al., 2006a), while nuclear DVL localization is required for canonical Wnt signaling (Itoh et al., 2005). In absence of Wnt ligand, cytoplasmic β-catenin is constitutively phosphorylated by several kinases, including Casein Kinase I (CKI) and glycogen synthase kinase 3β (GSK3β), a component of the β-catenin destruction complex, and thus targeted for proteasomal degradation (Aberle et al., 1997). At the same time, β-catenin-responsive transcriptional elements are bound by a complex of T-cell transcription factor (TCF)/lymphoid enhancer-binding factor 1 (LEF1) and Groucho, a group of Wnt repressor proteins. Upon activation of the canonical Wnt signaling pathway, the β-catenin destruction complex dissociates and active β-catenin enters the nucleus, where it displaces Groucho and activates transcription of β-catenin/TCF/LEF1 targets (Fig. 7.1A and B).

Unlike canonical Wnt signaling, transcriptional targets of noncanonical Wnt signaling have yet to be identified. Among the signaling cascades putatively activated by noncanonical Wnt signals are Ras homology gene family member A (RhoA)/Rho-associated, coiled-coil-containing protein kinase (ROCK), c-jun N-terminal kinase (JNK), as well as Ca^{2+} and G protein signaling (Fig. 7.1C; Veeman et al., 2003a). In the absence of a direct reporter of activated noncanonical Wnt signaling, the observation of characteristic phenotypes in animal models is a crucial tool to establish defects in noncanonical Wnt signaling. Hallmarks of disrupted planar cell polarity signaling (PCP, a subset of noncanonical Wnt signals) are defective convergence and extension (CE) movements in the developing embryo (Solnica-Krezel, 2005), misoriented stereociliary bundles in the organ of Corti in the mouse inner ear, open eyelids, neural tube defects, as well as a randomization of stroke direction in fur (mice)/wing hairs (flies) which display characteristic swirls (Seifert and Mlodzik, 2007; Veeman et al., 2003a).

Although the categorization into two branches of Wnt signaling is helpful in our general understanding of the role of the cilium and basal

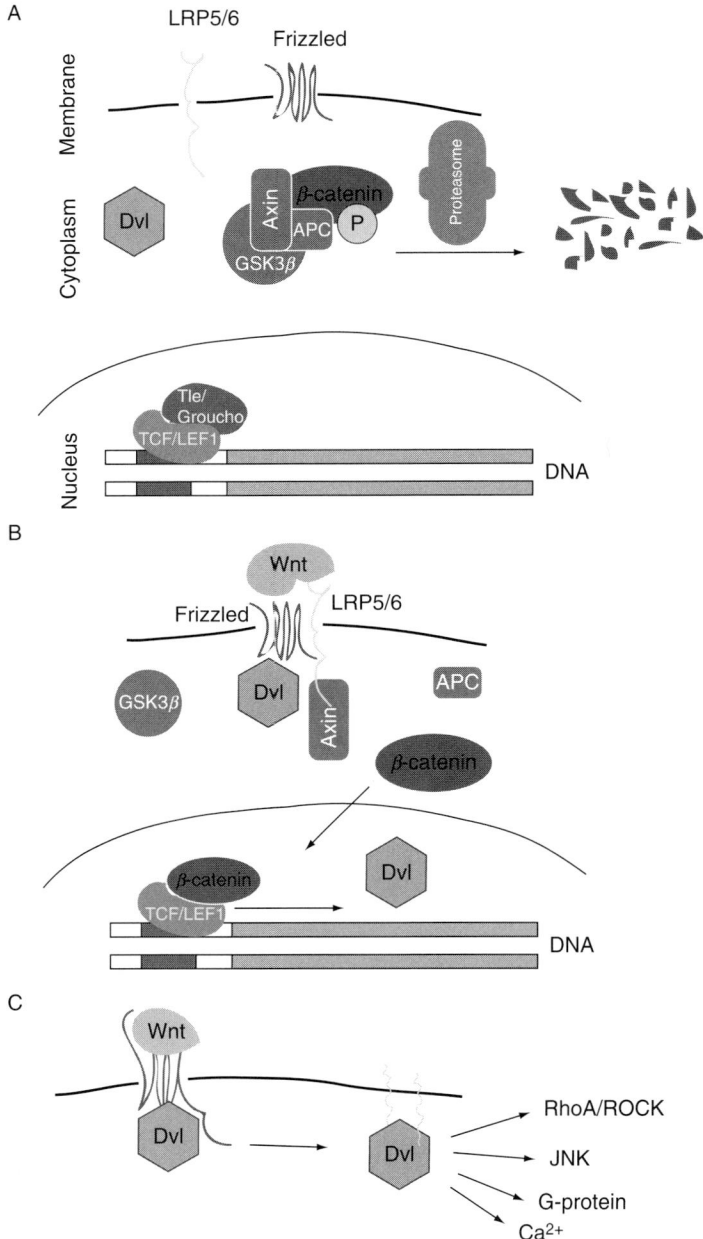

Figure 7.1 Wnt signaling overview. (A) In absence of extracellular Wnt ligand, β-catenin is constitutively phosphorylated by the β-catenin destruction complex and thus targeted for proteasomal degradation. Transcription factors TCF/LEF1 are complexed by Tle/Groucho protein to repress transcription of β-catenin/TCF/LEF1 targets. (B) In canonical β-catenin-dependent Wnt signaling, external Wnt ligand binds to a

body, it is likely going to prove to be grossly oversimplified. For example, some Wnt ligands, as well as Fz receptors, can trigger both β-catenin-dependent and β-catenin-independent signaling cascades depending on the context. Even Wnt5a, which acts as a noncanonical Wnt ligand, has been implicated in the activation of β-catenin-dependent signaling in presence of Fz4 (Mikels and Nusse, 2006). Bearing this in mind, we will continue to broadly distinguish between canonical and noncanonical Wnt signaling and focus on the differential role of ciliary and basal body proteins with respect to β-catenin-dependent and β-catenin-independent signaling activity.

3. EMERGING ROLES OF THE PRIMARY CILIUM AND BASAL BODY IN WNT SIGNALING MODULATION

3.1. Suppression of canonical Wnt signaling

Over the last few years, a number of independent lines of evidence showing an involvement of the primary cilium in Wnt signaling have emerged. Most of the data have suggested that the cilium and its anchoring structure, the basal body, exert an inhibitory role in β-catenin Wnt signaling, with potentially reciprocal, but not proportionate, activating/promoting roles for noncanonical Wnt signals. The first hint of such a role was obtained from studies of *Inversin/NPHP2*, a gene mutated in nephronophthisis (Otto *et al.*, 2003). Simons *et al.* (2005) demonstrated that overexpression of Inversin (INV) reduces cytoplasmic, but not membrane-bound DVL levels, thus differentiating between canonical and noncanonical Wnt signaling pathways. In *Xenopus laevis* embryos, INV inhibits formation of a secondary body axis induced by excessive canonical Wnt signaling, supporting the notion that it has an inhibitory effect on canonical Wnt signaling. Concomitantly, reporter assays for β-catenin activity showed that excessive INV blocks DVL but not β-catenin-stimulated canonical Wnt signaling *in vitro*, indicating that it acts upstream of β-catenin. Suppression of *inversin* in

---

heterodimer of Fz receptor and LRP5/6 coreceptor, leading to dissociation of the β-catenin destruction complex. Both DVL and β-catenin are localized to the nucleus in the context of canonical Wnt signaling, where β-catenin displaces Tle/Groucho and leads to transcriptional activation of β-catenin targets. (C) In β-catenin-independent signaling, an external Wnt ligand binds to Fz receptor and activates DVL, which is commonly associated with the membrane in the context of noncanonical Wnt signaling. DVL then activates putative noncanonical Wnt signaling components such as RhoA/ROCK, JNK, G protein or Ca^{2+} signaling. (See Color Insert.)

zebrafish led to cyst formation in the pronephric glomerulus at 4 days postfertilization, a phenotype that is potentially linked to the sensing of renal flow, because, in inner medullary collecting ducts (IMCD) 3 cells, protein levels of INV increased in response to a flow rate equivalent to urine flow, whereas levels of β-catenin slightly decreased (Simons et al., 2005). However, in addition to the basal body and ciliary axoneme, INV pools have also been seen in other cellular sites, such as cell junctions and the nucleus (Nurnberger et al., 2002, 2004). As such, the formal possibility remained that the observed INV activities originated from a nonciliary site.

Two more recent studies provided data that might help resolve this issue. The first focused on the role of the Bardet-Biedl syndrome (BBS) gene products in Wnt signaling. BBS (OMIM 209900) is a multisystemic disorder characterized, among others, by retinal degeneration, obesity, polydactyly, and renal cystic disease and has proven to be a useful ciliopathy model (reviewed in Badano et al., 2006). All known BBS proteins examined to date localize preferentially or exclusively to the basal body and cilium and some have been proposed to assemble in a multisubunit complex termed the BBSome that localizes to nonmembranous centriolar satellites in the cytoplasm, but also to the membrane of the cilium (Nachury et al., 2007). A recent study showed that suppression of BBS1/4/6 in human embryonic kidney cells (HEK293T) leads to increased cytoplasmic and nuclear protein levels of DVL and β-catenin as well as increased canonical Wnt signaling (Gerdes et al., 2007). Importantly, this phenotype could be phenocopied by suppression of KIF3a, a subunit of the axonemal kinesin motor complex necessary for ciliogenesis (Marszalek et al., 1999), but not by treatment with subeffective doses of vinblastine, which disrupts the microtubular network but leaves the ciliary axoneme intact. This supports the notion that the observed Wnt signature is linked to the ciliary, but not cytoplasmic role of the BBS proteins. The stabilization of β-catenin protein is potentially caused by impaired proteasomal targeting; BBS4 interacts with the RPN10 subunit of the proteasome and this interaction could implicate the basal body in proteasomal targeting of a subset of proteins. Another study investigated the Wnt signaling pathway in ciliary ($Kif3a^{-/-}$, $Ift88^{orpk/orpk}$) and basal body mouse mutants ($Ofd1^{-/-}$) (Corbit et al., 2008) and reached similar conclusions. In $Kif3a^{-/-}$ mice, canonical Wnt signaling was increased compared to age-matched wt mice. On the cellular level, β-catenin-dependent signaling activity was found to be increased in $Kif3a^{-/-}$ and $Ift88^{orpk/orpk}$ mouse embryonic fibroblasts (MEFs), as well as in $Ofd1^{-/-}$ mouse embryonic stem cells.

In contrast to the cilia-mediated mechanism(s) of Shh signaling, where two key SHH receptors, PTCH and SMO, have been localized to the ciliary axoneme (Corbit et al., 2005; Rohatgi et al., 2007), the role of the cilium and basal body to Wnt signal propagation is unclear. Nonetheless, several lines of evidence have indirectly shown a connection, including the fact that a number of key components of the Wnt signaling cascade localize to the primary cilium or basal body: pools of adenomatous polyposis coli (APC), a member of the β-catenin destruction complex, and β-catenin localize to the primary cilium, and the catalytic subunit α of protein phosphatase 1 (PPP1C), implicated in β-catenin dephosphorylation, Casein Kinase I (CKI) δ as well as phosphorylated β-catenin are present at the basal body and/or centrosome (Corbit et al., 2008; Gerdes et al., 2007). Kif3a, but not Ift88 or Ofd1, inhibits CKI-mediated phosphorylation of DVL, suggesting that there might be several mechanisms by which the primary cilium can suppress canonical Wnt signaling. Importantly, both in vivo and in vitro data indicate that ciliary and basal body mutants are hyperresponsive to, but still require, a Wnt stimulus, indicating that Wnt signaling is not globally misregulated (Fig. 7.2; Corbit et al., 2008; Gerdes et al., 2007).

Recently, the identification of a nonciliary gene that regulates the balance between canonical and noncanonical Wnt signaling has added to our understanding of the cilia-specific Wnt signaling pathway: disruption of seahorse (sea), the zebrafish homolog of human Leucine-rich repeat-containing protein 6 (LRRC6), causes renal cysts and laterality defects typical for ciliary genes (Kishimoto et al., 2008). In a proteomics analysis, murine Lrrc6 was found in the photoreceptor-sensory complex consisting of the outer segment and its cytoskeleton, including the ciliary rootlet (Liu et al., 2007). Although it is highly expressed in ciliated tissues, seahorse does not localize to the cilium and does not affect ciliogenesis or ciliary architecture in the sea morphants. Instead, sea morphants are dorsalized—implicating excessive β-catenin signaling activity—and show CE defects indicative of impaired noncanonical Wnt signaling. Taken together, these data suggest that sea might represent a downstream component of cilia-specific Wnt signaling.

3.2. Transmission of noncanonical Wnt signaling

While some studies focused on the role of the cilium in canonical Wnt signaling exclusively, others have addressed the impact of ciliary dysfunction on noncanonical Wnt signaling. Suppression of Inv in X. laevis embryos causes CE defects indicative of impaired PCP/noncanonical Wnt signaling (Simons et al., 2005). At the same time, loss of Inv also leads to excessive β-catenin-dependent canonical Wnt signaling, which is consistent with the notion that both branches of the Wnt signaling pathway are interconnected in a reciprocal, but not proportionate, relationship (Habas and Dawid, 2005; Malbon and Wang, 2006; Wallingford and Habas, 2006). Inversin interacts with DVL and, because it is required for degradation of nuclear (canonical) but not membrane-associated (noncanonical) DVL, it has been proposed to act as a switch between the two arms of the Wnt pathway. PCP defects such as exencephaly, open eyelids, or misarranged stereocilia in the organ of

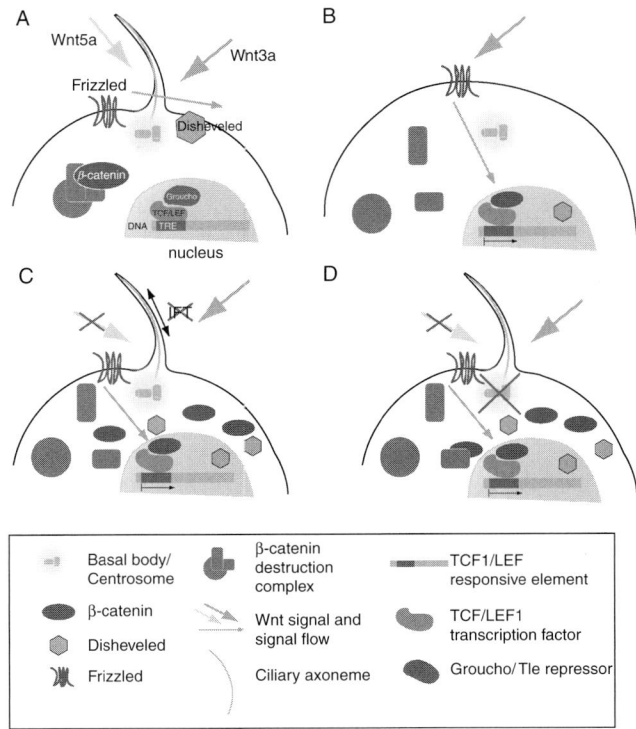

Figure 7.2 The role of the primary cilium and basal body in Wnt signaling. (A) In the presence of a primary cilium, noncanonical Wnt signals (such as Wnt5a) antagonize canonical signals (such as Wnt3a) and the Wnt signal is transmitted through membrane-associated DVL to downstream targets. (B) In unciliated cells, the external Wnt stimulus is transmitted through Fz receptors and leads to stabilization of β-catenin and subsequent transcriptional activation of β-catenin targets. (C) and (D) If the ciliary axoneme is disrupted (such as by suppression of KIF3a or IFT88), or the basal body is compromised (by suppression of BBS4 or Ofd1), antagonism of noncanonical and canonical Wnt signals and thus inhibition of β-catenin-dependent signaling is lost; both cytoplasmic β-catenin and DVL are stabilized and accumulate in the cytoplasm and nucleus, leading to transcriptional activation of β-catenin responsive elements. (See Color Insert.)

Corti in *Bbs* mouse mutants provided more evidence for an involvement of the basal body and cilium in noncanonical Wnt signaling (Ross *et al.*, 2005). Moreover, although *Bbs1* and *Bbs6* heterozygotes where phenotypically indistinguishable from wt littermates, the introduction of a heterozygous null mutation for *Vangl2*, encoding a core PCP component (Montcouquiol *et al.*, 2003), yielded stereociliary misorganization phenotypes reminiscent of the *Bbs1* and *Bbs6* homozygous mutant (Ross *et al.*, 2005).

One can reasonably argue that the above data suggest, but do not prove, the role of the basal body in regulating PCP, not least because it is theoretically possible that the BBS proteins investigated have hitherto unknown nonciliary roles. However, data from other *bona fide* ciliary mutants reinforce the above findings by showing that stereociliary bundle defects are not restricted to the BBS proteins. Kinocilia are absent from the organ of Corti in the inner ear of conditional *Ift88* knockout mice ($Ift88^{CKO/CKO}$) starting at E14.5, and the stereociliary bundles are misoriented compared to that of $Ift88^{CKO/+}$ heterozygous littermates (Jones *et al.*, 2008). Moreover, *Ift88* interacts genetically with *Vangl2* in the orientation of mouse inner hair cells, but $Ift88^{CKO/CKO}$ mice present with loss of the V-shaped formation of stereociliary bundles which are now arranged in a circle with the basal body in the center in ~10% of the cells, a phenotype not observed for *Vangl2* or other core PCP components. Because core PCP proteins such as VANGL2 or FZ3 partition normally to the apical cellular membrane, the role of the basal body/cilium seems to be not in the core PCP signaling pathway, but in the translation of polarized morphological cues. However, VANGL2 has been observed at the axoneme of both motile and primary cilia (human respiratory epithelium and IMCD3 cells, respectively; Ross *et al.*, 2005), and although it is not known whether VANGL2 is mislocalized in *Bbs* mutants, or indeed whether its ciliary localization is necessary for proper PCP signaling, it suggests a role for the cilium in core PCP signaling. One possible explanation is that the role of basal body/ciliary proteins is tissue-specific and that the orientation of the kinocilium and its basal body is a specialized readout of noncanonical Wnt signaling. In addition, it cannot be excluded that the basal body/primary cilium has separate roles in core PCP and downstream effector signaling, implicating that the cilium has been co-opted by the noncanonical Wnt signaling pathway in more than one way.

In zebrafish, suppression of *bbs* message gives rise to CE defects and abnormal posterior development, both hallmarks of impaired noncanonical Wnt signaling (Marlow *et al.*, 2004; Solnica-Krezel, 2005). Consistent with observations in mammalian systems, *bbs1,bbs4*, and *bbs6* interact genetically with *silberblick/Wnt11,pipetail/Wnt5b*, and *trilobite/Vangl2*, providing more evidence that bbs proteins participate in the noncanonical Wnt signaling pathway (Gerdes *et al.*, 2007; Ross *et al.*, 2005). Moreover, in the *bbs* morphants, expression of β-catenin transcriptional targets *axin2* and *sp5 transcription factor like* (*sp5l*) is upregulated, indicating that loss of noncanonical Wnt signaling is concomitant with gain of canonical Wnt signaling. Interestingly, *sea* has a similar role in constraining the canonical and facilitating the noncanonical Wnt response in ciliated cells, although it does not localize to the basal body or cilium (Kishimoto *et al.*, 2008). In HEK293T cells, the Wnt3a/Wnt5a antagonism (Mikels and Nusse, 2006; Topol *et al.*, 2003) is lost upon depletion of BBS4, suggesting a possible mechanism how the balance between both branches might be regulated (Gerdes *et al.*, 2007).

4. WNT AND CILIOGENESIS

In addition to the roles of the mature cilium and basal body in Wnt signal interpretation, another series of experiments has suggested that noncanonical Wnt signaling is also critical for ciliogenesis of both multi- and uniciliated cells; two *X. laevis* homologs of planar polarity effectors (PPE), *inturned* (*int*) and *fuzzy* (*fy*), play a role in cilia formation (Park et al., 2006). PPE proteins are not components of the core PCP signaling pathway, but are required for the correct translation of the morphological cues of planar cell polarization (Adler, 2002). Suppression of either *Xint* or *Xfy* results in Hedgehog signaling as well as CE defects consistent with ciliary dysfunction. The CE defects were exacerbated by a dominant negative DVL mutant, *Xdd1*, implicating a genetic interaction of *Xint/Xfy* with a Wnt signaling component (Park et al., 2006). Although DVL is required for both canonical and noncanonical Wnt signaling, CE defects and the interaction with PPE genes strongly suggest that PCP signaling is required for proper cilia formation in multiciliated epidermal cells.

Ciliogenesis is preceded by apical organization of an actin network and basal body docking to the apical membrane, a process that is controlled by *RhoA* (Pan et al., 2007) and *Dvl* (Park et al., 2008) in multiciliated cells. Remarkably, DVL has dual roles in this process: basal body docking, and subsequent polarization of the basal bodies to facilitate coordinated beating of the motile cilia are controlled by DVL (Park et al., 2008). This involvement has not yet been observed for and it remains to be seen whether DVL has been co-opted for formation of all cilia or is restricted to motile cilia and/or multiciliated cells.

The zebrafish gene *duboraya* (*dub*), an ortholog of human *CapZ-interacting protein* (*CapZIP.*), an interactor of actin filament capping proteins, supports a role for the actin network in ciliogenesis (Oishi et al., 2006): *Dub* morphants show CE and laterality defects, indicative of impaired PCP signaling and ciliogenesis. *Frizzled-2* (*fz2*), a predominantly noncanonical Wnt signaling component (Kuehl et al., 2000), is required for dub phosphorylation, indicating that *fz2* acts upstream of *dub*. Indeed, *fz2* morphants phenocopy the *dub* ciliogenesis defects; *dub* likely facilitates ciliogenesis by organizing the actin cytoskeleton in Kupffer's vesicle (KV) and pronephric epithelial cells. Overexpression of *DvlΔPDZ*, a construct that activates the canonical, but not the noncanonical Wnt signaling pathway, also impaired cilia formation in KV (Oishi et al., 2006). Dub is required for the organization of the actin network in ciliated cells and it is tempting to speculate that FZ2 activates a signaling cascade involving DVL and RhoA leading to reorganization of the actin cytoskeleton.

5. WNT DYSFUNCTION AND CILIOPATHY PHENOTYPES

While ciliary perturbations result in aberrant Wnt signaling, one recent example showed that perturbation of a Wnt component produced phenotypes common for ciliary dysfunction. In medaka fish (*oryzias latipes*), inhibition of the Wnt repressor gene family *Groucho/Tle* leads to left–right asymmetry and abnormal KV development (Bajoghli *et al.*, 2007). Overexpression of *amino-terminal enhancer of split* (*aes*), a dominant negative form of the *Tle* genes, revealed dual roles of Groucho protein: one in Brachyury signaling (KV development) and another in Wnt signaling (left–right asymmetry). Cilia in KVs of *Aes* overexpressants formed normally, although they seemed misoriented and reduced in numbers, a finding consistent with the observably smaller and dysmorphic KVs. Importantly, upregulation of the canonical Wnt signaling activity by overexpression of a gain-of-function β-catenin mutant replicated the observed laterality defects, but not the impairment in KV development. These data reiterate an emerging hypothesis, where modest overactivity of the canonical Wnt signaling pathway could account for a portion of the observed ciliary phenotypes (Gerdes *et al.*, 2007).

6. WHAT DRIVES THE PHENOTYPE: GAIN OF CANONICAL WNT SIGNALING OR LOSS OF NONCANONICAL WNT SIGNALING?

Perturbations of Wnt signaling have several distinctive sets of phenotypes commonly associated with either canonical or noncanonical Wnt signaling. We propose that the ciliary phenotype is predominantly driven by loss of noncanonical Wnt signaling for the following reasons: first, if excessive β-catenin signaling was a major outcome of ciliary dysfunction, one would expect to find a higher incidence of cancer or other proliferative phenotypes (Grigoryan *et al.*, 2008; Harada *et al.*, 1999; Nusse, 2005) in ciliopathy patients; however, there is no overtly increased cancer risk in such cohorts or in fact mouse models, with the possible exception of Von Hippel–Lindau disease (OMIM 193300). Indeed, evidence suggests that, while cells with disrupted basal bodies or cilia are hyperresponsive to canonical Wnt stimuli, the pathway is not activated in absence of any Wnt ligand (Corbit *et al.*, 2008; Gerdes *et al.*, 2007). Moreover, several targeted β-catenin gain-of-function mutations have informed us about tissue-specific phenotypes associated with excessive canonical Wnt signaling (for a recent review, see Grigoryan *et al.*, 2008). We will focus our discussion on three tissue types that are perturbed in many ciliopathies: eye/retina, ear, and kidney (Badano *et al.*, 2006; Fliegauf *et al.*, 2006).

Retinal degeneration is one of the most common phenotypes of all ciliopathies. However, gain of function in different compartments of the eye leads to developmental phenotypes never observed in the context of ciliopathies. Excessive β-catenin activity in the retina, for example, is associated with a loss of proper retinal structure and an apparent transdifferentiation of neural cells into pigmented cells (Fu et al., 2006). In the lens, on the other hand, constitutively activated β-catenin leads to a reduction of lens-specific marker expression concomitant with an increase of neural-specific marker expression, which in turn results in abnormal lens morphologies and loss of lens structure (Miller et al., 2006; Smith et al., 2005). Neither of these phenotypes has been observed in ciliopathy mouse models such as $Bbs4^{-/-}$ (Eichers et al., 2006; Mykytyn et al., 2004), or in retinal-specific conditional $Kif3a^{RHO-cre/RHO-cre}$ knockout mice (Jimeno et al., 2006); instead, mice show retinal degeneration, a phenotype that has been linked to a transport defect along the connecting cilium of mammalian photoreceptors as indicated by mislocalization of rhodopsin in $Bbs2^{-/-}$ mouse retinas (Nishimura et al., 2004). With respect to a role of ciliary and basal body proteins in Wnt signaling and the known implication of aberrant Wnt signaling in neurodegenerative phenotypes (Logan and Nusse, 2004), however, we cannot exclude a role for β-catenin-independent signaling in the observed retinal degeneration phenotype.

In mouse embryonic development, the ectodermal field that gives rise to both the otic placode and epidermis is characterized by *Pax2* expression, and the otic placode is the origin of the composite sensory structure of the inner ear. Expression of constitutively active β-catenin under the control of Pax2 responsive elements leads to expansion of the otic placode at the expense of epidermis (Ohyama et al., 2006). A similar phenotype has not been observed in any of the ciliopathy mouse models; rather, abnormalities in the mouse inner ear involve the organ of Corti and the misorganization of the stereociliary bundles of the outer hair cells that implicate aberrant noncanonical Wnt/PCP signaling (see above; Jones et al., 2008; Ross et al., 2005). In the organ of Corti, the orientation of the V-shaped stereociliary bundles is under the control of the kinocilium which is required for the translation of morphological cues from PCP signaling. In ciliary mutants, these directional cues are lost and kinocilia and basal bodies do not migrate properly, resulting in misarranged stereociliary bundles in the outer hair cells of the organ of Corti (see Chapter 8).

Finally, the potential implication of aberrant Wnt signaling in cystic kidney phenotypes associated with many ciliopathies has generated much discussion in recent years. Early studies demonstrated that a transgenic mouse model expressing constitutively active β-catenin in renal epithelium develops cystic kidneys (Saadi-Kheddouci et al., 2001). In addition, elevated β-catenin levels were observed in cystic kidneys of targeted $Kif3a^{CKO/CKO}$ mice (Lin et al., 2003), suggesting that excessive β-catenin-dependent signaling might be causally related to the formation of renal cysts. However, two conditional models expressing constitutively active β-catenin either in the kidney mesenchyme or in the epithelium do not develop cystic kidneys (Marose et al., 2008; Park et al., 2007). Instead, excessive β-catenin activity in renal mesenchyme led to enlarged kidneys that showed reduced branching and a lack of epithelial structures compared to controls, while in renal epithelium, gain of β-catenin activity suppressed tissue differentiation. Because the aforementioned targeted mouse models were generated using different regulatory elements, it is possible that the discrepancy between these findings reflects the differential role of β-catenin-dependent signaling throughout renal development. This argument is supported by the finding that the renal pathology related to impaired ciliary function is dependent on the time at which ciliary function is lost (Davenport et al., 2007; Patel et al., 2008). The area and incidence of cyst formation correlate with proliferative activity in kidney tissue but, importantly, loss of cilia does not induce proliferative activity (Patel et al., 2008). Rather, loss of ciliary/basal body function leads to abnormalities in oriented cell division, potentially implicating PCP signaling (Gong et al., 2004). In *polycystic kidney* (*pck*) rats, a model for autosomal recessive polycystic kidney disease (Lager et al., 2001), misoriented cell division had been implicated in cyst formation (Fischer et al., 2006). Recently, loss of *Fat4*, a murine ortholog of the *Drosophila melanogaster* PCP component *Fat*, has also been implicated in the etiology of cystic kidney disease (Saburi et al., 2008). $Fat4^{-/-}$ mice have misorganized stereociliary bundles in the organ of Corti, suggesting the role of FAT4 in PCP signaling is conserved between *Drosophila* and mouse; however, there is no apparent genetic interaction of *Fat4* and the core PCP component *Vangl2* (Saburi et al., 2008), possibly indicating that FAT4 and VANGL2 are components of two parallel but independent PCP signaling pathways. The data are suggestive that randomization of oriented cell division and thus loss of noncanonical Wnt signaling might be driving renal cyst formation, although it has not been unequivocally shown.

7. OTHER PHENOTYPES

One prominent symptom of the ciliopathy Meckel–Gruber syndrome (MKS, OMIM 249000) are abnormalities of the central nervous system (CNS), typically occipital encephalocele in combination with cleft palate, cystic kidneys, hepatic fibrosis, and polydactyly (Badano et al., 2006). Neural tube closure defects are also observed in $Bbs4^{-/-}$ mice and provided evidence that PCP might be impaired by basal body and ciliary dysfunction (Ross et al., 2005). Several core compounds have been implicated in neurulation and neural tube closure, including VANGL2, FZ3, and FZ6 as well as

DVL (Ciruna et al., 2006; Wallingford and Harland, 2002; Wang et al., 2006b), and it is suggestive that this phenotype is associated with defective PCP signaling. However, impairment of Hedgehog signaling can also lead to neural tube closure defects (Copp et al., 2003) and it is not clear which pathways contribute to the observed phenotypes. Constitutively active β-catenin in the CNS leads to an enlarged spinal cord and brain as well as a shift in the ratio of neural progenitor cells to differentiated neurons (Grigoryan et al., 2008), neither phenotypes is observed in ciliary mutants.

Many ciliopathies present with a high comorbidity with pancreatic abnormalities (10% of ADPKD patients have pancreatic cysts; Cano et al., 2004) or diabetes (Badano et al., 2006), suggesting that the cilium plays a role in pancreatic function. Oregon Ridge polycystic kidney (orpk) mice with a hypomorphic allele of Ift88 show a loss of acinar cells as well as expanded ductal epithelium in the exocrine compartment of the pancreas (Cano et al., 2004; Zhang et al., 2005), a finding that is replicated in pancreas specific $Kif3a^{-/-}$ mice (Cano et al., 2006). In pancreatic tissue of orpk mice, levels of Wnt signaling components such as β-catenin, LEF1, and TCF3 are elevated. However, constitutively active β-catenin during pancreas development leads to a lack of pancreatic branching and reduced size of both exocrine and endocrine compartments, whereas activation of β-catenin signaling at later stages leads to an expansion of β-cells in the endocrine compartment, indicating that excessive canonical Wnt signaling is unlikely to explain the observed phenotype. Moreover, although the acini are the structures most strongly affected by ciliary disruption, they are not known to be ciliated; in fact, in the exocrine pancreas only the ductal epithelial cells have been shown to emanate cilia, along with the β-cells of the endocrine pancreas (Zhang et al., 2005). Because of considerable fibrosis of the pancreas of ciliopathy models, the transforming growth factor (TGF) β signaling pathway has been implicated in the etiology of pancreatic abnormalities as observed in ciliopathy models and, indeed, expression of TGF-β2, TGF-β3, and connective tissue growth factor (CTGF) are elevated in the pancreas of the targeted $Kif3a^{-/-}$ mice (Cano et al., 2006). Shh signaling also plays a role in pancreas development and homeostasis (Nielsen et al., 2008) and it will be important to see how the pancreatic phenotype observed in orpk mice relates to the defects associated with the various candidate signaling pathways.

It is likely that the phenotype observed for ciliary dysfunction is tissue- as well as time-specific. For example, the skin phenotype varies depending on which ciliopathy gene is disrupted: The coats of inv mice do not have a uniform stroke direction but display swirls reminiscent of mice with a Frizzled mutation (Simons et al., 2005), providing the first clue of the interconnection of aberrant PCP signaling and ciliary function. K14 cre-driven gain-of-function mutations of β-catenin targeting the epidermis and teeth have a quite different phenotype, showing de novo hair generation and tumor formation after wounding in the epidermis as well as an excessive number of teeth. While these phenotypes have not been observed in *inv*, $Bbs2^{-/-}$, or $Bbs4^{-/-}$ mice, *orpk* mice develop an additional molar tooth and show follicular dysplasias reminiscent of K14 cre-driven constitutively active β-catenin (Lehman *et al.*, 2008).

8. Discussion

The involvement of primary cilia with the transmission of paracrine signals such as Shh and Wnt potentially represent both a significant expansion of the functions attributable to primary cilia and an illustration of how ancient structures have been co-opted to new utility, since there is ample evidence that both Shh and Wnt signaling are cilia-independent in invertebrates, such as *D. melanogaster* (Fanto and McNeill, 2004; Jacob and Lum, 2007; Seto and Bellen, 2004). At the same time, despite substantial progress in a short period of time, the precise role(s) of cilia in propagating and/or interpreting these signals are far from clear. Focusing on Wnt (since Shh is discussed elsewhere in this volume), there remain important questions.

Critically, definitive proof that defects in noncanonical, rather than canonical signaling, are driving cilia-mediated Wnt phenotypes is still lacking. The strongest evidence in support of a β-catenin-independent defect is the absence of typical β-catenin phenotypes including cancer in a variety of organs in both ciliopathy patients and animal models. At the same time, whether the defects observed represent core Wnt phenotypes or downstream surrogates is still debatable; examination of stereociliary bundle defects, neural tube defects and others reveal many similarities between ciliary mutant and *bona fide* PCP mutants, but there are also important differences. For example, both ciliopathy patients and mouse mutants exhibit neural tube defects, but the craniorachischisis phenotype characteristic of the *Vangl2* mutation in loop-tail mice (van Abeelen and Raven, 1968; Wilson and Finta, 1980) is never seen.

Central to that difficulty is the fact that a reliable biochemical assay for measuring PCP dysfunction remains elusive. While β-catenin defects can be quantified *in vitro* with the TOPFlash assay (Korinek *et al.*, 1997) and *in vivo* by reporter mice such as TOPGAL (DasGupta and Fuchs, 1999) or BAT GAL (Maretto *et al.*, 2003) that operate under a similar premise, the equivalent tools for β-catenin-independent signaling are yet to be developed. It is notable that loss of *BBS4* in HEK293 cells was concomitant with 40% reduction in JNK signaling (Gerdes *et al.*, 2007), a potential noncanonical Wnt output, suggesting that JNK signal output might prove to be a useful readout. Likewise, the quantification of Rho/ROCK as well as Ca^{2+} signaling in ciliary mutants might represent useful experimental tools to dissect the effect at the biochemical level on a tissue-by-tissue level of resolution (Boutros *et al.*, 1998; Kim and Han, 2005; Sheldahl *et al.*, 2003; Veeman *et al.*, 2003b; Yamanaka *et al.*, 2002), although this approach is hampered by the lack of specificity of these signaling cascades to PCP signaling.

In parallel, it will be important to determine the cilia-mediated effector molecules that regulate Wnt signal transmission. In part because this signal is not known and in part because of the aforementioned difficulties in measuring β-catenin-independent signal defects, the evidence implicating the primary cilium in this process remains indirect and associative. Unlike Shh signaling, where both PTCH and SMO have been shown to be required to localize to the primary cilium for efficient GLI processing, a ciliary Wnt receptor or coreceptor has not yet been discovered. In addition to testing of the known components of the Wnt receptor complex, such as the various receptors and classical and nonclassical coreceptors, such as FZ, LRP, or RYK receptor-like tyrosine kinase (RYK), we speculate that the primary cilium might contain a set of receptors with specialized noncanonical Wnt roles. Some candidates already exist: Meckelin (MKS3), for example, a seven-pass transmembrane protein with a cystein-rich domain shares topological similarity to the Fz proteins (Dawe *et al.*, 2007; Smith *et al.*, 2006). Moreover, loss of MKS3 *in vivo* was shown recently to cause PCP-like phenotypes in zebrafish (Leitch *et al.*, 2008), raising the possibility that Meckelin/MKS3 might bind to a subset of Wnt ligands and/or interact with Fz and LRP (co)receptors. Moreover, a cursory examination of the ciliary proteome (http://www.ciliaproteome.orgGherman *et al.*, 2006) showed an abundance of multipass transmembrane proteins with predicted receptor functions that might serve as useful candidates. Alternatively, the formal possibility also exists that the disruption of the balance between β-catenin-dependent and β-catenin-independent Wnt signaling could be a secondary effect to the disruption of another pathway, such as Shh. There is considerable crosstalk between Shh and Wnt signaling pathways (Ulloa *et al.*, 2007) and it is thus possible that the documented loss of Shh in ciliary mutants might contribute to the Wnt signaling phenotype. Finally, the Wnt signal could converge at the basal body after being processed by apically localized receptors; the accumulation of the proteasome in the vicinity of the basal body and centrosome (Wigley *et al.*, 1999) could reflect this function as a processing and relay station for a variety of different signaling cascades.

ACKNOWLEDGMENTS

We thank Edwin Oh for helpful discussions about the manuscript. This work was funded by R01HD04260 from the National Institute of Child Health and Development, and R01DK072301 and R01DK075972 from the National Institute of Diabetes, Digestive and Kidney Disorders, NIH (NK), and by a grant from the Macular Vision Research Foundation (NK).

REFERENCES

Abeelen, J. V., and Raven, S. (1968). Enlarged ventricles in the cerebrum of loop-tail mice. *Experientia* **24,** 191–192.

Aberle, H., Bauer, A., Stappert, J., Kispert, A., and Kemler, R. (1997). Beta-catenin is a target for the ubiquitin-proteasome pathway. *EMBO J.* **16,** 3797–3804.

Adler, P. N. (2002). Planar signaling and morphogenesis in *Drosophila*. *Dev. Cell* **2,** 525–535.

Afzelius, B. A. (1976). A human syndrome caused by immotile cilia. *Science* **193,** 317–319.

Badano, J. L., Mitsuma, N., Beales, P. L., and Katsanis, N. (2006). The ciliopathies: An emerging class of human genetic disorders. *Ann. Rev. Genomics Hum. Genet.* **7,** 125–148.

Bajoghli, B., Aghaallaei, N., Soroldoni, D., and Czerny, T. (2007). The roles of Groucho/Tle in left-right asymmetry and Kupffer's vesicle organogenesis. *Dev. Biol.* **303,** 347–361.

Boutros, M., Paricio, N., Strutt, D. I., and Mlodzik, M. (1998). Dishevelled activates JNK and discriminates between JNK pathways in planar polarity and wingless signaling. *Cell* **94,** 109–118.

Cano, D., Sekine, S., and Hebrok, M. (2006). Primary cilia deletion in Pancreatic Epithelial Cells results in cyst formation and pancreatitis. *Gastroenterology* **131,** 1856–1869.

Cano, D. A., Murcia, N. S., Pazour, G. J., and Hebrok, M. (2004). Orpk mouse model of polycystic kidney disease reveals essential role of primary cilia in pancreatic tissue organization. *Development* **131,** 3457–3467.

Ciruna, B., Jenny, A., Lee, D., Mlodzik, M., and Schier, A. (2006). Planar Cell Polarity Signaling couples cell division and morphogenesis during neurulation. *Nature* **439,** 220–224.

Clevers, H. (2006). Wnt/beta-catenin signaling in development and disease. *Cell* **127,** 469–480.

Copp, A., Greene, N., and Murdoch, J. (2003). The genetic basis of mammalian neurulation. *Nat. Rev. Genet.* **4,** 784–793.

Corbit, K., Shyer, A., Dowdle, W., Gaulden, J., Singla, V., and Reiter, J. (2008). Kif3a constrains beta-catenin-dependent Wnt signalling through dual ciliary and non-ciliary mechanisms. *Nat. Cell Biol.* **10,** 70–76.

Corbit, K. C., Aanstad, P., Singla, V., Norman, A. R., Stainier, D. Y., and Reiter, J. F. (2005). Vertebrate Smoothened functions at the primary cilium. *Nature* **437,** 1018–1021.

DasGupta, R., and Fuchs, E. (1999). Multiple roles for activated LEF/TCF transcription complexes during hair follicle development and differentiation. *Development* **126,** 4557–4568.

Davenport, J., Watts, A., Roper, V., Croyle, M., Groen, T. V., Wyss, J., Nagy, T., Kesterson, R., and Yoder, B. (2007). Disruption of intraflagellar transport in adult mice leads to obesity and slow-onset cystic kidney disease. *Curr. Biol.* **17,** 1586–1594.

Davis, E., Brueckner, M., and Katsanis, N. (2006). The emerging complexity of the vertebrate cilium: New functional roles for an ancient organelle. *Dev. Cell* **11,** 9–19.

Dawe, H., Smith, U., Cullinane, A., Gerrelli, D., Cox, P., Badano, J., Blair-Reid, S., Sriram, N., Katsanis, N., Attie-Bitach, T., Afford, S., Copp, A., *et al.* (2007). The Meckel-Gruber syndrome proteins MKS1 and meckelin interact and are required for primary cilium formation. *Hum. Mol. Genet.* **16,** 173–186.

Eichers, E., Abd-El-Barr, M., Paylor, R., Lewis, R., Weimin, B., Xiaodi, X., Meehan, T., Stockton, D., Wu, S., Lindsay, E., Justice, M., Beales, P., *et al.* (2006). Phenotypic characterization of Bbs4 null mice reveals age-dependent penetrance and variable expressivity. *Hum. Genet.* **120,** 211–226.

Fanto, M., and McNeill, H. (2004). Planar Polarity from flies to vertebrates. *J. Cell Sci.* **117,** 527–533.

Fischer, E., Legue, E., Doyen, A., Nato, F., Nicolas, J., Torres, V., Yaniv, M., and Ponotglio, M. (2006). Defective planar cell polarity in polycystic kidney disease. *Nat. Genet.* **38**, 21.

Fliegauf, M., Benzing, T., and Omran, H. (2006). When cilia go bad: Cilia defects and ciliopathies. *Nat. Rev. Mol. Cell Biol.* **8**, 880–893.

Fu, X., Sun, H., Klein, W., and Mu, X. (2006). Beta-catenin is essential for lamination but not neurogenesis in mouse retinal development. *Dev. Biol.* **299**, 424–437.

Gerdes, J. M., Liu, Y., Zaghloul, N. A., Leitch, C. C., Lawson, S., Kato, M., Beachy, P., Beales, P., DeMartino, G., Fisher, S., Badano, J. L., and Katsanis, N. (2007). Disruption of the basal body compromises proteasomal function and perturbs intracellular Wnt response. *Nat. Genet.* **39**, 1350–1360.

Gherman, A., Davis, E. E., and Katsanis, N. (2006). The ciliary proteome database: An integrated community resource for the genetic and functional dissection of cilia. *Nat. Genet.* **38**, 961–962.

Gong, Y., Mo, C., and Fraser, S. (2004). Planar cell polarity signalling controls cell division orientation during zebrafish gastrulation. *Nature* **430**, 689–693.

Grigoryan, T., Wend, P., Klaus, A., and Birchmeier, W. (2008). Deciphering the function of canonical Wnt signaling in development and disease: Conditional loss- and gain-of-function mutations of beta-catenin in mice. *Genes Dev.* **22**, 2308–2341.

Habas, R., and Dawid, I. (2005). Dishevelled and Wnt signaling: Is the nucleus the final frontier? *J. Biol.* **4**, 2.

Harada, N., Tamai, Y., Ishikawa, T., Sauer, B., Takaku, K., Oshima, M., and Taketo, M. (1999). Intestinal polyposis in mice with a dominant stable mutation of the beta-catenin gene. *EMBO J.* **18**, 5931–5942.

Huangfu, D., Liu, A., Rakeman, A. S., Murcia, N. S., Niswander, L., and Anderson, K. V. (2003). Hedgehog signalling in the mouse requires intraflagellar transport proteins. *Nature* **426**, 83–87.

Itoh, K., Brott, B. K., Bae, G.-U., Ratcliffe, M. J., and Sokol, S. Y. (2005). Nuclear localization is required for Dishevelled function in Wnt/beta-catenin signaling. *J. Biol.* **4**.

Jacob, L., and Lum, L. (2007). Hedgehog signaling pathway. *Sci. STKE.* cm6.

Jimeno, D., Feiner, L., Lillo, C., Teofilo, K., Goldstein, L., Pierce, E., and Williams, D. (2006). Analysis of kinesin-2 function in photoreceptor cells using synchronous Cre-loxP knockout of Kif3a with RHO-Cre. *Invest. Ophthalmol. Vis. Sci.* **47**, 5039–5046.

Jones, C., Roper, V. C., Foucher, I., Qian, D., Banizs, B., Petit, C., Yoder, B. K., and Chen, P. (2008). Ciliary proteins link basal body polarization to planar cell polarity. *Nature Genet.* **40**, 69–77.

Kim, G., and Han, J. (2005). JNK and ROKalpha function in the noncanonical Wnt/RhoA signaling pathway to regulate Xenopus convergent extension movements. *Dev. Dyn.* **232**, 958–968.

Kishimoto, N., Cao, Y., Park, A., and Sun, Z. (2008). Cystic kidney gene *seahorse* regulates cilia-mediated processes and Wnt pathways. *Dev. Cell* **14**, 954–961.

Korinek, V., Barker, N., Morin, P. J., Wichen, D. V., Weger, R. D., Kinzler, K. W., Vogelstein, B., and Clevers, H. (1997). Constitutive transcriptional activation by a beta-catenin-tcf complex in APC-/- colon carcinoma. *Science* **275**, 1784–1787.

Kuehl, M., Sheldahl, L. C., Park, M., Miller, J. R., and Moon, R. T. (2000). The Wnt/Ca2+ pathway: A new vertebrate Wnt signalling takes shape. *Trends Genet.* **16**, 279–283.

Lager, D., Qian, Q., Bengal, R., Ishibashi, M., and Torres, V. (2001). The pck rat: A new model that resembles human autosomal dominant polycystic kidney and liver disease. *Kidney Int.* **59**, 126–136.

Lehman, J., Michaud, E., Schoeb, T., Aydin-Son, Y., Miller, M., and Yoder, B. (2008). The oak ridge polycystic kidney mouse: Modeling ciliopathies of mice and men. *Dev. Dyn.* **237**, 1960–1971.

Leitch, C., Zaghloul, N., Davis, E., Stoetzel, C., Diaz-Font, A., Rix, S., Alfadhel, M., Lewis, R., Eyaid, W., Banin, E., Dollfus, H., Beales, P., et al. (2008). Hypomorphic mutations in syndromic encephalocele genes are associated with Bardet-Biedl syndrome. *Nat. Genet.* **40,** 443–448.

Lin, F., Hiesberger, T., Cordes, K., Sinclair, A. M., Goldstein, L. S., Somlo, S., and Igarashi, P. (2003). Kidney-specific inactivation of the KIF3A subunit of kinesin-II inhibits renal ciliogenesis and produces polycystic kidney disease. *Proc. Natl. Acad. Sci. USA* **100**.

Liu, Q., Tan, G., Levenkova, N., Li, T., EN Pugh, J., Rux, J., Speicher, D., and Pierce, E. (2007). The proteome of the mouse photoreceptor sensory cilium complex. *Mol. Cell Proteomics* **6,** 1299–1317.

Logan, C. Y., and Nusse, R. (2004). The Wnt signaling pathway in development and disease. *Ann. Rev. Cell Dev. Biol.* **20,** 781–810.

Malbon, C., and Wang, H. (2006). Dishevelled: A mobile scaffold catalyzing development. *Curr. Top. Dev. Biol.* **72,** 153–166.

Maretto, S., Cordenonsi, M., Dupont, S., Braghetta, P., Broccoli, V., Hassan, A., Volpin, D., Bressan, G., and Piccolo, S. (2003). Mapping Wnt/beta-catenin signaling during mouse development and in colorectal tumors. *Proc. Natl. Acad. Sci. USA* **100,** 3299–3304.

Marlow, F., Gonzalez, E. M., Yin, C. Y., Rojo, C., and Solnica-Krezel, L. (2004). No tail co-operates with non-canonical Wnt signaling to regulate posterior body morphogenesis in zebrafish. *Development* **131,** 203–216.

Marose, T., Merkel, C., McMahon, A., and Carroll, T. (2008). Beta-catenin is necessary to keep cells of ureteric bud/Wolffian duct epithelium in a precursor state. *Dev. Biol.* **314,** 112–126.

Marshall, W., and Kintner, C. (2008). Cilia orientation and the fluid mechanics of development. *Curr. Opin. Cell. Biol.* **20,** 48–52.

Marszalek, J., Ruiz-Lozano, P., Roberts, E., Chien, K., and Goldstein, L. (1999). Situs inversus and embryonic ciliary morphogenesis defects in mouse mutants lacking the KIF3A subunit of kinesin-II. *Proc. Natl. Acad. Sci. USA* **96,** 5043–5048.

Mikels, A., and Nusse, R. (2006). Purified Wnt5a protein activates or inhibits beta-catenin–TCF signaling depending on receptor context. *PLoS Biol.* **4,** e115.

Miller, L., Smith, A., Taketo, M., and Lang, R. (2006). Optic cup and facial patterning defects in ocular ectoderm beta-catenin gain-of-function mice. *BMC Dev. Biol.* **6,** 14.

Montcouquiol, M., Rachel, R. A., Lanford, P. J., Copeland, N. G., Jenkins, N. A., and Kelley, M. W. (2003). Identification of Vangl2 and Scrb1 as planar polarity genes in mammals. *Nature* **423,** 173–177.

Mykytyn, K., Mullins, R., Andrews, M., Chiang, A., Swiderski, R., Yang, B., Braun, T., Casavant, T., Stone, E., and Sheffield, V. (2004). Bardet-Biedl syndrome type 4 (BBS4)-null mice implicate Bbs4 in flagella formation but not global cilia assembly. *Proc. Natl. Acad. Sci. USA* **101,** 8664–8669.

Na, J., Lykke-Andersen, K., Padilla, M. T., and Zernicka-Goetz, M. (2006). Dishevelled proteins regulate cell adhesion in mouse blastocyst and serve to monitor changes in Wnt signaling. *Dev. Biol.* **302,** 40–49.

Nachury, M., Loktev, A., Zhang, Q., Westlake, C., Peränen, J., Merdes, A., Slusarski, D., Scheller, R., Bazan, J., Sheffield, V., and Jackson, P. (2007). A core complex of BBS proteins cooperates with the GTPase Rab8 to promote ciliary membrane biogenesis. *Cell* **129,** 1201–1203.

Nielsen, S., Mollgard, K., Clement, C., Veland, I., Awan, A., Yoder, B., Novak, I., and Christensen, S. (2008). Characterization of primary cilia and hedgehog signaling during development of the human pancreas and in human pancreatic ducts cancer cell lines. *Dev. Dyn.* **237,** 2039–2052.

Nishimura, D., Fath, M., Mullins, R., Searby, C., Andrews, M., Davis, R., Andorf, J., Mykytyn, K., Swiderski, R., Yang, B., Carmi, R., Stone, E., et al. (2004). Bbs2-null mice have neurosensory deficits, a defect in social dominance, and retinopathy associated with mislocalization of rhodopsin. Proc. Natl. Acad. Sci. USA **101**, 16588–16593.

Nurnberger, J., Bacallao, R. L., and Phillips, C. L. (2002). Inversin forms a complex with catenins and N-cadherin in polarized epithelial cells. Mol. Biol. Cell **13**, 3096–3106.

Nurnberger, J., Kribben, A., Saez, A. O., Heusch, G., Philipp, T., and Phillips, C. (2004). The Invs gene encodes a microtubule-associated protein. J. Am. Soc. Nephrol. **15**, 1700–1710.

Nusse, R. (2005). Wnt signaling in disease and in development. Cell Res. **15**, 28–32.

Ohyama, T., Mohamed, O., Taketo, M., Dufort, D., and Groves, A. (2006). Wnt signals mediate a fate decision between otic placode and epidermis. Development **133**, 865–875.

Oishi, I., Kawakami, Y., Raya, A., Callol-Massot, C., and Bemonte, J. I. (2006). Regulation of primary cilia formation and left-right patterning in zebrafish by a noncanonical Wnt signaling mediator, duboraya. Nat. Genet. **38**, 1316–1322.

Otto, E. A., Schermer, B., Obara, T., O'Toole, J. F., Hiller, K. S., Mueller, A. M., Ruf, R. G., Hoefele, J., Beekmann, F., Landau, D., Foreman, J. W., Goodship, J. A., et al. (2003). Mutations in INVS encoding inversin cause nephronophthisis type 2, linking renal cystic disease to the function of primary cilia and left-right axis determination. Nat. Genet. **34**, 413–420.

Pan, J., You, Y., Huang, T., and Brody, S. (2007). RhoA-mediated apical actin enrichment is required for ciliogenesis and promoted by Foxj1. J. Cell Sci. **120**, 1868–1876.

Park, J., Valerius, M., and McMahon, A. (2007). Wnt/beta-catenin signaling regulates nephron induction during mouse kidney development. Development **134**.

Park, T., Haigo, S., and Wallingford, J. (2006). Ciliogenesis defects in embryos lacking inturned or fuzzy function are associated with failure of planar cell polarity and Hedgehog signaling. Nat. Genet. **38**, 303–311.

Park, T., Mitchell, B., Abitua, P., Kintner, C., and Wallingford, J. (2008). Dishevelled controls apical docking and planar polarization of basal bodies in ciliated epithelial cells. Nat. Genet. **40**, 871–879.

Patel, V., Li, L., Cobo-Stark, P., Shao, X., Somlo, S., Lin, F., and Igarashi, P. (2008). Acute kidney injury and aberrant planar cell polarity induce cyst formation in mice lacking renal cilia. Hum. Mol. Genet. **17**, 1578–1590.

Rohatgi, R., Milenkovic, L., and Scott, M. (2007). Patched1 regulates hedgehog signaling at the primary cilium. Science **317**, 372–376.

Ross, A. J., May-Simera, H., Eichers, E. R., Kai, M., Hill, J., Jagger, D. J., Leitch, C. C., Chapple, J. P., Munro, P. M., Fisher, S., Tan, P. L., Phillips, H. M., et al. (2005). Disruption of Bardet-Biedl syndrome ciliary proteins perturbs planar cell polarity in vertebrates. Nat. Genet. **37**, 1135–1140.

Saadi-Kheddouci, S., Berrebi, D., Romagnolo, B., Cluzeaud, F., Peuchmauer, M., Kahn, A., Vanderwalle, A., and Perret, C. (2001). Early development of polycystic kidney disease in transgenic mice expressing an activated mutant of the beta-catenin gene. Oncogene **20**, 5972–5981.

Saburi, S., Hester, I., Fischer, E., Pontoglio, M., Eremina, V., Gessler, M., Quaggin, S., Harrison, R., Mount, R., and McNeill, H. (2008). Loss of Fat4 disrupts PCP signaling and oriented cell division and leads to cystic kidney disease. Nat. Genet. **40**, 1010–1015.

Schneider, L., Clement, C. A., Teilmann, S. C., Pazour, G. J., Hoffmann, E. K., Satir, P., and Christensen, S. T. (2005). PDGFRalpha signaling is regulated through the primary cilium in fibroblasts. Curr. Biol. **15**, 1861–1866.

Seifert, J., and Mlodzik, M. (2007). Frizzled/PCP signalling: A conserved mechanism regulating cell polarity and directed motility. Nat. Rev. Genet. **8**, 126–138.

Seto, E. S., and Bellen, H. J. (2004). The ins and outs of Wingless signaling. *Trends in Cell Biology* **14**, 45–53.

Sheldahl, L., Slusarski, D., Pandur, P., Miller, J., Kühl, M., and Moon, R. (2003). Dishevelled activates Ca2+ flux, PKC, and CamKII in vertebrate embryos. *J. Cell. Biol.* **161**, 769–777.

Simons, M., Gloy, J., Ganner, A., Bullerkotte, A., Bashkurov, M., Kronig, C., Schermer, B., Benzing, T., Cabello, O. A., Jenny, A., Mlodzik, M., Polok, B., et al. (2005). Inversin, the gene product mutated in nephronophthisis type II, functions as a molecular switch between Wnt signalling pathways. *Nat. Genet.* **37**, 537–543.

Smith, A., Miller, L., Song, N., Taketo, M., and Lang, R. (2005). The duality of beta-catenin function: A requirement in lens morphogenesis and signaling suppression of lens fate in periocular ectoderm. *Dev. Biol.* **285**, 477–489.

Smith, U. M., Consugar, M., Tee, L. J., McKee, B. M., Maina, E. N., Whelan, S., Morgan, N. V., Goranson, E., Gissen, P., Lilliquist, S., Aligianis, I. A., Ward, C. J., et al. (2006). The transmembrane protein meckelin (MKS3) is mutated in Meckel-Gruber syndrome and the wpk rat. *Nat. Genet.* **38**, 191–196.

Solnica-Krezel, L. (2005). Conserved patterns of cell movements during vertebrate gastrulation. *Curr. Biol.* **15**, R213–R228.

Topol, L., Jiang, X., Choi, H., Garrett-Beal, L., Carolan, P., and Yang, Y. (2003). Wnt-5a inhibits the canonical Wnt pathway by promoting GSK-3-independent beta-catenin degradation. *J. Cell. Biol.* **162**, 899–908.

Ulloa, F., Itasaki, N., and Briscoe, J. (2007). Inhibitory Gli3 activity negatively regulates Wnt/beta-catenin signaling. *Curr. Biol.* **17**, 545–550.

Veeman, M. T., Axelrod, J. D., and Moon, R. T. (2003a). A second canon. Functions and mechanisms of beta-catenin-independent Wnt signaling. *Dev. Cell* **5**, 367–377.

Veeman, M. T., Slusarski, D. C., Kaykas, A., Louie, S. H., and Moon, R. T. (2003b). Zebrafish prickle, a modulator of noncanonical Wnt/Fz signaling, regulates gastrulation movements. *Curr. Biol.* **13**, 680–685.

Wallingford, J., and Habas, R. (2006). The developmental biology of Dishevelled: An enigmatic protein governing cell fate and cell polarity. *Development* **132**, 4421–4436.

Wallingford, J., and Harland, R. (2002). Neural tube closure requires Dishevelled-dependent convergent extension of the midline. *Development* **129**, 5815–5825.

Wang, J., Hamblet, N., Mark, S., Dickinson, M., Brinkman, B., Segil, N., Fraser, S., Chen, P., Wallingford, J., and Wynshaw-Boris, A. (2006a). Dishevelled genes mediate a conserved mammalian PCP pathway to regulate convergent extension during neurulation. *Development* **133**, 1767–1778.

Wang, Y., Guo, N., and Nathans, J. (2006b). The role of Frizzled3 and Frizzled6 in neural tube closure and in the planar polarity of inner-ear sensory hair cells. *J. Neurosci.* **26**, 2147–2156.

Wigley, W., Fabunmi, R., Lee, M., Marino, C., Muallem, S., DeMartino, G., and Thomas, P. (1999). Dynamic association of proteasomal machinery with the centrosome. *J. Cell. Biol.* **145**, 481–490.

Wilson, D., and Finta, L. (1980). Early development of the brain and spinal cord in dysraphic mice: A transmission electron microscopic study. *J. Comp. Neurol.* **190**, 363–371.

Yamanaka, H., Moriguchi, T., Masuyama, N., Kusakabe, M., Hanafusa, H., Takada, R., Takada, S., and Nishida, E. (2002). JNK functions in the non-canonical Wnt pathway to regulate convergent extension movements in vertebrates. *EMBO Rep.* **3**, 69–75.

Zhang, Q., Davenport, J., Croyle, M., Haycraft, C., and Yoder, B. (2005). Disruption of IFT results in both exocrine and endocrine abnormalities in the pancreas of Tg737(orpk) mice. *Lab. Invest.* **85**, 45–64.

CHAPTER EIGHT

Primary Cilia in Planar Cell Polarity Regulation of the Inner Ear

Chonnettia Jones *and* Ping Chen

Contents

1. Introduction	198
2. Planar Cell Polarity and Ciliogenesis of the Inner Ear Sensory Organs	200
2.1. Planar cell polarity of the mammalian inner ear sensory organs	200
2.2. Kinocilia and development of the stereociliary bundle	202
2.3. Physiological function of primary cilia in the inner ear sensory organs	205
3. Planar Cell Polarity Regulation for the Development of the Inner Ear	206
3.1. Planar cell polarity pathway in orienting the sensory cells of the inner ear	206
3.2. Planar cell polarity pathway in patterning and growth of the inner ear sensory organs	209
4. Cilia and PCP Regulation	209
4.1. Evidence linking cilia and PCP signaling in vertebrates	209
4.2. Cilia in regulating PCP of the vertebrate inner ear	211
4.3. Crosstalk between cilia and conserved PCP proteins	213
5. Cilia and Determination of the Intrinsic Cellular Polarity of Inner Ear Cells	214
5.1. The molecular machinery for building the intrinsically polarized hair bundles	214
5.2. Cilia determine intrinsic cellular polarity	215
6. Conclusions and Perspectives	217
References	219

Abstract

Primary cilia are essential components of diverse cellular processes. Many of the requirements can be linked to the apparent signaling function of primary cilia. Recent studies have also uncovered a role for primary cilia in planar cell polarity (PCP) signaling. PCP refers to the coordinated orientation of cells along

Department of Cell Biology, Emory University School of Medicine, Atlanta, Georgia an axis parallel to the plane of the cell sheet. In vertebrates, the inner ear sensory organs display distinctive forms of PCP. One of the inner ear PCP characteristics is the coordinated positioning of a primary cilium eccentrically in every sensory hair cell within each organ. The inner ear, therefore, provides an opportunity to explore the cellular role of primary cilia in PCP signaling. In this chapter, we will introduce the PCP of the inner ear sensory organs, describe the conserved mechanism underlying the establishment of the planar polarity axis in invertebrates and vertebrates, and highlight a unique requirement for primary cilia in PCP regulation in vertebrates. Additionally, we will discuss a potentially ubiquitous role for cilia in cellular polarization in general.

1. Introduction

Cellular polarity underlies many complex developmental processes in higher organisms, such as the generation of different cell lineages from a single fertilized egg, the establishment of the asymmetric body axes from the blastula, the directional migration of cells, the specification of the axon and dendrites in a neuron, the compartmentalization of apical–basal domains of epithelial cells, and the precise orientation of cells within a tissue. The acquisition of each of these polarities often involves common molecular and cellular components that act in a cell context-dependent manner (Lawrence et al., 2007).

Primary cilia are microtubule-based organelles assembled from the centriole-derived basal body and extend from the surfaces of cells (Davis et al., 2006). The assembly of primary cilia and transport of proteins along the length of the ciliary axonemes depends upon intraflagellar transport (IFT) (Davis et al., 2006; Scholey and Anderson, 2006; Taulman et al., 2001). Consistent with their antennae-like morphology, cilia can apparently receive remote information and transduce the signal to the cell body largely through IFT proteins. The signaling function of cilia has been linked to many cellular processes, including mechanotransduction, cellular proliferation and differentiation, cell migration, and signaling transduction for Hedgehog (Hh) and platelet-derived growth factor receptor (PDGFR) (Christensen and Ott, 2007; Christensen et al., 2007; Corbit et al., 2005; Davenport et al., 2007; Davis et al., 2006; Haycraft et al., 2005, 2007; Liu et al., 2005; Michaud and Yoder, 2006; Ou et al., 2007; Schneider et al., 2005; Scholey and Anderson, 2006; Singla and Reiter, 2006; Yoder et al., 2002). Recent studies have also revealed links between cilia and planar cell polarity (PCP) signaling in the mouse inner ear and in the zebrafish (Jones et al., 2008; Park et al., 2006, 2008; Ross et al., 2005; Wallingford, 2006).

In the inner ear sensory organs, each sensory hair cell consists of a hair bundle that is made up of specialized F-actin-filled microvilli, known as stereocilia, and a single primary cilium, known as the kinocilium. On the apical surface of each hair cell, stereocilia are arranged with graded heights into a staircase that is distinctively shaped for each inner ear sensory organ. The kinocilium sits snugly near the tallest stereocilia (Fig. 8.1). The position

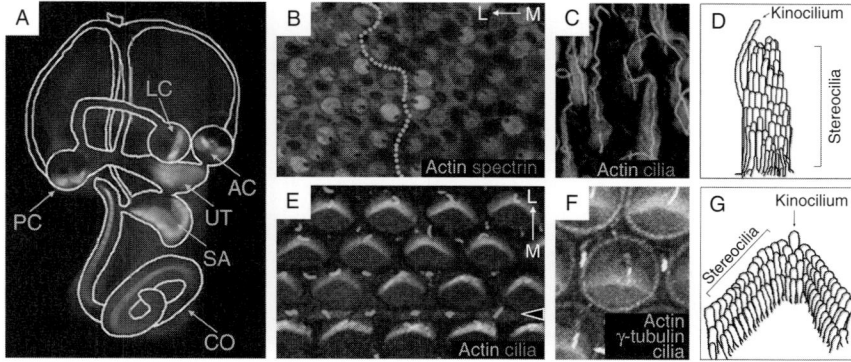

Figure 8.1 Planar cell polarity of inner ear sensory epithelia. (A) Mouse inner ear isolated at E18.5. The white tracing outlines the fluid-filled labyrinth that connects the vestibule and the cochlea of the inner ear. A Math1-GFP transgene (green) is expressed in hair cells and highlights the sensory epithelia of the six sensory organs: the organ of Corti of the cochlea (CO); the maculae of the utricle (UT) and the saccule (SA); and the three perpendicular cristae: the posterior crista (PC), anterior crista (AC), and lateral crista (LC). (B)–(G) The planar cell polarity of vestibular (B)–(D) and cochlear (E)–(G) sensory organs isolated from a mouse embryo at E18.5 viewed by immunohistochemistry and confocal microscopy (B), (C), (E), (F), and the intrinsic polarity of vestibular (D) and cochlear (G) hair cells illustrated by schematic diagrams. The hair cells are polarized across the sensory epithelium along the medial-to-lateral axis (M, medial; L, lateral). (B) The utricle viewed at a level just beneath the cell cortex. Actin (green) is enriched at the cell membranes. Spectrin (red) accumulates in the cuticular plates of hair cells, is excluded from the pericentriolar region surrounding the basal body and serves as a convenient marker of hair cell polarity. The dotted line (purple) demarcates the line of polarity reversal, where the polarities of the hair cells are reversed in the utricle. (E) The surface of the organ of Corti. Actin (green) accentuates the microvilli that make up the stereociliary bundles. Tubulin (red) marks both the cilia associated with stereociliary bundles atop the sensory hair cells and the cilia of intervening nonsensory supporting cells. The black arrowhead marks the separation of the inner hair cells from the outer hair cells, which are uniformly oriented toward the abneural, or lateral, edge of the sensory epithelium. (C) and (F) Confocal projections of vestibular (C) and cochlear (F) hair cell bundles at high magnification. The kinocilia (C, red; F, yellow) are assembled from the eldest of two centrioles (C, not shown; F, red) that make up the basal bodies, and are closely apposed to the hair cell bundles (C, F, green). Kinocilia and stereocilia of the vestibule (C) are considerably longer than those of the cochlea (F). (D) and (G) Diagrams of a typical vestibular (D) and cochlear (G) hair cell bundle showing the array of stereocilia rows increasing in height and the eccentrically placed kinocilium which protrudes from behind the tallest row of stereocilia. Note the round morphology of the vestibular hair cell bundle in comparison to the "V"-shaped morphology of the cochlear hair cell bundle. (See Color Insert.)

of the kinocilium and the organization of the stereociliary bundle mark an intrinsic polarity and the orientation of each hair cell. All of the hair cells in each inner ear sensory organ are coordinately oriented, conferring a precise polarity within the plane of the sensory epithelium (Fig. 8.1) (Lewis and Davies, 2002). The regulation of PCP therefore requires directional information, the establishment of the planar polarity axis that is relayed to all of the cells across the entire tissue, and transformation of polarity signals to the formation of the polarized structure of hair bundles. Genetic studies identified a set of conserved genes encoding the so-called core PCP proteins that are asymmetrically partitioned to define the planar polarization axis and to coordinate the orientation of hair cells across the sensory epithelium (Fig. 8.3) (Curtin *et al.*, 2003; Lu *et al.*, 2004; Montcouquiol *et al.*, 2003, 2006; Wang *et al.*, 2005, 2006a,b). However, the molecular and cellular apparatus that receives directional information is obscure (Dabdoub *et al.*, 2003; Qian *et al.*, 2007), and the precise mechanism that determines the intrinsic polarity of hair cells has not been determined.

During development, the positioning of the primary cilium of the hair cell, the kinocilium, leads and predicts the polarity of each stereociliary bundle, supporting the hypothesis that kinocilia direct the polarization of hair cells (Cotanche and Corwin, 1991; Denman-Johnson and Forge, 1999; Tilney *et al.*, 1992). This review summarizes recent results from inner ear studies that revealed a surprising but likely ubiquitous role for primary cilia in cellular polarity regulation, through the positioning of the basal body to direct the polarity of cells (Absalon *et al.*, 2007; Benzing and Walz, 2006; Boisvieux-Ulrich and Sandoz, 1991; Boisvieux-Ulrich *et al.*, 1985; Jones *et al.*, 2008; Mitchell *et al.*, 2007).

2. Planar Cell Polarity and Ciliogenesis of the Inner Ear Sensory Organs

2.1. Planar cell polarity of the mammalian inner ear sensory organs

The mammalian inner ear is comprised of two distinct regions: the cochlea, which regulates auditory function, and the vestibule, which perceives motion and balance (Fig. 8.1). There are six sensory organs within the inner ear that are largely responsible for regulating these specialized functions (Fig. 8.1). The auditory sensory organ, called the organ of Corti, runs along the length of the spiraling cochlea (Fig. 8.1). The two macular sensory organs, the saccule and the utricle, reside in the center of the inner ear and detect gravity and linear acceleration (Fig. 8.1). Each of three cristae lies at the base of three perpendicular semicircular canals of the vestibule to detect rotational motion and contribute to the maintenance of balance (Fig. 8.1).

The auditory and vestibular sensory organs contain polarized epithelial cells of two general types: the mechanosensory hair cells and the nonsensory supporting cells. The hair cells have elongated cell bodies whose lateral membranes are tightly connected to the surrounding supporting cells by adherens and tight junctions. Together, these cells of the sensory epithelia make up a cellular mosaic, such that hair cells are separated from adjacent hair cells by supporting cells. In the organ of Corti, the mechanosensory hair cells are arranged into four parallel rows, one row of inner hair cells (IHCs) and three rows of outer hair cells (OHCs) (Fig. 8.1). The mechanosensory hair cells also alternate with supporting cells to form a characteristically shaped patch in each of the vestibular sensory epithelia (Fig. 8.1).

The mammalian cochlea and vestibule can process sound and positional signals, respectively, with remarkable resolution and sensitivity. This ability depends largely on an extraordinary mechanotransduction apparatus present atop the mechanosensory hair cells. At the apical surfaces of the mechanosensory hair cells are a bundle of modified actin-rich microvilli called stereocilia, often referred to as stereociliary bundles (Fig. 8.1). Each bundle is made up of tightly packed rows of stereocilia that are connected by extracellular linkages and increase in height to form a staircase-like structure (Fig. 8.1). Behind and central to the tallest row of stereocilia sits a single true cilium, or kinocilium, which is linked to the stereociliary bundle via extracellular kinocilial links (Fig. 8.1). The stereocilia are anchored beneath the cell cortex to the cuticular plate, a meshwork of crosslinked actin filaments and associated cytoskeletal proteins. The basal body at the base of the kinocilium is closely apposed to the cuticular plate and connects the filaments to apical microtubules that radiate along the periphery of the cell cortex and along the length of the lateral cell membranes.

The eccentric position of the kinocilium and the organization of the stereociliary bundle mark the intrinsic polarity and the orientation of the hair cell. The orientations of all of the stereociliary bundles are coordinately aligned across the plane of the epithelium of each sensory organ, displaying a distinct cellular synchrony called planar cell polarity (Jones and Chen, 2007; Kelly and Chen, 2007; Lewis and Davies, 2002). In the cochlea, the stereociliary bundle has a "V" shape, and the vertices of the "V"-shaped stereociliary bundles invariantly point toward the abneural, or lateral, edge of the epithelium (Fig. 8.1). The vestibular hair cell bundles of the cristae are uniformly oriented toward one side of the epithelia. In the macular sensory organs of the vestibule, the stereociliary bundles are also oriented along the mediolateral axis of the epithelia (Fig. 8.1). However, the cells on either side of a line of polarity reversal, which lies lateral to a physiologically and morphologically distinct striolar region, have opposite orientations

(Fig. 8.1) (Deans et al., 2007; Denman-Johnson and Forge, 1999; Kelly and Chen, 2007).

PCP of the inner ear sensory organs is continuous across each sensory epithelium and contributed by both hair cells and supporting cells. The supporting cells of the inner ear sensory organs that interdigitate between hair cells are also polarized. In the cochlea, the supporting cells are characterized by long phalangeal processes, which are oriented coordinately across the sensory epithelium (reviewed by Jones and Chen, 2007). In the vestibule, asymmetric subcellular localization of proteins reveals the intrinsic polarity and coordinated orientation of supporting cells (Deans et al., 2007; Wang et al., 2006b).

2.2. Kinocilia and development of the stereociliary bundle

The hair cell bundles are intrinsically polarized, bilaterally symmetrical structures (Figs. 8.1 and 8.2). In the organ of Corti, the kinocilium sitting precisely at the vertex of each "V"-shaped stereociliary bundle. Moreover, the bundle is oriented so that the kinocilium and tallest stereocilia row face the cell periphery; while the shortest stereocilia row faces the cell interior. The vestibular hair cell bundles exhibit a similar staircase structure, and the placement of the kinocilium in relation to the bundle is preserved. However, the vestibular hair bundle morphology is circular in shape, and the lengths of stereocilia and kinocilia are considerably longer (Fig. 8.1).

The polarized structures of the stereociliary bundles have functional significance. Deflection of the bundle either toward or away from the tallest stereocilia row (parallel to the planar polarity axis) leads to opposite mechanosensory responses. Deflections of the bundle perpendicular to the planar polarity axis have no effect, supporting the directional sensitivity of the hair cell bundle (Hudspeth, 1989, 2000). Moreover, mutations in genes that disrupt the morphology of hair cell bundles cause impaired hearing and vestibular dysfunction (El-Amraoui and Petit, 2005). It is conceivable that the coordinated orientation of hair cell bundles across the epithelium of each inner ear sensory organ is essential for their physiological functions and contributes to the sensitivity of the inner ear.

Temporal observations of stereociliary bundles by scanning electron microscopy during mammalian ear development have revealed that the development of the stereociliary bundle is carefully orchestrated during hair cell maturation (Fig. 8.2) (Frolenkov et al., 2004). In the mouse, the microvilli that will give rise to the stereocilia are apparent on the apical surfaces of newly differentiated hair cells by embryonic day 13.5 (E13.5) in the vestibule and E15.5 in the cochlea. Initially, the microvilli are of a uniform height at the cell cortex. Near the center of the immature bundle sits a single kinocilium, easily distinguishable at this stage because it is longer than the microvilli surrounding it (Fig. 8.2) (Frolenkov et al., 2004; Kikuchi

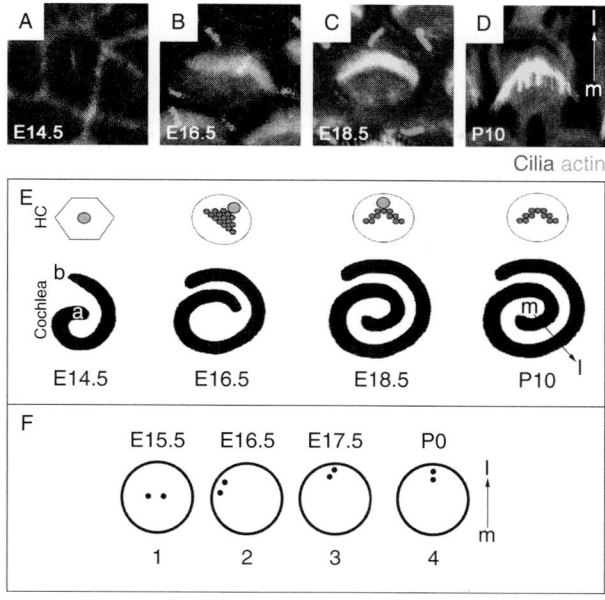

Figure 8.2 Kinocilium and basal body polarization. Confocal images of cochlear whole mounts (A)–(D) and a schematic diagram (E) focusing on a single cochlear hair cell bundle at successive stages in mouse ear development. Extension of the cochlear spiral unidirectionally from the base (b) to the distal tip, or apex (a), occurs concurrently with hair cell (HC) differentiation in the organ of Corti. In an immature hair cell, the kinocilium (red) projects centrally from the apical cell membrane. The kinocilium proceeds to polarize the growing microvilli (green) toward the lateral edge of the cell surface. By the end of embryogenesis, the polarized V-shaped morphology of the stereociliary bundle becomes apparent. The orientation of the hair cell bundle is refined postnatally until the kinocilium associated with the vertex of the stereociliary bundle points distally along the medial-to-lateral axis. The medial (neural) and the lateral (abneural) sides of the sensory epithelium face the interior and periphery of the cochlea, respectively. The kinocilium regenerates postnatally prior to the onset of hearing in mammalian cochleae. (F) Diagram illustrating the positioning of the ciliary basal body during cochlear hair cell bundle development. The pair of centrioles that make up the basal body undergoes similar polarization movements during hair cell maturation. The movements can be described as a four-step process: (1) the pair of centrioles is positioned centrally beneath the cell cortex in newly differentiating hair cells; (2) the pair of centrioles migrates to the lateral cell surface; (3) one of the centrioles leads the other toward the distal edge of the sensory epithelium; (4) the pair of centrioles go through a presumed period of refinement as they align themselves along the medial-to-lateral axis, such that the daughter centriole is ultimately positioned distally (laterally), while the elder maternal centriole, from which the kinocilium is assembled, is positioned proximally (medially) and is closely apposed to the stereociliary bundle (not shown) (M, medial; L, lateral). (See Color Insert.)

and Hilding, 1965; Kikuchi et al., 1988). By E15.5 in the vestibule and E16.5 in the cochlea, the kinocilium migrates to the periphery of the bundle toward the lateral side of the cell surface membrane. Over the next couple of days, the microvilli nearest the kinocilium begin to elongate, followed by microvilli progressively a distance away from the kinocilium. Stereocilia elongation ceases in a defined order leading to the formation of the staircase pattern. Some of the stereocilia filaments extend basally into the apical cytoplasm and become anchored to the developing cuticular plate. By E18.5, the kinocilia and tightly associated stereociliary bundles are relatively close to their final orientations. Furthermore, the characteristic "V"-shape morphology of the hair bundles is well defined in most of the cochlea. Postnatally, the orientation of the hair cell bundle goes through a period of refinement as the cuticular plate locks the bundle in place (Dabdoub and Kelley, 2005). Prior to the onset of hearing at postnatal day 10 (P10), the kinocilia in the cochlea are resorbed (Leibovici et al., 2005; Sobkowicz et al., 1995). In contrast, the kinocilia in the vestibular sensory organs persist throughout the life of the animal. Together, these observations provided three lines of evidence suggest that the kinocilium may play an initial role in stereociliary bundle development and orientation. The first evidence is the observation that the polarization of the kinocilium precedes the polarization of the stereociliary bundle during hair cell bundle development (Fig. 8.2) (Frolenkov et al., 2004). Secondly, the direction of kinocilium migration during bundle development is not random, but is biased toward the final orientation of the bundle (Dabdoub et al., 2003). Thirdly, the kinocilia in the mammalian cochlea degenerate prior to the onset of hearing (Leibovici et al., 2005; Sobkowicz et al., 1995).

Hair cell regeneration studies in which mechanically induced injury was applied to cochlear explants obtained from newborn mice also provided some insightful observations pertaining to a potential role of the kinocilium in stereociliary bundle formation during regeneration of hair cells (Sobkowicz et al., 1995). The authors observed that regenerating kinocilia reformed prior to hair cell recovery, even when the injury was induced at 10 days in vitro, at which time the kinocilia are normally absent in culture, as in the intact animal. Furthermore, the regenerating kinocilium often assumed a central position in the recovering hair cells and proceeded to polarize the newly formed hair cell bundle through the typical developmental sequence, supporting a morphogenetic role for the kinocilium in hair cell development.

Notably, not only are mechanosensory hair cells ciliated, but also the nonsensory supporting cells (Fig. 8.1). The supporting cells in the cochlea can be subdivided into three subclasses based upon their distinct cellular morphology: inner pillar cells, inner phalangeal cells, and Dieters cells (Fig. 8.3). Their cilia also appear to be developmentally transient. In the vestibule, the cilia of supporting cells persist alongside their hair cell counterparts, but are much shorter than the kinocilia of hair cells. However, ciliogenesis in the supporting cells has not been examined, nor is their relationship to the establishment of PCP known.

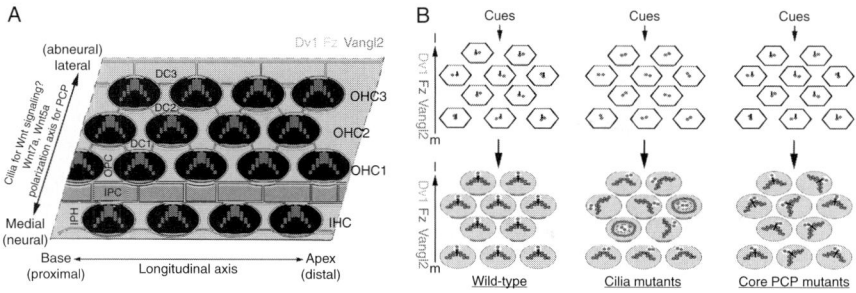

Figure 8.3 Models of PCP regulation in the ear. (A) Schematic drawing of the organ of Corti viewed from its surface. Hair cells (black) are arranged into four parallel rows and are separated from each other by nonsensory supporting cells (gray). Hair cells are synchronously aligned across the sensory epithelium, displaying planar cell polarity. The apical and asymmetric distribution of conserved PCP proteins, Dvl (green), Fz (yellow), and Vangl2 (magenta) along the planar polarity axis is required to direct planar cell polarity of the hair cells in the cochlea. Wnt7a and Wnt5a have been implicated in regulating planar cell polarity of hair cells. The role of primary cilia in planar cell polarity has recently been confirmed, and cilia may act as receivers for Wnts or an unknown polarizing signal for PCP regulation (OHC, outer hair cell; IPC, inner pillar cell; IPHC, inner phalangeal cell; DC, Deiters cell). (B) Model of the role of cilia in the morphological polarization of hair cells in the organ of Corti. An unknown polarization signal directs the asymmetric membrane sorting of core PCP proteins Dvl (green), Fz (blue), and Vangl2 (magenta) in wild-type animals and ciliary mutants. Cell–cell interactions reinforce the distribution of PCP proteins along the planar polarity axis. In contrast, the PCP proteins fail to organize properly in core PCP mutants. Consequently, in wild-type cells, the kinocilia (black lines) and hair cell bundles (green) are polarized and all the hair cells are uniformly aligned along the medial-to-lateral, or planar polarization, axis. Like wild-type animals, PCP mutants exhibit normally polarized kinocilia and hair cell bundles, but the bundles are randomly oriented. Ciliary mutants lack kinocilia and exhibit both randomly polarized hair cell bundles and bundles that have lost their intrinsic polarity. Therefore, the distribution of core PCP proteins is necessary but not sufficient to direct polarization of the kinocilia and the stereociliary bundle. Basal bodies, through the positioning and alignment of the centrioles (red), may direct the organization of the cytoskeletal network to build the polarized structure of the stereociliary bundles. It is likely that the interactions between core PCP proteins and the basal bodies, together with their associated cytoskeletal components, cooperate to direct the coordination of stereociliary bundles across the sensory epithelium. (See Color Insert.)

2.3. Physiological function of primary cilia in the inner ear sensory organs

A mechanosensory role has been proposed for the vestibular kinocilia, which remain tightly associated to the stereociliary bundle (Eatock et al., 1998). Kinocilia of vestibular hair cells may participate partially or wholly in mechanotransduction. It has been proposed that the kinocilia may relay fluid-induced movements of the overlying otoconial membrane to the deflection of stereociliary bundles in the macular sensory organs. Because of the extracellular kinocilial links that connect the kinocilium to the stereociliary bundle, kinocilial deflection affects deflection of the stereociliary bundle, making it difficult to experimentally distinguish a separate mechanosensory role for the kinocilia. Both the kinocilia and stereocilia of vestibular hair cells of mice are considerably longer than those in the cochlea and vary in length amongst the different vestibular sensory organs. The variations in the length of both vestibular kinocilia and stereocilia presumably contribute to the transduction sensitivity of these cells in their respective organs. A mechanosensory role for the kinocilium in the nonmammalian cochlea where it persists has not been ruled out.

3. Planar Cell Polarity Regulation for the Development of the Inner Ear

3.1. Planar cell polarity pathway in orienting the sensory cells of the inner ear

PCP is displayed in many forms. In *Drosophila*, the best known examples are the uniformly oriented fly wing bristles and the regular arrangement of ommatidia in the adult fly eye (Adler, 1992; Gubb and Garcia-Bellido, 1982). In these tissues that exhibit PCP, there is a defined polarity both in terms of the structure of each unit, such as the pointed bristle that sits at the vertex of each wing cell or the precise arrangement of photoreceptors within each ommatidium, and in terms of the arrangement of the different units relative to each other, such as the uniform orientation of all the wing bristles that point distally and the regular orientation of ommatidia in the fly compound eye. Therefore, the establishment of PCP requires a signaling network consisting of multiple regulatory components: (1) global guidance cues that provide directional information, (2) cellular factors to interpret the directional signals and establish the PCP axis, and (3) tissue-specific effectors necessary to build the intrinsic polarity of each individual cell (Klein and Mlodzik, 2005; Lawrence *et al.*, 2007; Ma *et al.*, 2003; Strutt and Strutt, 2005; Tree *et al.*, 2002).

In *Drosophila*, it appears that a network of adhesive molecules may propagate and amplify an initial molecular gradient to provide the directional information (Lawrence *et al.*, 2007; Strutt and Strutt, 2005); a set of transmembrane and membrane-associated proteins, known as core PCP proteins, form complexes at opposing sides of the cell to define the PCP axis (Axelrod, 2001; Klein and Mlodzik, 2005; Ma *et al.*, 2003); and several cytoskeleton-binding proteins act downstream of core PCP genes to affect cellular morphological polarization (Adler *et al.*, 1990, 1994; Collier *et al.*, 2005; Park *et al.*, 1994, 1996; Turner and Adler, 1998; Yun *et al.*, 1999).

In vertebrates, PCP was first described for a process known as convergent extension (CE) (Wallingford et al., 2000). CE occurs during gastrulation and neurulation (Keller and Tibbetts, 1989; Keller et al., 1985) and involves the mediolateral polarization of cells that drives the convergence of cells toward the midline and the extension of the tissue along the perpendicular anterior–posterior axis (Brodland, 2006; Brodland and Veldhuis, 2006; Zajac et al., 2003; Zallen and Wieschaus, 2004). Studies in *Xenopus* and zebrafish identified a vertebrate PCP pathway that consists of a similar cassette of conserved core genes, including *Celsr1*, *Frizzled* (*Fz*), *Dishevelled* (*Dvl*), and *Ltap/Vangl2*, required for CE during gastrulation and neurulation (Ciruna et al., 2006; Djiane et al., 2000; Goto and Keller, 2002; Jessen et al., 2002; Jiang et al., 2005; Mlodzik, 2006; Moeller et al., 2006; Park and Moon, 2002; Schwarz-Romond et al., 2002; Sokol, 1996; Veeman et al., 2003; Wallingford et al., 2000). Two members of the Rho family of GTPases that are capable of modifying cytoskeletal components, RhoA and Rac, are implicated in vertebrate PCP signaling downstream of Dvl during CE in *Xenopus* (Habas and He, 2006; Habas et al., 2001, 2003; Marlow et al., 2002; Phillips et al., 2005). The vertebrate homologs of fly PCP effector proteins, Inturned and Fuzzy, regulate PCP signaling downstream of core PCP genes and participate in CE regulation in *Xenopus* and zebrafish (Park et al., 2006).

Protein localization and loss-of-function studies established essential roles for conserved core PCP genes in coordinated cellular orientation in inner ear sensory organs. During the establishment of epithelial PCP in the cochlea, Fz3 and Fz6, Dishevelled 2 (Dvl2), and Vangl2, display polarized subcellular localization and are distributed asymmetrically across the cochlear epithelium along the mediolateral axis of the cochlear epithelium (Fig. 8.3) (Montcouquiol et al., 2006; Wang et al., 2005, 2006b). Mutations in core PCP genes lead to the misorientation of stereociliary bundles (Curtin et al., 2003; Lu et al., 2004; Montcouquiol et al., 2006; Wang et al., 2005, 2006b). These data together suggest that the vertebrate PCP pathway may utilize a similar molecular mechanism in establishing the PCP axis in the inner ear sensory epithelia.

Notably, mutations in mouse core PCP genes do not affect the formation of the polarized stereociliary bundles and their association with the kinocilium but only the coordinated orientation of all the stereociliary bundles in the sensory epithelia (Fig. 8.3) (Curtin et al., 2003; Lu et al., 2004; Montcouquiol et al., 2003; Wang et al., 2005, 2006b), indicating that the cell–cell coordination but not the intrinsic polarity of hair cells is affected. This idea is consistent with the observed subcellular asymmetry of some of the core PCP proteins in reference to the polarity of hair cells. One of the mammalian homologs of *Drosophila* core PCP protein Prickle (Pk), Pk2, is preferentially localized to the medial side of cells on both sides of the line of polarity reversal in the utricle (Deans et al., 2007), regardless of the reverse polarity of hair cells across the line of polarity reversal (Fig. 8.1). Similar to Pk2, the subcellular localization of Fz6 on the lateral side of cells does not appear to differ in hair cells with opposite polarity on either sides of the line of polarity reversal (Deans et al., 2007). There are three implications from the observations of core PCP protein localization and the phenotypes of core PCP mutants: (1) the polarity of core PCP protein localization does not determine the direction of the cellular polarity, (2) a cellular mechanism independent of core PCP proteins is responsible for the formation of the intrinsic polarity of hair cells, and (3) the cellular mechanism for cell-intrinsic polarity must interact with the core PCP proteins for coordinated alignment of cells across the tissue in a cell context-dependent manner.

The molecules that translate the polarity of core PCP proteins to the orientation of hair cells in a cell context-dependent manner have not been identified. The role of any of the genes known to act downstream of core PCP genes in regulating vertebrate CE and *Drosophila* PCP has yet to be examined in the inner ear. As discussed in Section 4 below, recent studies of primary cilia have identified a crucial requirement for ciliary and basal body genes in PCP regulation in vertebrates and expose them as attractive candidates underlying cellular morphological polarization (Jones et al., 2008; Ross et al., 2005). In particular, the asymmetric localization of core PCP proteins in the auditory sensory epithelium is maintained in ciliary mutants, although the hair cells are misoriented or completely lose their intrinsic polarity in ciliary mutants (Fig. 8.3) (Jones et al., 2008). These observations suggest that cilia are essential components in regulating cell-intrinsic polarity and must interact with core PCP proteins for coordinated orientation of all the hair cells (Fig. 8.3). However, the direct molecular link between cilia or the basal body to core PCP proteins is missing.

In addition, the extracellular cues that provide directional information to core PCP proteins to establish the PCP axis remain unresolved. In contrast to an apparently dispensable role for Wnts in PCP regulation in *Drosophila*, ligands for Fz receptors, Wnt5 and Wnt11, were shown to play essential roles in convergent extension in *Xenopus* and zebrafish (Heisenberg et al., 2000; Kilian et al., 2003; Ohkawara et al., 2003; Smith et al., 2000; Tada and Smith, 2000). Several Wnts are expressed in the cochlea (Dabdoub et al., 2003; Qian et al., 2007). Inactivation of one of them, Wnt5a, results in characteristic PCP phenotypes in the cochlea (Qian et al., 2007). However, it is yet to be determined whether Wnts play an instructive role in generating directional information or merely a permissive role for vertebrate PCP signaling. Furthermore, recent identification of the primary cilium as a signaling apparatus that represses β-catenin-dependent canonical Wnt signaling supports the possibility that primary cilia may promote Wnts for the noncanonical PCP signaling pathway (Corbit et al., 2008; Gerdes et al., 2007; He, 2008; Kishimoto et al., 2008).

3.2. Planar cell polarity pathway in patterning and growth of the inner ear sensory organs

During ear development, the precursor cells that give rise to mechanosensory hair cells and nonsensory supporting cells of the mouse auditory sensory organ form a zone of nonproliferating cells in the cochlea (Chen and Segil, 1999; Chen *et al.*, 2002; Ruben, 1967). Accompanying the acquisition of the highly polarized structure of hair cells and supporting cells during terminal cellular differentiation and the establishment of PCP in the cochlea, the length of the organ of Corti is more than doubled from the shorter and thicker nonproliferating sensory primordium. Such an elongation of the sensory organ in the absence of additional cell divisions suggests cellular rearrangements characteristic of convergent extension during terminal differentiation of the organ of Corti (Chen and Segil, 1999; Chen *et al.*, 2002).

The morphological defects observed in PCP mutants support the concurrence of CE and establishment of PCP in the cochlea. Associated with misorientation of stereociliary bundles, the cochlear duct is shortened and widened in *Wnt5a* and several core PCP mutants (Montcouquiol *et al.*, 2003; Qian *et al.*, 2007; Wang *et al.*, 2005, 2006b). Instead of the invariable four rows of hair cells, there are additional rows of hair cells, especially toward the distal end of the cochlear duct (Montcouquiol *et al.*, 2003; Qian *et al.*, 2007; Wang *et al.*, 2005, 2006b). These growth defects of the cochlear duct and inappropriate patterning of its sensory organ are consistent with a defect in CE.

As discussed below, apparent cochlear CE defects were also observed in ciliary mutants (Jones *et al.*, 2008). The association of CE and cellular orientation defects in the cochlea of both PCP mutants and ciliary mutants suggests shared molecular components by both processes. However, it is not clear what the molecular roles are for *Wnt5a* and core PCP genes in CE of the organ of Corti, nor is it known whether the same cellular polarization process simultaneously drives CE and causes the coordinated polarization of hair cells and supporting cells in the cochlea.

4. Cilia and PCP Regulation

4.1. Evidence linking cilia and PCP signaling in vertebrates

Much of the evidence that links cilia and PCP signaling has come from studies of polarized cellular movements during convergent extension of the vertebrate body axis during *Xenopus* and zebrafish gastrulation. Mutations in conserved PCP genes or ciliary genes independently lead to typical CE phenotypes, such as shorter and wider body axes. The relative ease of *Xenopus* and zebrafish embryo manipulations, morpholino-based knockdown approaches and the accessibility of ciliated tissues have made them leading vertebrate models in exploring the relationship between cilia and PCP signaling.

The coupling of cilia and PCP has predictably included the examination of whether ciliary protein functions involve PCP signaling, also known as noncanonical Wnt signaling activity. A recurring theme appears to be that a common function of ciliary genes is to constrain canonical β-catenin-dependent Wnt signaling activity while simultaneously promoting the noncanonical Wnt signaling activity (Corbit et al., 2008; Gerdes et al., 2007; He, 2008; Kishimoto et al., 2008). Analyses of various ciliary mutants have led to the proposal that ciliary proteins may participate as switches between canonical and noncanonical Wnt signaling pathways to interpret the Wnt signals into appropriate cellular responses during development.

For example, knockdown of the ciliary gene *Inversin* in gastrulating zebrafish and *Xenopus* embryos causes defects in convergent extension, which are enhanced in the *trilobite* mutants, mutants of core PCP gene, *Vangl2* (Ross et al., 2005; Simons et al., 2005). *Inversin* morphants also manifest renal defects that can be rescued by Diversin, a putative vertebrate homolog of the fly PCP protein, Diego. Inversin may indirectly promote PCP processes by negatively regulating canonical Wnt signaling through the targeting of Dvl for proteasome-mediated degradation (Simons et al., 2005).

Many of the genes implicated in Bardet-Biedl syndrome (BBS) have been linked to cilia and/or basal body assembly and function. *bbs1*, *bbs4*, and *bbs6* morphants exhibit gastrulation defects consistent with defects in convergent extension (Gerdes et al., 2007; Ross et al., 2005). Like inversin, convergent extension defects are enhanced when *bbs4* morpholinos are injected into the *trilobite* mutant background (Ross et al., 2005). Moreover, *bbs1*, *bbs4*, and *bbs6* interact with noncanonical *Wnt11* and *Wnt5b* in convergent extension movements (Gerdes et al., 2007). Suppression of bbs gene expression leads to the stabilization of β-catenin, suggesting that they normally inhibit canonical Wnt signaling. The stabilization of β-catenin and Dvl is phenocopied by suppressing another ciliary gene, *kif3a*, the anterograde kinesin required for ciliogenesis (Corbit et al., 2008; Gerdes et al., 2007).

The zebrafish gene *duboraya* is required for ciliogenesis, left–right body patterning and convergent extension (Oishi et al., 2006). Morphants of the PCP gene, *frizzled-2*, could phenocopy the laterality and ciliogenesis defects of *duboraya* morphants; and *frizzled-2* genetically interacts with *duboraya* to regulate left–right patterning. Duboraya function is regulated by phosphorylation, which is mediated by frizzled-2-dependent noncanonical Wnt signaling activity (Oishi *et al.*, 2006). Whether Duboraya can inhibit canonical Wnt activity has not been reported.

Zebrafish gene *seahorse* is enriched in ciliated tissues, and *seahorse* morphants phenocopy ciliary mutants in the manifestations of kidney cysts and left–right asymmetry defects (Kishimoto *et al.*, 2008). In addition, *seahorse* morphants show dorsalization phenotypes and ectopic Wnt target transcription, indicative of ectopic β-catenin-dependent signaling activity. The ectopic activity can be suppressed by *seahorse* mRNA injection, suggesting that seahorse normally inhibits canonical Wnt signaling. Unlike *seahorse* morphants alone, knockdown of seahorse simultaneously with either the ciliary gene *inversin* or the PCP gene *prickle* induces gastrulation phenotypes consistent with PCP defects (Kishimoto *et al.*, 2008). In addition, Seahorse can be found in a complex with Dvl, which is shared by both Wnt pathways, consistent with the idea that Seahorse may play a role in modulating signaling down either pathway.

4.2. Cilia in regulating PCP of the vertebrate inner ear

Many parallels between PCP regulation in flies and mice have been reported. However, there is no evidence that the mechanosensory cilia in flies play any role in PCP establishment or maintenance. It is possible that the role of primary cilia in PCP regulation may have evolved in vertebrates. Yet, the exact nature of the relationship of primary cilia to PCP signaling has been obscure. The first genetic evidence that the relationship between cilia and PCP signaling was conserved in mammalian ear development was revealed through studies of *bbs* mouse mutants. Not only were there abnormal stereociliary bundle morphologies in the cochleae of *bbs*-deficient mice, but the kinocilia lost their close associations with the stereociliary bundles. Furthermore, mice that were simultaneously mutated for *bbs* genes and the PCP gene, *Vangl2*, showed an enhancement of the cochlear phenotype. However, since kinocilia remained in the *bbs* mutants, it was difficult to assess the specific role of kinocilia in planar polarity of the hair cells.

To test the role of cilia in PCP regulation in the ear, Cre-mediated recombination was used to inactivate a floxed allele of IFT protein 88 (IFT88) in the mouse ear at approximately E10.5, thereby disrupting kinocilia formation in the inner ear primordium several days prior to hair cell differentiation in the organ of Corti (Jones *et al.*, 2008). The authors observed misoriented stereociliary bundles in the cochleae of *IFT88* mutant mice and cochlear extension defects consistent with aberrant convergent extension. In addition, there were both a significant increase in the number of misrotated bundles in mice mutated for *IFT88* and *Vangl2* and an enhancement of the cochlear extension defect in the double mutants. Similar observations were made with mice lacking *Kif3a*, the anterograde kinesin required for ciliogenesis. Together, these findings provided compelling evidence for a direct role for cilia in planar polarization processes and a genetic relationship between the ciliogenic and PCP pathways.

A direct role for the kinocilium in the development of the intrinsic polarity and orientation of the stereociliary bundle was also revealed in the study of *IFT88* mutant cochleae. A proportion of the hair cell bundles in *IFT88* mutants exhibited abnormal morphologies, most of which were circular in shape. The circular bundles were perhaps reminiscent of immature bundles of the cochlea that had arrested in their development, indicating that the cilium is essential for the polarized structure of the hair cell bundle. The circular bundle morphology exhibited low penetrance in the mutant animals. One explanation could be delayed or variable Cre-mediated recombination at the floxed sites and therefore remnant IFT88 activity. Alternatively, the orientations of the stereociliary bundles were established prior to the time in which the Cre recombinase was active, and aberrant bundle morphologies were consequently indicative of the failure to maintain the intrinsic polarity of the hair bundle.

Normally, the centrioles, integral components of the basal body, are aligned along the planar polarity axis in the cochleae of wild-type mice (Jones *et al.*, 2008). In addition, the position of the maternal centriole, the template for kinocilium assembly, is positioned more proximally to the daughter centriole and thus is closely associated to the stereociliary bundle (Fig. 8.1). Ablation of kinocilia in the *IFT88* mutant cochleae left the basal bodies intact. Interestingly, there was a strong correlation between the positions of the basal bodies and the orientation of the stereociliary bundles; that is, the bundles were consistently misoriented in the same direction as the basal bodies in the ciliary mutants. This observation suggested that the basal body, through the placement and alignment of the centrioles, is likely the cell-intrinsic regulator of stereociliary bundle polarity and orientation. The basal body is an ideal candidate for the intrinsic regulator of bundle polarity because of its own intrinsic chirality, both in terms of the asymmetric structure of the centrioles and the molecular and structural distinction between the two centrioles. Consistent with this suggestion is the observation that some circularly shaped hair bundles in *IFT88* mutants had centrally positioned basal bodies. The positioning of basal bodies underlying planar polarization processes in multiciliated cells has also been reported, and may be a general intracellular mechanism that is coordinated to polarize cells across a tissue (Marshall and Kintner, 2008; Mitchell *et al.*, 2007). Interestingly, core PCP protein Dvl is required for the apical positioning and planar polarization of basal bodies in multiciliated mucociliary epithelia of the *Xenopus* epidermis (Park *et al.*, 2008). Furthermore, the loss of either Dvl or PCP effector protein Inturned leads to a failure to apically localize basal bodies and consequently affects apical cytoskeletal organization (Park *et al.*, 2008).

4.3. Crosstalk between cilia and conserved PCP proteins

How the ciliogenic and PCP signaling pathways interact to coordinate the alignment of the stereociliary bundles is not known. Surprisingly, the study of *IFT88* mutants (Jones et al., 2008) revealed that core PCP proteins, Fz3 and Vangl2, were polarized normally along the planar polarity axis in *IFT88* mutants, suggesting that IFT88, or the cilium, is not required for the PCP proteins to receive and interpret cues to affect their polarized distribution in these *IFT88* mutant animals. However, the study did not exclude the possibility that kinocilia may be involved in transducing directional cues to core PCP proteins in the mouse prior to the time that IFT88 was inactivated, nor the possibility that a polarization signal was provided non-autonomously from cells in adjacent tissues where the Cre recombinase may not have been active. In addition, the structures of kinocilia and their tight associations with stereociliary bundles in all PCP mutants examined thus far appear normal. However, it is possible that while their structures appear normal, the sensory functions of kinocilia may be perturbed in PCP mutants and have not yet been determined.

The identities of the molecular players involved in the crosstalk between the ciliogenic and PCP pathways are not known. The basal body may serve as an organizing center for ciliogenic and PCP signaling proteins. The accumulation of PCP proteins Dvl and Vangl2 at the basal body in *Xenopus in vivo* (Park et al., 2006, 2008; Simons et al., 2005) and cultured cells *in vitro* was observed (Ross et al., 2005). However, neither the localization of PCP proteins to the basal bodies in cochlear or vestibular cells nor the direct association between these proteins in cochlear or vestibular extracts has been determined. The normal polarization of PCP proteins in *IFT88* mutants suggests that the IFT88 and/or cilia may function downstream of the PCP signaling pathway. It is possible therefore that the PCP complexes recruit ciliary proteins transiently. For example, the PCP proteins may recruit PCP effector proteins to the basal body to cause the changes in the cytoskeleton necessary to orient the cells across the entire epithelium. Studies conducted in other vertebrate models support this idea. For example, knockdown of PCP effector genes, *Inturned* and *Fuzzy*, in *Xenopus* affects both convergent extension movements and cilia morphogenesis (Park et al., 2006). The ciliogenic defects in *Inturned* and *Fuzzy* morphants are attributed to a failure to orient ciliary microtubules during ciliogenesis (Park et al., 2006). Similar cytoskeletal disorganization was reported in *duboraya* morphants (Oishi et al., 2006). Furthermore, knockdown of Dvl leads to decreased apical actin accumulation, reduced ciliogenesis in multi-ciliated cells, and misorientation of cilia rootlets, supporting an upstream role for core PCP genes in cytoskeletal alterations for the formation and polarity of cilia (Park et al., 2008).

5. Cilia and Determination of the Intrinsic Cellular Polarity of Inner Ear Cells

5.1. The molecular machinery for building the intrinsically polarized hair bundles

The intrinsic polarity of hair cells is marked by the position of the kinocilium and the polarity of the stereociliary bundle. The molecular components of the machinery that builds the stereociliary bundle in hair cells have been partially identified from genetic studies (El-Amraoui and Petit, 2005). Usher syndrome (USH) is the most frequent cause of hereditary deafblindness in humans. USH can also be associated with vestibular dysfunction, reduced odor identification, and sperm motility (Kremer et al., 2006). There are three clinical subtypes, USH1–3, according to the severity and onset of the hearing impairment, vestibular dysfunction, and retinitis pigmentosa. Each USH subtype is genetically heterogeneous, and at least 12 chromosomal loci are assigned to USH. Five USH1 genes have been cloned, including those encoding an unconventional myosin VIIa for USH1B (el-Amraoui et al., 1996; Weil et al., 1996), a PDZ domain-containing protein harmonin for USH1C (Verpy et al., 2000), cadherin 23 for USH1D (Bolz et al., 2001; Bork et al., 2001; Di Palma et al., 2001), a cadherin-related protein protocadherin 15 for USH1F (Ahmed et al., 2001; Alagramam et al., 2001b), and a putative scaffold protein with three ankyrin repeats and a sterile alpha motif (SAM) domain and a C-terminal PDZ domain, Sans, for USH1G (Weil et al., 2003). Also, three USH2 and one USH3 genes were identified that encode transmembrane protein usherin (USH2A) (Adato et al., 2000; Dreyer et al., 2000; Rivolta et al., 2000; Weston et al., 2000), a G protein-coupled seven-transmembrane receptor VLGR1b (USH2C) (Weston et al., 2004), a PDZ domain- and proline-rich domain-containing scaffold protein whirlin (USH2D) (Adato et al., 2005a), and a four-transmembrane domain protein clarin-1 (USH3A) (Adato et al., 2002). Mouse models for USH1 and USH2D linked USH1 and USH2D genes to the development of hair cell stereociliary bundles.

Mice that carry mutated loci for USH1B gene *myosin VIIa* (shaker-1) (El-Amraoui et al., 1996; Weil et al., 1997), USH1C gene *harmonin* (deaf circler) (Verpy et al., 2000), USH1D gene *cadherin 23* (waltzer) (Di Palma et al., 2001), USH1F gene *protocadherin 15* (Ames waltzer) (Alagramam et al., 2001a), USH1G gene *Sans* (Jackson shaker) (Kikkawa et al., 2003), or USH2D gene *whirlin* (whirler) (Mburu et al., 2003) all display stereocilia defects as revealed by scanning electron microscope (SEM) analysis. Stereocilia contain up to 2000 parallel actin filaments that are anchored in an actin meshwork of the cuticular plate at the apical cortex of hair cells and are arranged in three to four rows of increasing height toward the kinocilium.

The stereocilia are interconnected by ankle links, lateral and tip links, and to the nearby kinocilium by fibrous extracellular links. The USH proteins are often present in multiple isoforms in hair cells, and show dynamic and overlapping expression patterns to various regions of the stereociliary bundle or the kinocilium during development of hair bundles. Protein–protein interaction assays *in vitro* further suggest that USH proteins function in a protein network and affect actin dynamics. In particular, scaffold proteins harmonin and whirlin can integrate other USH proteins and alter their association with transmembrane adhesion molecules and actin cytoskeleton (Adato *et al.*, 2005b; El-Amraoui and Petit, 2005). Consistent with their localization and protein–protein interactions, mutations in any of the USH1 and USH2D genes leads to common stereocilia defects, including disorganization of stereocilia and misshapening of the hair bundle, loss of the graded heights of stereocilia, shortened or enlarged and fused stereocilia, and displacement of the kinocilium (Di Palma *et al.*, 2001; Holme and Steel, 2002; Johnson *et al.*, 2003; Kazmierczak *et al.*, 2007; Lagziel *et al.*, 2005; Lefevre *et al.*, 2008; Libby *et al.*, 2003; Mogensen *et al.*, 2007; Mustapha *et al.*, 2007). Although it has not been determined how exactly the formation of the polarized structure of hair bundles is regulated by USH genes, the genetic, biochemical, and cell biological studies together identified USH protein complexes as components of a protein network that builds the precise structure of the hair bundles.

5.2. Cilia determine intrinsic cellular polarity

Despite the shared defect of the loss of the uniform orientation of stereociliary bundles in the cochlea of both core PCP and ciliary mutants, the mechanisms underlying their defect differ. In known core PCP mutants, the polarized structure of the hair bundle, including graded heights of stereocilia, the staircase arrangement of stereocilia, the shape of the stereociliary bundle, and the placement of the kinocilium and the basal body near the tallest stereocilia, are not affected (Curtin *et al.*, 2003; Lu *et al.*, 2004; Montcouquiol *et al.*, 2003; Wang *et al.*, 2005, 2006b). In contrast, circular stereociliary bundles are present (Jones *et al.*, 2008), and remaining kinocilia are often shorter and displaced from their association with the tallest stereocilia in ciliary mutants (Jones *et al.*, 2008; Ross *et al.*, 2005). This difference between core PCP mutants and ciliary mutants indicates that known core PCP genes are not essential for the intrinsic polarity of hair cells while ciliary genes are required.

The examination of the cellular alignment in core PCP mutants and ciliary mutants revealed an additional fundamental difference in the two types of mutants. In *looptail* mutants that carry a loss-of-function allele for core PCP gene *Vangl2*, the polarized subcellular distributions of Dvl2, Fz, and Pk2 are affected (Deans *et al.*, 2007; Wang *et al.*, 2005, 2006b), suggesting a failure in normal cell–cell alignment that is communicated through core PCP proteins. In contrast, core PCP proteins, Vangl2 and Fz3, are normally partitioned along the mediolateral axis of the cochlea and show polarized subcellular localization across the cochlear sensory epithelium in ciliary mutants (Fig. 8.3) (Jones et al., 2008), indicating normal cell–cell communication by core PCP proteins but a failure in responding to the polarization signals from the asymmetrically partitioned core PCP proteins in ciliary mutants.

As morphological studies of the inner ear and studies of basal body and ciliary genes in inner ear development implicated a role for cilia in positioning of the basal body and as well as the basal body as a determinant of the intrinsic polarity of hair cells, three critical issues emerged. (1) What is the signal for cilia to position the basal body? (2) How does the basal body regulate the intrinsic polarity of the cell? (3) How do cilia and/or the basal body fulfill an essential role in transducing the polarity signals from core PCP proteins?

It is possible that cilia respond to an extracellular directional cue, either the same cue as that received by the core PCP proteins or an entirely independent ocue, for the positioning of cilia and the basal body. Another possibility is that ciliogenesis is first required to acquire or maintain a particular configuration of the pair of centrioles or the basal body to respond directly to a directional cue transduced from asymmetrically partitioned core PCP proteins. Alternatively, although the results obtained by conditionally knocking out ciliary genes revealed normal partitioning of core PCP proteins, there may be remnant cilia activity in these conditional ciliary mutants that is sufficient for an early role in transducing a directional signal for planar polarization, including the positioning of cilia and basal body, and the polarization of core PCP proteins. No experimental data is available to support or exclude any of the above possibilities for the positioning of the basal body.

The advancement of USH gene studies, on the other hand, has implicated a common genetic pathway for ciliary genes and USH genes in the determination of intrinsic cellular polarity. In particular, a recent study by Lefevre et al. (2008) systematically examined hair bundle polarity in mutant mice for five USH1 genetic forms. Together with several earlier studies, a core cochlear phenotype that shares many similarities with ciliary mutants was revealed. The stereocilia phenotype in USH1 mutant mice includes the replacement of the V-shaped stereociliary bundle by clusters of stereocilia, the presence of circularly shaped stereocilia, the loss of graded heights of stereocilia, and the reduction of lateral links between stereocilia. In particular, in mice mutated for cadherin 23, protocadherin 15, or Sans, some stereociliary bundles lack polarity and become clustered or circular, and the kinocilium is frequently dissociated from the stereociliary bundle or cluster. The localization of cadherin 23, protocadherin 15, and Sans further implicates a potential interaction between the basal body and USH protein complexes. During development, the full-length protocadherin 15 protein is enriched at the base of the tallest row of stereocilia that are adjacent to the base of the kinocilium (Haywood-Watson et al., 2006; Lefevre et al., 2008). Cadherin 23 has been localized along the kinocilia in the vestibule, and has also been detected in the kinocilia in the distal tip of the cochlear duct (Lagziel et al., 2005). Prior to the morphological polarization of hair cells and the formation of distinct stereocilia at E14 in mice, cadherin 23 is initially detected in the centrosome in cells of the developing organ of Corti. Sans is enriched in the immediate vicinity of the basal body of the auditory kinocilia in neonatal mice (El-Amraoui and Petit, 2005), and shows a pattern that is very similar to pericentriolar γ-tubulin staining (P. Chen, unpublished data).

As USH syndrome affects hearing and vision, studies of USH proteins in vertebrate photoreceptor cells provide new insight regarding the molecular role for USH proteins, and support a direct interaction between the basal body and USH protein complexes. USH proteins are highly expressed in photoreceptors, another type of cells with specialized cilia, and are collectively localized at the periciliary region (Maerker et al., 2008). In particular, USH1G Sans, which is concentrated at the proximity of the basal body of the outer hair cell kinocilium, is also localized to the basal body and the connecting cilium in mouse and *Xenopus* photoreceptor cells. In NIH3T3 cells, Sans is detected in the centrosome. In addition, Sans is associated with microtubules and its subcellular distribution can be altered by destabilizing microtubules in NIH3T3 and photoreceptor cells. Together, these observations implicated a direct interaction between USH proteins and the basal body and the microtubules.

It is conceivable that to act as a cellular polarity determinant, the basal body must interact with the machinery that builds the asymmetric structure of stereociliary bundles in hair cells. Sans, Cadherin 23, or protocadherin 15 directly interact with the basal body, which may serve as a compass to orient cytoplasmic microtubules and coordinate the activity of USH proteins. Sans, cadherin 23, and protocadherin 15 may then integrate the positional information provided by the basal body to the other USH proteins to direct the formation of the polarized structure of stereociliary bundles in hair cells.

6. Conclusions and Perspectives

The arrangement of the primary cilium of the inner ear sensory hair cell, the kinocilium, and the stereociliary bundle atop each inner ear sensory hair cell marks an intrinsic polarity of the hair cell, and the coordinated orientation of all of the hair cells in each of the six inner ear sensory organs confers a polarity within the plane of the epithelium of each inner ear sensory organ. The establishment of PCP in the inner ear sensory organs requires a set of conserved core PCP proteins that response to directional cues and undergo asymmetrical partitioning to define the PCP axis for downstream morphological polarization. The involvement of a ciliary and basal body genes in mediating Wnt signaling (Corbit et al., 2008; He, 2008; Kishimoto et al., 2008; Simons et al., 2005; Singla and Reiter, 2006), the functional complementation of the loss of ciliary function by PCP proteins (Simons et al., 2005), and the role of several Wnt transducers, including Fz and Dvl, in establishing the planar polarization axis have implicated a role for cilia in PCP regulation. Since cilia can function as specialized subcellular compartments for localizing and concentrating membrane bound signal receptors and complexes in order to relay signals from the extracellular environment to the cell interior, it was thought that the primary cilium of the inner ear sensory hair cell, the kinocilium, may function as a signaling apparatus to transduce polarization signals. Indeed, two recent studies on BBS genes and IFT genes in inner ear development have also established an essential role for ciliary and basal body genes in vertebrate PCP signaling.

Surprisingly, the study of ciliary mutant cochleae showed that the proper positioning of ciliary basal bodies and the formation of the polarized cellular structure of inner ear sensory hair cells, the stereociliary bundles, are disrupted in ciliary mutants, whereas core PCP proteins are partitioned normally along the polarization axis (Jones et al., 2008). These data uncover a distinct requirement for ciliary genes downstream of core PCP proteins in basal body positioning and a role for the basal body in the determination of the intrinsic polarity, possibly the intrinsic polarized organization of cytoskeleton, of hair cells. Although the putative polarization signal remains elusive, the positioning of basal bodies may be a general mechanism underlying other cellular polarization processes involving cilia (Absalon et al., 2007; Boisvieux-Ulrich et al., 1985; Feldman et al., 2007; Mitchell et al., 2007).

However, despite of the abundance of genes known to be involved in stereocilia development, the dearth of information on proteins that act downstream of core PCP proteins to affect cytoskeletal polarization leaves open the question how do cilia through the positioning of the basal body determine the intrinsic polarity of the hair cell? Two of the major challenges that remain are (1) to identify the molecular link that connects the basal body and the protein network that builds the polarized structure of the stereociliary bundle and (2) to illustrate how the basal body communicates with polarized core PCP proteins to coordinately orient all of the hair cells within each inner ear sensory epithelium. In addition, the existing data do not rule out the possibility that the kinocilium may also have a role in transducing a signal for core PCP proteins. While the conditional inactivation of IFT genes in inner ear sensory epithelia have effectively bypassed the requirement of these genes in early embryonic processes and targeted studies on cellular polarization processes in ear development, conditional knockout approaches may not sufficiently inactivate cilia function at all times required to affect polarization processes during ear development. More effective control of temporal IFT gene inactivation in the inner ear will help resolve this issue. Until then, it is tempting to speculate that the kinocilium acts as a signaling apparatus and transduces polarization signals to core PCP proteins, which in turn regulate the positioning and polarization of the basal body that collaborate with an intracellular protein network to build and orient the intrinsically polarized structure of stereociliary bundles in hair cells.

REFERENCES

Absalon, S., et al. (2007). Basal body positioning is controlled by flagellum formation in Trypanosoma brucei. PLoS ONE **2**, e437.

Adato, A., et al. (2000). Three novel mutations and twelve polymorphisms identified in the USH2A gene in Israeli USH2 families. Hum. Mutat. **15**, 388.

Adato, A., et al. (2002). USH3A transcripts encode clarin-1, a four-transmembrane-domain protein with a possible role in sensory synapses. Eur. J. Hum. Genet. **10**, 339–350.

Adato, A., et al. (2005a). Usherin, the defective protein in Usher syndrome type IIA, is likely to be a component of interstereocilia ankle links in the inner ear sensory cells. Hum. Mol. Genet. **14**, 3921–3932.

Adato, A., et al. (2005b). Interactions in the network of Usher syndrome type 1 proteins. Hum. Mol. Genet. **14**, 347–356.

Adler, P. N. (1992). The genetic control of tissue polarity in Drosophila. BioEssays **14**, 735–741.

Adler, P. N., et al. (1990). Molecular structure of frizzled, a Drosophila tissue polarity gene. Genetics **126**, 401–416.

Adler, P. N., et al. (1994). The Drosophila tissue polarity gene inturned functions prior to wing hair morphogenesis in the regulation of hair polarity and number. Genetics **137**, 829–836.

Ahmed, Z. M., et al. (2001). Mutations of the protocadherin gene PCDH15 cause Usher syndrome type 1F. Am. J. Hum. Genet. **69**, 25–34.

Alagramam, K. N., et al. (2001a). The mouse Ames waltzer hearing-loss mutant is caused by mutation of Pcdh15, a novel protocadherin gene. Nat. Genet. **27**, 99–102.

Alagramam, K. N., et al. (2001b). Mutations in the novel protocadherin PCDH15 cause Usher syndrome type 1F. Hum. Mol. Genet. **10**, 1709–1718.

Axelrod, J. D. (2001). Unipolar membrane association of Dishevelled mediates Frizzled planar cell polarity signaling. Genes Dev. **15**, 1182–1187.

Benzing, T., and Walz, G. (2006). Cilium-generated signaling: A cellular GPS? Curr. Opin. Nephrol. Hypertens. **15**, 245–249.

Boisvieux-Ulrich, E., and Sandoz, D. (1991). Determination of ciliary polarity precedes differentiation in the epithelial cells of quail oviduct. Biol. Cell **72**, 3–14.

Boisvieux-Ulrich, E., et al. (1985). The orientation of ciliary basal bodies in quail oviduct is related to the ciliary beating cycle commencement. Biol. Cell **55**, 147–150.

Bolz, H., et al. (2001). Mutation of CDH23, encoding a new member of the cadherin gene family, causes Usher syndrome type 1D. Nat. Genet. **27**, 108–112.

Bork, J. M., et al. (2001). Usher syndrome 1D and nonsyndromic autosomal recessive deafness DFNB12 are caused by allelic mutations of the novel cadherin-like gene CDH23. Am. J. Hum. Genet. **68**, 26–37.

Kikuchi, K., and Hilding, D. (1965). The development of the organ of Corti in the mouse. *Acta Otolaryngol.* **60,** 207–222.

Kikuchi, T., et al. (1988). Development of apical-surface structures of mouse otic placode. *Acta Otolaryngol.* **106,** 200–207.

Kilian, B., et al. (2003). The role of Ppt/Wnt5 in regulating cell shape and movement during zebrafish gastrulation. *Mech. Dev.* **120,** 467–476.

Kishimoto, N., et al. (2008). Cystic kidney gene *seahorse* regulates cilia-mediated processes and Wnt pathways. *Dev. Cell* **14,** 954–961.

Klein, T. J., and Mlodzik, M. (2005). Planar cell polarization: An emerging model points in the right direction. *Annu. Rev. Cell Dev. Biol.* **21,** 155–176.

Kremer, H., et al. (2006). Usher syndrome: Molecular links of pathogenesis, proteins and pathways. *Hum. Mol. Genet.* **15**(Spec No. 2), R262–R270.

Lagziel, A., et al. (2005). Spatiotemporal pattern and isoforms of cadherin 23 in wild type and waltzer mice during inner ear hair cell development. *Dev. Biol.* **280,** 295–306.

Lawrence, P. A., et al. (2007). Planar cell polarity: One or two pathways? *Nat. Rev. Genet.* **8,** 555–563.

Lefevre, G., et al. (2008). A core cochlear phenotype in USH1 mouse mutants implicates fibrous links of the hair bundle in its cohesion, orientation and differential growth *Development* **135,** 1427–1437.

Leibovici, M., et al. (2005). Initial characterization of kinocilin, a protein of the hair cell kinocilium. *Hear. Res.* **203,** 144–153.

Lewis, J., and Davies, A. (2002). Planar cell polarity in the inner ear: How do hair cells acquire their oriented structure? *J. Neurobiol.* **53,** 190–201.

Libby, R. T., et al. (2003). Cdh23 mutations in the mouse are associated with retinal dysfunction but not retinal degeneration. *Exp. Eye Res.* **77,** 731–739.

Liu, A., et al. (2005). Mouse intraflagellar transport proteins regulate both the activator and repressor functions of Gli transcription factors. *Development* **132,** 3103–3111.

Lu, X., et al. (2004). PTK7/CCK-4 is a novel regulator of planar cell polarity in vertebrates. *Nature* **430,** 93–98.

Ma, D., et al. (2003). Fidelity in planar cell polarity signalling. *Nature* **421,** 543–547.

Maerker, T., et al. (2008). A novel Usher protein network at the periciliary reloading point between molecular transport machineries in vertebrate photoreceptor cells. *Hum. Mol. Genet.* **17,** 71–86.

Marlow, F., et al. (2002). Zebrafish Rho kinase 2 acts downstream of Wnt11 to mediate cell polarity and effective convergence and extension movements. *Curr. Biol.* **12,** 876–884.

Marshall, W. F., and Kintner, C. (2008). Cilia orientation and the fluid mechanics of development. *Curr. Opin. Cell Biol.* **20,** 48–52.

Mburu, P., et al. (2003). Defects in whirlin, a PDZ domain molecule involved in stereocilia elongation, cause deafness in the whirler mouse and families with DFNB31. *Nat. Genet.* **34,** 421–428.

Michaud, E. J., and Yoder, B. K. (2006). The primary cilium in cell signaling and cancer. *Cancer Res.* **66,** 6463–6467.

Mitchell, B., et al. (2007). A positive feedback mechanism governs the polarity and motion of motile cilia. *Nature* **447,** 97–101.

Mlodzik, M. (2006). A GAP in convergent extension scores PAR. *Dev. Cell* **11,** 2–4.

Moeller, H., et al. (2006). Diversin regulates heart formation and gastrulation movements in development. *Proc. Natl Acad. Sci. USA* **103,** 15900–15905.

Mogensen, M. M., et al. (2007). The deaf mouse mutant whirler suggests a role for whirlin in actin filament dynamics and stereocilia development. *Cell Motil. Cytoskeleton* **64,** 496–508.

Montcouquiol, M., et al. (2003). Identification of Vangl2 and Scrb1 as planar polarity genes in mammals. *Nature* **423,** 173–177.

Montcouquiol, M., et al. (2006). Asymmetric localization of Vangl2 and Fz3 indicate novel mechanisms for planar cell polarity in mammals. *J. Neurosci.* **26**, 5265–5275.

Mustapha, M., et al. (2007). Whirler mutant hair cells have less severe pathology than shaker 2 or double mutants. *J. Assoc. Res. Otolaryngol.* **8**, 329–337.

Ohkawara, B., et al. (2003). Role of glypican 4 in the regulation of convergent extension movements during gastrulation in *Xenopus laevis*. *Development* **130**, 2129–2138.

Oishi, I., et al. (2006). Regulation of primary cilia formation and left–right patterning in zebrafish by a noncanonical Wnt signaling mediator, *duboraya*. *Nat. Genet.* **38**, 1316–1322.

Ou, G., et al. (2007). Sensory ciliogenesis in *Caenorhabditis elegans*: Assignment of IFT components into distinct modules based on transport and phenotypic profiles. *Mol. Biol. Cell* **18**, 1554–1569.

Park, M., and Moon, R. T. (2002). The planar cell-polarity gene stbm regulates cell behaviour and cell fate in vertebrate embryos. *Nat. Cell Biol.* **4**, 20–25.

Park, W. J., et al. (1994). The frizzled gene of *Drosophila* encodes a membrane protein with an odd number of transmembrane domains. *Mech. Dev.* **45**, 127–137.

Park, W. J., et al. (1996). The *Drosophila* tissue polarity gene inturned acts cell autonomously and encodes a novel protein. *Development* **122**, 961–969.

Park, T. J., et al. (2006). Ciliogenesis defects in embryos lacking inturned or fuzzy function are associated with failure of planar cell polarity and Hedgehog signaling. *Nat. Genet.* **38**, 303–311.

Park, T. J., et al. (2008). Dishevelled controls apical docking and planar polarization of basal bodies in ciliated epithelial cells. *Nat. Genet.* **40**, 871–879.

Phillips, H. M., et al. (2005). Vangl2 acts via RhoA signaling to regulate polarized cell movements during development of the proximal outflow tract. *Circ. Res.* **96**, 292–299.

Qian, D., et al. (2007). Wnt5a functions in planar cell polarity regulation in mice. *Dev. Biol.* **306**, 121–133.

Rivolta, C., et al. (2000). Missense mutation in the USH2A gene: Association with recessive retinitis pigmentosa without hearing loss. *Am. J. Hum. Genet.* **66**, 1975–1978.

Ross, A. J., et al. (2005). Disruption of Bardet-Biedl syndrome ciliary proteins perturbs planar cell polarity in vertebrates. *Nat. Genet.* **37**, 1135–1140.

Ruben, R. J. (1967). Development of the inner ear of the mouse: A radioautographic study of terminal mitoses. *Acta Otolaryngol.* **220**(Suppl.), 1–44.

Schneider, L., et al. (2005). PDGFRalphaalpha signaling is regulated through the primary cilium in fibroblasts. *Curr. Biol.* **15**, 1861–1866.

Scholey, J. M., and Anderson, K. V. (2006). Intraflagellar transport and cilium-based signaling. *Cell* **125**, 439–442.

Schwarz-Romond, T., et al. (2002). The ankyrin repeat protein Diversin recruits Casein kinase Iepsilon to the beta-catenin degradation complex and acts in both canonical Wnt and Wnt/JNK signaling. *Genes Dev.* **16**, 2073–2084.

Simons, M., et al. (2005). Inversin, the gene product mutated in nephronophthisis type II, functions as a molecular switch between Wnt signaling pathways. *Nat. Genet.* **37**, 537–543.

Singla, V., and Reiter, J. F. (2006). The primary cilium as the cell's antenna: Signaling at a sensory organelle. *Science* **313**, 629–633.

Smith, J. C., et al. (2000). Xwnt11 and the regulation of gastrulation in *Xenopus*. *Philos. Trans. R. Soc. Lond. B Biol. Sci.* **355**, 923–930.

Sobkowicz, H. M., et al. (1995). The kinocilium of auditory hair cells and evidence for its morphogenetic role during the regeneration of stereocilia and cuticular plates. *J. Neurocytol.* **24**, 633–653.

Sokol, S. Y. (1996). Analysis of Dishevelled signalling pathways during *Xenopus* development. *Curr. Biol.* **6**, 1456–1467.

Strutt, H., and Strutt, D. (2005). Long-range coordination of planar polarity in *Drosophila*. *BioEssays* **27**, 1218–1227.

Tada, M., and Smith, J. C. (2000). Xwnt11 is a target of *Xenopus* Brachyury: Regulation of gastrulation movements via Dishevelled, but not through the canonical Wnt pathway. *Development* **127**, 2227–2238.

Taulman, P. D., et al. (2001). Polaris, a protein involved in left–right axis patterning, localizes to basal bodies and cilia. *Mol. Biol. Cell* **12**, 589–599.

Tilney, L. G., et al. (1992). Actin filaments, stereocilia and hair cells of the bird cochlea. VI. How the number and arrangement of stereocilia are determined. *Development* **116**, 213–226.

Tree, D. R., et al. (2002). A three-tiered mechanism for regulation of planar cell polarity. *Semin. Cell Dev. Biol.* **13**, 217–224.

Turner, C. M., and Adler, P. N. (1998). Distinct roles for the actin and microtubule cytoskeletons in the morphogenesis of epidermal hairs during wing development in *Drosophila*. *Mech. Dev.* **70**, 181–192.

Veeman, M. T., et al. (2003). A second canon. Functions and mechanisms of beta-catenin-independent Wnt signaling. *Dev. Cell* **5**, 367–377.

Verpy, E., et al. (2000). A defect in harmonin, a PDZ domain-containing protein expressed in the inner ear sensory hair cells, underlies Usher syndrome type 1C. *Nat. Genet.* **26**, 51–55.

Wallingford, J. B. (2006). Planar cell polarity, ciliogenesis and neural tube defects. *Hum. Mol. Genet.* **15**(Spec No. 2), R227–R234.

Wallingford, J. B., et al. (2000). Dishevelled controls cell polarity during *Xenopus* gastrulation. *Nature* **405**, 81–85.

Wang, J., et al. (2005). Regulation of polarized extension and planar cell polarity in the cochlea by the vertebrate PCP pathway. *Nat. Genet.* **37**, 980–985.

Wang, J., et al. (2006a). Dishevelled genes mediate a conserved mammalian PCP pathway to regulate convergent extension during neurulation. *Development* **133**, 1767–1778.

Wang, Y., et al. (2006b). The role of Frizzled3 and Frizzled6 in neural tube closure and in the planar polarity of inner-ear sensory hair cells. *J. Neurosci.* **26**, 2147–2156.

Weil, D., et al. (1996). Human myosin VIIA responsible for the Usher 1B syndrome: A predicted membrane-associated motor protein expressed in developing sensory epithelia. *Proc. Natl Acad. Sci. USA* **93**, 3232–3237.

Weil, D., et al. (1997). The autosomal recessive isolated deafness, DFNB2, and the Usher 1B syndrome are allelic defects of the myosin-VIIA gene. *Nat. Genet.* **16**, 191–193.

Weil, D., et al. (2003). Usher syndrome type I G (USH1G) is caused by mutations in the gene encoding SANS, a protein that associates with the USH1C protein, harmonin. *Hum. Mol. Genet.* **12**, 463–471.

Weston, M. D., et al. (2000). Genomic structure and identification of novel mutations in usherin, the gene responsible for Usher syndrome type IIa. *Am. J. Hum. Genet.* **66**, 1199–1210.

Weston, M. D., et al. (2004). Mutations in the VLGR1 gene implicate G-protein signaling in the pathogenesis of Usher syndrome type II. *Am. J. Hum. Genet.* **74**, 357–366.

Yoder, B. K., et al. (2002). The polycystic kidney disease proteins, polycystin-1, polycystin-2, polaris, and cystin, are co-localized in renal cilia. *J. Am. Soc. Nephrol.* **13**, 2508–2516.

Yun, U. J., et al. (1999). The inturned protein of *Drosophila melanogaster* is a cytoplasmic protein located at the cell periphery in wing cells. *Dev. Genet.* **25**, 297–305.

Zajac, M., et al. (2003). Simulating convergent extension by way of anisotropic differential adhesion. *J. Theor. Biol.* **222**, 247–259.

Zallen, J. A., and Wieschaus, E. (2004). Patterned gene expression directs bipolar planar polarity in *Drosophila*. *Dev. Cell* **6**, 343–355.

CHAPTER NINE

THE PRIMARY CILIUM: AT THE CROSSROADS OF MAMMALIAN HEDGEHOG SIGNALING

Sunny Y. Wong *and* Jeremy F. Reiter

Contents

1. Hh Signals Pattern Diverse Developing Tissues	226
2. Hh Signal Transduction: Activating an Activator and Repressing a Repressor	229
3. Functions of Gli Proteins in Development	233
4. Defective Intraflagellar Transport Disrupts Mammalian Hh Signaling	236
5. The Primary Cilium as a Cellular Signaling Center	239
6. Mammalian Smo Activates the Hh Pathway at the Primary Cilium	241
7. Proper Ciliary Function is Required for Processing of Gli Proteins	242
8. Additional Ciliary Components Function in Hh Signal Transduction	243
9. Are Cilia Involved in Hh Pathway-Mediated Tumorigenesis?	246
10. Lingering Questions	248
Acknowledgments	249
References	249

Abstract

Cilia function as critical sensors of extracellular information, and ciliary dysfunction underlies diverse human disorders including situs inversus, polycystic kidney disease, retinal degeneration, and Bardet-Biedl syndrome. Importantly, mammalian primary cilia have recently been shown to mediate transduction of Hedgehog (Hh) signals, which are involved in a variety of developmental processes. Mutations in several ciliary components disrupt the patterning of the neural tube and limb bud, tissues that rely on precisely coordinated gradients of Hh signal transduction. Numerous components of the Hh pathway, including Patched, Smoothened, and the Gli transcription factors, are present within primary cilia, indicating that key steps of Hh signaling may occur within the cilium.

Department of Biochemistry and Biophysics, Cardiovascular Research Institute, University of California, San Francisco, California

Because dysregulated Hh signaling promotes the development of a variety of human tumors, cilia may also have roles in cancer. Together, these findings have shed light on one mechanism by which primary cilia transduce signals critical for both development and disease.

1. HH SIGNALS PATTERN DIVERSE DEVELOPING TISSUES

Hh proteins are secreted lipoproteins that specify cell patterning in many metazoa and in various developmental contexts (Huangfu and Anderson, 2006; Lum and Beachy, 2004; Wang et al., 2007b). Investigations into Hh signal transduction have proved to be a rich source of insight into fundamental mechanisms by which cells coordinate their functions during development and homeostasis, and have also revealed that dysregulated Hh signaling underlies a variety of human congenital diseases and cancer (Hooper and Scott, 2005; McMahon et al., 2003; Ruiz i Altaba et al., 2002). Indeed, studies on Hh signaling have continued to yield unanticipated surprises, one of which, the involvement of the primary cilium in vertebrate Hh signaling, is the focus of this review.

Hh was initially identified by Nüsslein-Volhard and Wieschaus (1980) in a systematic screen for larval segmentation defects in *Drosophila melanogaster*. Whereas flies possess a single Hh gene, mammals have three, the best studied of which encodes Sonic hedgehog (Shh), an important regulator of both embryonic development and postnatal homeostasis (Chiang et al., 1996; Echelard et al., 1993). The other two members of this family, Indian hedgehog and Desert hedgehog, participate in bone development and spermatogenesis, respectively (Bitgood et al., 1996; St-Jacques et al., 1999; Vortkamp et al., 1996).

All Hh proteins are initially synthesized as inactive precursors that possess an amino-terminal signaling domain and a carboxy-terminal intein-like domain which is later removed by autocatalytic cleavage (Perler, 1998). In the choanoflagellate *Monosiga brevicollis*, separate proteins containing domains homologous to the Hh signaling domain and the intein-like domain have recently been identified, suggesting that Hh proteins evolved through shuffling of domains found in our unicellular antecedents (King et al., 2008; Snell et al., 2006). However, some organisms, most notably *Caenorhabditis elegans*, lack proteins with homology to the Hh amino-terminal signaling domain, suggesting that the Hh pathway has subsequently diverged in some metazoan phyla (Bürglin, 1996).

In addition to its role in segment polarity, *Drosophila* Hh regulates the development of a variety of other tissues, including the wing, eye, muscle, and gut (Hooper and Scott, 2005; McMahon et al., 2003).

Similarly, vertebrate Shh is important for proper morphogenesis of organs such as the skin, eye, lung, muscle, and pancreas (McMahon et al., 2003). Perhaps the best understood roles of Shh are in the patterning of the neural tube and limb bud.

In the developing mouse embryo, *Shh* is first expressed at embryonic day (E) 7.5 in the ventral node (Echelard et al., 1993). By E9.5, *Shh* is expressed by cells in the ventral forebrain, notochord, floor plate of the neural tube, and posterior limb buds. In the mouse neural tube, Shh is critical for specifying neuronal cell fates in a concentration- and time-dependent manner (Dessaud et al., 2007; Ericson et al., 1997; Martí et al., 1995; Roelink et al., 1995). Shh is initially produced by cells in the notochord and is essential for the specification of cells in the overlying neural tube as floor plate (or, in zebrafish, as lateral floor plate) (Chiang et al., 1996; Martí et al., 1995; Roelink et al., 1995; Schauerte et al., 1998). Once specified, the floor plate also expresses Shh, becoming a second source of axial midline Shh that patterns the remainder of the ventral neural tube.

Shh produced by the notochord, the floor plate and possibly the gut moves away from these sources, establishing a dorsoventral gradient in the neural tube that specifies the fate of five additional neuronal subtypes (Fig. 9.1A) (Ericson et al., 1997). The ventral-most cells in the neural tube—those of the floor plate and the V3 interneurons—are normally exposed to and specified by the highest concentrations of Shh, whereas more dorsal cells are exposed to successively lower concentrations of Shh. Thus, *Shh* mutants display a loss of the floor plate and almost all ventral neural subtypes, with a concomitant expansion of more dorsal neural subtypes into the ventral neural tube (Fig. 9.1B) (Chiang et al., 1996). Conversely, ectopic placement of Shh sources is sufficient to specify ventral neural cell types in the lateral neural tube (Roelink et al., 1994).

In addition to concentration of Shh, duration of Hh signaling is also an important factor in determining neural cell fates. Neural explants can be induced to become progressively more ventral neural subtypes either by exposing these cells to increasing concentrations of Shh or by lengthening the time of exposure (Dessaud et al., 2007). Similarly, ventral neural cells that eventually become V3 interneurons initially express a marker of the more dorsal motor neuron fate, suggesting that ventral cell fates are sequentially determined *in vivo* by the duration of Hh pathway activity (Dessaud et al., 2007). Together, these results provide evidence that both concentration and duration of Hh signaling generate the diversity of ventral neural tube precursors.

A similar scenario occurs in the limb buds, where Shh specifies digit identity and number in a concentration- and time-dependent manner. During embryonic development, three organizing centers are crucial for establishing the axes of the nascent limb bud: the apical ectodermal ridge, which is necessary for the outgrowth of the limb and patterning of the

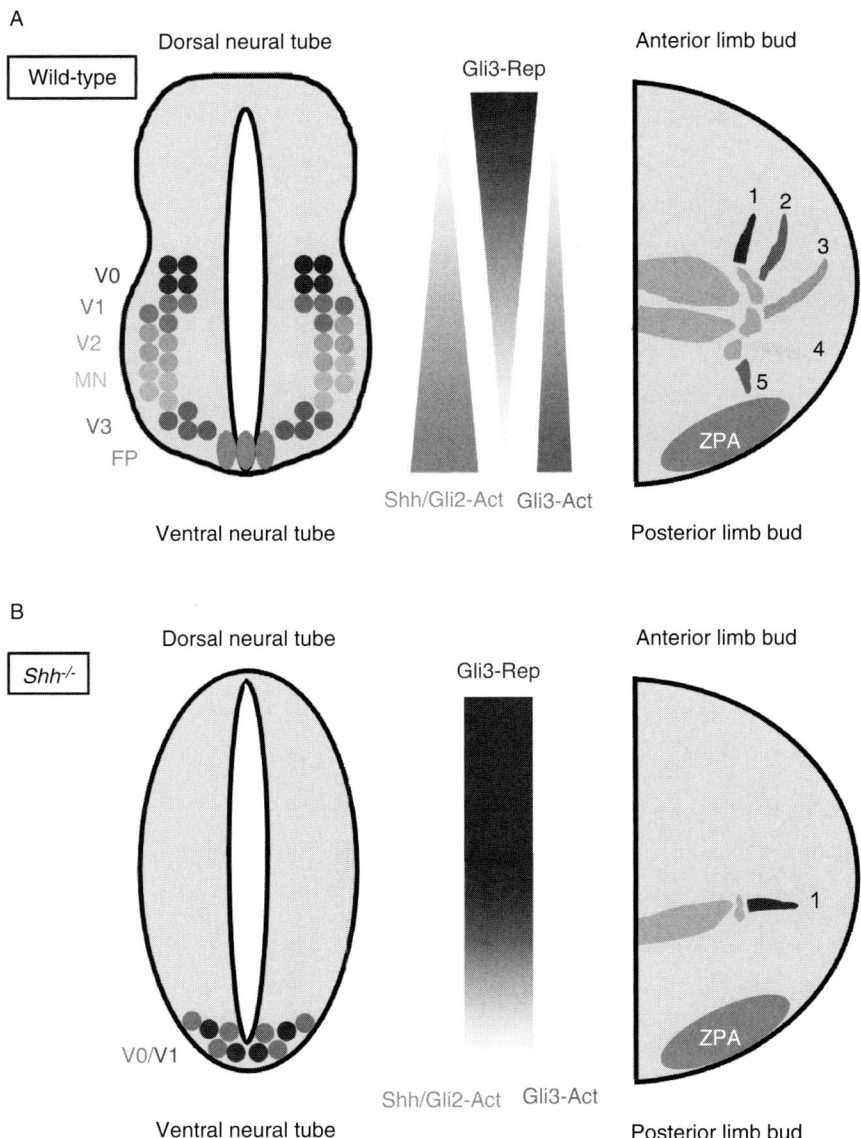

Figure 9.1 Proper patterning of the neural tube and limb buds is mediated by a gradient of Shh. (A) In the neural tubes of wild-type embryos, Shh produced by the notochord, floor plate (FP) and possibly gut forms a gradient along the dorsal–ventral axis. Similarly, Shh released by cells in the zone of polarizing activity (ZPA) in the limb buds forms a gradient along the anterior–posterior axis. Increased Shh concentration and duration of action are associated with increased Gli2 activator (Gli2-Act) and Gli3 activator (Gli3-Act) functions, and reduced Gli3 repressor (Gli3-Rep) formation. High-level Hh signaling is critical for specifying the ventral-most cell types in the neural tube, including FP and V3 interneurons (V3). More dorsal cell types (V0, V1, V2, and motor neurons (MN)) are specified by lower levels of Hh signaling. In limb buds, the Shh and

proximal–distal axis; the surface ectoderm, which regulates dorsal–ventral patterning; and the zone of polarizing activity (ZPA), which specifies digit number and identity. The ZPA is located in the posterior mesenchyme of the limb bud and secretes Shh, which forms a gradient as it moves anteriorly (Fig. 9.1A) (Riddle *et al.*, 1993). Posterior cells within the ZPA are exposed to high concentrations of Shh for the longest period of time and are specified to form digit 5 (the little finger in the human hand), whereas anterior cells in the limb bud are exposed to little or no Shh, and are specified into digit 1 (thumb) (Ahn and Joyner, 2004; Harfe *et al.*, 2004; Scherz *et al.*, 2007). The rest of the digits are specified by intermediate levels and/or durations of Shh signaling. In *Shh* mutant embryos, only a single digit similar to digit 1 is formed (Fig. 9.1B) (Chiang *et al.*, 1996).

Although vertebrate Hh signaling impinges on many aspects of development, the importance of this pathway later in life appears more limited. During adulthood, the Hh pathway maintains normal tissue homeostasis and plays regulatory roles for stem cells found, for instance, in the central nervous system and hair follicle (Han *et al.*, 2008; Lai *et al.*, 2003; Levy *et al.*, 2005). Improper activation of the Hh pathway may lead to cancer, and in two tumor types in particular—basal cell carcinoma (BCC) and medulloblastoma—dysregulated Hh signaling appears to play a causative role.

2. Hh Signal Transduction: Activating an Activator and Repressing a Repressor

Hh initiates signaling by binding to its receptor, the 12-transmembrane protein Patched (Ptch). Unlike receptors in many signaling pathways, in the absence of Hh, Ptch tonically inhibits the downstream pathway by repressing the function of Smoothened (Smo), a seven-transmembrane protein that acts as the central positive mediator of Hh signaling (Hooper and Scott, 2005; McMahon *et al.*, 2003). Vertebrates have two homologs of Ptch. Consistent with its function as a negative regulator of the Hh pathway, *Ptch1* mutant embryos display increased and ectopic Hh pathway activity, leading to an expansion of ventral neural cell types and polydactyly (Goodrich *et al.*, 1997; Milenkovic *et al.*, 1999; Motoyama *et al.*, 2003).

Gli3-Rep gradients are interpreted for the specification of digit identity (digits 1–5). (B) In *Shh* mutant embryos, the gradient of Hh pathway activation is absent. Gli3 is mostly converted into Gli3-Rep along the dorsal–ventral axis of the neural tube and the anterior–posterior axis of the limb buds, although Ihh may still activate the pathway to some degree. In mutant neural tubes, only the V0 and V1 ventral neuronal subtypes are specified, and in mutant limb buds, only the anterior-most digit (digit 1) is formed. (See Color Insert.)

In *Drosophila*, Ptch normally retains Smo in cytoplasmic vesicles. Upon binding of Hh to Ptch, inhibition of Smo is relieved, leading to phosphorylation and plasma membrane accumulation of Smo (Denef et al., 2000; Jia et al., 2004; Zhang et al., 2004). Activated Smo at the cell surface subsequently transduces signals through Cubitus interruptus (Ci), a zinc finger transcription factor that serves as a pivotal switch for downstream pathway activity.

In an intriguing demonstration of biological parsimony, the *Drosophila* Hh signaling pathway uses Ci both to repress and activate the Hh transcriptional program. In the absence of Hh signals, Ci is processed into a smaller repressor form (CiR) that inhibits the expression of Hh target genes (Aza-Blanc et al., 1997). This processing is facilitated by a kinesin-related protein known as Costal-2 (Cos2), which complexes with Ci and helps exclude it from the nucleus (Farzan et al., 2008; Robbins et al., 1997; Sisson et al., 1997; Wang and Holmgren, 1999; Zhang et al., 2005). This Ci–Cos2 complex is also bound by additional proteins in the absence of Hh signals, including unphosphorylated Smo, the serine/threonine kinase Fu, and Sufu (Lum et al., 2003). Importantly, Cos2 recruits several kinases, including protein kinase A (PKA), casein kinase I (CKI), and glycogen synthase kinase 3 (GSK3), which phosphorylate full-length Ci (Price and Kalderon, 1999, 2002). The sequential action of these kinases creates phosphopeptide-binding motifs that recruit the Supernumerary limbs (Slimb) subunit of the Skp1/Cullin1/F-box (SCF) E3 ubiquitin ligase complex, which ubiquitinates Ci to direct it for processing into CiR (Jiang and Struhl, 1998; Smelkinson and Kalderon, 2006). The proteolytic processing required for formation of CiR is mediated by the proteasome, which acts in a regulated manner to degrade only the carboxy-terminal portion of Ci containing the transcriptional activator domain (Chen et al., 1999; Wang and Price, 2008). Consistent with their roles in mediating Ci processing, loss of Cos2 or any one of the kinases that phosphorylate Ci can lead to accumulation of full-length Ci and increased activity of the Hh pathway (Price and Kalderon, 2002; Wang and Holmgren, 1999; Zhang et al., 2005).

In the presence of Hh, *Drosophila* Smo traffics to the membrane and induces phosphorylation and activation of Fu, leading to subsequent phosphorylation of Cos2 and Sufu (Lum et al., 2003; Ruel et al., 2003). Phosphorylation of Cos2 weakens its interactions with Ci and Smo, permitting dissociation of Ci from the complex (Liu et al., 2007; Ruel et al., 2007). Sufu, which suppresses the Hh pathway through its direct binding to Ci, is also inhibited by phosphorylation (Dussillol-Godar et al., 2006). Once freed of its interactions with Cos2 and Sufu, an activator form of Ci can enter the nucleus, where it induces expression of downstream target genes (Chen et al., 1999). In addition to its role in promoting CiR formation, PKA also appears to have a positive role in Hh pathway activation (Zhou et al., 2006). However, how PKA mediates these positive effects and whether additional post-translational events activate Ci are currently unknown.

Vertebrates possess three homologs of Ci, called Gli transcription factors (Gli1–3). Like Ci, Gli2 and Gli3 possess an amino-terminal repressor domain and a carboxy-terminal activator domain flanking the central DNA-binding zinc fingers (Ruiz i Altaba, 1999; Sasaki et al., 1999). Gli1, however, lacks the amino-terminal repressor domain and functions only as a transcriptional activator. Gli2 similarly functions mainly as an activator of Hh signaling, whereas Gli3 functions primarily as a repressor. Ultimately, it is the balance of the collective activator and repressor functions of these Gli transcription factors that determines the status of the Hh transcriptional program.

As with Ci, proteasome-mediated proteolysis regulates the abundance and transcriptional activity of the vertebrate Gli proteins. In the absence of Hh signals, Gli3 is phosphorylated by PKA, CKI, and GSK3, generating a phosphopeptide motif bound by β-transducin repeat-containing protein (β-TrCP), the vertebrate homolog of Slimb (Fig. 9.2A) (Wang and Li, 2006; Wang et al., 2000). β-TrCP recruits the SCF ubiquitin ligase complex, which targets Gli3 to the proteasome for partial proteolysis into a truncated form that functions as a repressor (Tempé et al., 2006). Gli1 and Gli2 are also phosphorylated by these kinases prior to binding by β-TrCP, but are mainly targeted for complete degradation by the proteasome (Kaesler et al., 2000; Pan et al., 2006). In some cases, Gli2 has been observed to be processed into truncated forms that might serve as transcriptional repressors (Pan et al., 2006; Ruiz i Altaba, 1999), although the significance of these less abundant species is currently unclear.

SCF is not the only ubiquitin ligase complex that regulates Ci and Gli activity. In *Drosophila*, a second complex which includes HIB/Roadkill and Cullin3 (Cul3) also binds full-length Ci proteins in the eye and wing discs to target them for proteasome-mediated degradation (Kent et al., 2006). A mammalian homolog of HIB, SPOP, can functionally substitute for HIB in flies, suggesting that the same complex may also regulate Gli stability (Zhang et al., 2006). Additionally, Gli1, but not Gli2 or Gli3, can interact with Numb, which recruits another E3 ubiquitin ligase, Itch, leading to degradation (Marcotullio et al., 2006).

Upon activation of Hh signaling, both degradation of Gli2 and proteolytic processing of Gli3 into its repressive form are inhibited (Pan et al., 2006; Wang et al., 2000), thereby permitting Gli2 to function as a strong activator of the Hh transcriptional program and allowing full-length Gli3 to serve as a activator in some circumstances (Fig. 9.2B) (Bai et al., 2004; Wang et al., 2007a). Overexpression of full-length Gli2 causes weak activation of the Hh pathway, whereas overexpression of a truncated form of Gli2 lacking its amino-terminal repressor domain leads to constitutive activation of Hh signaling both *in vitro* and *in vivo* (Meyer and Roelink, 2003; Sasaki et al., 1999; Sheng et al., 2002). Overexpression of a truncated Gli3 lacking its carboxy-terminal activator domain causes constitutive repression of

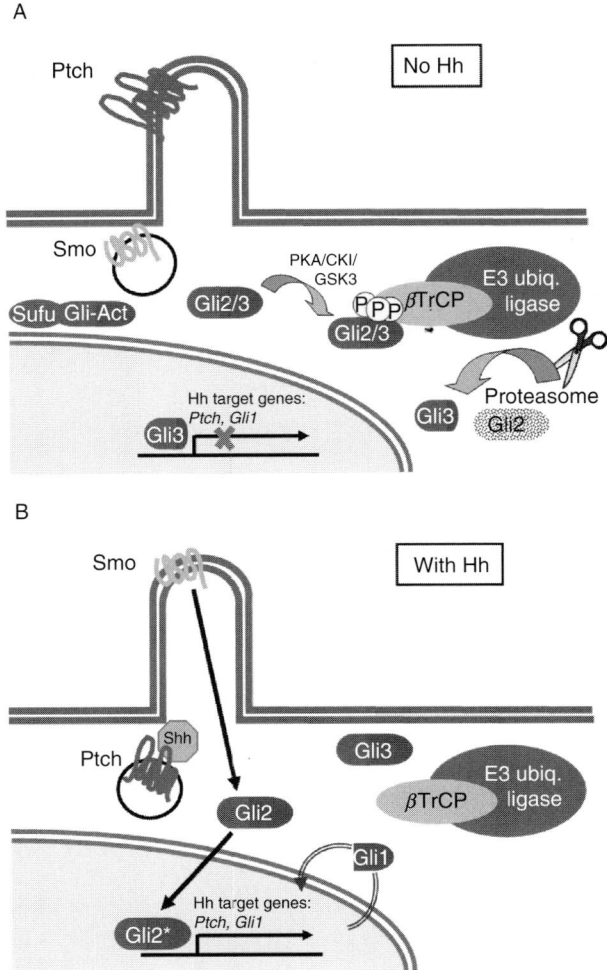

Figure 9.2 A model of mammalian Hh signal transduction. (A) In the absence of Shh, Ptch localizes to the primary cilium and, through an unknown mechanism, prevents Smo from entering the cilium. Gli2 and Gli3 are phosphorylated by kinases, including PKA, CKI, and GSK3, to generate phosphopeptide-binding motifs for β-TrCP, an important substrate recognition component of the E3 SCF ubiquitin ligase complex. Ubiquitination of Gli2 and Gli3 targets these proteins to the proteasome, which processes Gli3 into a carboxy-terminal-truncated repressor form and degrades Gli2. Gli3 subsequently enters the nucleus and inhibits transcription of downstream Hh target genes such as *Ptch1* and *Gli1*. (B) In the presence of Shh, Ptch is internalized, allowing Smo to traffic into the primary cilium. It is unclear whether Smo traffics directly to the cilium, or accumulates first in the plasma membrane. At the cilium, Smo inhibits the formation of Gli3 repressor and activates Gli2, which enters the nucleus to promote transcription of *Ptch1* and *Gli1*, whose protein products negatively and positively feed back on this pathway, respectively. (See Color Insert.)

downstream signaling (Meyer and Roelink, 2003; Ruiz i Altaba, 1999). Although these artificial constructs may not faithfully recapitulate the normal post-translational regulation of these transcription factors, these results suggest that the presence of activator and repressor domains critically determines the functions of Gli2 and Gli3, and that modulating the activity of these domains may be one mechanism by which the same proteins may exert both positive and negative effects on the Hh transcriptional program.

3. FUNCTIONS OF GLI PROTEINS IN DEVELOPMENT

Of the three Gli transcription factors, Gli2 plays the most important role in mediating the effects of Hh signaling in the neural tube. *Gli2*-deficient mouse embryos display loss of ventral neuronal subtypes normally specified by high levels of Hh signaling (the floor plate and some V3 interneurons), and exhibit an expansion of more dorsal cell types normally specified by lower levels of Hh signaling (Fig. 9.3A) (Ding *et al.*, 1998; Matise *et al.*, 1998). In contrast, *Gli1*-deficient mice manifest no obvious developmental defects (Park *et al.*, 2000), and *Gli3*-deficient animals exhibit only a subtle expansion of V0, V1, and V2 neural subtypes (Fig. 9.3B) (Persson *et al.*, 2002). These differences suggest that, in the neural tube, cells rely primarily on Gli2 to activate the Hh pathway in response to Shh, whereas Gli1 and Gli3 play supportive roles. Consistent with this, both *Gli1;Gli2*, and *Gli2;Gli3* double mutant embryos display more severe deficits in ventral neuronal subtype specification than do embryos lacking *Gli2* alone (Bai *et al.*, 2004; Lei *et al.*, 2004; Motoyama *et al.*, 2003; Park *et al.*, 2000).

A different situation occurs in the limb buds, where absence of *Gli2* has no effect on digit patterning, but loss of *Gli3* leads to polydactyly (Hui and Joyner, 1993), a phenotype also caused by increased or ectopic Shh in the limb bud (Fig. 9.3A and B) (Riddle *et al.*, 1993). Embryos lacking both *Shh* and *Gli3* exhibit polydactyly similar to mutants lacking only *Gli3*, suggesting that Shh patterns the limb bud by suppressing Gli3 repressor formation (Litingtung *et al.*, 2002; te Welscher *et al.*, 2002). Indeed, a gradient of the repressor form of Gli3 inverse to that of Shh forms in the limb bud (Litingtung *et al.*, 2002; Wang *et al.*, 2000). At posterior sites proximal to the source of Shh secretion (the future site of digit 5), Gli3 repressor levels are lowest, whereas in the anterior limb bud farthest from the source of Shh (the future site of digit 1), Gli3 repressor levels are highest. Thus, tissues can respond to Hh signals in two fundamentally different ways. In the first way, exemplified by the neural tube, the principle action of Shh is to induce the activator function of Gli2. In the second way, exemplified by the limb bud, Shh acts primarily to inhibit the formation of Gli3 repressor.

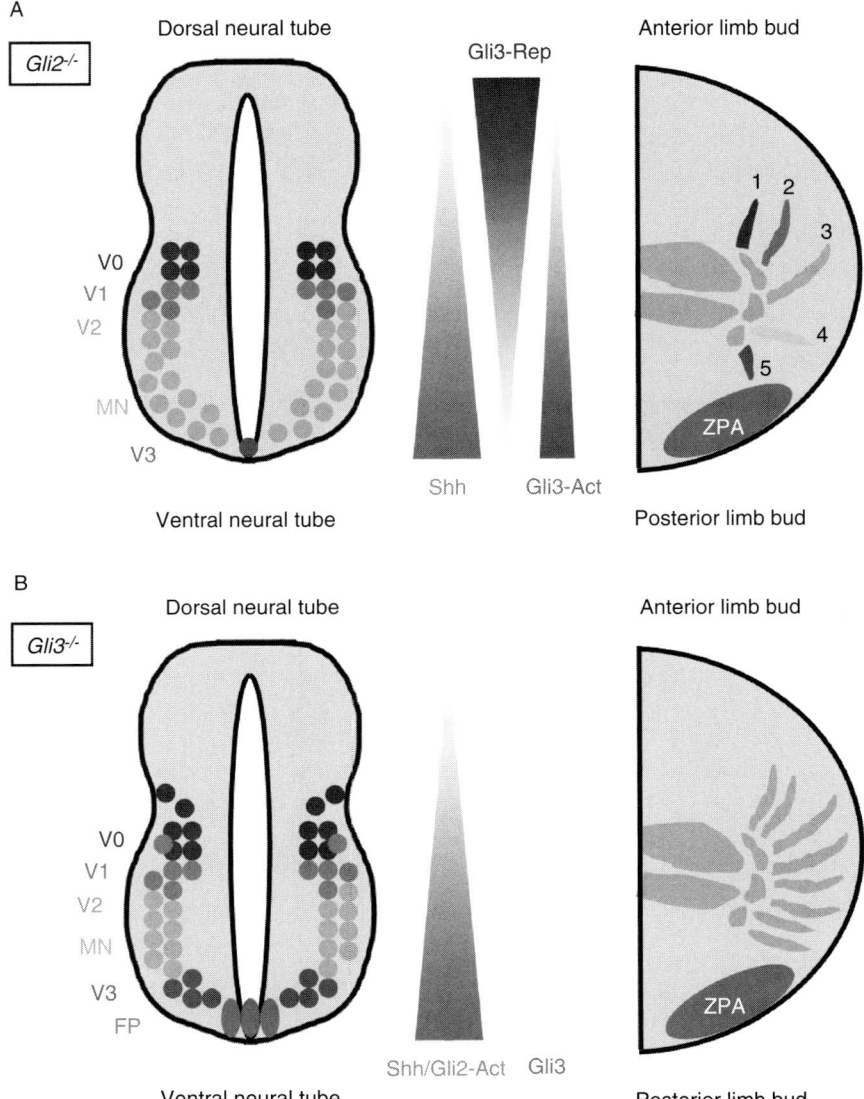

Figure 9.3 The neural tube and limb buds differ in their reliance on Gli2 and Gli3 for proper patterning. (A) In *Gli2* mutant embryos, gradients of Shh and Gli3-Rep/Act are established, but downstream signaling is not properly activated. In the neural tube, which relies predominantly on Gli2 activator function, specification of neuronal subtypes that require the highest levels of Hh pathway activity (FP and V3) is deficient. Because proper patterning of the limb bud depends upon a gradient of Gli3-Act/-Rep, and not on Gli2-Act, digit specification is normal in *Gli2* mutants. (B) In *Gli3* mutant embryos, gradients of Shh and Gli2-Act are established, but formation of Gli3-Rep is absent. *Gli3*-deficient neural tubes display only a subtle expansion of V0 and V1 neuronal subtypes. Because proper limb patterning depends on a gradient of Gli3-Rep, *Gli3* mutant limb buds upregulate genes normally repressed by Gli3-Rep, including *Gremlin* and *Hoxd13*, and exhibit polydactyly. (See Color Insert.)

In many tissues, Hh signaling activates the expression of *Ptch1* and *Gli1*, which, in addition to being integral components of the Hh signal transduction pathway, are direct transcriptional targets of Hh signaling (Bai *et al.*, 2002; Goodrich *et al.*, 1996; Marigo and Tabin, 1996). As mentioned previously, the developmental role of Gli1 is limited, although double mutant, gain-of-function and gene replacement experiments indicate that Gli1 can act as a positive mediator of the Hh transcriptional program (Bai and Joyner, 2001; Bai *et al.*, 2002; Park *et al.*, 2000). Induction of *Ptch1* expression has more complex effects on Hh patterning. Increased Ptch limits the movement of Hh proteins in a developmental field, presumably by directly binding and internalizing Hh proteins (Hepker *et al.*, 1997; Jeong and McMahon, 2005). Given its role in Smo inhibition, upregulation of Ptch may also be an important feedback by which cells become desensitized to Hh signals (Casali and Struhl, 2004; Dessaud *et al.*, 2007).

It is currently unclear the degree to which the functions of the *Drosophila* Ci interacting proteins Fu, Cos2, and Sufu are conserved in vertebrates. In zebrafish, inhibition of the Cos2 homolog Kif7 induces ectopic Hh signaling in the myotome and neural tube, consistent with data from flies that Cos2 acts as a suppressor of the Hh pathway (Tay *et al.*, 2005). The role of Cos2 homologs in mammals, however, remains unclear, as inhibition of Kif7 and its paralog Kif27 in cell culture assays did not affect Hh signaling (Varjosalo *et al.*, 2006). Mice deficient for *Fu* also do not display overt Hh-related defects (Chen *et al.*, 2005; Merchant *et al.*, 2005), although inhibition of *Fu* in zebrafish suppresses the specification of myotome subtypes that rely on high-level Hh signaling (Wolff *et al.*, 2003). In contrast, Sufu functions as a negative regulator of Hh signaling in both zebrafish and mice (Cooper *et al.*, 2005; Koudijs *et al.*, 2005; Svärd *et al.*, 2006). This is somewhat unexpected, given that Sufu plays only a minor role in *Drosophila* Hh signaling (Ohlmeyer and Kalderon, 1998; Préat, 1992). The molecular mechanism by which Sufu inhibits vertebrate Hh signaling is unclear, although it may both restrict Gli nuclear import and inhibit the activity of Gli transcriptional activators in the nucleus (Cheng and Bishop, 2002; Ding *et al.*, 1999; Paces-Fessy *et al.*, 2004; Pearse *et al.*, 1999; Stone *et al.*, 1999).

Thus, the homologs of the Ci interacting proteins appear to manifest varied functions in vertebrate Hh signaling. Fu, a critical regulator of Hh signaling in *Drosophila*, is important for Hh signaling in zebrafish but not in mouse. On the other hand, Sufu, a minor regulator of Hh signaling in fly, is a critical repressor of Hh signaling in both zebrafish and mouse. Together, these observations indicate that whereas much of the Hh signal transduction pathway is remarkably conserved, the mechanisms by which Hh signals are transduced between Smo and Gli have diverged considerably during metazoan evolution (Huangfu and Anderson, 2006).

4. Defective Intraflagellar Transport Disrupts Mammalian Hh Signaling

A recent screen for N-ethyl-N-nitrosourea (ENU)-induced mouse mutants that display neural tube defects yielded the surprising observation that mammalian Hh signaling relies on intraflagellar transport (IFT) (Huangfu et al., 2003). First visualized over a decade ago in Chlamydomonas flagella, IFT consists of large, multisubunit complexes that commute bidirectionally between the flagellar base and tip (Kozminski et al., 1993). IFT proteins are conserved among ciliated organisms, and IFT movement along cilia has also been observed in C. elegans chemosensory neuronal cilia (Orozco et al., 1999; Ou et al., 2005; Snow et al., 2004) and in mammalian primary cilia (Tran et al., 2008).

IFT plays a critical role in assembling flagella and cilia, as protein synthesis does not occur within these organelles (Rosenbaum and Witman, 2002). Thus, the building blocks required for generating these structures must be transported from the cell cortex to the distal end of the cilium. In Chlamydomonas, IFT is mediated by at least 17 proteins organized into two complexes, A and B (Cole et al., 1998). IFT from the base of the cilium to its tip (anterograde IFT) is powered by Kinesin II (Cole et al., 1998; Kozminski et al., 1995), whereas IFT from the ciliary tip to the base (retrograde IFT) is mediated by a dynein motor (Pazour et al., 1998; Porter et al., 1999). Disruption of either Chlamydomonas Kinesin II or one of several IFT complex B proteins prevents flagellar assembly (Deane et al., 2001; Pazour et al., 2000). In contrast, mutation of genes encoding components of the Chlamydomonas retrograde dynein motor or several IFT complex A proteins do not completely abrogate flagellum formation. Instead, these mutants have short flagella that accumulate flagellar proteins at the distal end due to defective retrograde IFT (Pazour et al., 1998; Piperno et al., 1998; Porter et al., 1999).

Consistent with their involvement in Chlamydomonas flagellum biogenesis, mutations that inhibit the function of mouse IFT also disrupt ciliogenesis (Marszalek et al., 1999; Nonaka et al., 1998; Pazour et al., 2000). The first cilia known to form during mouse development are those of the ventral node. These motile cilia generate a leftward flow in the node that is essential for proper left–right axis determination (Nonaka et al., 2002). Disruption of cilia blocks establishment of this flow, leading to left–right axis defects (Marszalek et al., 1999; Nonaka et al., 1998). Similarly, defects in ciliogenesis in the zebrafish embryo disrupt flow in Kupffer's vesicle, the ventral node equivalent, randomizing left–right axis formation (Essner et al., 2005). Paralysis of motile cilia also presumably causes the situs inversus that often accompanies human primary ciliary dyskinesia (Carlén and Stenram, 2005; Storm van's Gravesande and Omran, 2005).

In addition, mutations in ciliary proteins are linked to human polycystic kidney disease (PKD), a common genetic disorder characterized by the formation of fluid-filled renal cysts (Pan et al., 2005; Simons and Walz, 2006). Although disruption of ciliogenesis in mice can cause midgestation lethality, preventing analysis of kidney phenotypes, animals bearing a hypomorphic allele of *Ift88*, which encodes a complex B protein, survive longer and develop stunted primary cilia in the kidneys and PKD (Pazour et al., 2000; Yoder et al., 2002). Similarly, kidney-specific deletion of *Kif3a*, which encodes an essential component of the Kinesin II motor required for ciliogenesis, also causes PKD (Lin et al., 2003). Together, these results establish links between defects in IFT function and the pathogenesis of important human diseases.

In addition to left–right axis defects, mutant mouse embryos lacking Kif3a display an open neural tube, as do embryos lacking the IFT complex B components Ift188 or Ift172 (Huangfu and Anderson, 2005; Huangfu et al., 2003; Marszalek et al., 1999). *Kif3a, Ift88*, and *Ift172* mutants also display loss of ventral cell types in the neural tube, including loss of the floor plate, V3 interneurons, and motor neurons (Fig. 9.4A). These deficiencies are accompanied by an expansion of lateral neural markers into more ventral domains. Unlike *Shh* mutant embryos (Chiang et al., 1996), neural subtypes such as the V2 precursors are specified in *Ift88* and *Ift172* mutants, and the dorsal neural progenitor marker Pax7 is not expanded (Huangfu et al., 2003). Thus, the neural tube phenotypes of *Kif3a, Ift88*, and *Ift172* mutants are reminiscent of those caused by abrogated Hh signaling, but are more severe than those caused by loss of Gli2 and less severe than those observed in the absence of *Shh*.

Analyses of additional mice defective for various complex B components have yielded concordant results. For example, disruption of *Ift57* or *Ift52* causes ventral neural tube patterning defects resembling those observed in *Ift88* and *Ift172* mutants (Houde et al., 2006; Liu et al., 2005). In the cases in which it has been examined, loss of complex B proteins is associated with an absence of cilia in the ventral node. These results suggest that all IFT complex B proteins may be essential for mammalian Hh signaling and that their loss perturbs this pathway through a common mechanism.

Disruption of retrograde IFT also affects mammalian Hh signaling. Loss of the atypical dynein heavy chain Dnchc2 (also known as Dynch1), a component of the retrograde IFT motor, causes neural tube closure defects and loss of the floor plate and V3 precursors (Huangfu and Anderson, 2005; May et al., 2005). Loss of a different dynein subunit, D2lic, also likely interferes with retrograde IFT and is associated with phenotypes indicative of impaired Hh signaling (Rana et al., 2004). In neither case are cilia completely lost; as with *Chlamydomonas* mutations affecting retrograde IFT transport, the *Dnchc2* and *D2lic* mutant embryos both display stunted, bulged cilia. Defects in retrograde IFT in *C. elegans* have also been observed to cause a similar bulging (Schafer et al., 2003).

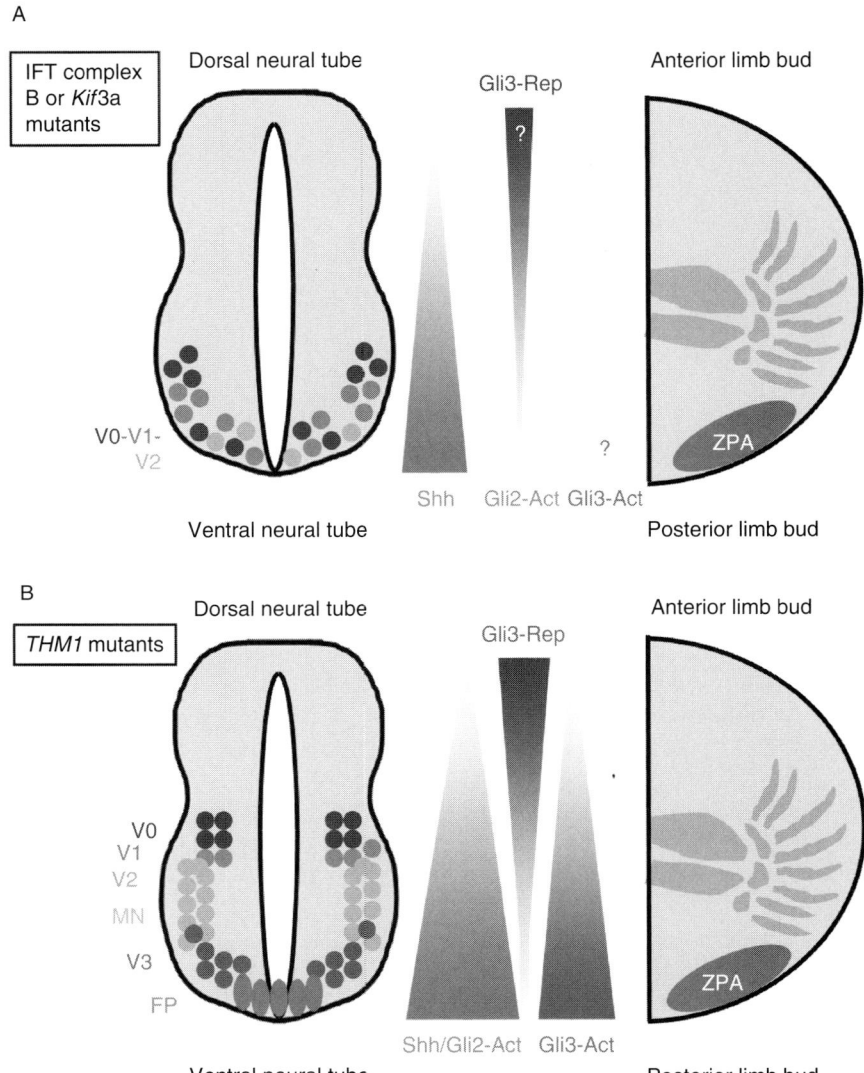

Figure 9.4 Many IFT mutations cause polydactyly and loss of ventral neural tube fates. (A) In embryos lacking IFT components such as IFT88, IFT172, and Kif3a, a gradient of Shh is established, but signaling through Gli2-Act and formation of Gli3-Rep are disrupted. Consequently, these mutants do not properly activate the Hh transcriptional targets, and display loss of ventral cell types (FP, V3, MN) and expansion of more dorsal subtypes (V0, V1, V2) in the neural tube. In IFT mutant limb buds, disruption of Gli3-Rep formation leads to polydactyly. (B) In contrast to other IFT mutants, *THM1* mutants display increased Hh signaling mediated by Gli2-Act or Gli3-Act, leading to an increase in ventral subtypes (FP, V3, MN) in the neural tube. THM1 is also essential for Gli3 processing, defects in which result in polydactyly. (See Color Insert.)

The involvement of IFT proteins in mammalian Hh signaling is surprising, given that mutations in homologous genes do not perturb *Drosophila* Hh signaling (Han *et al.*, 2003; Sarpal *et al.*, 2003). Indeed, many *Drosophila* tissues patterned by Hh are not known to possess cilia. It is interesting that epistatic analyses have suggested that IFT functions at a point downstream of Smo and upstream of Gli3 (Huangfu and Anderson, 2005; Huangfu *et al.*, 2003). As noted previously, analyses of the vertebrate homologs of Sufu, Fu, and Cos2 have suggested that this part of the pathway has diverged most extensively between arthropods and mammals. The contrasting reliance of *Drosophila* and mouse Hh signaling on cilia appears to be part of this evolutionary divergence. Analysis of whether Hh signals are also transduced through cilia in organisms such as zebrafish, tunicates, and Cnidaria will help reveal whether *Drosophila* have lost ciliary involvement in Hh signaling or whether mammals have gained this capability.

In addition to neural tube defects, disruption of mouse IFT proteins Dnchc2, Ift88 or Ift52 causes preaxial polydactyly, which is typically associated with increased Hh signaling in the limb bud (Fig. 9.4A) (Haycraft *et al.*, 2005; Liu *et al.*, 2005; May *et al.*, 2005). As discussed previously, *Shh* mutants form only a single digit (Chiang *et al.*, 1996; Litingtung *et al.*, 2002; te Welscher *et al.*, 2002). In contrast, *Shh; Ift88* double mutants display polydactyly similar to that of *Ift88* single mutants and *Shh;Gli3* double mutants (Liu *et al.*, 2005). Thus, as in the neural tube, IFT functions downstream of Shh in limb patterning. However, in contrast to neural tube patterning, loss of IFT results in a limb bud phenotype characteristic of Hh gain of function. How do IFT proteins regulate Hh signaling in this seemingly paradoxical manner? Is it their role in ciliogenesis that mediates Hh signaling, or some alternate activity? Fortunately, recent insights into ciliary function have shed light on these important questions.

5. THE PRIMARY CILIUM AS A CELLULAR SIGNALING CENTER

First described by Zimmerman (1898) over a hundred years ago, cilia are found in many tissues across diverse phyla. The recent renewed interest in primary cilia comes from the recognition that these organelles function as critical signaling centers in the cell (Eggenschwiler and Anderson, 2007). Indeed, the varied metaphors used to describe cilia in the literature over the past few years—as "cellular antennae," as "watchtowers," as "global positioning systems"—underscore the prominent role primary cilia are now regarded to play in the detection and integration of a variety of signaling cues (Benzing and Walz, 2006; Mans *et al.*, 2008; Marshall and Nonaka, 2006; Singla and Reiter, 2006).

Unlike the more familiar motile cilia that abundantly line such surfaces as the airway and oviduct, primary cilia are solitary microtubule-based extensions that protrude from the plasma membrane of most cells and, with the exception of cilia found in the embryonic node, are thought to be nonmotile (Satir and Christensen, 2007). They are also dynamic structures that are extended during interphase and later resorbed during cell cycle progression (Quarmby and Parker, 2005). This regulated assembly and disassembly is thought to be necessary for proper mitosis, as the base of the cilium, known as the basal body, contains a centriole that must be released from the cell membrane to move to the spindle pole.

The formation of a primary cilium initiates during interphase when the centrioles move to the plasma membrane. Vesicles budding from the Golgi deliver proteins that encapsulate the distal end of the mother centriole, leading to an accumulation of pericentriolar material that promotes microtubule polymerization (Salisbury, 2004; Satir and Christensen, 2007). These microtubules, which exist as triplets in the basal body, extend as doublets in a $9 + 0$ configuration along the length of the cilium, forming a core structure known as the ciliary axoneme. This microtubule configuration differs from that of most motile cilia, which possess an additional central microtubule pair in a $9 + 2$ arrangement.

At the interface between the axoneme and the basal body are the terminal plate and transition fibers, structures which may function as gateways that regulate the trafficking of proteins into and out of the cilium (Arima et al., 1984). Within the overlying plasma membrane, a condensed lipid zone encircles the base of the cilium, which may also serve to restrict the entry of membrane proteins (Gilula and Satir, 1972; Vieira et al., 2006). The basal body, in coordination with these gatekeeper structures, is thought to regulate the loading of ciliary proteins onto IFT particles and their entry into the cilium.

A variety of human diseases are associated with dysfunctional primary cilia. In addition to left–right axis defects and PKD, as discussed above, dysfunctional cilia have been linked to polymorphic diseases such as Bardet-Biedl syndrome (BBS) (Ansley et al., 2003). BBS is characterized by a range of phenotypes including polycystic kidney disease, polydactyly, diabetes, obesity, situs inversus, retinal degeneration, anosmia, sensory defects, and cognitive deficits. To date, mutations in at least 12 different *Bbs* genes have been linked to this disorder, and many of these genes encode proteins that localize to cilia and basal bodies (Nachury et al., 2007; Tobin and Beales, 2007). Defective ciliary function may also underlie other multisystem disorders, including Meckel-Gruber syndrome and Joubert syndrome (Adams et al., 2007). Interestingly, like mouse embryos with defects in IFT or Gli3, many BBS- or Meckel-Gruber syndrome-affected individuals exhibit polydactyly. Whether this is a manifestation of altered Hh signaling is currently unclear.

6. MAMMALIAN SMO ACTIVATES THE HH PATHWAY AT THE PRIMARY CILIUM

As discussed previously, double mutant analyses have indicated that IFT proteins function downstream of Smo and upstream of Gli proteins in the Hh pathway. Mechanistic insights into how IFT mediates Hh signal transduction have been provided by recent studies which show that both Ptch and Smo can localize to primary cilia, and that localization of these proteins is mutually exclusive (Corbit et al., 2005; Rohatgi et al., 2007). Thus, in the absence of Hh ligand, Ptch localizes to cilia and inhibits the Hh pathway by keeping Smo out of the cilium. To activate signaling, Hh binds Ptch at the cilium (Rohatgi et al., 2007) and becomes localized to a juxta-ciliary region (Chamberlain et al., 2008). Consequently, Ptch is internalized, allowing Smo to move into the cilium, where it promotes downstream pathway activation by shifting the processing of the Gli transcription factors in favor of activator forms (Fig. 9.2B). How Ptch regulates Smo localization, and how Smo, once in the cilium, transduces signals to the Gli transcription factors remain unclear.

Loss of *Ptch1* is sufficient to confer constitutive ciliary localization for Smo (Ocbina and Anderson, 2008; Rohatgi et al., 2007). Recent work has focused on determining whether Ptch inhibits Smo localization and activity through small molecule intermediates (Rohatgi et al., 2007; Taipale et al., 2002). Suggestive evidence for this hypothesis comes from the fact that Ptch acts catalytically to inhibit Smo, and is structurally related to the resistance-nodulation-cell division (RND) family of bacterial pumps and the Niemann–Pick C1 cholesterol transporter (Taipale et al., 2002). Thus, Ptch may modulate the concentration of a small molecule, possibly related to cholesterol, which affects Smo trafficking into and/or out of the cilium.

Further support for this hypothesis comes from the fact that sterols affect cellular response to Shh. Treatment of cultured fibroblasts with specific oxysterol derivatives of cholesterol promotes localization of Smo into the cilium (Rohatgi et al., 2007) and can induce differentiation of pluripotent mesenchymal cells into osteoblasts, an effect also observed when these cells are exposed to Shh (Dwyer et al., 2007). Conversely, depleting cholesterol by cyclodextrin or HMG-CoA reductase inhibitor can inhibit pathway response to Shh (Cooper et al., 2003). Furthermore, human genetic disorders such as Smith–Lemli–Opitz syndrome and lathosterolosis are caused by deficiencies in cholesterol biosynthesis, and are characterized by features such as holoprosencephaly and polydactyly, phenotypes that can result from alterations in Hh signaling (Chiang et al., 1996; Kelley et al., 1996; Krakowiak et al., 2003). Pharmacological inhibitors of the sterol synthesis pathway have also been shown to limit the Hh pathway activity and growth of murine $Ptch1^{+/-}$; $p53^{-/-}$ medulloblastoma cells (Corcoran and Scott, 2006). Together, these data suggest that Ptch suppresses Hh signaling either by transporting inhibitory sterols to Smo, or by preventing activating sterols from reaching Smo (Bijlsma et al., 2006; Corcoran and Scott, 2006).

It is tempting to speculate that sterols may also affect Smo activity indirectly by altering the membrane dynamics that regulate the trafficking of proteins into and out of the cilium. Another possibility is that oxysterols might act directly on Ptch to inhibit its function. It is also currently unclear how Smo reaches the cilium upon activation of Hh signaling. In the presence of Shh, Smo can be phosphorylated by G protein-coupled receptor kinase 2 (GRK2), which induces Smo binding to β-arrestin and Kif3a in NIH-3T3 cells (Chen et al., 2004; Kovacs et al., 2008; Meloni et al., 2006). Both β-arrestin and Kif3a are required for Smo to traffic into the cilium, suggesting that Kinesin II may directly transport Smo along the axoneme (Kovacs et al., 2008).

7. Proper Ciliary Function is Required for Processing of Gli Proteins

The finding that Smo localization to the cilium is required for activating mammalian Hh signal transduction has provided mechanistic insights into the embryonic phenotypes observed upon disruption of IFT components. In the absence of IFT complex B components, cilia are lost or structurally compromised, likely preventing Smo from signaling through Gli2 and activating the Hh transcriptional program. Although this explanation can account for why disruption of IFT complex B components leads to Hh loss-of-function phenotypes observed in the neural tube, it does not explain why polydactyly is also observed in the limb buds of IFT mutants.

How can IFT mediate Hh pathway activation in certain developmental contexts and Hh pathway inhibition in others? One possible explanation for these dual functions comes from the finding that, in addition to their role in potentiating Gli2 activator formation, IFT components also promote Gli3 repressor activity. This is evidenced by the observation that embryos mutant for genes encoding several IFT components, including Ift88, Ift172, Kif3a, and Dnchc2, display decreased processing of Gli3 into its repressor form (Haycraft et al., 2005; Huangfu and Anderson, 2005; Liu et al., 2005; May et al., 2005).

As mentioned previously, the principle role of Shh during limb bud patterning is to inhibit formation of Gli3 repressor (Wang et al., 2000). In the absence of Gli3 repressor formation, Shh-dependent genes such as *Gremlin* and *Hoxd13* are ectopically expressed in the limb (Litingtung et al., 2002; te Welscher et al., 2002). Consistent with the observed defects in Gli3 repressor formation, Gremlin and *Hoxd13* are also ectopically expressed in the limb buds of embryos lacking Ift88 or Dnchc2 (Liu et al., 2005; May et al., 2005). In addition, overexpression of full-length Gli3 in IFT-deficient limb bud cells is incapable of suppressing Gli1-mediated Hh pathway activation, although overexpression of a truncated Gli3 lacking its carboxy-terminal activator domain can still function as a repressor in the absence of IFT (Haycraft et al., 2005). Together, these results indicate that IFT plays a critical role in Gli3 repressor formation and activity both *in vitro* and *in vivo*.

As also previously mentioned, proper patterning of the neural tube depends primarily on Gli2 activator function, with Gli3 activator playing a supportive role, as evidenced by the observation that *Gli2;Gli3* double mutant embryos display more severe ventral neural patterning defects than do *Gli2* single mutants (Bai et al., 2004; Lei et al., 2004; Motoyama et al., 2003). IFT complex B mutants display loss of ventral cell types that largely resemble the phenotypes observed in *Gli2;Gli3* double mutant embryos (Huangfu and Anderson, 2005; Huangfu et al., 2003). However, loss of Gli3 partially mitigates the neural patterning defects observed in either *Ift88-* or *Ift172-*deficient embryos by restoring formation of motor neurons (Huangfu et al., 2003). Given the phenotypes observed in the limb bud and neural tube, the most parsimonious explanation for reconciling these results is that, while the cilium is essential for full Gli3 repressor activity, some residual repressor activity remains even in the absence of the cilium.

Finally, all three Gli proteins, as well as Sufu, can localize to the cilium (Haycraft et al., 2005), suggesting that key Gli processing steps do not simply require cilia, but occur within the cilium itself. Interestingly, Gli1, which can localize to cilia, does not require IFT for activity, unlike Gli2 and Gli3 (Haycraft et al., 2005). Proteasomes, which are critical for regulating Gli2 protein levels and for generating Gli3 repressor, have been observed to accumulate near basal bodies (Wigley et al., 1999). Thus, it is possible that Gli2 and Gli3 may be post-translationally modified at the cilium in an Hh-dependent manner, and that proteasomes at the ciliary base may subsequently interpret these cues, prior to Gli2/3 entry into the nucleus.

8. Additional Ciliary Components Function in Hh Signal Transduction

If cilia are in fact integral to the transduction of Hh signals during mammalian development, inhibition of critical ciliary proteins such as those involved with proper functioning of the basal body should also be expected to disrupt Hh signal transduction. One such basal body protein is Ofd1, disruption of which causes oral–facial–digital type I (Ofd1), a human ependymal and kidney epithelial cilia to be short and reduced in number (Town et al., 2008). Despite these defects in ciliary morphology, *Stumpy* mutants lack recognized Hh-associated phenotypes.

Additional proteins that have previously been found to modulate the Hh pathway may also have roles in ciliary function. Among these is Rab23, a Rab family GTPase that negatively regulates Hh signaling downstream of Smo and upstream of Gli2 (Eggenschwiler et al., 2001). The immunophilin family member FK506-binding protein 8 (FKBP8) is another intriguing negative regulator of Hh signaling that functions downstream of Smo and upstream of Gli2 in neural tissues (Bulgakov et al., 2004; Cho et al., 2008). Sil1, a cytosolic protein of unknown function, is a positive mediator of Hh signaling that acts downstream of Ptch (Izraeli et al., 2001). Evc, which localizes to the base of the cilium, potentiates Indian Hh-regulated bone development (Ruiz-Perez et al., 2007). Finally, genetic analyses of Tectonic have suggested that this protein may potentiate Hh signaling in the presence of Hh ligands, but may also suppress the pathway in the absence of upstream signals (Reiter and Skarnes, 2006). In summary, these studies have provided tantalizing clues that suggest the involvement of additional novel components in ciliogenesis and Hh signaling.

9. Are Cilia Involved in Hh Pathway-Mediated Tumorigenesis?

During many developmental processes, activation of the Hh pathway is mediated through paracrine interactions between the epithelium and mesenchyme. In the adult, however, Hh signaling plays a more limited role, although upregulation of Hh pathway activity, either through autocrine signaling or mutations that act cell autonomously, can give rise to cancer. Indeed, a diversity of human tumors express targets of the Hh pathway, and there is evidence that inappropriate activation of Hh signaling may contribute to the development of lung cancer, pancreatic cancer, prostate cancer, and glioma (Pasca di Magliano and Hebrok, 2003; Rubin and de Sauvage, 2006; Ruiz i Altaba et al., 2002). However, for two human cancers in particular—BCC and medulloblastoma—there is extensive evidence that increased Hh signaling causes these malignancies (Athar et al., 2006; Tang et al., 2007).

Humans diagnosed with Basal cell nevus syndrome (also known as Gorlin's syndrome) are prone to developing widespread BCCs. These patients have been found to be heterozygous for *Ptch1*, and *Ptch1$^{+/-}$* mice exhibit many features of this disease, including a propensity for developing medulloblastomas and sarcomas, and BCCs after irradiation (Aszterbaum *et al.*, 1999; Hahn *et al.*, 1998). Mutations in the *Ptch1* locus are also detected in 10–20% of sporadic medulloblastomas in humans, and in 50–60% of sporadic BCCs (Tang *et al.*, 2007). BCCs without *Ptch1* mutations often harbor mutations in *Smo* (Crowson, 2006; Lam *et al.*, 1999; Xie *et al.*, 1998). These mutations likely render Smo insensitive to the inhibitory action of Ptch, leading to constitutive activation of downstream signaling.

A homolog of at least one oncogenic form of Smo isolated from a human patient with BCC localizes constitutively to the cilium in the absence of Hh ligand (Corbit *et al.*, 2005). This is in contrast to wild-type Smo, which moves to the cilium only upon Hh stimulation. In addition, cyclopamine, which has been used as a chemotherapeutic agent for BCC and medulloblastoma in animal models, removes Smo from the cilium (Corbit *et al.*, 2005). Given the preponderance of data indicating that cilia are required for proper Hh pathway activity downstream of Smo, one important question might be, Are cilia necessary for the growth of tumors, particularly those initiated and sustained by Smo activation? And are human tumors ciliated?

Although there are currently no published reports that have examined ciliary distribution across a panel of different tumor types, we have observed that cilia are frequently absent from established human cancer cell lines (S. Y. Wong and J. F. Reiter, unpublished results). However, these cells have been selected for growth under tissue culture conditions and may not be representative of tumors arising in human patients. Indeed, we have recently examined biopsies from human BCCs and found that cells from these tumors are frequently ciliated (Wong *et al.*, manuscript submitted). In addition, we have observed that cilia are also present on cells from BCC-like lesions that arise in transgenic mice induced to express a constitutively active form of Smo (SmoM2) specifically in the skin. Others have recently observed that cells from a subset of human medulloblastomas can also be ciliated (Han *et al.*, manuscript submitted).

Because cilia play critical roles in modulating Hh signaling during development, it is likely that these organelles also serve important functions during tumorigenesis, particularly for Hh-dependent cancers such as BCC and medulloblastoma. As discussed previously, while cilia are necessary for propagating signals initiated by Smo and transduced downstream through Gli activators, these organelles are also required for Gli3 repressor formation, which may inhibit the expression of a host of genes, including those that promote cell proliferation. Thus, the functions of cilia during tumorigenesis are likely to be complex (Fig. 9.5). In addition, cilia may have roles in modulating a multitude of other signaling pathways important for cancer, including Wnt and PDGF (Corbit *et al.*, 2008; Gerdes *et al.*, 2007; Schneider *et al.*, 2005).

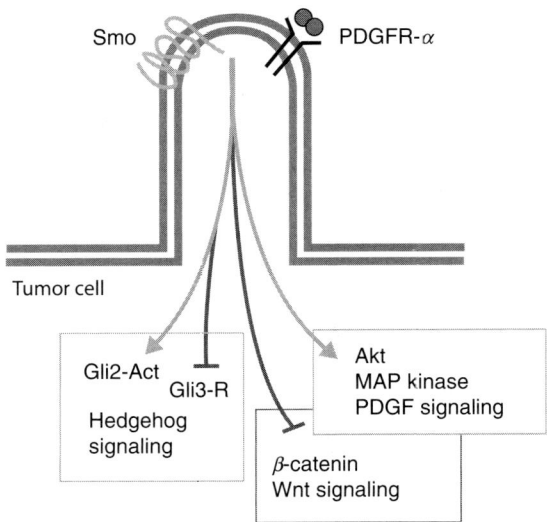

Figure 9.5 The primary cilium likely coordinates a variety of signaling pathways important for cancer. In Hh-dependent cancers such as BCC and medulloblastoma, the cilium may modulate the activity of signaling pathways in addition to Hh. For instance, localization of PDGF receptor-α to the cilium is important for transducing downstream signals mediated by Akt and MAP kinases. The cilium may also normally restrain Wnt pathway activity by preventing the accumulation of β-catenin. The balance of the downstream effects of these pathways likely governs cellular decisions for proliferation and apoptosis. (See Color Insert.)

10. Lingering Questions

The identification of the primary cilium as a critical modulator of both positive and negative signals of the Hh pathway has placed this organelle at the crossroads of Hh signal transduction. While many new questions have been raised, many old ones remain to be addressed. For instance, it remains unclear how Ptch restrains Smo activity or how Smo triggers Gli activation. The central involvement of the cilium in these processes suggests that understanding how Ptch alters the cilium, what proteins move Smo into the cilium, and what Gli processing components function within the cilium may provide significant insights into these difficult problems.

It is also currently unknown how general a role cilia play in Hh signaling in other organisms. Although it is clear that mouse Hh signaling relies on cilia and *Drosophila* Hh signaling does not, the involvement of cilia in Hh signaling in other organisms has not been well explored. For example, does Hh signaling act through cilia in ascidians or Cnidaria? Are cilia important for Hh signaling in fish? Although loss of IFT components causes kidney cysts in zebrafish (Sun *et al.*, 2004), Hh-related phenotypes have not yet been observed, possibly because cilia persist in these mutants through early development, presumably due to maternal deposition of wild-type IFT mRNA or protein (Tsujikawa and Malicki, 2004). Thus, it will be interesting in future studies to assess the phenotypes of fish lacking both maternal and zygotic IFT contributions.

Even in mammals, it is currently unclear whether all Hh signaling is transduced through the cilium, or whether some cell types or developmental events respond to Hh signals through alternative mechanisms. One possible example of cilium-independent Hh signaling might be in the guidance of commissural axons. Shh functions as a midline signal that guides commissural axons toward their intermediate target, the floor plate (Charron et al., 2003). As Shh presumably acts at the growth cone to induce rapid cytoskeletal changes, it is difficult to imagine how the cilium could mediate this guidance. Intriguingly, the correct projection of commissural axons depends on Boc, a cell surface protein that binds Shh (Okada et al., 2006). Could Ptch mediate cilium-dependent Hh activity, while Boc, or a similar protein, Cdo, mediate cilium-independent Hh activity?

Perhaps a more philosophical question is, Why is the cilium required for Hh signaling at all? As noted above, work from *Drosophila* has shown that organisms can clearly solve the problem of Hh signal transduction through other means. Cilia have evolutionarily ancient roles in sensing and transducing environmental information, extending back to our unicellular ancestors. Upon achieving multicellularity, there would have been new demands for coordinating cellular behaviors. Perhaps there was a low barrier for adapting the signaling machinery already present in the cilium for these purposes. If so, ciliary transduction of Hh signals may represent a truly ancient form of intercellular communication likely to be well represented throughout the animal kingdom.

ACKNOWLEDGMENTS

The authors acknowledge the support of grants from the NIH (RO1AR054396), the Burroughs Wellcome Fund, the Packard Foundation, and the Sandler Family Supporting Foundation to J.F.R. S.Y.W. acknowledges the support of the A. P. Giannini Foundation, the Herbert W. Boyer Fund, and the American Cancer Society.

REFERENCES

Adams, N. A., Awadein, A., and Toma, H. S. (2007). The retinal ciliopathies. *Ophthalmic Genet.* **28,** 113–125.

Ahn, S., and Joyner, A. L. (2004). Dynamic changes in the response of cells to positive hedgehog signaling during mouse limb patterning. *Cell* **118,** 505–516.

Ansley, S. J., Badano, J. L., Blacque, O. E., Hill, J., Hoskins, B. E., Leitch, C. C., Kim, J. C., Ross, A. J., Eichers, E. R., Teslovich, T. M., Mah, A. K., Johnsen, R. C., et al. (2003). Basal body dysfunction is a likely cause of pleiotropic Bardet-Biedl syndrome. *Nature* **425,** 628–633.

Arima, T., Shibata, Y., and Yamamoto, T. (1984). A deep-etching study of the guinea pig tracheal cilium with special reference to the ciliary transitional region. *J. Ultrastruct. Res.* **89,** 34–41.

Aszterbaum, M., Epstein, J., Oro, A., Douglas, V., LeBoit, P. E., Scott, M. P., and Epstein, E. H., Jr. (1999). Ultraviolet and ionizing radiation enhance the growth of BCCs and trichoblastomas in patched heterozygous knockout mice. *Nat. Med.* **5,** 1285–1291.

Athar, M., Tang, X., Lee, J. L., Kopelovich, L., and Kim, A. L. (2006). Hedgehog signalling in skin development and cancer. *Exp. Dermatol.* **15,** 667–677.

Aza-Blanc, P., Ramírez-Weber, F. A., Laget, M. P., Schwartz, C., and Kornberg, T. B. (1997). Proteolysis that is inhibited by hedgehog targets Cubitus interruptus protein to the nucleus and converts it to a repressor. *Cell* **89,** 1043–1053.

Bai, C. B., and Joyner, A. L. (2001). Gli1 can rescue the *in vivo* function of Gli2. *Development* **128,** 5161–5172.

Bai, C. B., Auerbach, W., Lee, J. S., Stephen, D., and Joyner, A. L. (2002). Gli2, but not Gli1, is required for initial Shh signaling and ectopic activation of the Shh pathway. *Development* **129,** 4753–4761.

Bai, C. B., Stephen, D., and Joyner, A. L. (2004). All mouse ventral spinal cord patterning by Hedgehog is Gli dependent and involves an activator function of Gli3. *Dev. Cell* **6,** 103–115.

Benzing, T., and Walz, G. (2006). Cilium-generated signaling: A cellular GPS? *Curr. Opin. Nephrol. Hypertens.* **15,** 245–249.

Bijlsma, M. F., Spek, C. A., Zivkovic, D., van de Water, S., Rezaee, F., and Peppelenbosch, M. P. (2006). Repression of Smoothened by Patched-dependent (pro-) vitamin D3 secretion. *PLoS Biol.* **4,** 1397–1410.

Bitgood, M. J., Shen, L., and McMahon, A. P. (1996). Sertoli cell signaling by Desert hedgehog regulates the male germline. *Curr. Biol.* **6,** 298–304.

Bonnafe, E., Touka, M., AitLounis, A., Baas, D., Barras, E., Ucla, C., Moreau, A., Flamant, F., Dubruille, R., Couble, P., Collignon, J., Durand, B., et al. (2004). The transcription factor RFX3 directs nodal cilium development and left–right asymmetry specification. *Mol. Cell. Biol.* **24,** 4417–4427.

Bulgakov, O. V., Eggenschwiler, J. T., Hong, D. H., Anderson, K. V., and Li, T. (2004). FKBP8 is a negative regulator of mouse sonic hedgehog signaling in neural tissues. *Development* **131,** 2149–2159.

Bürglin, T. R. (1996). Warthog and groundhog, novel families related to hedgehog. *Curr. Biol.* **6,** 1047–1050.

Carlén, B., and Stenram, U. (2005). Primary ciliary dyskinesia: A review. *Ultrastruct. Pathol.* **29,** 217–220.

Casali, A., and Struhl, G. (2004). Reading the Hedgehog morphogen gradient by measuring the ratio of bound to unbound Patched protein. *Nature* **431,** 76–80.

Caspary, T., Larkins, C. E., and Anderson, K. V. (2007). The graded response to Sonic Hedgehog depends on cilia architecture. *Dev. Cell* **12,** 767–778.

Chamberlain, C. E., Jeong, J., Guo, C., Allen, B. L., and McMahon, A. P. (2008). Notochord-derived Shh concentrates in close association with the apically positioned basal body in neural target cells and forms a dynamic gradient during neural patterning. *Development* **135,** 1097–1106.

Charron, F., Stein, E., Jeong, J., McMahon, A. P., and Tessier-Lavigne, M. (2003). The morphogen sonic hedgehog is an axonal chemoattractant that collaborates with netrin-1 in midline axon guidance. *Cell* **113,** 11–23.

Chen, C. H., von Kessler, D. P., Park, W., Wang, B., Ma, Y., and Beachy, P. A. (1999). Nuclear trafficking of Cubitus interruptus in the transcriptional regulation of Hedgehog target gene expression. *Cell* **98,** 305–316.

Chen, W., Ren, X. R., Nelson, C. D., Barak, L. S., Chen, J. K., Beachy, P. A., de Sauvage, F., and Lefkowitz, R. J. (2004). Activity-dependent internalization of smoothened mediated by beta-arrestin 2 and GRK2. *Science* **306,** 2257–2260.

Chen, M. H., Gao, N., Kawakami, T., and Chuang, P. T. (2005). Mice deficient in the fused homolog do not exhibit phenotypes indicative of perturbed hedgehog signaling during embryonic development. *Mol. Cell. Biol.* **25,** 7042–7053.

Cheng, S. Y., and Bishop, J. M. (2002). Suppressor of Fused represses Gli-mediated transcription by recruiting the SAP18–mSin3 corepressor complex. *Proc. Natl Acad. Sci. USA* **99,** 5442–5447.

Chiang, C., Litingtung, Y., Lee, E., Young, K. E., Corden, J. L., Westphal, H., and Beachy, P. A. (1996). Cyclopia and defective axial patterning in mice lacking Sonic hedgehog gene function. *Nature* **383,** 407–413.

Cho, A., Ko, H. W., and Eggenschwiler, J. T. (2008). FKBP8 cell-autonomously controls neural tube patterning through a Gli2- and Kif3a-dependent mechanism. *Dev. Biol.* **321,** 27–39.

Cole, D. G., Diener, D. R., Himelblau, A. L., Beech, P. L., Fuster, J. C., and Rosenbaum, J. L. (1998). *Chlamydomonas* kinesin-II-dependent intraflagellar transport (IFT): IFT particles contain proteins required for ciliary assembly in *Caenorhabditis elegans* sensory neurons. *J. Cell Biol.* **141,** 993–1008.

Cooper, M. K., Wassif, C. A., Krakowiak, P. A., Taipale, J., Gong, R., Kelley, R. I., Porter, F. D., and Beachy, P. A. (2003). A defective response to Hedgehog signaling in disorders of cholesterol biosynthesis. *Nat. Genet.* **33,** 508–513.

Cooper, A. F., Yu, K. P., Brueckner, M., Brailey, L. L., Johnson, L., McGrath, J. M., and Bale, A. E. (2005). Cardiac and CNS defects in a mouse with targeted disruption of suppressor of fused. *Development* **132,** 4407–4417.

Corbit, K. C., Aanstad, P., Singla, V., Norman, A. R., Stainier, D. Y., and Reiter, J. F. (2005). Vertebrate Smoothened functions at the primary cilium. *Nature* **437,** 1018–1021.

Corbit, K. C., Shyer, A. E., Dowdle, W. E., Gaulden, J., Singla, V., Chen, M. H., Chuang, P. T., and Reiter, J. F. (2008). Kif3a constrains beta-catenin-dependent Wnt signalling through dual ciliary and non-ciliary mechanisms. *Nat. Cell Biol.* **10,** 70–76.

Corcoran, R. B., and Scott, M. P. (2006). Oxysterols stimulate Sonic hedgehog signal transduction and proliferation of medulloblastoma cells. *Proc. Natl Acad. Sci. USA* **103,** 8408–8413.

Crowson, A. N. (2006). Basal cell carcinoma: Biology, morphology and clinical implications. *Mod. Pathol.* **19,** S127–S147.

Deane, J. A., Cole, D. G., Seeley, E. S., Diener, D. R., and Rosenbaum, J. L. (2001). Localization of intraflagellar transport protein IFT52 identifies basal body transitional fibers as the docking site for IFT particles. *Curr. Biol.* **11,** 1586–1590.

Denef, N., Neubüser, D., Perez, L., and Cohen, S. M. (2000). Hedgehog induces opposite changes in turnover and subcellular localization of patched and smoothened. *Cell* **102,** 521–531.

Dessaud, E., Yang, L. L., Hill, K., Cox, B., Ulloa, F., Ribeiro, A., Mynett, A., Novitch, B. G., and Briscoe, J. (2007). Interpretation of the sonic hedgehog morphogen gradient by a temporal adaptation mechanism. *Nature* **450,** 717–720.

Ding, Q., Motoyama, J., Gasca, S., Mo, R., Sasaki, H., Rossant, J., and Hui, C. C. (1998). Diminished Sonic hedgehog signaling and lack of floor plate differentiation in Gli2 mutant mice. *Development* **125,** 2533–2543.

Ding, Q., Fukami, S., Meng, X., Nishizaki, Y., Zhang, X., Sasaki, H., Dlugosz, A., Nakafuku, M., and Hui, C. (1999). Mouse suppressor of fused is a negative regulator of sonic hedgehog signaling and alters the subcellular distribution of Gli1. *Curr. Biol.* **9,** 1119–1122.

Dussillol-Godar, F., Brissard-Zahraoui, J., Limbourg-Bouchon, B., Boucher, D., Fouix, S., Lamour-Isnard, C., Plessis, A., and Busson, D. (2006). Modulation of the suppressor of fused protein regulates the Hedgehog signaling pathway in Drosophila embryo and imaginal discs. *Dev. Biol.* **291,** 53–66.

Dwyer, J. R., Sever, N., Carlson, M., Nelson, S. F., Beachy, P. A., and Parhami, F. (2007). Oxysterols are novel activators of the hedgehog signaling pathway in pluripotent mesenchymal cells. *J. Biol. Chem.* **282,** 8959–8968.

Echelard, Y., Epstein, D. J., St-Jacques, B., Shen, L., Mohler, J., McMahon, J. A., and McMahon, A. P. (1993). Sonic hedgehog, a member of a family of putative signaling molecules, is implicated in the regulation of CNS polarity. *Cell* **75,** 1417–1430.

Eggenschwiler, J. T., and Anderson, K. V. (2007). Cilia and developmental signaling. *Annu. Rev. Cell Dev. Biol.* **23,** 345–373.

Eggenschwiler, J. T., Espinoza, E., and Anderson, K. V. (2001). Rab23 is an essential negative regulator of the mouse Sonic hedgehog signalling pathway. *Nature* **412,** 194–198.

Ericson, J., Briscoe, J., Rashbass, P., van Heyningen, V., and Jessell, T. M. (1997). Graded sonic hedgehog signaling and the specification of cell fate in the ventral neural tube. *Cold Spring Harb. Symp. Quant. Biol.* **62,** 451–466.

Essner, J. J., Amack, J. D., Nyholm, M. K., Harris, E. B., and Yost, H. J. (2005). Kupffer's vesicle is a ciliated organ of asymmetry in the zebrafish embryo that initiates left–right development of the brain, heart and gut. *Development* **132,** 1247–1260.

Farzan, S. F., Ascano, M., Jr., Ogden, S. K., Sanial, M., Brigui, A., Plessis, A., and Robbins, D. J. (2008). Costal2 functions as a kinesin-like protein in the Hedgehog signal transduction pathway. *Curr. Biol.* **18,** 1215–1220.

Ferrante, M. I., Zullo, A., Barra, A., Bimonte, S., Messaddeq, N., Studer, M., Dollé, P., and Franco, B. (2006). Oral–facial–digital type I protein is required for primary cilia formation and left–right axis specification. *Nat. Genet.* **38,** 112–117.

Gerdes, J. M., Liu, Y., Zaghloul, N. A., Leitch, C. C., Lawson, S. S., Kato, M., Beachy, P. A., Beales, P. L., DeMartino, G. N., Fisher, S., Badano, J. L., and Katsanis, N. (2007). Disruption of the basal body compromises proteasomal function and perturbs intracellular Wnt response. *Nat. Genet.* **39,** 1350–1360.

Gilula, N. B., and Satir, P. (1972). The ciliary necklace. A ciliary membrane specialization. *J. Cell Biol.* **53,** 494–509.

Goodrich, L. V., Johnson, R. L., Milenkovic, L., McMahon, J. A., and Scott, M. P. (1996). Conservation of the hedgehog/patched signaling pathway from flies to mice: Induction of a mouse patched gene by Hedgehog. *Genes Dev.* **10,** 301–312.

Goodrich, L. V., Milenković, L., Higgins, K. M., and Scott, M. P. (1997). Altered neural cell fates and medulloblastoma in mouse patched mutants. *Science* **277,** 1109–1113.

Hahn, H., Wojnowski, L., Zimmer, A. M., Hall, J., Miller, G., and Zimmer, A. (1998). Rhabdomyosarcomas and radiation hypersensitivity in a mouse model of Gorlin syndrome. *Nat. Med.* **4,** 619–622.

Han, Y. G., Kwok, B. H., and Kernan, M. J. (2003). Intraflagellar transport is required in Drosophila to differentiate sensory cilia but not sperm. *Curr. Biol.* **13,** 1679–1686.

Han, Y. G., Spassky, N., Romaguera-Ros, M., Garcia-Verdugo, J. M., Aguilar, A., Schneider-Maunoury, S., and Alvarez-Buylla, A. (2008). Hedgehog signaling and primary cilia are required for the formation of adult neural stem cells. *Nat. Neurosci.* **11,** 277–284.

Han, Y. G., Dlugosz, A. A., and Alvarez-Buylla, A. Double-edged roles of primary cilia in oncogenic hedgehog signaling in medulloblastoma. Manuscript submitted.

Harfe, B. D., Scherz, P. J., Nissim, S., Tian, H., McMahon, A. P., and Tabin, C. J. (2004). Evidence for an expansion-based temporal Shh gradient in specifying vertebrate digit identities. *Cell* **118,** 517–528.

Haycraft, C. J., Banizs, B., Aydin-Son, Y., Zhang, Q., Michaud, E. J., and Yoder, B. K. (2005). Gli2 and Gli3 localize to cilia and require the intraflagellar transport protein polaris for processing and function. *PLoS Genet.* **1,** 480–488.

Hepker, J., Wang, Q. T., Motzny, C. K., Holmgren, R., and Orenic, T. V. (1997). *Drosophila* cubitus interruptus forms a negative feedback loop with patched and regulates expression of Hedgehog target genes. *Development* **124,** 549–558.

Hong, D. H., Yue, G., Adamian, M., and Li, T. (2001). Retinitis pigmentosa GTPase regulator (RPGR)-interacting protein is stably associated with the photoreceptor ciliary axoneme and anchors RPGR to the connecting cilium. *J. Biol. Chem.* **276,** 12091–12099.

Hooper, J. E., and Scott, M. P. (2005). Communicating with Hedgehogs. *Nat. Rev. Mol. Cell Biol.* **6,** 306–317.

Houde, C., Dickinson, R. J., Houtzager, V. M., Cullum, R., Montpetit, R., Metzler, M., Simpson, E. M., Roy, S., Hayden, M. R., Hoodless, P. A., and Nicholson, D. W. (2006). Hippi is essential for node cilia assembly and Sonic hedgehog signaling. *Development* **300,** 523–533.

Huangfu, D., and Anderson, K. V. (2005). Cilia and hedgehog responsiveness in the mouse. *Proc. Natl Acad. Sci. USA* **102,** 11325–11330.

Huangfu, D., and Anderson, K. V. (2006). Signaling from Smo to Ci/Gli: Conservation and divergence of Hedgehog pathways from *Drosophila* to vertebrates. *Development* **133,** 3–14.

Huangfu, D., Liu, A., Rakeman, A. S., Murcia, N. S., Niswander, L., and Anderson, K. V. (2003). Hedgehog signalling in the mouse requires intraflagellar transport proteins. *Nature* **426,** 83–87.

Hui, C. C., and Joyner, A. L. (1993). A mouse model of greig cephalopolysyndactyly syndrome: The extra-toesJ mutation contains an intragenic deletion of the Gli3 gene. *Nat. Genet.* **3,** 241–246.

Izraeli, S., Lowe, L. A., Bertness, V. L., Campaner, S., Hahn, H., Kirsch, I. R., and Kuehn, M. R. (2001). Genetic evidence that Sil is required for the Sonic Hedgehog response pathway. *Genesis* **31,** 72–77.

Jeong, J., and McMahon, A. P. (2005). Growth and pattern of the mammalian neural tube are governed by partially overlapping feedback activities of the hedgehog antagonists patched 1 and Hhip1. *Development* **132,** 143–154.

Jia, J., Tong, C., Wang, B., Luo, L., and Jiang, J. (2004). Hedgehog signalling activity of Smoothened requires phosphorylation by protein kinase A and casein kinase I. *Nature* **432,** 1045–1050.

Jiang, J., and Struhl, G. (1998). Regulation of the Hedgehog and Wingless signalling pathways by the F-box/WD40-repeat protein Slimb. *Nature* **391,** 493–496.

Kaesler, S., Lüscher, B., and Rüther, U. (2000). Transcriptional activity of GLI1 is negatively regulated by protein kinase A. *Biol. Chem.* **381,** 545–551.

Kelley, R. L., Roessler, E., Hennekam, R. C., Feldman, G. L., Kosaki, K., Jones, M. C., Palumbos, J. C., and Muenke, M. (1996). Holoprosencephaly in RSH/Smith–Lemli–Opitz syndrome: Does abnormal cholesterol metabolism affect the function of Sonic Hedgehog? *Am. J. Med. Genet.* **66,** 478–484.

Kent, D., Bush, E. W., and Hooper, J. E. (2006). Roadkill attenuates Hedgehog responses through degradation of Cubitus interruptus. *Development* **133,** 2001–2010.

King, N., Westbrook, M. J., Young, S. L., Kuo, A., Abedin, M., Chapman, J., Fairclough, S., Hellsten, U., Isogai, Y., Letunic, I., Marr, M., Pincus, D., *et al.* (2008). The genome of the choanoflagellate *Monosiga brevicollis* and the origin of metazoans. *Nature* **451,** 783–788.

Koudijs, M. J., den Broeder, M. J., Keijser, A., Wienholds, E., Houwing, S., van Rooijen, E. M., Geisler, R., and van Eeden, F. J. (2005). The zebrafish mutants dre, uki, and lep encode negative regulators of the hedgehog signaling pathway. *PLoS Genet.* **1,** e19.

Kovacs, J. J., Whalen, E. J., Liu, R., Xiao, K., Kim, J., Chen, M., Wang, J., Chen, W., and Lefkowitz, R. J. (2008). Beta-arrestin-mediated localization of Smoothened to the primary cilium. *Science* **320**, 1777–1781.

Kozminski, K. G., Johnson, K. A., Forscher, P., and Rosenbaum, J. L. (1993). A motility in the eukaryotic flagellum unrelated to flagellar beating. *Proc. Natl Acad. Sci. USA* **90**, 5519–5523.

Kozminski, K. G., Beech, P. L., and Rosenbaum, J. L. (1995). The *Chlamydomonas* kinesin-like protein FLA10 is involved in motility associated with the flagellar membrane. *J. Cell Biol.* **131**, 1517–1527.

Krakowiak, P. A., Wassif, C. A., Kratz, L., Cozma, D., Kovárová, M., Harris, G., Grinberg, A., Yang, Y., Hunter, A. G., Tsokos, M., Kelley, R. I., and Porter, F. D. (2003). Lathosterolosis: An inborn error of human and murine cholesterol synthesis due to lathosterol 5-desaturase deficiency. *Hum. Mol. Genet.* **12**, 1631–1641.

Lai, K., Kaspar, B. K., Gage, F. H., and Schaffer, D. V. (2003). Sonic hedgehog regulates adult neural progenitor proliferation *in vitro* and *in vivo*. *Nat. Neurosci.* **6**, 21–27.

Lam, C. W., Xie, J., To, K. F., Ng, H. K., Lee, K. C., Yuen, N. W., Lim, P. L., Chan, L. Y., Tong, S. F., and McCormick, F. (1999). A frequent activated smoothened mutation in sporadic basal cell carcinomas. *Oncogene* **18**, 833–836.

Lei, Q., Zelman, A. K., Kuang, E., Li, S., and Matise, M. P. (2004). Transduction of graded Hedgehog signaling by a combination of Gli2 and Gli3 activator functions in the developing spinal cord. *Development* **131**, 3593–3604.

Levy, V., Lindon, C., Harfe, B. D., and Morgan, B. A. (2005). Distinct stem cell populations regenerate the follicle and interfollicular epidermis. *Dev. Cell* **9**, 855–861.

Lin, F., Hiesberger, T., Cordes, K., Sinclair, A. M., Goldstein, L. S., Somlo, S., and Igarashi, P. (2003). Kidney-specific inactivation of the KIF3A subunit of kinesin-II inhibits renal ciliogenesis and produces polycystic kidney disease. *Proc. Natl Acad. Sci. USA* **100**, 5286–5291.

Litingtung, Y., Dahn, R. D., Li, Y., Fallon, J. F., and Chiang, C. (2002). Shh and Gli3 are dispensable for limb skeleton formation but regulate digit number and identity. *Nature* **418**, 979–983.

Liu, A., Wang, B., and Niswander, L. A. (2005). Mouse intraflagellar transport proteins regulate both the activator and repressor functions of Gli transcription factors. *Development* **132**, 3103–3111.

Liu, Y., Cao, X., Jiang, J., and Jia, J. (2007). Fused–Costal2 protein complex regulates Hedgehog-induced Smo phosphorylation and cell-surface accumulation. *Genes Dev.* **21**, 1949–1963.

Lum, L., and Beachy, P. A. (2004). The Hedgehog response network: Sensors, switches, and routers. *Science* **304**, 1755–1759.

Lum, L., Zhang, C., Oh, S., Mann, R. K., von Kessler, D. P., Taipale, J., Weis-Garcia, F., Gong, R., Wang, B., and Beachy, P. A. (2003). Hedgehog signal transduction via Smoothened association with a cytoplasmic complex scaffolded by the atypical kinesin, Costal-2. *Mol. Cell* **12**, 1261–1274.

Mans, D. A., Voest, E. E., and Giles, R. H. (2008). All along the watchtower: Is the cilium a tumor suppressor organelle? *Biochim. Biophys. Acta* (in press).

Marcotullio, L. D., Ferretti, E., Greco, A., Smaele, E. D., Po, A., Sico, M. A., Alimandi, M., Giannini, G., Maroder, M., Screpanti, I., and Gulino, A. (2006). Numb is a suppressor of Hedgehog signalling and targets Gli1 for Itch-dependent ubiquitination. *Nat. Cell Biol.* **8**, 1415–1423.

Marigo, V., and Tabin, C. J. (1996). Regulation of patched by sonic hedgehog in the developing neural tube. *Proc. Natl Acad. Sci. USA* **93**, 9346–9351.

Marshall, W. F., and Nonaka, S. (2006). Cilia: Tuning in to the cell's antenna. *Curr. Biol.* **16**, R604–R614.

Marszalek, J. R., Ruiz-Lozano, P., Roberts, E., Chien, K. R., and Goldstein, L. S. (1999). Situs inversus and embryonic ciliary morphogenesis defects in mouse mutants lacking the KIF3A subunit of kinesin-II. *Proc. Natl Acad. Sci. USA* **96,** 5043–5048.

Martí, E., Bumcrot, D. A., Takada, R., and McMahon, A. P. (1995). Requirement of 19K form of Sonic hedgehog for induction of distinct ventral cell types in CNS explants. *Nature* **375,** 322–325.

Matise, M. P., Epstein, D. J., Park, H. L., Platt, K. A., and Joyner, A. L. (1998). Gli2 is required for induction of floor plate and adjacent cells, but not most ventral neurons in the mouse central nervous system. *Development* **125,** 2759–2770.

May, S. R., Ashique, A. M., Karlen, M., Wang, B., Shen, Y., Zarbalis, K., Reiter, J., Ericson, J., and Peterson, A. S. (2005). Loss of the retrograde motor for IFT disrupts localization of Smo to cilia and prevents the expression of both activator and repressor functions of Gli. *Dev. Biol.* **287,** 378–389.

McMahon, A. P., Ingham, P. W., and Tabin, C. J. (2003). Developmental roles and clinical significance of hedgehog signaling. *Curr. Top. Dev. Biol.* **53,** 1–114.

Meloni, A. R., Fralish, G. B., Kelly, P., Salahpour, A., Chen, J. K., Wechsler-Reya, R. J., Lefkowitz, R. J., and Caron, M. G. (2006). Smoothened signal transduction is promoted by G protein-coupled receptor kinase 2. *Mol. Cell. Biol.* **26,** 7550–7560.

Merchant, M., Evangelista, M., Luoh, S. M., Frantz, G. D., Chalasani, S., Carano, R. A., van Hoy, M., Ramirez, J., Ogasawara, A. K., McFarland, L. M., Filvaroff, E. H., French, D. M., *et al.* (2005). Loss of the serine/threonine kinase fused results in postnatal growth defects and lethality due to progressive hydrocephalus. *Mol. Cell. Biol.* **25,** 7054–7068.

Meyer, N. P., and Roelink, H. (2003). The amino-terminal region of Gli3 antagonizes the Shh response and acts in dorsoventral fate specification in the developing spinal cord. *Dev. Biol.* **257,** 343–355.

Milenkovic, L., Goodrich, L. V., Higgins, K. M., and Scott, M. P. (1999). Mouse patched1 controls body size determination and limb patterning. *Development* **126,** 4431–4440.

Motoyama, J., Milenkovic, L., Iwama, M., Shikata, Y., Scott, M. P., and Hui, C. C. (2003). Differential requirement for Gli2 and Gli3 in ventral neural cell fate specification. *Dev. Biol.* **259,** 150–161.

Nachury, M. V., Loktev, A. V., Zhang, Q., Westlake, C. J., Peränen, J., Merdes, A., Slusarski, D. C., Scheller, R. H., Bazan, J. F., Sheffield, V. C., and Jackson, P. K. (2007). A core complex of BBS proteins cooperates with the GTPase Rab8 to promote ciliary membrane biogenesis. *Cell* **129,** 1201–1213.

Nonaka, S., Tanaka, Y., Okada, Y., Takeda, S., Harada, A., Kanai, Y., Kido, M., and Hirokawa, N. (1998). Randomization of left–right asymmetry due to loss of nodal cilia generating leftward flow of extraembryonic fluid in mice lacking KIF3B motor protein. *Cell* **95,** 829–837.

Nonaka, S., Shiratori, H., Saijoh, Y., and Hamada, H. (2002). Determination of left–right patterning of the mouse embryo by artificial nodal flow. *Nature* **418,** 96–99.

Nüsslein-Volhard, C., and Wieschaus, E. (1980). Mutations affecting segment number and polarity in *Drosophila*. *Nature* **287,** 795–801.

Ocbina, P. J., and Anderson, K. V. (2008). Intraflagellar transport, cilia, and mammalian Hedgehog signaling: Analysis in mouse embryonic fibroblasts. *Dev. Dyn.* **237,** 2030–2038.

Ohlmeyer, J. T., and Kalderon, D. (1998). Hedgehog stimulates maturation of Cubitus interruptus into a labile transcriptional activator. *Nature* **396,** 749–753.

Okada, A., Charron, F., Morin, S., Shin, D. S., Wong, K., Fabre, P. J., Tessier-Lavigne, M., and McConnell, S. K. (2006). Boc is a receptor for sonic hedgehog in the guidance of commissural axons. *Nature* **444,** 369–373.

Orozco, J. T., Wedaman, K. P., Signor, D., Brown, H., Rose, L., and Scholey, J. M. (1999). Movement of motor and cargo along cilia. *Nature* **398,** 674.

Ou, G., Blacque, O. E., Snow, J. J., Leroux, M. R., and Scholey, J. M. (2005). Functional coordination of intraflagellar transport motors. *Nature* **436,** 583–587.

Paces-Fessy, M., Boucher, D., Petit, E., Paute-Briand, S., and Blanchet-Tournier, M. F. (2004). The negative regulator of Gli, Suppressor of fused (Sufu), interacts with SAP18, Galectin3 and other nuclear proteins. *Biochem. J.* **378,** 353–362.

Pan, J., Wang, Q., and Snell, W. J. (2005). Cilium-generated signaling and cilia-related disorders. *Lab. Invest.* **85,** 452–463.

Pan, Y., Bai, C. B., Joyner, A. L., and Wang, B. (2006). Sonic hedgehog signaling regulates Gli2 transcriptional activity by suppressing its processing and degradation. *Mol. Cell. Biol.* **26,** 3365–3377.

Park, H. L., Bai, C., Platt, K. A., Matise, M. P., Beeghly, A., Hui, C. C., Nakashima, M., and Joyner, A. L. (2000). Mouse Gli1 mutants are viable but have defects in SHH signaling in combination with a Gli2 mutation. *Development* **127,** 1593–1605.

Pasca di Magliano, M., and Hebrok, M. (2003). Hedgehog signalling in cancer formation and maintenance. *Nat. Rev. Cancer* **3,** 903–911.

Pazour, G. J., Wilkerson, C. G., and Witman, G. B. (1998). A dynein light chain is essential for the retrograde particle movement of intraflagellar transport (IFT). *J. Cell Biol.* **141,** 979–992.

Pazour, G. J., Dickert, B. L., Vucica, Y., Seeley, E. S., Rosenbaum, J. L., Witman, G. B., and Cole, D. G. (2000). Chlamydomonas IFT88 and its mouse homologue, polycystic kidney disease gene tg737, are required for assembly of cilia and flagella. *J. Cell Biol.* **151,** 709–718.

Pearse, R. V., II, Collier, L. S., Scott, M. P., and Tabin, C. J. (1999). Vertebrate homologs of *Drosophila* suppressor of fused interact with the Gli family of transcriptional regulators. *Dev. Biol.* **212,** 323–336.

Perler, F. B. (1998). Protein splicing of inteins and hedgehog autoproteolysis: Structure, function, and evolution. *Cell* **92,** 1–4.

Persson, M., Stamataki, D., te Welscher, P., Andersson, E., Böse, J., Rüther, U., Ericson, J., and Briscoe, J. (2002). Dorsal–ventral patterning of the spinal cord requires Gli3 transcriptional repressor activity. *Genes Dev.* **16,** 2865–2878.

Piperno, G., Siuda, E., Henderson, S., Segil, M., Vaananen, H., and Sassaroli, M. (1998). Distinct mutants of retrograde intraflagellar transport (IFT) share similar morphological and molecular defects. *J. Cell Biol.* **143,** 1591–1601.

Porter, M. E., Bower, R., Knott, J. A., Byrd, P., and Dentler, W. (1999). Cytoplasmic dynein heavy chain 1b is required for flagellar assembly in *Chlamydomonas*. *Mol. Cell. Biol.* **10,** 693–712.

Préat, T. (1992). Characterization of suppressor of fused, a complete suppressor of the fused segment polarity gene of *Drosophila melanogaster*. *Genetics* **132,** 725–736.

Price, M. A., and Kalderon, D. (1999). Proteolysis of cubitus interruptus in *Drosophila* requires phosphorylation by protein kinase A. *Development* **126,** 4331–4339.

Price, M. A., and Kalderon, D. (2002). Proteolysis of the Hedgehog signaling effector Cubitus interruptus requires phosphorylation by Glycogen Synthase Kinase 3 and Casein Kinase 1. *Cell* **108,** 823–835.

Quarmby, L. M., and Parker, J. D. (2005). Cilia and the cell cycle? *J. Cell Biol.* **169,** 707–710.

Rana, A. A., Barbera, J. P., Rodriguez, T. A., Lynch, D., Hirst, E., Smith, J. C., and Beddington, R. S. (2004). Targeted deletion of the novel cytoplasmic dynein mD2LIC disrupts the embryonic organiser, formation of the body axes and specification of ventral cell fates. *Development* **131,** 4999–5007.

Reiter, J. F., and Skarnes, W. C. (2006). Tectonic, a novel regulator of the Hedgehog pathway required for both activation and inhibition. *Genes Dev.* **20,** 22–27.

Riddle, R. D., Johnson, R. L., Laufer, E., and Tabin, C. (1993). Sonic hedgehog mediates the polarizing activity of the ZPA. *Cell* **75,** 1401–1416.

Robbins, D. J., Nybakken, K. E., Kobayashi, R., Sisson, J. C., Bishop, J. M., and Thérond, P. P. (1997). Hedgehog elicits signal transduction by means of a large complex containing the kinesin-related protein costal2. *Cell* **90,** 225–234.

Roelink, H., Augsburger, A., Heemskerk, J., Korzh, V., Norlin, S., Ruiz i Altaba, A., Tanabe, Y., Placzek, M., Edlund, T., Jessell, T. M., and Dodd, J. (1994). Floor plate and motor neuron induction by vhh-1, a vertebrate homolog of hedgehog expressed by the notochord. *Cell* **76,** 761–775.

Roelink, H., Porter, J. A., Chiang, C., Tanabe, Y., Chang, D. T., Beachy, P. A., and Jessell, T. M. (1995). Floor plate and motor neuron induction by different concentrations of the amino-terminal cleavage product of sonic hedgehog autoproteolysis. *Cell* **81,** 445–455.

Rohatgi, R., Milenkovic, L., and Scott, M. P. (2007). Patched1 regulates Hedgehog signaling at the primary cilium. *Science* **317,** 372–376.

Romio, L., Fry, A. M., Winyard, P. J., Malcolm, S., Woolf, A. S., and Feather, S. A. (2004). OFD1 is a centrosomal/basal body protein expressed during mesenchymal–epithelial transition in human nephrogenesis. *J. Am. Soc. Nephrol.* **15,** 2556–2568.

Rosenbaum, J. L., and Witman, G. B. (2002). Intraflagellar transport. *Nat. Rev. Mol. Cell Biol.* **3,** 813–825.

Rubin, L. L., and de Sauvage, F. J. (2006). Targeting the Hedgehog pathway in cancer. *Nat. Rev. Drug Discov.* **5,** 1026–1033.

Ruel, L., Rodriguez, R., Gallet, A., Lavenant-Staccini, L., and Thérond, P. P. (2003). Stability and association of Smoothened, Costal2 and Fused with Cubitus interruptus are regulated by Hedgehog. *Nat. Cell Biol.* **5,** 907–913.

Ruel, L., Gallet, A., Raisin, S., Truchi, A., Staccini-Lavenant, L., Cervantes, A., and Thérond, P. P. (2007). Phosphorylation of the atypical kinesin Costal2 by the kinase Fused induces the partial disassembly of the Smoothened–Fused–Costal2–Cubitus interruptus complex in Hedgehog signalling. *Development* **134,** 3677–3689.

Ruiz i Altaba, A. (1999). Gli proteins encode context-dependent positive and negative functions: Implications for development and disease. *Development* **126,** 3205–3216.

Ruiz i Altaba, A., Sanchez, P., and Dahmane, N. (2002). Gli and Hedgehog in cancer: Tumours, embryos and stem cells. *Nat. Rev. Cancer* **2,** 361–372.

Ruiz-Perez, V. L., Blair, H. J., Rodriguez-Andres, M. E., Blanco, M. J., Wilson, A., Liu, Y. N., Miles, C., Peters, H., and Goodship, J. A. (2007). Evc is a positive mediator of Ihh-regulated bone growth that localises at the base of chondrocyte cilia. *Development* **134,** 2903–2912.

Salisbury, J. L. (2004). Primary cilia: Putting sensors together. *Curr. Biol.* **14,** R765–R767.

Sarpal, R., Todi, S. V., Sivan-Loukianova, E., Shirolikar, S., Subramanian, N., Raff, E. C., Erickson, J. W., Ray, K., and Eberl, D. F. (2003). *Drosophila* KAP interacts with the kinesin II motor subunit KLP64D to assemble chordotonal sensory cilia, but not sperm tails. *Curr. Biol.* **13,** 1687–1696.

Sasaki, H., Nishizaki, Y., Hui, C. C., Nakafuku, M., and Kondoh, H. (1999). Regulation of Gli2 and Gli3 activities by an amino-terminal repression domain: Implication of Gli2 and Gli3 as primary mediators of Shh signaling. *Development* **126,** 3915–3924.

Satir, P., and Christensen, S. T. (2007). Overview of structure and function of mammalian cilia. *Annu. Rev. Physiol.* **69,** 377–400.

Schafer, J. C., Haycraft, C. J., Thomas, J. H., Yoder, B. K., and Swoboda, P. (2003). XBX-1 encodes a dynein light intermediate chain required for retrograde intraflagellar transport and cilia assembly in *Caenorhabditis elegans*. *Mol. Biol. Cell* **14,** 2057–2070.

Schauerte, H. E., van Eeden, F. J., Fricke, C., Odenthal, J., Strähle, U., and Haffter, P. (1998). Sonic hedgehog is not required for the induction of medial floor plate cells in the zebrafish. *Development* **125,** 2983–2993.

Scherz, P. J., McGlinn, E., Nissim, S., and Tabin, C. J. (2007). Extended exposure to Sonic hedgehog is required for patterning the posterior digits of the vertebrate limb. *Dev. Biol.* **308,** 343–354.

Schneider, L., Clement, C. A., Teilmann, S. C., Pazour, G. J., Hoffmann, E. K., Satir, P., and Christensen, S. T. (2005). PDGFRalphaalpha signaling is regulated through the primary cilium in fibroblasts. *Curr. Biol.* **15,** 1861–1866.

Sheng, H., Goich, S., Wang, A., Grachtchouk, M., Lowe, L., Mo, R., Lin, K., de Sauvage, F. J., Sasaki, H., Hui, C. C., and Dlugosz, A. A. (2002). Dissecting the oncogenic potential of Gli2: Deletion of an NH2-terminal fragment alters skin tumor phenotype. *Cancer Res.* **62,** 5308–5316.

Simons, M., and Walz, G. (2006). Polycystic kidney disease: Cell division without a c(l)ue? *Kidney Int.* **70,** 854–864.

Singla, V., and Reiter, J. F. (2006). The primary cilium as the cell's antenna: Signaling at a sensory organelle. *Science* **313,** 629–633.

Sisson, J. C., Ho, K. S., Suyama, K., and Scott, M. P. (1997). Costal2, a novel kinesin-related protein in the Hedgehog signaling pathway. *Cell* **90,** 235–245.

Smelkinson, M. G., and Kalderon, D. (2006). Processing of the *Drosophila* hedgehog signaling effector Ci-155 to the repressor Ci-75 is mediated by direct binding to the SCF component Slimb. *Curr. Biol.* **16,** 110–116.

Snell, E. A., Brooke, N. M., Taylor, W. R., Casane, D., Philippe, H., and Holland, P. W. (2006). An unusual choanoflagellate protein released by Hedgehog autocatalytic processing. *Proc. Biol. Sci.* **273,** 401–407.

Snow, J. J., Ou, G., Gunnarson, A. L., Walker, M. R., Zhou, H. M., Brust-Mascher, I., and Scholey, J. M. (2004). Two anterograde intraflagellar transport motors cooperate to build sensory cilia on *C. elegans* neurons. *Nat. Cell Biol.* **6,** 1109–1113.

St-Jacques, B., Hammerschmidt, M., and McMahon, A. P. (1999). Indian hedgehog signaling regulates proliferation and differentiation of chondrocytes and is essential for bone formation. *Genes Dev.* **13,** 2072–2086.

Stone, D. M., Murone, M., Luoh, S., Ye, W., Armanini, M. P., Gurney, A., Phillips, H., Brush, J., Goddard, A., de Sauvage, F. J., and Rosenthal, A. (1999). Characterization of the human suppressor of fused, a negative regulator of the zinc-finger transcription factor Gli. *J. Cell Sci.* **112,** 4437–4448.

Storm van's Gravesande, K., and Omran, H. (2005). Primary ciliary dyskinesia: Clinical presentation, diagnosis and genetics. *Ann. Med.* **37,** 439–449.

Sun, Z., Amsterdam, A., Pazour, G. J., Cole, D. G., Miller, M. S., and Hopkins, N. (2004). A genetic screen in zebrafish identifies cilia genes as a principal cause of cystic kidney. *Development* **131,** 4085–4093.

Svärd, J., Heby-Henricson, K., Persson-Lek, M., Rozell, B., Lauth, M., Bergström, A., Ericson, J., Toftgård, R., and Teglund, S. (2006). Genetic elimination of Suppressor of fused reveals an essential repressor function in the mammalian Hedgehog signaling pathway. *Dev. Cell* **10,** 187–197.

Taipale, J., Cooper, M. K., Maiti, T., and Beachy, P. A. (2002). Patched acts catalytically to suppress the activity of Smoothened. *Nature* **418,** 892–897.

Tang, J. T., So, P. L., and Epstein, E. H., Jr. (2007). Novel Hedgehog pathway targets against basal cell carcinoma. *Toxicol. Appl. Pharmacol.* **224,** 257–264.

Tay, S. Y., Ingham, P. W., and Roy, S. (2005). A homologue of the *Drosophila* kinesin-like protein Costal2 regulates Hedgehog signal transduction in the vertebrate embryo. *Development* **132,** 625–634.

Tempé, D., Casas, M., Karaz, S., Blanchet-Tournier, M. F., and Concordet, J. P. (2006). Multisite protein kinase A and glycogen synthase kinase 3beta phosphorylation leads to Gli3 ubiquitination by SCFbetaTrCP. *Mol. Cell. Biol.* **26,** 4316–4326.

te Welscher, P., Zuniga, A., Kuijper, S., Drenth, T., Goedemans, H. J., Meijlink, F., and Zeller, R. (2002). Progression of vertebrate limb development through Shh-mediated counteraction of Gli3. *Science* **298,** 827–830.

Tobin, J. L., and Beales, P. L. (2007). Bardet-Biedl syndrome: Beyond the cilium. *Pediatr. Nephrol.* **22,** 926–936.

Town, T., Breunig, J. J., Sarkisian, M. R., Spilianakis, C., Ayoub, A. E., Liu, X., Ferrandino, A. F., Gallagher, A. R., Li, M. O., Rakic, P., and Flavell, R. A. (2008). The stumpy gene is required for mammalian ciliogenesis. *Proc. Natl Acad. Sci. USA* **105,** 2853–2858.

Tran, P. V., Haycraft, C. J., Besschetnova, T. Y., Turbe-Doan, A., Stottmann, R. W., Herron, B. J., Chesebro, A. L., Qiu, H., Scherz, P. J., Shah, J. V., Yoder, B. K., and Beier, D. R. (2008). THM1 negatively modulates mouse sonic hedgehog signal transduction and affects retrograde intraflagellar transport in cilia. *Nat. Genet.* **40,** 403–410.

Tsujikawa, M., and Malicki, J. (2004). Intraflagellar transport genes are essential for differentiation and survival of vertebrate sensory neurons. *Neuron* **42,** 703–716.

Varjosalo, M., Li, S. P., and Taipale, J. (2006). Divergence of hedgehog signal transduction mechanism between *Drosophila* and mammals. *Dev. Cell* **10,** 177–186.

Vieira, O. V., Gaus, K., Verkade, P., Fullekrug, J., Vaz, W. L., and Simons, K. (2006). FAPP2, cilium formation, and compartmentalization of the apical membrane in polarized Madin-Darby canine kidney (MDCK) cells. *Proc. Natl Acad. Sci. USA* **103,** 18556–18561.

Vierkotten, J., Dildrop, R., Peters, T., Wang, B., and Ruther, U. (2007). Ftm is a novel basal body protein of cilia involved in Shh signalling. *Development* **134,** 2569–2577.

Vortkamp, A., Lee, K., Lanske, B., Segre, G. V., Kronenberg, H. M., and Tabin, C. J. (1996). Regulation of rate of cartilage differentiation by Indian hedgehog and PTH-related protein. *Nature* **273,** 613–622.

Wang, Q. T., and Holmgren, R. A. (1999). The subcellular localization and activity of *Drosophila* cubitus interruptus are regulated at multiple levels. *Development* **126,** 5097–5106.

Wang, B., and Li, Y. (2006). Evidence for the direct involvement of {beta}TrCP in Gli3 protein processing. *Proc. Natl Acad. Sci. USA* **103,** 33–38.

Wang, Y., and Price, M. A. (2008). A unique protection signal in Cubitus interruptus prevents its complete proteasomal degradation. *Mol. Cell. Biol.* **28,** 5555–5568.

Wang, B., Fallon, J. F., and Beachy, P. A. (2000). Hedgehog-regulated processing of Gli3 produces an anterior/posterior repressor gradient in the developing vertebrate limb. *Cell* **100,** 423–434.

Wang, C., Ruther, U., and Wang, B. (2007a). The Shh-independent activator function of the full-length Gli3 protein and its role in vertebrate limb digit patterning. *Dev. Biol.* **305,** 460–469.

Wang, Y., McMahon, A. P., and Allen, B. L. (2007b). Shifting paradigms in Hedgehog signaling. *Curr. Opin. Cell Biol.* **19,** 159–165.

Wigley, W. C., Fabunmi, R. P., Lee, M. G., Marino, C. R., Muallem, S., DeMartino, G. N., and Thomas, P. J. (1999). Dynamic association of proteasomal machinery with the centrosome. *J. Cell Biol.* **145,** 481–490.

Wolff, C., Roy, S., and Ingham, P. W. (2003). Multiple muscle cell identities induced by distinct levels and timing of hedgehog activity in the zebrafish embryo. *Curr. Biol.* **13,** 1169–1181.

Wong, S. Y., So, P. L., Dluogsz, A. A., Epstein, Jr., E. H., and Reiter, J. F. Dual roles for the ciliary motor Kinesin-II in hedgehog-mediated skin cancer. Manuscript submitted.

Xie, J., Murone, M., Luoh, S. M., Ryan, A., Gu, Q., Zhang, C., Bonifas, J. M., Lam, C. W., Hynes, M., Goddard, A., Rosenthal, A., Epstein, E. H., Jr., et al. (1998). Activating Smoothened mutations in sporadic basal-cell carcinoma. *Nature* **391**, 90–92.

Yoder, B. K., Tousson, A., Millican, L., Wu, J. H., Bugg, C. E., Jr., Schafer, J. A., and Balkovetz, D. F. (2002). Polaris, a protein disrupted in orpk mutant mice, is required for assembly of renal cilium. *Am. J. Physiol. Renal Physiol.* **282**, F541–F552.

Zhang, C., Williams, E. H., Guo, Y., Lum, L., and Beachy, P. A. (2004). Extensive phosphorylation of Smoothened in Hedgehog pathway activation. *Proc. Natl Acad. Sci. USA* **101**, 17900–17907.

Zhang, W., Zhao, Y., Tong, C., Wang, G., Wang, B., Jia, J., and Jiang, J. (2005). Hedgehog-regulated Costal2–kinase complexes control phosphorylation and proteolytic processing of Cubitus interruptus. *Dev. Cell* **8**, 267–278.

Zhang, Q., Zhang, L., Wang, B., Ou, C. Y., Chien, C. T., and Jiang, J. (2006). A hedgehog-induced BTB protein modulates hedgehog signaling by degrading Ci/Gli transcription factor. *Dev. Cell* **10**, 719–729.

Zhou, Q., Apionishev, S., and Kalderon, D. (2006). The contributions of protein kinase A and smoothened phosphorylation to hedgehog signal transduction in *Drosophila melanogaster*. *Genetics* **173**, 2049–2062.

Zimmerman, H. (1898). Beitrage zur kenntniss einiger drusen und epithelien. *Arch. Mikr. Anat.* **52**, 552–706.

CHAPTER TEN

THE PRIMARY CILIUM COORDINATES SIGNALING PATHWAYS IN CELL CYCLE CONTROL AND MIGRATION DURING DEVELOPMENT AND TISSUE REPAIR

Søren T. Christensen,* Stine F. Pedersen,* Peter Satir,[†] Iben R. Veland,* and Linda Schneider*

Contents

1. Introduction	262
2. Cell Cycle Entry Regulated by PDGFR$\alpha\alpha$ Signaling in the Primary Cilium	267
2.1. Regulation of PDGFR$\alpha\alpha$ signaling in the primary cilium	270
3. Directional Cell Migration is Regulated by PDGFRα in the Primary Cilium	272
3.1. Ciliary PDGFRα signaling regulates directed cell migration in fibroblasts	273
3.2. Primary cilia in wound healing	276
3.3. Roles of the Na$^+$/H$^+$ exchanger NHE1 in PDGFRα-mediated control of cell migration	277
3.4. Downstream mechanisms by which NHE1 controls directed cell migration and invasion	279
4. The Extracellular Matrix and the Primary Cilium	280
5. Polarization, Cell Migration, Cell Cycle Control and Wnt Signaling in the Primary Cilium	282
5.1. Wnt signaling in human health and disease	282
5.2. Wnt signaling and the primary cilium	283
6. Conclusions and Perspectives	286
Acknowledgments	288
References	288

* Department of Biology, Section of Cell and Developmental Biology, The August Krogh Building, University of Copenhagen, Universitetsparken 13, DK-2100 Copenhagen OE, Denmark
[†] Department of Anatomy and Structural Biology, Albert Einstein College of Medicine of Yeshiva University, Bronx, NY 10461

Current Topics in Developmental Biology, Volume 85   © 2008 Elsevier Inc.
ISSN 0070-2153, DOI: 10.1016/S0070-2153(08)00810-7   All rights reserved.

Abstract

Cell cycle control and migration are critical processes during development and maintenance of tissue functions. Recently, primary cilia were shown to take part in coordination of the signaling pathways that control these cellular processes in human health and disease. In this review, we present an overview of the function of primary cilia and the centrosome in the signaling pathways that regulate cell cycle control and migration with focus on ciliary signaling via platelet-derived growth factor receptor alpha (PDGFRα). We also consider how the primary cilium and the centrosome interact with the extracellular matrix, coordinate Wnt signaling, and modulate cytoskeletal changes that impinge on both cell cycle control and cell migration.

1. Introduction

The successful coordination of embryonic development and maintenance of tissues and organs highly depends on the concerted action of complex signal transduction pathways that orchestrate the interaction between individual cells and their extracellular environment. It is now generally appreciated that primary cilia play a unique role in the coordination of signaling pathways that control cell survival, proliferation, migration, and differentiation—processes, which are central to development and tissue homeostasis. Consequently, defects in assembly or sensory function of primary cilia in the embryo and in the adult are tightly coupled to developmental defects, diseases, and disorders. Primary cilia coordinate a series of signal transduction pathways, including Hedgehog (Hh), Wnt, and platelet-derived growth factor alpha (PDGFRα) pathways. Moreover, they function as mechano- and osmosensing units that probe the extracellular environment and transmit signals to the cell (Christensen *et al.*, 2007). The primary cilium has been given a variety of names to designate its function, including cybernetic probe (Poole *et al.*, 1985), cellular GPS (Benzing and Walz, 2006), and environmental rheostat (Plotnikova *et al.*, 2008) to give a few examples. Defects in primary cilia have been linked to many pathologies such as hypoplasia, skeletal malformation, cancer, obesity, diabetes, as well as a number of other diseases and developmental defects, now jointly referred as to ciliopathies, discussed further below (Fliegauf *et al.*, 2007; Lehman *et al.*, 2008; Satir and Christensen, 2007).

The primary cilium emanates from the distal end of the mother centriole of the centrosome. During growth arrest (G_0) the centrosome translocates to the cell surface to form the cilium as an antennal-like structure that protrudes into the extracellular environment (Blacque *et al.*, 2008; Pedersen *et al.*, 2008). Assembly of the cilium takes place after docking of the centrosome at the plasma membrane and formation of the ciliary necklace by a process known as intraflagellar transport (IFT) that requires molecular motors, kinesins for anterograde transport and a special dynein for retrograde transport. IFT moves ciliary precursors from the base of the cilium along the axoneme to its distal tip and *vice versa* (Pedersen et al., 2008; Rosenbaum and Witman, 2002; Chapter 2, this volume), adding components to the axoneme and depositing receptors and channels in the ciliary membrane or removing components for replacement or ciliary shortening. The primary cilium usually resorbs at this point in the cell cycle where duplicated centrosomes begin to form the mitotic spindle. The cilium reassembles in postmitotic cells that reenter growth arrest (Fig. 10.1).

Since various signal transduction components essential for cell cycle control and migration are integrated into the primary cilium and the centrosome, and sometimes move between them, it has been suggested that this axis of cellular compartments comprises a unique signaling platform that guides developmental processes and tissue homeostasis, as originally hypothesized by Tucker et al. (1979). Regulatory proteins and transcription factors generated by ciliary signaling coordinate the trafficking of signaling components in and out of the cilium, as in polycystin-1, Hh and PDGFRα signaling (Christensen and Ott, 2007; Christensen et al., 2007; Kolb and Nauli, 2008). Signaling molecules enter the cilium at an IFT loading zone at or near the centriole; signaling molecules leaving the cilium arrive at the centrosome, which functions as a major docking station for cell cycle regulators (Schatten, 2008).

The activity and turnover of regulatory proteins in the centrosome change during the different phases of the cell cycle such that specific proteins and transcription factors regulated by ciliary signaling control the G_1/S phase transition, the disassembly of the cilium and the apical microtubule cytoskeleton and the duplication of the centrosomes upon cell cycle entrance. Other proteins may control assembly of the mitotic spindle during mitosis and the formation of a new primary cilium upon reentrance into growth arrest. Finally, the turnover of ciliary and centrosomal signaling molecules may be part of the dynamic process that controls differential processes, allocating different signal systems to different cell types to determine cell fate and function. The dynamic signaling interactions between the cilium and the centrosome must come together with other cellular signaling compartments and regulatory cell systems, including those of the plasma membrane, the cytoskeleton, Golgi, and the nucleus, to control the continuous homeostasis of the cell. Consistent with a role for cilia in growth control, defective primary cilia have been associated with proliferative diseases such as cancer (Mans et al., 2008; Plotnikova et al 2008; Wheatley 1996) and polycystic kidney disease (PKD) (Pazour et al., 2000; Yoder, 2007)—the first disease to be connected to defective primary cilia. Several types of cancer cells are recognized by a significantly lower frequency of

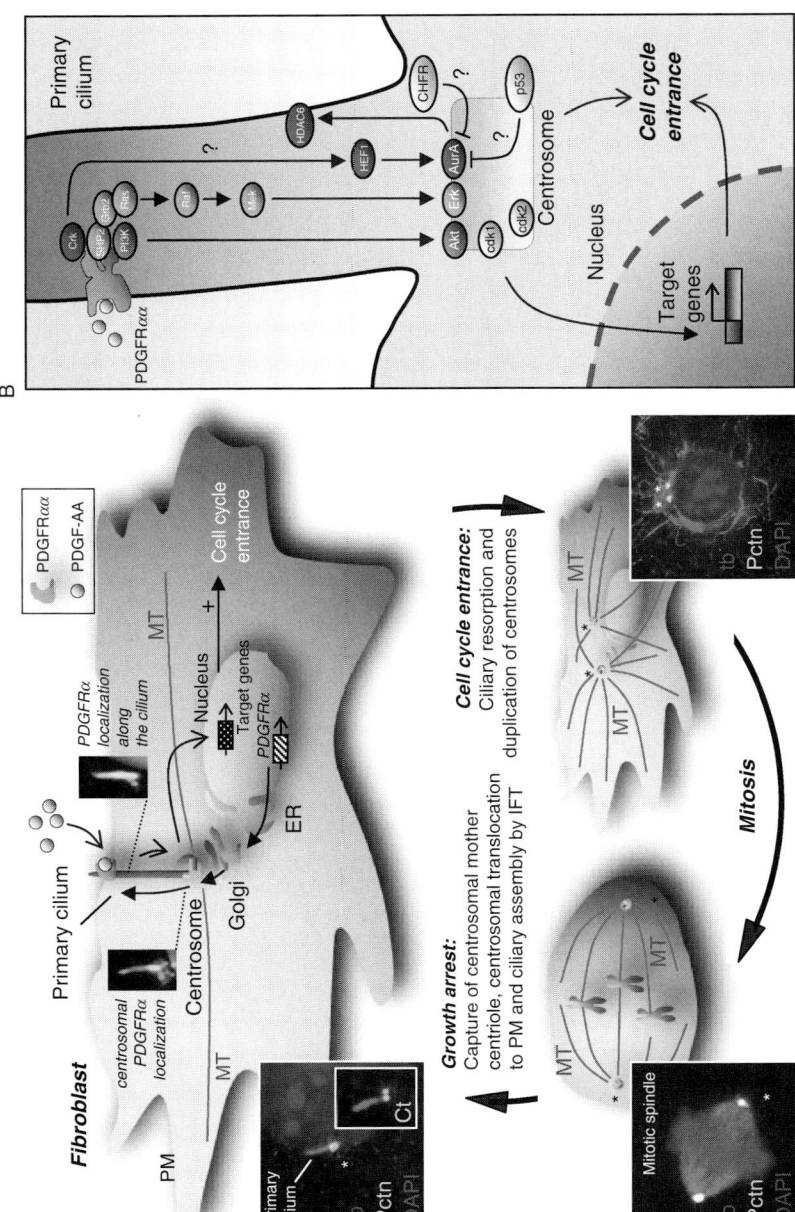

primary cilia (Nielsen et al 2008; Wheatley et al., 1996), indicating that the cilium may function as a tumor suppressor organelle, which when defect leads to aberrant PDGFRα, Hh, Wnt, or von Hippel–Lindau (VHL) tumor suppressor signaling (Mans et al., 2008).

Some specific examples of ciliary signal control of cell cycle progression have been demonstrated. Ciliary PDGFRα signaling activates cyclin-dependent kinase 4 (cdk4)-mediated phosphorylation of retinoblastoma protein (Rb) and cdk1/cdc2 that are essential for G1/S and G2/M phase transitions (Schneider et al., 2005). Hh signaling promotes G1/S phase and G2/M phase progression via cyclin D1 and B1 (Adolphe et al., 2006). In most cases, however, we still know little as to the mechanisms by which ciliary receptors transmit their signal to the regulators in cell cycle control. Nonetheless, we know that aberrant activation or absence of ciliary signaling is correlated with the occurrence of many types of human cancers (Kuehn et al., 2007; Mans et al., 2008; Michaud and Yoder, 2006; Nielsen et al., 2008; Plotnikova et al., 2008).

Many other players in the cilium/centrosome axis have been implicated in signal generation and cell regulation. These include (1) the NIMA-related

---

Figure 10.1 *Cell cycle control regulated by PDGFRαα in the primary cilium of fibroblasts.* (A) Upon growth arrest PDGFRα expression is upregulated and the receptor is translocated to the centrosome/ciliary base and then into the cilium. Activation of ciliary PDGFRαα by, e.g., PDGF-AA turns on a series of signal transduction pathways in the cilium and to the cell. Ciliary localization of PDGFRα in NIH3T3 fibroblasts is shown by immunofluorescence microscopy analysis (IF) with antiacetylated a-tubulin (tb, cilium, red) and anti-PDGFRα (green). The nucleus is stained with DAPI (blue). From Schneider et al. (2005) with permission. PDGFRαα activation causes ciliary resorption, duplication of the centrosome and cell cycle entrance followed by mitosis where the two centrosomes form the mitotic spindle. After cytokinesis the daughter cells may reenter growth arrest by capturing of the centrosomal mother centriole and translocation of the centrosome to the plasma membrane (PM) at which the primary cilium is formed by intraflagellar transport (IFT). The three major inset images show IF of NIH3T3 fibroblasts in interphase growth (lower right image), mitosis (lower image left) and growth arrest (upper left image). The primary cilium, cytoskeletal microtubules (MT) networks, and mitotic spindle are stained with acetylated a-tubulin (tb, red), centrosomes are stained with antipericentrin (Pctn, light blue, asterisks) and the nucleus/chromosomes is stained with DAPI (dark blue). The centrioles at the base of the primary cilium were further stained with anticentrin (Ct, light blue) (upper left image). (B) Current model on some of the signal transduction pathways in the ciliary/centrosome axis that controls growth arrest and cell cycle entrance. Activation of PDGFRαα in the primary cilium leads to ciliary resorption and cell cycle entrance via activation of the PI3K-Akt, Mek1/2-Erk1/2 and hypothetically Crk/Hef1 pathways in the cilium and/or at the centrosome. This results in altered gene expression and induces ciliary resorption via Aurora A kinase (AurA) and tubulin deacetylase (HDAC6). Other hypothetical players in cell cycle control at the ciliary/centrosome axis may include cyclin-dependent kinases, cdk1 and cdk2, and tumor suppressor proteins, p53 and CHFR, which we suggest to inhibit ciliary resorption and cell cycle entrance by blocking the activity of AurA. (See Color Insert.)

kinases, Neks (Bradley and Quarmby, 2005; O'Connell et al., 2003; Otto et al., 2008; White and Quarmby, 2008), (2) the human enhancer of filamentation (HEF1), Aurora A kinase (AurA) (Pugacheva et al., 2007), (3) VHL and Glycogen synthase kinase 3 beta (GSK-3β) (Schermer et al., 2006; Thoma et al., 2007), the Par3-Par6-aPKC polarity complex (Fan et al., 2004), (4) nephrocystin-2/Inversin (Morgan et al., 2002), (5) the microtubule-end binding protein, EB1 (Schrøder et al., 2007), (6) Bardet–Biedl syndrome (BBS) proteins (Blaque and Leroux, 2006; Nachury et al., 2007; Tobin and Beales, 2007), (7) outer dense fiber, ODF2/cenexin (Ishikawa et al., 2005), and (8) small G proteins, such as Rab and Arf-like GTPases, and their corresponding guanine exchange factors, GEFs (Nachury et al., 2007; Omori et al., 2008; Qin et al., 2007; Yoshimura et al., 2007). In vertebrates, Nek8 affects ciliary and centrosomal localization (Otto et al., 2008) and Nek1 plays a role in centrosome integrity affecting both ciliogenesis and centrosome stability (White and Quarmby, 2008). HEF1 is a scaffolding protein that activates the centrosomal protein, AurA, which subsequently directs disassembly of the primary cilium possibly by phosphorylating tubulin deacetylase, HDAC6, in the cilium (Pugacheva et al., 2007). The CALK protein in *Chlamydomonas* is a homologue of mammalian AurA and was the first direct connection between ciliary disassembly and a cell cycle protein (Pan and Snell, 2003). Similarly, *Chlamydomonas* IFT27, which is a Rab-like small G-protein, may act as a checkpoint that holds up cell cycle entry until the flagellum is resorbed (Qin et al., 2007). Mammalian Rab GTPases function at primary cilia and in ciliogenesis, Rab8a being localized to the cilium, and specifically interacting with ODF2/cenexin (Ishikawa et al., 2005; Yoshimura et al., 2007). The Arl-13b protein, hennin, is probably involved in transport of Hh signaling components (Caspary et al., 2007).

The centrosomal proteins CP110 and CEP97 are also potential players in ciliary disassembly and cell cycle control. CP110 interacts with CEP97, which recruits CP110 to the centriole to form a stabilizing CP110/CEP97 interaction. Inhibition of CEP97 and CP110 increases the number of cells displaying primary cilia in several cell types (Spektor et al., 2007), and expression of CEP97 and CP110 in nonproliferating cells inhibits ciliary formation, suggesting that these two centrosomal proteins function to suppress ciliary assembly (Santos and Reiter, 2008; Spektor et al., 2007). CP110 expression is highly induced during G_1 but is shown to be inhibited by phosphorylation via Cdk2 (Chen et al., 2002), which in turn is bound to nuclear Cyclin A and E, expressed predominantly in dividing cells. Cdk2 is also abundant in growth-arrested cells, but is not bound to cyclins under these conditions (Bresnahan et al., 1996). Thus, inhibition of CP110 by Cdk2-mediated phosphorylation may allow the generation of primary cilia during G_1/G_0, whereas activation of AurA, mediated by ciliary mitogenic signaling potentially accounts for ciliary disassembly upon cell cycle entrance. Additional functions of some of these and other players in ciliogenesis and in coordination of the cell cycle are reviewed in Pedersen *et al.* (2008), Plotnikova *et al.* (2008) and Santos and Reiter (2008).

A number of observations link the cilium/centrosome axis to cell migration and chemotaxis in mammalian cells. Some of the above-mentioned signaling elements involved in cell cycle control also impinge on cell migration and/or chemotaxis. These include PDGFRα (Schneider *et al.*, 2008a), Wnt signaling components (Ciani and Salinas, 2007; Kawasaki *et al.*, 2007; Kim *et al.*, 1999, 2007, Kurayoshi *et al.*, 2006, Yun *et al.*, 2005), and HEF1 and AurA (Guan *et al.*, 2007; Pugacheva *et al.*, 2007). Originally, Albrecht-Buehler (1977) discovered that primary cilia in migrating 3T3 fibroblasts were oriented predominantly parallel to the substrate and to the current movement direction, suggesting that the cilium has a role in the directional control of cell migration. More recently, *in vitro* wound healing assays showed that primary cilia emanating from the centrosome locate predominantly in front of the nucleus and orient in the direction of the leading lamellae in 3Y1 (Katsumoto *et al.*, 1994) and NIH3T3 cells (Schneider *et al.*, 2008a) as well as in vascular smooth muscle cells (VSMCs) (Lu *et al.*, 2008) (Fig. 10.2). Benzing and Walz (2006) hypothesized that the primary cilium works as a cellular GPS that guides directional movement. Conceivably, signaling through the cilium may regulate cytoskeletal changes essential in the formation of lamellipodia and other protrusive structures, and in local changes of cell–matrix interactions and cell volume, which collectively enable the cell to move forward. Indeed, as will be described below, the primary cilium may also coordinate responsiveness to chemokines such as PDGF-AA and organize interactions with the extracellular matrix (ECM) in directional cell migration. Both processes are critical in wound repair and developmental processes and when aberrantly activated may be a factor in oncogenesis and tumor invasiveness, in congruence with the central role of the primary cilium in regulating cell cycle progression.

2. Cell Cycle Entry Regulated by PDGFRαα Signaling in the Primary Cilium

Signaling via PDGF and their receptors (PDGFRs) control cell survival, proliferation, and directional cell migration during gastrulation and embryonic and fetal development as well as in tissue homeostasis in the adult organism. Consequently, defects in PDGFR signaling are a causal factor in a range of diseases, including cancer, vascular disorders and fibrosis (Andrae *et al.*, 2008). PDGFs are synthesized and secreted by a number of different cell types and their receptors are found on a wide range of cell types (Ross, 1987), including, e.g., fibroblasts, smooth muscle cells, epithelial cells,

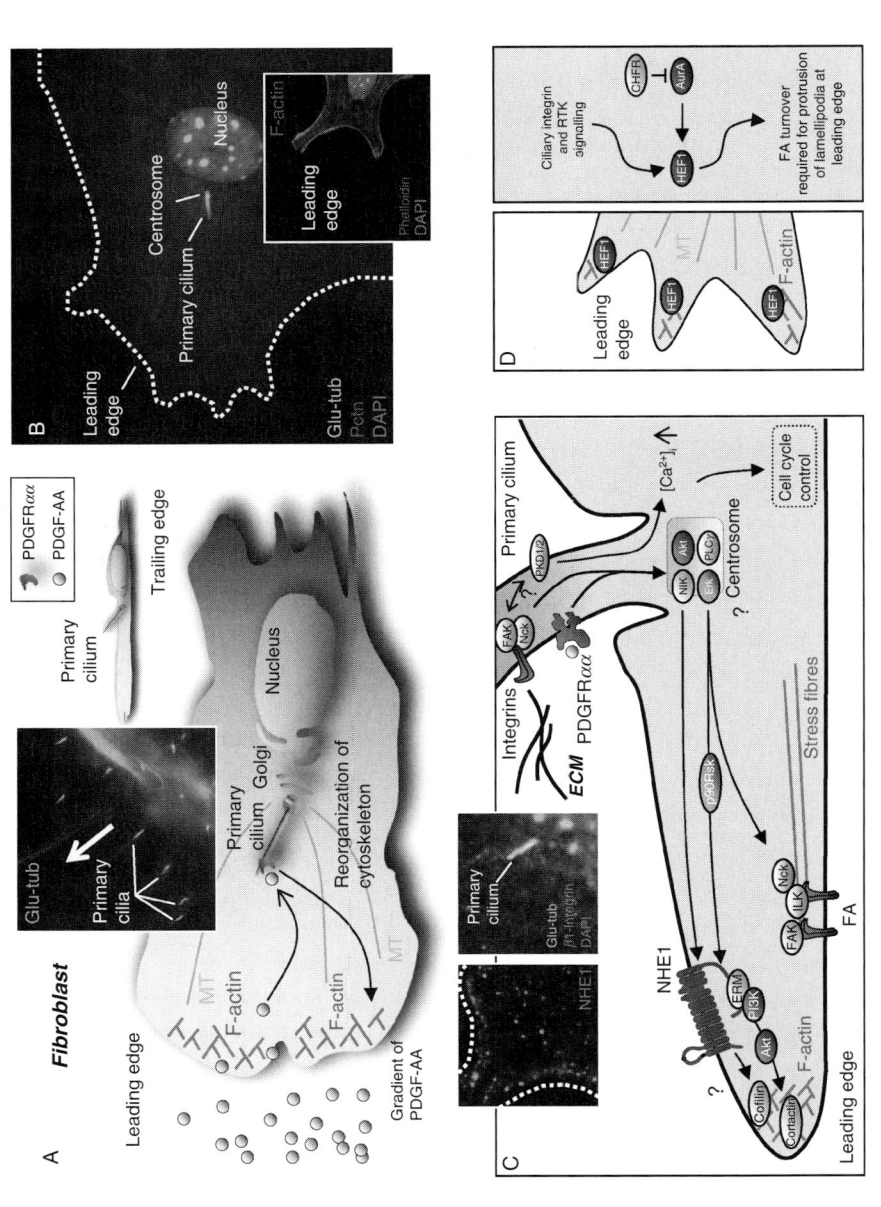

macrophages, and glia cells (Heldin and Westermark, 1999). The PDGF family consists of five homo- and heterodimers built up by disulphide linked polypeptide chains encoded by four different genes. Expression of PDGFs is controlled by external stimuli such as wounding and inflammation on both the translational and the posttranslational level (Andrae et al., 2008; Fredriksson et al., 2004). The five PDGF isoforms, PDGF-AA, PDGF-AB, PDGF-BB, PDGF-CC, and PDGF-DD require proteolytic processing for their activation and biological function (Bergsten et al., 2001; LaRochelle et al., 2001; Li et al., 2000). All isoforms operate via receptor tyrosine kinases, PDGFRα and PDGFRβ, which dimerize into three different homo- or heterodimer complexes, PDGFR$\alpha\alpha$, PDGFR$\alpha\beta$, and PDGFR$\beta\beta$ (Fredriksson et al., 2004). PDGF isoforms bind the receptors with different binding specificity: PDGF-AA, PDGF-AB, PDGF-BB and

Figure 10.2 *Cell migration regulated by PDGFR$\alpha\alpha$ and integrins in the primary cilium.* (A) In migrating cells the primary cilium functions as a cellular GPS that locates in front of the nucleus and orients in the direction of the leading edge to coordinate signaling (e.g., PDGFR$\alpha\alpha$ signaling) in reorganization of microtubules (MT) and F-actin in lamellipodia. The inset shows IF of primary cilia stained with antidetyrosinated tubulin (Glu-tub, green) orienting to the leading edge of growth-arrested NIH3T3 cells in an *in vitro* scratch assay. (B) IF of a single migrating fibroblast with the primary cilium (Glu-tub, green) orienting toward the leading edge and emerging from the centrosome stained with antipericentrin (Pctn, red) and in front of the nucleus (DAPI, blue). (C) Our model of some of the signal transduction pathways in the ciliary/centrosome axis that coordinate cell migration via PDGF-AA-mediated activation of PDGFR$\alpha\alpha$ and ECM-mediated activation of integrins in the primary cilium. Activation of the ciliary receptors leads to activation of signal transduction molecules at the centrosome (e.g., Akt, Erk1/2, phospholipase C-gamma, PLCγ, and Nck-interacting kinase, NIK). In integrin signaling these molecules are activated by focal adhesion kinase (FAK) and Nck. Also, polycystin-2 (PKD2) may be involved in both cell migration and cell cycle control by interaction with integrin signaling and mechanical deflection of the cilium to increase $[Ca^{2+}]_i$. Downstream of the ciliary/centrosome axis, a series of critical molecules in cell motility are activated to control formation of focal adhesions (FA) and lamellipodia at the leading edge of migrating cells. These include FAK, Nck, and NIK at FA and the plasma membrane Na^+/H^+ exchanger (NHE1) at leading edge lamellipodia. As examples, NHE1 directly interacts with ezrin/radixin/moesin (ERM) proteins to regulate cortactin via the PI3K-Akt pathway, and NHE1 regulates cofilin activity at the leading edge. IF insets show NHE1 (blue) localizing to the leading edge lamellipodia (dashed line, upper left IF) and β1-integrin (magenta) localizing to the primary cilium (anti-Glu-tub, green) (upper right IF) in NIH3T3 fibroblasts. While many of these signaling pathways are well described in the literature, little is known on how signaling from the cilium/centrosome axis is transmitted to NHE1 and FA. Some of the hypothetical pathways are also indicated with a question mark. (D) shows in two panels our hypothesis on the role of ciliary integrin and receptor tyrosine kinase (e.g., PDGFR$\alpha\alpha$) signaling in activation of human enhancer of filamentation 1 (HEF1) in control of protrusion of lamellipodia at the leading edge. In this pathway tumor suppressor, CHFR, may restrain excessive cell migration (as seen in invasive cancers) by inhibiting the activity of AurA, which activates HEF1. (See Color Insert.)

PDGF-CC bind PDGFRαα; PDGF-BB and PDGF-DD activate PDGFRββ, whereas PDGF-AB, PDGF-BB and PDGF-CC have high binding affinity for PDGFRαβ (Fredriksson et al., 2004; Heldin and Westermark, 1999). PDGFRαα is activated only by PDGF-AA. In all three receptor complexes, intracellular tyrosine residues are autophosphorylated to recruit SH2/PTB domain-containing adaptor- and effector proteins that ultimately regulate a variety of signal transduction pathways eliciting cytoskeletal reorganization and stimulation of cell growth, proliferation, motility, and chemotaxis. Two well-characterized such adaptor- and effector proteins are the tyrosine phosphatase Src homology phosphotyrosyl phosphatase 2 (SHP2) that binds to PDGFRα when the receptor is phosphorylated on Y^{720} or Y^{754}, and Phoshophatidyl Inositol 3-OH kinase (PI3K) that binds to the receptor phosphorylated on Y^{731} or Y^{742}. SHP2 connects the receptor to the adaptor molecule Grb2 activating Ras-Mek1/2-Erk1/2 pathway through binding of Sos, a positive regulator of Ras activity. Docking of PI3K to the receptor leads to activation/production of various downstream factors, e.g., phosphatidylinositol(4,5)bisphosphate (PtdIns(4,5)P$_2$), members of the Rho family of small GTPases, protein kinase C (PKC)s, Ras and the multifunctional protein kinase Akt/protein kinase B (PKB).

2.1. Regulation of PDGFRαα signaling in the primary cilium

In 1996 Lih et al. showed that PDGFRα mRNA is upregulated during serum starvation of cultured fibroblasts, i.e., PDGFRα is encoded by a growth arrest specific gene in these cells. This led Schneider et al. (2005) to investigate the potential relationship between PDGFRα signaling and the fibroblast primary cilium, which is formed exclusively in quiescent cells. We confirmed that upregulation of PDGFRα in fibroblasts correlates with the formation of the cilium during serum starvation. Further, PDGFRα localized to the primary cilium (Fig. 10.1) and was activated by either PDGF-AA or PDGF-BB only when the cells had entered growth arrest. In contrast, PDGFRβ localized to the plasma membrane, was expressed at a similar level in interphase and growth-arrested cells, and was activated by PDGF-BB in both interphase and growth-arrested cells (Schneider et al., 2005). These results suggested that PDGFRαα signaling is closely associated with the primary cilium. Indeed, PDGF-AA stimulation of growth-arrested cells activated PDGFRαα in the primary cilium by tyrosine phosphorylation of the receptor in positions 720, 742, and 754. This was followed by activation of the Mek1/2-Erk1/2 and Akt pathways in and at the base of the cilium, in addition to phosphorylation of Rb and Cdc2 (Fig. 10.1), which marks cell cycle entry and progression (Schneider et al., 2005). Embryonic fibroblasts from the ORPK ($IFT88^{Tg737Rpw}$) mouse, whose fibroblasts form no or very short cilia (hereafter referred as to $Tg737^{orpk}$

MEFs), do not show PDGFRα upregulation and signaling through the receptor is inhibited. In contrast, PDGFRβ is not targeted to the primary cilium, and signaling through PDGFRβ is not altered in $Tg737^{orpk}$ mutants, further supporting the conclusion that the primary cilium specifically regulates PDGFRαα activation and function during cell cycle entry in fibroblasts (Schneider et al., 2005) (Fig. 10.1).

The mechanisms by which PDGFRα is upregulated during growth arrest and targeted to the primary cilium (Fig. 10.1) are unknown, although data from the $Tg737^{orpk}$ MEFs suggest that upregulation is controlled by one or more players associated with formation of the primary cilium. Targeting of ciliary proteins has been suggested to be controlled, e.g., by lipid modifications, protein phosphorylation, and specific ciliary peptide targeting motifs that may coordinate the translocation of membrane proteins from the Golgi to the ciliary transition zone and ultimately to the primary cilium (Christensen et al., 2007; Pedersen et al., 2008). As an example, the motif RVXP located in the amino terminal region of mammalian polycystin-2 (PKD-2) is necessary and sufficient for ciliary localization (Geng et al., 2006). Interestingly, PDGFRα (but not PDGFRβ) has a similar motif (RLVP), hence, it may be hypothesized that the differential targeting of PDGFRα to the ciliary membrane and PDGFRβ to the plasma membrane could in part be coordinated by this motif. The regulated upregulation and ciliary targeting of PDGFRα, however, imply that the receptor is enriched in the cilia under normal conditions to ensure a strong signal in cell cycle control, since the basal level of the receptor in nonarrested cells does not elicit detectable activation (Schneider et al., 2005).

How is mitogenic signaling in the primary cilium coordinated with disassembly of the cilium? Clearly, activation of, e.g., PDGFRα in the primary cilium must be converted into a mitogenic response by activation of immediate early genes controlling cell cycle entry and potentially cell cycle progression, but it is not known how signaling from the receptor impinges on players in ciliary disassembly. Activated PDGFRαα binds and activates adapter proteins of the Crk family (CrkI, CrkII and CrkL) via docking of their SH2 domain to PDGFRα when the receptor is phosphorylated on Y^{762} (Yokote et al., 1998). Since HEF1 is a substrate for Crk, one possible sequence of events is that upon PDGF-AA stimulation, Crk proteins are activated in the primary cilium, which subsequently permits HEF1 to activate AurA, which then leads to ciliary disassembly. It is notable that activated PDGFRβ apparently does not bind Crks (Yokote et al., 1998), signifying that ciliary disassembly is highly coordinated with mitogenic signaling pathways that are associated with the primary cilium. Another potential player in ciliary disassembly and cell cycle control is the recently discovered tumor suppressor and early mitotic checkpoint protein, *Checkpoint with FHA and RING finger domains* (CHFR) (Scolnick and Halazonetis, 2000), which inhibits AurA by ubiquitination and degradation

(Yu et al., 2005) and when mutated causes 30–50% of all human cancers (Privette and Petty, 2008). Interestingly, both HEF1 and CHFR contain motifs similar to that of the PKD-2 ciliary targeting motif: R(402)LVP in HEF1 and R(504)VAP in CFHR, which indicate that to function as regulators in cell cycle control these proteins could localize to cilia, although such localization has not yet been demonstrated. Another central tumor suppressor protein is p53, which suppresses AurA activity. Genetic ablation of p53 in MEFs leads to abnormal centrosome duplication (Oda-Sato and Tanaka, 2007). We previously reported that activation of p53 by phosphorylation on Ser 15 in serum-starved and quiescent wt MEFs is inhibited in $Tg737^{orpk}$ MEFs (Christensen et al., 2006), suggesting that in the absence of externally provided stimuli, the primary cilium is required for activation of p53 to maintain the cells in growth arrest—potentially via inhibition of AurA activity.

Most recently, we and others showed that PDGFRα localizes to primary cilia of human embryonic stem cells (hESCs) (Awan et al., 2008) and in nerve stem cells (NSCs) and neuroblasts in the rat brain (Danilov et al., 2008). PDGFRαα signaling was proposed to control NCS proliferation in a process associated with disassembly of the primary cilium and redistribution of the receptor to the mitotic spindle; the progeny giving rise to neuroblasts (Danilov et al., 2008). Similarly, epidermal growth factor receptor (EGFR) signaling in NSC (Danilov et al., 2008) and Hh signaling in hESC (Kiprilov et al., 2008) may be coordinated by the primary cilium. In hESCs PDGFRαα signaling is known to induce cell proliferation and assist maintaining cells in an undifferentiated state (Pébay et al., 2005), indicating that the hESC primary cilium controls embryonic stem cell pluripotentiality. These observations support the conclusion that PDGFRαα signaling in cell cycle control is coordinated by the primary cilium in various cell types and tissues to control developmental processes and maintain tissue homeostasis.

3. Directional Cell Migration is Regulated by PDGFRα in the Primary Cilium

Cell motility and migration are necessary for many physiological processes including wound healing, immunity, neuronal patterning, embryogenesis, and angiogenesis, and for the development of pathologies such as fibrosis and tumor metastasis (Kay et al., 2008; Parent and Devreotes, 1999). For example, neural crest cells migrate extensively during embryonic development, giving rise to a wide variety of cell types, and consequently contributing to organ physiology (Begbie, 2008; Graham et al., 2004; Kuriyama and Mayor, 2008). Immune cells migrate to the location of inflammation as part of the immune defense, and migration of stem cells, epithelial cells and fibroblasts plays central roles in tissue homeostasis, wound healing processes and regeneration of a variety of tissues (Cahalan and Gutman, 2006; Dignass, 2001; Laird et al., 2008; Schwab et al., 2007). Defective migratory signaling causes severe malignancies including neuronal disorders, embryonic developmental malfunctions and cancer (Kuriyama and Mayor, 2008; O'Hayre et al., 2008; Raman et al., 2007).

Cell migration involves a highly complex series of events culminating in the formation of lamellipodia and other protrusive structures, and in local changes of cell–matrix interactions and cell volume, which collectively enable the cell to move forward. Cell migration requires extensive reorganization of elements of the actin-based cytoskeleton, including F-actin, cortactin, cofilin, and ezrin/radixin/moesin (ERM) proteins (Elliott et al., 2005; Yamaguchi and Condeelis, 2007). Cofilin, which is regulated both by phosphorylation and by changes in pH_i (Srivastava et al., 2007), is central to both actin polymerization and cancer metastasis (Yamaguchi and Condeelis, 2007). Cortactin localizes to the plasma membrane, lamellipodia/pseudopodia, and invadopodia, and is an important player in migration (van Rossum et al., 2006). In addition to the cytoskeletal rearrangement, cell migration is also critically dependent on activation of multiple ion transport proteins (Schwab, 2001a,b; Schwab et al., 2007).

Recently it has become apparent that many migratory malfunctions may be directly or indirectly related to defects in primary cilia or primary cilia signaling (Lu et al., 2008; Schneider et al., 2008a; Tobin et al., 2008). For instance, primary cilia are required for correct Hh and noncanonical Wnt signaling and subsequently for the specific cellular locomotion of neural crest cells and proper neuronal patterning (De Calisto et al., 2005; Haycraft et al., 2005; Liu et al., 2005; Tobin et al., 2008; Wada et al., 2005). In the context of this article, it is relevant to know whether signaling from the primary cilium influences the reorganization of the actin cytoskeleton and transport proteins in fibroblasts, for example, to control migration speed and direction of lamellipod formation. If so, the mechanisms through which ciliary signaling controls occur might be generally applicable to other cell types with primary cilia, including embryonic stem cells and neurons.

3.1. Ciliary PDGFRα signaling regulates directed cell migration in fibroblasts

Cell migration is known to be regulated through external stimuli, including growth factors such as PDGF. While PDGF-BB is well established as a chemoattractant, the effect of PDGF-AA on cell motility and chemotaxis (i.e., directed cell migration) remains somewhat controversial. Earlier studies suggest that PDGF-AA and PDGF-BB regulate cell migration diversely with different cell type specificity (Hayashi et al., 1995; Koyama et al., 1992; Siegbahn et al., 1990; Yokote et al., 1996, Yu et al., 2001). More recent studies have emphasized the importance of PDGF-AA as a unique chemoattractant during directed cell movement (Bornfeldt et al., 1995; Brockmann et al., 2003; Forsberg-Nilsson et al., 1998; Frost et al., 2008; Kato et al., 1998; Lind et al., 1995; Uren et al., 1994; Yokote et al., 1996). For example, PDGF-AA is an essential chemoattractant during CNS development and important for normal distribution of oligodendrocytes progenitors (OP) cells throughout the developing CNS *in vivo* (Frost et al., 2008).

Recently, our laboratory showed that the primary cilium is required for PDGF-AA mediated chemotaxis in growth arrested cultured fibroblasts, which are thought to model most tissue fibroblasts *in situ* (Fig. 10.2). Fibroblasts from wt mice with normal primary cilia acquire chemotactic behaviour when exposed to a PDGF-AA gradient, characterized by an increase in migration speed and directional cell movement up the gradient. In contrast, $Tg737^{orpk}$ MEFs with no cilia do not respond to PDGF-AA and do not display chemotaxis, strongly suggesting a significant role of primary cilia signaling in regulation of cell migration (Schneider et al., 2008a). These results imply that loss of the cilium, and potentially the lack of a cilium orienting parallel to the direction of migration in quiescent fibroblasts lead to a lack of regulatory control of the migratory speed and displacement of cells, which is characteristic to fibrosis and tumour cell invasion. Consequently, signals from the cilium may restrain excessive cell migration to prevent uncontrolled and incorrect displacement of cells as seen in $Tg737^{orpk}$ MEFs.

Ciliary PDGFRα activation by PDGF-AA leads to increased levels of phospho-Akt at the base of the cilium, i.e., at the mother centriole in fibroblasts. This increase is blocked in $Tg737^{orpk}$ MEFs (Schneider et al., 2008a), suggesting that PtdIns(3,4,5)P_3 and Akt lie downstream of ciliary PDGFRα signaling in the primary cilium. PDGFRα-mediated signaling via the PI3K-Akt pathway partially controls microtubule stabilization and actin dynamics at the leading edge during directional cell movement (Nishio et al., 2007). On the other hand, PtdIns(3,4,5)P_3 depletion or inhibition of PI3K does not completely block the directed migration. This indicates that alternative signaling pathways, including some involving PDGFRβ and PDGF-BB or ligands such as EGF for nonciliary cell membrane localized receptors, also control cell movement (Ferguson et al., 2007; Franca-Koh et al., 2007; Kay et al., 2008; Nishio et al., 2007). In some cells, PDGFRα signaling can negatively regulate migratory signaling induced by PDGFRβ (Koyama et al., 1992; Yokote et al., 1996).

The downstream effectors of activated PDGFRα and PI3K, PtdIns(3,4,5)P_3 and Akt play significant roles in chemotaxis and directional sensing by affecting a variety of molecular mechanisms, including transcriptional regulation of motility genes, actin cytoskeleton reorganisation and dynamics, promotion of microtubule cytoskeleton stabilization at the leading edge in migrating fibroblasts (Onishi et al., 2007), and control of cellular interactions with the ECM (Harvey and Lonial, 2007; Stambolic and Woodgett, 2006). PtdIns(3,4,5)P$_3$ regulates the activity of Rac and Cdc42 (Franca-Koh and Devreotes, 2004; Franca-Koh et al., 2007; Li et al., 2003; Srinivasan et al., 2003; Welch et al., 2002). The Akt isoforms regulate cell migration downstream of PtdIns(3,4,5)P$_3$ by multiple mechanisms (Arboleda et al., 2003; Hutchinson et al., 2004; Li et al., 2003; Manning and Cantley, 2007; Zhou et al., 2006). An interplay between Akt phosphorylation and increased AurA activity may enhance cell migration while leading to eventual ciliary resorption, as inhibition of AurA can decrease Akt1 phosphorylation and block cell migration (Guan et al., 2007). In line with these findings, stable knockdowns of CHFR significantly increased migration speed (Privette et al., 2007), indicating that accumulation of active AurA at the basal body may be required for increase of migratory movement.

In addition to the PI3K-Akt pathway, PDGF-AA-induced chemotaxis requires activation of several signaling molecules including MAP kinases, Src kinase and Phospholipase C gamma (PLCγ), which may work additively to modulate cell migration (McKinnon et al., 2005, Rosenkranz et al., 1999). These signaling systems act on the actin cytoskeleton and on focal adhesions to facilitate, initiate and stabilize movement. PDGFRs promote reorganization of the actin cytoskeletal network through activation of intracellular molecules such as Rho-ROCK, PKC, Rac1 and Wiscott–Aldrich syndrome protein (WASP), and the WASP-interacting protein, WIP (Anton et al., 2003; Chen and Guan, 1994; Di et al., 2002; Ishikura et al., 2005; Maddala et al., 2003; McKinnon et al., 2005). PDGFRα activation of PLCγ indirectly increases actin turnover by releasing actin-regulating subunits like gelsolin, cofilin, and profilin from the membrane. PLCγ activation also increases the level of free intracellular $Ca^{2+}[Ca^{2+}]_i$, supporting Ca^{2+}-dependent migratory events (Fabian et al., 2008; McKinnon et al., 2005). PDGFRα mediates efficient focal adhesion kinase (FAK) phosphorylation at focal adhesions (FAs), through Src or integrin induced regulation of RhoGTPases activity and consequently increased FA turnover (Carloni et al., 2000, 2002; Ding et al., 2003; Rosenkranz et al., 1999). In addition, PDGFRα-induced localization of HEF1 at FAs plays a potential role in regulating FA assembly/disassembly and correct migratory movement (Natarajan et al., 2006; van Seventer et al., 2001). PDGFRα signaling also controls FA dynamics by increasing the expression of Matrix Metallo Proteases (MMPs), which facilitate migration by activating a proteolytic remodeling of ECM (Laurent et al., 2003; Vu and Werb, 2000). Finally, PDGFRs regulate the activity of the ubiquitously expressed plasma membrane Na^+/H^+ exchanger, NHE1 (Yan et al., 2001), a key player in cell motility and migration. Defects in migratory PDGF-AA signaling cause serious malignancies such as various types of cancers, developmental malfunctions, atherosclerosis, and different forms of fibrosis (Andrae et al., 2008; Carvalho et al., 2005; Dell'albani, 2008; Matei et al., 2006; Reis et al., 2005), all diseases that resemble many of the physiological abnormalities caused by defects in primary cilia assembly or cilia signaling.

3.2. Primary cilia in wound healing

Wound healing is a dynamic interactive and distinctively regulated process controlled by different chemical stimuli, the various PDGFs being important chemoattractants in the earlier stages of wound repair *in vivo* (Martin, 1997; Parent and Devreotes, 1999; Singer and Clark, 1999). PDGF is secreted from aggregated platelets, neutrophils and macrophages located in the blood clot right after tissue injury. PDGF contributes to attracting immune cells for the inflammatory processes and to the initiation of proliferation and crawling of fibroblasts from the adjacent tissue around the wound into the wound space (for reviews see Singer and Clark, 1999), presumably by mechanisms which we have just elucidated in the above. Indeed, *in vivo* studies showed that $Tg737^{orpk}$ mutant mice that lack the primary cilium have a significantly reduced rate of wound repair and defects in wound closure compared to wt mice (Schneider et al., 2008a), accentuating an essential role of primary cilia in tissue repair. In the adult rat brain, PDGFRα is similarly localized to the primary cilium of NSCs and blastocysts, indicating that the cilium in these cells is involved in injury-induced neurogenesis in the brain (Danilov et al., 2008).

Other migratory factors related to ciliary PDGFRα signaling may be involved in correct wound healing. Disruption of PDGF-AA signaling decreases Transforming Growth Factor-β (TGFβ) induced wound repair, which suggests that there are feedback interactions between TGFβ and ciliary PDGF-AA signaling during wound healing/tumorigenesis (Gotzmann et al., 2006; Martin, 1997) and also in inflammatory processes during early tissue repair (Soma et al., 2002; Xie et al., 1994; Yeh et al., 1993). Migrating fibroblasts are responsible for remodeling the ECM by expressing ECM receptors including integrin receptors and producing large amounts of collagen matrix required for new cells to move into the wound during wound retraction (Clark et al., 1995; Martin, 1997; Singer and Clark, 1999; Xu and Clark, 1996). Thus, primary cilia are likely to contribute to the PDGF-mediated processes during late wound closure.

As will be discussed below, both ECM receptor components and integrin receptors become localized to primary cilia and interact directly with collagen fibers through electron-opaque-plagues linking the cilia directly to ECM (Jensen et al., 2004; Lu et al., 2008; McGlashan et al., 2006; Praetorius et al., 2004). Integrins and PDGFRα associate and perhaps together regulate part of the migratory machinery during wound healing (Baron et al., 2002; Ding et al., 2003). Taken together, these findings suggest that primary cilia are essential cellular compartments necessary for correct sensing of chemokines from the extracellular environment.

3.3. Roles of the Na$^+$/H$^+$ exchanger NHE1 in PDGFRα-mediated control of cell migration

The plasma membrane Na$^+$/H$^+$ exchanger, NHE1, is found in essentially all cells in the body and plays important roles in cellular pH and volume homeostasis and in control of cell proliferation and cell survival (Pedersen, 2006; Putney et al., 2002). NHE1 has moreover been assigned a central role in control of cell migration and chemotaxis in a wide range of cell types (Chiang et al., 2008; Denker and Barber, 2002; Lagana et al., 2000; Reshkin et al., 2000; Stock et al., 2005). Notably, NHE1 contributes not only to cell migration, but also to the invasive/metastatic capacity of tumor cells (Cardone et al., 2005). NHE1 is enriched in leading edge lamellipodia (Cardone et al., 2005; Frantz et al., 2007b; Lagana et al., 2000; Patel and Barber, 2005), where it interacts directly with ERM proteins (Denker et al., 2000), which are central players in cell migration and cancer metastasis (Elliott et al., 2005). Inhibition of NHE1 leads to retraction of pseudopodia and reduced motility, an effect most pronounced in highly invasive cells (Lagana et al., 2000). Fibroblasts deficient in NHE1, or expressing a mutant NHE1 unable to bind ERM proteins, cannot establish a primary leading edge lamellipodium, and their Golgi apparatus fails to localize to the leading-edge face of the nucleus (Denker and Barber, 2002). NHE1 also localizes to FAs, is activated by RhoA-ROCK signaling downstream of integrin activation, is required for integrin-mediated stress fiber formation (Putney et al., 2002; Tominaga and Barber, 1998), and plays a major role in FA remodeling during fibroblast migration (Denker and Barber, 2002).

We recently demonstrated that in fibroblasts, chemotaxis in response to PDGFRα activation is dependent on activation of NHE1 (Schneider et al. 2008b). During growth arrest, NHE1 mRNA and protein levels were upregulated in NIH3T3 cells and in wt and $Tg737^{orpk}$ MEFs independent of primary cilium formation. However, NHE1 activity was highly downregulated compared to interphase cells. Activity could be partially restored by PDGF-AA addition. In wound-healing assays on growth-arrested NIH3T3 cells and wt MEFs, inhibition of NHE1 potently reduced PDGF-AA-mediated displacement (net distance covered over 4 h) and directional migration (speed of movement into the wound). These effects were markedly reduced in interphase cells and essentially abolished in growth-arrested $Tg737^{orpk}$ MEFs without cilia. Therefore, directional responses to ciliary PDGFRα signals operate in part via upregulation of NHE1 activity to control migration of growth-arrested fibroblasts. Interestingly, NHE1 was more important for PDGFRα-stimulated than for basal migration, and more important for directional migration than for the overall migratory speed (Schneider *et al.*, 2008b). This indicates that NHE1 is important for transmitting ciliary PDGFRα stimulation into control of directionality.

How does ciliary PDGFRα signaling lead to the activation of NHE1 at the leading edge? Our analyses of cellular pH recovery suggest that regulation of directional migration may involve ciliary PDGFRα signaling to a local pool of NHE1. An obvious question in this regard is whether a pool of NHE1 may localize to the primary cilium. NHE1 contains two ciliary targeting motifs, including an N-terminal motif (R14IFP) similar to that required for ciliary targeting of polycystin-2 (R6VXP) (Geng *et al.*, 2006b). It is also notable in this regard that a novel NHE isoform, NHE10 (sNHE), localizes to the sperm flagellum, where it plays a major role in control of motility (Wang *et al.*, 2007). However, in fibroblasts, NHE1 staining in the primary cilium was generally not increased above the levels in the plasma membrane (Schneider *et al.*, 2008b), hence, while it remains a possibility that NHE1 plays a role in the cilium proper, the discussion below will focus on mechanisms through which the cilium may regulate NHE1 at the leading edge.

Given the known effectors of PDGFRα, at least five, not mutually exclusive, pathways linking ciliary PDGFRα activation to NHE1 activation and localization to leading edge lamellipodia can be envisaged (Fig. 10.2B). First, NHE1 is known to be activated via the ERK pathway, in a manner involving direct phosphorylation of NHE1 by the ERK effector p90ribosomal S kinase (Rsk) (Takahashi *et al.*, 1999). This is of particular interest since p90Rsk may play an important role in cell migration (see Anjum & Blenis, 2008). Second, the PLCγ-PKC pathway has been implicated in activation of NHE1 in response to EGF receptor activation (Maly *et al.*, 2002). Third, the Ste20-related kinase Nck-interacting kinase (NIK) activates NHE1 in response to PDGF stimulation (Yan *et al.*, 2001), although it is to our knowledge unknown whether specifically PDGFRα activates NIK. Fourth, NHE1 interacts directly with SHP-2, and SHP-2 overexpression increases NHE1 activity (Xue *et al.*, 2007). Finally, NHE1 may be both up- and downstream of PI3K activation in migrating cells. NHE1-mediated recruitment of PI3K through interaction with ERM proteins is required for Akt activation by stress stimuli in some (Wu *et al.*, 2004), although not all (Rasmussen *et al.*, 2008) cell types. In accordance with a role for NHE1 upstream of Akt, NHE1 appears to be required for leading edge PI3K signaling (Denker and Barber, 2002). On the other hand, PDGFRα signaling from primary cilia also activates Akt directly. As mentioned, Akt stabilizes microtubules at the developing leading edge of the cell. It seems possible that the PI3K-Akt pathway (and perhaps the other pathways discussed) could in turn stimulate vesicular transport or transport of NHE1 within the cell membrane along the stabilized microtubules. This could contribute to the redistribution of NHE1—and other transporters and receptors—from an unpolarized membrane distribution in resting cells, to a predominant localization in the leading edge lamellipodia in polarized, migrating cells after PDGF-AA stimulation in cells with primary cilia. While such a scenario has yet to be tested for NHE1, endocytic retrieval from other parts of the cell, vesicular transport through the cell, and exocytic insertion in the leading edge membrane has been proposed for other central components of the migration machinery, including integrins (Caswell and Norman, 2008) and K^+ channels (Nechyporuk-Zloy et al., 2008).

3.4. Downstream mechanisms by which NHE1 controls directed cell migration and invasion

Once present and activated at the leading edge, there are several possible ways for NHE1 to control the direction of cell migration (Fig. 10.2B). One likely role of NHE1 is to serve as an osmolyte influx pathway facilitating leading edge protrusion (not shown in Fig. 10.2B; for a recent discussion, see Mitchison et al., 2008). Another role of NHE1 could be to control actin-based cytoskeletal reorganization through direct interaction with ERM proteins (Denker and Barber, 2002). However, the interaction with ERM proteins is not sufficient for optimal cell migration, as ion translocation by NHE1 is also required (Denker and Barber, 2002). Other roles of NHE1 in migration likely reflect the intracellular alkalinization resulting from NHE1 activity, since increased pH_i is a prerequisite for cell polarity in a wide variety of cell types (Srivastava et al., 2007). Such effects include the regulation of strongly pH_i dependent cytoskeletal regulators with central roles in cell migration, including Cofilin, Villin, and Talin (Srivastava et al., 2007). Changes in pH_o also play an important role in NHE1-mediated regulation of cell migration and invasion (Stock et al., 2008). The extrusion of H^+ by NHE1 creates an extracellular acidic gradient along the axis of the migrating cell, with the most acidic environment found at the leading edge where NHE1 activity is highest. NHE1 was also recently shown to regulate another central factor in directional migration, namely leading edge Cdc42 activity, possibly by a mechanism involving pH-dependent binding of PtdIns(4,5)P_2 to a Cdc42 GEF (Frantz et al., 2007). Although the relative importance of these actions of NHE1 in affecting directional cell migration remains to be determined, we can now construct a general pathway by which signals generated by the primary cilium act through NHE1 to influence the formation and forward protrusion of the leading edge of the cell.

4. THE EXTRACELLULAR MATRIX AND THE PRIMARY CILIUM

The ECM is a critical mediator of communication between the cell and its surroundings during development and in tissue homeostasis. Components of ECM are produced intracellularly by resident cells and are then secreted into the extracellular environment where they aggregate and form an interlocking mesh of fibrous proteins and glycosaminoglycans. The cell interacts with the ECM via a series of receptors and adhesion molecules, including integrins that anchor the cells to the ECM and translate the interaction with ECM into cytosolic signaling events that regulate both physiological and pathological events, including angiogenesis, tumor growth, and metastasis (Stupack, 2007). Cell–matrix adhesion complexes are foci of cellular attachment to ECM that are mediated by integrins and adaptor proteins, providing the physical and regulatory links between ECM and the cytoskeleton. The adaptor proteins control the activity of a series of signal transduction pathways of which some are unique for integrin signaling—others are integrated with, e.g., Wnt and receptor tyrosine kinase signaling as outlined in Sections 2–4. Integrin signaling involves the activation of adaptor molecules such as FAK, Integrin-linked kinase (ILK), particularly interesting new cysteine-histidine rich protein (PINCH) and Noncatalytic (region of) tyrosine kinase adaptor protein (Nck) that via, e.g., Crk, PI3K/Akt, PLCγ, Grb7, ASAP1, and GRAF cooperatively interact with receptor tyrosine kinase signaling components to regulate survival, proliferation and differentiation as well as the adhesion, polarity and actin polymerization necessary for cell migration in development and vascular function (Hehlgans et al., 2007; Lock and Debnath, 2008; Parsons, 2003).

In 1985, Poole et al. hypothesized that formation and maintenance of a functionally effective matrix required the interaction between the primary cilium of connective tissue cells such as fibroblasts and ECM. It is now known that primary cilia in a series of different cell types embedded in ECM contain receptors for ECM components, including chondrocytes (McGlashan et al., 2006), VSMCs (Lu et al., 2008) and fibroblasts (Schneider and Christensen, unpublished), and that ciliary interaction with ECM is critical in tissue homeostasis and wound repair. However, receptors for ECM are also present in kidney tubule epithelial cells, where the primary cilium projects into the tubule lumen. In these cilia integrins potentiate fibronectin-induced and flow-independent Ca^{2+} signaling, a mechanism which was hypothesized to be involved in detecting damage to upstream renal tubules (Praetorius et al., 2004). VSMC primary cilia may be critical for cell–ECM interaction and mechanosensing that regulate vasculature (Lu et al., 2008). VSMC cilia contain PKD1, PKD2 and integrins and uniquely project from the front of VSMCs into the adjacent ECM, aligning at an approximately 60% angle in relation to the cross-sectional plane of the mouse aorta. The ciliary ECM receptors include $\alpha 3$ and $\beta 1$ integrins, and blocking of the latter by $\beta 1$-integrin antibodies inhibits positioning of cilia to the cell wound interface in *in vitro* scratch assays and blocks type 1 collagen-induced increase in $[Ca^{2+}]_i$ at a level comparable to that of non-ciliated cells. Further, growth-arrested VSMCs deciliated with chloral hydrate exhibit a reduced Ca^{2+} response induced by either mechanical stress or type 1 collagen stimulation and a delay in wound healing, supporting the conclusion that primary cilia are involved in the response to vessel injury (Lu et al., 2008). Similarly, we hypothesize that directional migration and PDGFR$\alpha\alpha$-mediated chemotaxis of fibroblasts is synchronized through ciliary integrins that connect to ECM for adhesion, cytoskeletal organization, and gene expression, where the cilium attaches to and pulls upon ECM as the cell moves, transmitting mechanical information from the outside milieu to the cell in embryonic patterning and adult tissue reorganization.

The *Tg737^{orpk}* mouse with defect primary cilia exhibits an array of skeletal patterning defects and shows stunted growth after birth (Lehman et al., 2008; Zhang et al., 2003). In chondrocytes, which produce and maintain ECM of cartilage, primary cilia directly interact with ECM components via specific ECM receptors, suggesting that mechanical stimuli may be transmitted through the cilium to control tissue development and to construct a mechanically robust skeletal system (Jensen et al., 2004; McGlashan et al., 2006). The chondrocyte ciliary ECM receptors include $\alpha 2$, $\alpha 3$, and $\beta 1$ integrins, which interact with collagens, fibronectin, and laminin, and the transmembrane proteoglycan, NG2, which binds collagen types V and VI in cell proliferation, adhesion, and cell spreading. More recently, McGlashan and co-workers showed that primary cilia of chondrocytes can be linked to defects in chondrocyte differentiation that results in delayed chondrocyte hypertrophy within the growth plate during development (McGlashan et al., 2007) as well as to articular cartilage degeneration in osteoarthritis (McGlashan et al., 2008). Probably, interaction between the primary cilium and ECM components is characteristic to cell types in deep tissues, where coordination of cell cycle control, migration and differentiation are regulated by the cilium in concert with chemokines, growth factors and other soluble signaling molecules, as partly shown in Fig. 10.2, to control embryonic patterning, developmental processes and adult tissue reorganization.

5. POLARIZATION, CELL MIGRATION, CELL CYCLE CONTROL AND WNT SIGNALING IN THE PRIMARY CILIUM

Above, we have outlined how ciliary PDGFRα and ECM signaling may coordinate cell cycle control and migration. However, multiple other signaling pathways in the cilium may also contribute. These include Hh and Wnt signaling, mechanisms in planar cell polarity (PCP) and signaling via TRP ion channels, which are extensively reviewed in this issue (Chen, 2008; Katsanis, 2008; Reiter, 2008). As an example, loss of primary cilia in the $Tg737^{orpk}$ mouse causes a series of abnormalities in the pancreas, including extensive cyst formation in ducts associated with defects in cell cycle control (Cano *et al.*, 2004; Zhang *et al.*, 2005) and spindle orientation (B. K. Yoder, personal communication). Recently, Hh signaling in the primary cilium was suggested to regulate normal development of the human pancreas and defects in ciliary assembly and/or aberrant ciliary Hh signaling could lead to human pancreatic cancers (Nielsen *et al.*, 2008). In the dilated ducts and cysts of the $Tg737^{orpk}$ mouse, PKD2 is mislocalized to intracellular compartments, the cytosolic localization of β-Catenin is increased and there is an increased expression of Tcf/Lef transcription, suggesting an alteration in the Wnt signaling pathway (Cano *et al.*, 2004). In this section, we will outline some of the critical pathways by which Wnt signaling may control cell migration and cell cycle control in relation to cell polarity and in concert with cytoskeletal changes and downstream effector molecules in ECM, PDGFRα, and NHE1 signaling modulated by the primary cilium.

5.1. Wnt signaling in human health and disease

Wnt signaling is a keystone in many developmental and homeostatic processes including gastrulation, organogenesis, bone and cartilage formation, hematopoeisis, sex determination, and tissue self-renewal (Clevers, 2006; Johnson and Rajamannan, 2006; Kikuchi and Yamamoto, 2008; Maatouk *et al.*, 2008; Merkel *et al.*, 2007; Nemeth *et al.*, 2007; Schambony *et al.*, 2004). Defects in this signaling system result in a variety of developmental disorders, degenerative diseases and cancers (Clevers, 2006; Johnson and Rajamannan, 2006). Traditionally, vertebrate Wnt signaling is divided into at least three distinct signaling pathways comprising the canonical pathway leading to cell fate determination and proliferation (Chen *et al.*, 2007; Logan and Nusse, 2004) and the noncanonical or β-catenin-independent pathways, Wnt/Ca^{2+} and PCP. The noncanonical pathway controls cell polarization and migration during processes that rely on convergence extension movements, such as gastrulation and neurulation (Montcouquiol *et al.*, 2006; Veeman et al., 2003; Wallingford, 2006; Wang et al., 2006). In broad strokes, all of these pathways are initiated by binding of a ligand of the Wnt family to a 7TM Frizzled receptor in the presence or absence of the co-receptors Lrp5/6, knypek, Ryk, or Ror2 (Gordon and Nusse, 2006; Heisenberg and Tada, 2002). Yet this representation might be too simplistic due to extensive crosstalk and, accordingly, Wnt signaling was recently described as a network rather than comprising individual pathways (Kestler and Kuhl, 2008). Moreover, many PCP proteins equally participate in Wnt/Ca^{2+} signaling (Veeman et al., 2003), hence the traditional distinction between canonical and noncanonical Wnt signaling may be the most suitable.

In canonical Wnt signaling, dishevelled (Dvl) stabilizes β-catenin by inhibiting its degradation by Axin, Adenomatous poliposis coli (APC), GSK-3β and Casein kinase I (CKI), thus permitting translocation of β-catenin to the nucleus. Here, it serves as a transcriptional co-activator of genes implicated in cell cycle progression, proliferation, cell fate determination, differentiation and regulation of embryonic development (Chen et al., 2007; Logan and Nusse, 2004; Miller et al., 1999; Vlad et al., 2008). Whereas both membrane-bound and cytoplasmic Dvl can participate in canonical Wnt signaling, only membrane-associated Dvl activates the noncanonical pathways, as has been demonstrated in Xenopus laevis embryos (Park et al., 2005). The microtubule (MT)-interacting and ciliary protein Inversin (Fig. 10.3C), which interacts with the anaphase promoting complex/Cyclosome (APC/C) (Morgan et al., 2002; Nurnberger et al., 2004) cooperates with Nephrocystin-3 to target cytoplasmic Dvl for proteasomal degradation, hereby favoring β-catenin-independent Wnt signaling (Bergmann et al., 2008; Simons et al., 2005). The noncanonical Wnt signaling proceeds through several effectors including PKC, Ca^{2+}/calmodulin-dependent kinase II (CamKII) and the Rho GTPases with downstream effectors to induce remodeling of the actin and MT cytoskeleton during the polarization of cells and tissues preceding cell migration and CE movements (Kikuchi and Yamamoto, 2008; Veeman et al., 2003). For instance, membrane-associated Dvl has been shown to modulate MT dynamics in fibroblasts and neurons by activating c-Jun N-terminal kinase (JNK) and concomitantly inactivate GSK-3β (Ciani and Salinas, 2007, Habas et al., 2003, Moriguchi et al., 1999). Further, Dvl can induce stress fibers in fibroblasts via the actin-regulating proteins Daam1 and Profilin (Habas et al., 2001; Sato et al., 2006) that are critical in cell migration.

5.2. Wnt signaling and the primary cilium

The primary cilium probably plays number of roles in regulating both canonical and noncanonical Wnt signaling in cell cycle regulation and migration. A series of signaling molecules in both pathways have been shown to localize to the cilium/centrosome axis. These include Vangl2

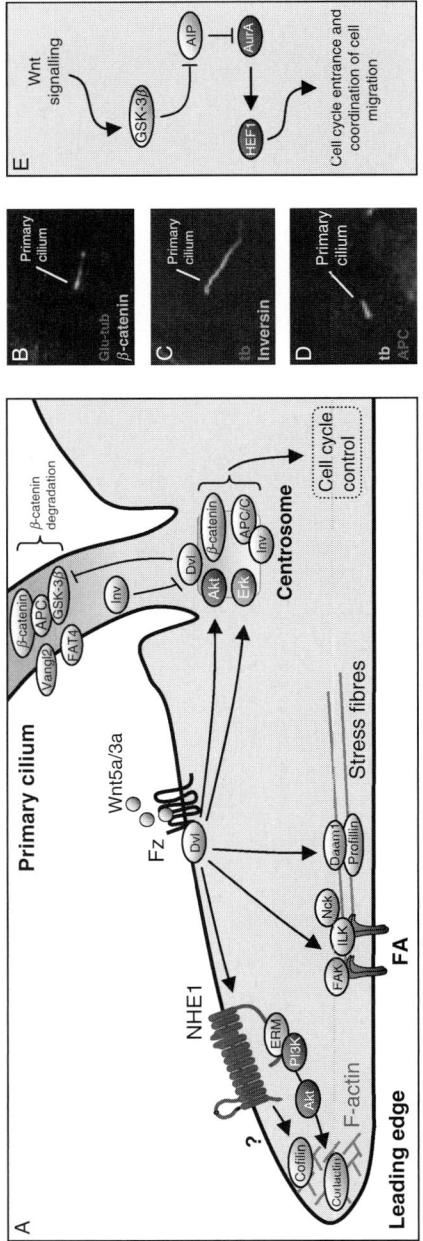

Figure 10.3 *Cell migration and cell cycle control regulated by Wnt signaling.* (A) Hypothesis on some of the Wnt signaling components at the plasma membrane and in the ciliary/centrosome axis that coordinate cell migration and cell cycle control. Ciliary Inversin impairs the canonical Wnt pathway by restricting Dvl-mediated impairment of β-catenin degradation by Adenomatous poliposis coli (ACP) and glycogen synthase kinase 3 beta (GSK-3β) in the primary cilium and potentiates degradation Wnt signaling through membrane-bound dishevelled (Dvl), which is activated by Wnt5a binding to Frizzled (Fz) 3. In turn, Dvl activates NHE1 and F-actin reorganization at the leading edge in addition to stress fiber formation and focal adhesion (FA) turnover. Dvl also mediates Wnt5a-induced activations of integrins via the PI3K-Akt pathway, in addition to Src- and PKC-mediated activation of focal adhesion kinase (FAK). Besides β-catenin-mediated transcription of genes in cell proliferation Wnt3a activates the Raf-Mek1/2-Erk1/2 pathway at the centrosome for cell cycle progression. Moreover, Inversin interacts with Anaphase Promoting Complex/Cyclosome (APC/C), which controls mitosis. (B–D) IF localization of β-catenin (green) (Corbit *et al.*, 2008; with permission) (B), Inversin (green) (Courtesy Prof. J. Goodship) (C) and APC (red) (Corbit *et al.*, 2008; with permission) (D) to the primary cilium stained with either detyrosinated tubulin (Glu-tub) or acetylated α-tubulin (tb). (E) Our hypothesis on the potential overlap between Wnt signaling and AurA-induced cell cycle control and migration. In addition to its role in β-catenin degradation complex, ciliary GSK-3β releases the inhibition of centrosomal AurA through AurA-Interacting protein (AIP), resulting in activation of human enhancer of filamentation 1 (HEF1). (See Color Insert.)

(Ross et al., 2005), Inversin (Morgan et al., 2002; Watanabe et al., 2003), Dvl, Inturned, Fuzzy (Park et al., 2006, 2008), APC and β-Catenin (Corbit et al., 2008), GSK-3β (Corbit et al., 2008; Thoma et al., 2007; Wilson and Lefebvre, 2004) and the protocadherin, Fat4 (Saburi et al., 2008). Surprisingly, Wnt signaling receptors of the Frizzled family have so far not been reported to localize to the cilium, which is in sharp contrast to receptors of, e.g., the PDGFR$\alpha\alpha$ and Hh signaling pathways that are activated by binding of their ligands within the primary cilium (Rohatgi et al., 2007; Schneider et al., 2005). In Wnt signaling, the primary cilium itself may not exclusively work as a sensory organelle in receptor activation but perhaps rather as a unique microdomain that coordinates the downstream molecules in canonical and noncanonical Wnt signaling, although we still know little as to the localization of Wnt receptors in mammalian cells. Indeed, it was proposed that Inversin controls the Dvl-dependent switch between the canonical and noncanonical Wnt pathways by selective degradation of cytosolic Dvl to inhibit the canonical and promote the noncanonical pathway in renal development (Simons et al., 2005). Further, defects in maintenance of the primary cilium and lack of normal signaling in the cilium tip the balance in favor of the canonical pathway followed by uncontrolled cell proliferation and differentiation (Cano et al., 2004; Corbit et al., 2008; Gerdes et al., 2007; Lin et al., 2003).

Many of the additional downstream signaling components in both the canonical and noncanonical Wnt pathways are identical to the effector molecules in ciliary PDGFR$\alpha\alpha$ and ECM signaling, including Mek1/2-Erk1/2, PI3K-Akt, and PKC. Both Wnt3a and Wnt5a can stimulate PI3K-mediated activation of Akt (Kawasaki et al., 2007; Kim et al., 2007) and Wnt3a activates the Raf1-Mek1/2-Erk1/2 pathway in, e.g., fibroblasts (Kawasaki et al., 2007; Yun et al., 2005). This implies that there is an extensive crosstalk between Wnt, PDGFR$\alpha\alpha$ and ECM signaling to regulate cell cycle control and cell migration (Christensen et al., 2007). In terms of cell cycle control, a link between Wnt and PDGFR$\alpha\alpha$ signaling is further indicated by regulation of AurA as outlined in Figs. 10.1 and 10.3. The canonical pathway partly regulates cell proliferation via transcriptional activation of genes involved in cell cycle progression and prevention of apoptosis (Vlad et al., 2008) and GSK-3β promotes mitosis as it binds and restricts the AurA interacting protein, AIP (Fumoto et al., 2008), thus preventing AIP-mediated downregulation of AurA levels (Kiat et al., 2002; Lim and Gopalan, 2007).

In terms of cell migration, Dvl may directly activate NHE1 (Putney et al., 2002), indicating that directional cell migration is highly coordinated by signaling events regulated by both the primary cilium via, e.g., PDGFRα and receptors in the plasma membrane. In turn these events set off a series of downstream signaling pathways, including the PI3K-Akt pathway, that partly interact with one another to control cytoskeletal rearrangements associated with cell migration. Noncanonical Wnt signaling affects various events in the process of cell migration, including filopodia formation, membrane ruffling, stress fiber persistence and FA turnover, as outlined in Fig. 10.3 (Ciani and Salinas, 2007; Kurayoshi et al., 2006). Most prominently, Wnt5a has been demonstrated to activate integrins via the PI3K-Akt pathway in fibroblasts (Kawasaki et al., 2007) and FAK, probably via Src and PKC, in gastric cancer cells (Kurayoshi et al., 2006). Integrins and cadherins are both important modulators and effectors in Wnt signaling and several vertebrate protocadherins are directly implicated in PCP (Nelson and Nusse, 2004; Saburi et al., 2008; Schambony et al., 2004). In turn, ILK stabilizes β-catenin at multiple levels (Schambony et al., 2004), including a downregulation of E-cadherins whereby junction-associated β-catenin is released to the cytosol (Oloumi et al., 2004).

A final link between the Wnt pathways and signaling through PDGFR$\alpha\alpha$ and ECM in the primary cilium comes from studies on orientation of the centrosome during cell migration and regulatory proteins in formation of the primary cilium. The recruitment of APC to MT plus ends is mandatory for centrosome reorientation in wound healing and requires the inactivation of GSK-3β by the Par6-aPKCζ complex in neurons (Etienne-Manneville and Hall, 2003; Etienne-Manneville et al., 2005). Wnt5a has been found to activate aPKCζ via Dvl during neuronal polarization (Zhang et al., 2007). In fibroblasts the process of centrosome reorientation involves the interplay between Wnt5a, Axin and Dvl, of which the latter interacts directly with aPKCζ in a Cdc42-dependent manner (Schlessinger et al., 2007). These data imply a Dvl-mediated interplay between the pathways controlling apicobasal and PCP, and there is reason to speculate that the primary cilium may be the uniting factor in this aspect, since the Par3–Par6–aPKCζ complex has been localized to primary cilia with substantial support of their implication in ciliogenesis (reviewed by Pedersen et al., 2008). As Cdc42-mediated inactivation of GSK-3β through the (Par3-)Par6–aPKCζ complex is also involved in β-catenin stabilization (Etienne-Manneville and Hall, 2003, Etienne-Manneville et al., 2005, Wu et al., 2006), it is possible that via the primary cilium aPKCζ is implicated in regulating the β-catenin degradation complex and canonical Wnt signaling. Consequently, defects in the coordination of Wnt signaling and apicobasal polarization may have dire consequences for PDGFR$\alpha\alpha$ and ECM signaling, since these signaling pathways rely on the correct formation of the cilium in which PDGFRα and ECM receptors are activated.

6. Conclusions and Perspectives

The primary cilium is a sensory organelle that coordinates many different signaling events during development and in tissue homeostasis. The cilium/centrosome axis is subjected to substantial trafficking and activation of receptors, ion channels, effector molecules, and transcription factors in and out of the primary cillim in response to changes in the extracellular environment to control pathways in cell cycle entrance, migration, polarity, differentiation, and apoptosis. In this scenario, the centrosome functions as a major docking station with multiple functions, while the cilium acts as specialized cellular compartment to concentrate and sequester signaling receptor and effector molecules. One function of the centrosome is to coordinate the assembly of the primary cilium, which is a prerequisite for ciliary signaling, and to deliver, receive, and organize the trafficking and turnover of signaling molecules in the cilium. In cell cycle regulation, signaling molecules from the cilium impinge on centrosomal constituents to monitor cell cycle progression, causing disassembly of the cilium in concert with events that control duplication of the centrosome and formation of the mitotic spindle during cell division. In addition, the dynamic exchange of different signaling molecules to/from the cilium is part of the process that controls migration, polarity and differential processes in determining cell fate and function.

There is still much to be learned about signaling pathways in the primary cilium and how these impinge on human physiology and pathophysiology. However, the available data clearly indicate the importance of different signal transduction pathways, including PDGFRα, integrin-ECM signaling, Hh signaling, TRP ion channels, and Wnt signaling in the cilium/centrosome signaling axis. We are just beginning to understand two fundamental questions: (1) how are diverse signaling pathways individually coordinated to control diverse cellular processes and (2) how do events initiated by ciliary signaling control cellular organization at a distance, for example, at the leading edge of a migrating cell. The first question is particularly relevant for those receptors which when activated in the primary cilium can promote quite different responses, such as in PDGFRαα signaling. We have indicated that this signaling coordinates both cell cycle entrance, in which the cilium is resorbed, and directional cell migration, in which the cilium is preserved, to play a role in cellular orientation and chemotaxis. Probably, the "decision" of a cell as to whether to migrate or to enter the cell cycle is dependent on the precise network of events operating at a given time point and in concert with other signaling pathways, such as Wnt and ECM signaling. As outlined in this review these other pathways crosstalk with PDGFRαα signaling and the fine-tuned balance between the activity, expression, concentration, and localization of the relevant molecular players from these systems as they interact at the centrosome with centrosomal proteins will determine the outcome. AurA, HEF1, and CHFR, all of which may regulate both cell migration and cell cycle entrance, are relevant centrosomal proteins for this process.

With respect to the second question, we have some specific information regarding PDGFRα signaling, and at least a hint as to how this signaling system could act to organize the actin cytoskeleton and the membrane at the leading edge of a cell in a gradient of PDGF-AA. Prior to migration in a gradient or after wounding, the MTOC and the primary cilium become positioned toward the leading edge in front of the nucleus. This spatial organization in fibroblasts with primary cilia moves the cilium to point in the direction of movement. The reorientation suggests that the cilium detects the spatial gradient of growth factor necessary to initiate migration and reasses it moment by moment while the cell is moving—that is that the cilium has a true GPS-like function in this system. In this way the gradient of growth factor acting on its ciliary receptor induces chemotaxis. Chemotaxis and movement depend in part on the activation of NHE1 at the leading edge of the cell. Activation of Akt at the ciliary basal body of the centrosome is a principal effect of PDGF-AA binding to its ciliary receptor. PI3K-Akt signaling is one of the factors that affect both NHE1 activity and MT stabilization at the leading edge. In turn, NHE1 activity, and perhaps MT stability, can affect the organization of the actin cytoskeleton and the generation of leading edge lamellipodia. So at least one pathway from ciliary receptor to cytoskeleton can be outlined, although the detailed mechanisms remain to be elucidated. This type of relay of information to control cytoskeletal positioning and membrane microdomain differentiation could be widely used in development, perhaps with different ciliary receptors and slightly different signaling pathways coupling the gradient detection with cytoskeletal response.

ACKNOWLEDGMENTS

This work was supported by the Lundbeck Foundation (STC), the Danish Science Research Council (STC, SFP), The Carlsberg Foundation (SFP), grants from NIDDK (PS), a Novo-Nordic Scholarship (IRV) and funds from the Department of Biology, University of Copenhagen (LS).

REFERENCES

Adolphe, C., Hetherington, R., Ellis, T., and Wainwright, B. (2006). Patched1 functions as a gatekeeper by promoting cell cycle progression. *Cancer Res.* **66,** 2081–2088.

Anjum, R., and Blenis, J. (2008). The RSK family of kinases: Emerging roles in cellular signalling. *Nat. Rev. Mol. Cell Biol.* **9,** 747–758.

Albrecht-Buehler, G. (1977). Phagokinetic tracks of 3T3 cells: Parallels between the orientation of track segments and of cellular structures which contain actin or tubulin. *Cell* **12,** 333–339.

Albrecht-Buehler, G., and Bushnell, A. (1979). The orientation of centrioles in migrating 3T3 cells. *Exp. Cell Res.* **120,** 111–118.

Andrae, J., Gallini, R., and Betsholtz, C. (2008). Role of platelet-derived growth factors in physiology and medicine. *Genes Dev.* **22,** 1276–1312.

Anton, I. M., Saville, S. P., Byrne, M. J., Curcio, C., Ramesh, N., Hartwig, J. H., and Geha, R. S. (2003). WIP participates in actin reorganization and ruffle formation induced by PDGF. *J. Cell Sci.* **116,** 2443–2451.

Arboleda, M. J., Lyons, J. F., Kabbinavar, F. F., Bray, M. R., Snow, B. E., Ayala, R., Danino, M., Karlan, B. Y., and Slamon, D. J. (2003). Overexpression of AKT2/protein kinase Bbeta leads to up-regulation of beta1 integrins, increased invasion, and metastasis of human breast and ovarian cancer cells. *Cancer Res.* **63,** 196–206.

Awan, A., Oliveri, R. S., Jensen, P. L., Christensen, S. T., and Andersen, C. Y. (2008). Characterization of Human Embryonic Stem Cells (hESCs) grown under feeder-free conditions. In "Methods in Molecular Biology" (K. Turksen, ed.). Humana Press, Totowa (*In press*).

Baron, W., Shattil, S. J., and ffrench-Constant, C. (2002). The oligodendrocyte precursor mitogen PDGF stimulates proliferation by activation of alpha(v)beta3 integrins. *EMBO J.* **21,** 1957–1966.

Begbie, J. (2008). Migration of neuroblasts from neurogenic placodes. *Dev. Neurosci.* **30,** 33–35.

Benzing, T., and Walz, G. (2006). Cilium-generated signaling: A cellular GPS? *Curr. Opin. Nephrol. Hypertens* **15,** 245–249.

Bergmann, C., Fliegauf, M., Bruchle, N. O., Frank, V., Olbrich, H., Kirschner, J., Schermer, B., Schmedding, I., Kispert, A., Kranzlin, B., Nurnberg, G., Becker, C., et al. (2008). Loss of nephrocystin-3 function can cause embryonic lethality, Meckel-Gruber-like syndrome, *situs inversus*, and renal-hepatic-pancreatic dysplasia. *Am. J. Hum. Genet.* **82,** 959–970.

Bergsten, E., Uutela, M., Li, X., Pietras, K., Ostman, A., Heldin, C. H., Alitalo, K., and Eriksson, U. (2001). PDGF-D is a specific, protease-activated ligand for the PDGF beta-receptor. *Nat. Cell Biol.* **3,** 512–516.

Blacque, O. E., and Leroux, M. R. (2006). Bardet–Biedl syndrome: An emerging pathomechanism of intracellular transport. *Cell Mol. Life Sci.* **18,** 2145–2161.

Blacque, O. E., Cevik, S., and Kaplan, O. I. (2008). Intraflagellar transport: from molecular characterisation to mechanism. *Front Biosci.* **13,** 2633–2652.

Bornfeldt, K. E., Raines, E. W., Graves, L. M., Skinner, M. P., Krebs, E. G., and Ross, R. (1995). Platelet-derived growth factor. Distinct signal transduction pathways associated with migration versus proliferation. *Ann. NY Acad. Sci.* **766,** 416–430.

Bradley, B. A., and Quarmby, L. M. (2005). A NIMA-related kinase, Cnk2p, regulates both flagellar length and cell size in Chlamydomonas. *J. Cell Sci.* **118,** 3317–3326.

Bresnahan, W. A., Boldogh, I., Ma, T., Albrecht, T., and Thompson, E. A. (1996). Cyclin E/Cdk2 activity is controlled by different mechanisms in the G0 and G1 phases of the cell cycle. *Cell Growth Differ.* **7,** 1283–1290.

Brockmann, M. A., Ulbricht, U., Gruner, K., Fillbrandt, R., Westphal, M., and Lamszus, K. (2003). Glioblastoma and cerebral microvascular endothelial cell migration in response to tumor-associated growth factors. *Neurosurgery* **52,** 1391–1399.

Cahalan, M. D., and Gutman, G. A. (2006). The sense of place in the immune system. *Nat. Immunol.* **7,** 329–332.

Cano, D. A., Murcia, N. S., Pazour, G. J., and Hebrok, M. (2004). Orpk mouse model of polycystic kidney disease reveals essential role of primary cilia in pancreatic tissue organization. *Development* **131,** 3457–3467.

Cardone, R. A., Bagorda, A., Bellizzi, A., Busco, G., Guerra, L., Paradiso, A., Casavola, V., Zaccolo, M., and Reshkin, S. J. (2005). Protein kinase A gating of a pseudopodial-located RhoA/ROCK/p38/NHE1 signal module regulates invasion in breast cancer cell lines. *Mol. Biol. Cell* **16,** 3117–3127.

Carloni, V., Pinzani, M., Giusti, S., Romanelli, R. G., Parola, M., Bellomo, G., Failli, P., Hamilton, A. D., Sebti, S. M., Laffi, G., and Gentilini, P. (2000). Tyrosine phosphorylation of focal adhesion kinase by PDGF is dependent on ras in human hepatic stellate cells. *Hepatology* **31,** 131–140.

Carloni, V., Defranco, R. M., Caligiuri, A., Gentilini, A., Sciammetta, S. C., Baldi, E., Lottini, B., Gentilini, P., and Pinzani, M. (2002). Cell adhesion regulates platelet-derived growth factor-induced MAP kinase and PI-3 kinase activation in stellate cells. *Hepatology* **36,** 582–591.

Carvalho, I., Milanezi, F., Martins, A., Reis, R. M., and Schmitt, F. (2005). Overexpression of platelet-derived growth factor receptor alpha in breast cancer is associated with tumour progression. *Breast Cancer Res.* **7,** R788–R795.

Caspary, T., Larkins, C. E., and Anderson, K. V. (2007). The graded response to Sonic Hedgehog depends on cilia architecture. *Dev. Cell* **12,** 767–778.

Caswell, P., and Norman, J. (2008). Endocytic transport of integrins during cell migration and invasion. *Trends. Cell Biol.* **18,** 257–263.

Chen, H. C., and Guan, J. L. (1994). Stimulation of phosphatidylinositol 3′-kinase association with foca adhesion kinase by platelet-derived growth factor. *J. Biol. Chem.* **269,** 31229–31233.

Chen, Z., Indjeian, V. B., McManus, M., Wang, L., and Dynlacht, B. D. (2002). CP110, a cell cycle-dependent CDK substrate, regulates centrosome duplication in human cells. *Dev. Cell* **3,** 339–350.

Chen, S., McLean, S., Carter, D. E., and Leask, A. (2007). The gene expression profile induced by Wnt 3a in NIH 3T3 fibroblasts. *J. Cell Commun. Signal.* **1,** 175–183.

Chiang, Y., Chou, C. Y., Hsu, K. F., Huang, Y. F., and Shen, M. R. (2008). EGF upregulates Na + /H + exchanger NHE1 by post-translational regulation that is important for cervical cancer cell invasiveness. *J. Cell Physiol.* **214,** 810–819.

Christensen, S. T., and Ott, C. M. (2007). Cell signaling. A ciliary signaling switch. *Science* **317,** 330–331.

Christensen, S. T., Schneider, L., Clement, C. A., Pazour, G., Hoffmann, E. K., and Satir, P. (2006). The primary cilium is a sensory organelle that regulates growth control and tissue homeostasis. *FASEB J.* **20,** A437

Christensen, S. T., Pedersen, L. B., Schneider, L., and Satir, P. (2007). Sensory cilia and integration of signal transduction in human health and disease. *Traffic* **8,** 97–109.

Ciani, L., and Salinas, P. C. (2007). c-Jun N-terminal kinase (JNK) cooperates with Gsk3beta to regulate Dishevelled-mediated microtubule stability. *BMC Cell Biol.* **8,** 27.

Clark, R. A., Nielsen, L. D., Welch, M. P., and McPherson, J. M. (1995). Collagen matrices attenuate the collagen-synthetic response of cultured fibroblasts to TGF-beta. *J. Cell Sci.* **108,** (Pt 3), 1251–1261.

Clevers, H. (2006). Wnt/beta-catenin signaling in development and disease. *Cell* **127,** 469–480.

Corbit, K. C., Shyer, A. E., Dowdle, W. E., Gaulden. J., Singla, V., Chen, M. H., Chuang, P. T., and Reiter, J. F. (2008). Kif3a constrains beta-catenin-dependent Wnt signalling through dual ciliary and non-ciliary mechanisms. *Nat. Cell Biol.* **10,** 70–76.

Danilov, A. I., Leal, W. G., Ahlenius, H., Kokaia, Z., Carlemalm, E., and Lindvall, O. (2008). Ultrastructural and antigenic properties of neural stem cells and their progeny in adult rat subventricular zone. Aug 15. *Glia* [Epub ahead of print].

De Calisto, C. J., Araya, C., Marchant, L., Riaz, C. F., and Mayor, R. (2005). Essential role of non-canonical Wnt signalling in neural crest migration. *Development* **132,** 2587–2597.

Dell'albani, P. (2008). Stem cell markers in Gliomas. *Neurochem. Res.* May 21. [Epub ahead of print].

Denker, S. P., and Barber, D. L. (2002). Cell migration requires both ion translocation and cytoskeletal anchoring by the Na-H exchanger NHE1. *J. Cell Biol.* **159,** 1087–1096.

Denker, S. P., Huang, D. C., Orlowski, J., Furthmayr, H., and Barber, D. L. (2000). Direct binding of the Na–H exchanger NHE1 to ERM proteins regulates the cortical cytoskeleton and cell shape independently of H(+) translocation. *Mol. Cell* **6,** 1425–1436.

abnormalities in $Tg737^{orpk}$ mice lacking the primary cilia protein polaris. *Matrix Biol.* **26**, 234–246.

McGlashan, S. R., Cluett, E. C., Jensen, C. G., and Poole, C. A. (2008). Primary cilia in osteoarthritic chondrocytes: From chondrons to clusters. *Dev. Dyn.* **237**, 2013–2020.

McKinnon, R. D., Waldron, S., and Kiel, M. E. (2005). PDGF alpha-receptor signal strength controls an RTK rheostat that integrates phosphoinositol 3′-kinase and phospholipase Cgamma pathways during oligodendrocyte maturation. *J. Neurosci.* **25**, 3499–3508.

Merkel, C. E., Karner, C. M., and Carroll, T. J. (2007). Molecular regulation of kidney development: Is the answer blowing in the Wnt? *Pediatr. Nephrol.* **22**, 1825–1838.

Michaud, E. J., and Yoder, B. K. (2006). The primary cilium in cell signaling and cancer. *Cancer Res.* **66**, 6463–6467.

Miller, J. R., Hocking, A. M., Brown, J. D., and Moon, R. T. (1999). Mechanism and function of signal transduction by the Wnt/beta-catenin and Wnt/Ca2+ pathways. *Oncogene* **18**, 7860–7872.

Mitchison, T. J., Charras, G. T., and Mahadevan, L. (2008). Implications of a poroelastic cytoplasm for the dynamics of animal cell shape. *Semin. Cell Dev. Biol.* **19**, 215–223.

Montcouquiol, M., Crenshaw, E. B., III, and Kelley, M. W. (2006). Noncanonical Wnt signaling and neural polarity. *Annu. Rev. Neurosci.* **29**, 363–386.

Morgan, D., Eley, L., Sayer, J., Strachan, T., Yates, L. M., Craighead, A. S., and Goodship, J. A. (2002). Expression analyses and interaction with the anaphase promoting complex protein Apc2 suggest a role for inversin in primary cilia and involvement in the cell cycle. *Hum. Mol. Genet.* **11**, 3345–3350.

Moriguchi, T., Kawachi, K., Kamakura, S., Masuyama, N., Yamanaka, H., Matsumoto, K., Kikuchi, A., and Nishida, E. (1999). Distinct domains of mouse dishevelled are responsible for the c-Jun N-terminal kinase/stress-activated protein kinase activation and the axis formation in vertebrates. *J. Biol. Chem.* **274**, 30957–30962.

Nachury, M. V., Loktev, A. V., Zhang, Q., Westlake, C. J., Peranen, J., Merdes, A., Slusarski, D. C., Scheller, R. H., Bazan, J. F., Sheffield, V. C., and Jackson, P. K. (2007). A core complex of BBS proteins cooperates with the GTPase Rab8 to promote ciliary membrane biogenesis. *Cell* **129**, 1201–1213.

Nechyporuk-Zloy, V., Dieterich, P., Oberleithner, H., Stock, C., and Schwab, A. (2008). Dynamics of single potassium channel proteins in the plasma membrane of migrating cells. *Am. J. Physiol. Cell Physiol.* **294**, C1096–C1102.

Natarajan, M., Stewart, J. E., Golemis, E. A., Pugacheva, E. N., Alexandropoulos, K., Cox, B. D., Wang, W., Grammer, J. R., and Gladson, C. L. (2006). HEF1 is a necessary and specific downstream effector of FAK that promotes the migration of glioblastoma cells. *Oncogene.* **25**, 1721–1732.

Nelson, W. J., and Nusse, R. (2004). Convergence of Wnt, beta-catenin, and cadherin pathways. *Science* **303**, 1483–1487.

Nemeth, M. J., Topol, L., Anderson, S. M., Yang, Y., and Bodine, D. M. (2007). Wnt5a inhibits canonical Wnt signaling in hematopoietic stem cells and enhances repopulation. *Proc. Natl. Acad. Sci. USA* **104**, 15436–15441.

Nielsen, S. K., Mollgard, K., Clement, C. A., Veland, I. R., Awan, A., Yoder, B. K., Novak, I., and Christensen, S. T. (2008). Characterization of primary cilia and Hedgehog signaling during development of the human pancreas and in human pancreatic duct cancer cell lines. *Dev. Dyn.* **237**, 2039–2052.

Nishio, M., Watanabe, K., Sasaki, J., Taya, C., Takasuga, S., Iizuka, R., Balla, T., Yamazaki, M., Watanabe, H., Itoh, R., Kuroda, S., Horie, Y., *et al.* (2007). Control of cell polarity and motility by the PtdIns(3,4,5)P3 phosphatase SHIP1. *Nat. Cell Biol.* **9**, 36–44.

Nurnberger, J., Bacallao, R. L., and Phillips, C. L. (2002). Inversin forms a complex with catenins and N-cadherin in polarized epithelial cells. *Mol. Biol. Cell* **13**, 3096–3106.

Nurnberger, J., Kribben, A., Opazo, S. A., Heusch, G., Philipp, T., and Phillips, C. L. (2004). The Invs gene encodes a microtubule-associated protein. *J. Am. Soc. Nephrol.* **15**, 1700–1710.

O'Connell, M. J., Krien, M. J., and Hunter, T. (2003). Never say never. The NIMA-related protein kinases in mitotic control. *Trends Cell Biol.* **13**, 221–228.

O'Hayre, M., Salanga, C. L., Handel, T. M., and Allen, S. J. (2008). Chemokines and cancer: Migration, intracellular signalling and intercellular communication in the microenvironment. *Biochem. J.* **409**, 635–649.

Oda-Sato, E., and Tanaka, N. (2007). Abnormal centrosome amplification and Aurora-A activation in p53-deficient cells. *J. Nippon Med. Sch.* **74**, 384–385.

Oloumi, A., McPhee, T., and Dedhar, S. (2004). Regulation of E-cadherin expression and beta-catenin/Tcf transcriptional activity by the integrin-linked kinase. *Biochim. Biophys. Acta.* **1691**, 1–15.

Omori, Y., Zhao, C., Saras, A., Mukhopadhyay, S., Kim, W., Furukawa, T., Sengupta, P., Veraksa, A., and Malicki, J. (2008). Elipsa is an early determinant of ciliogenesis that link the IFT particle to membrane-associated small GTPase Rab8. *Nat. Cell Biol.* **10**, 437–444.

Onishi, K., Higuchi, M., Asakura, T., Masuyama, N., and Gotoh, Y. (2007). The PI3K-Akt pathway promotes microtubule stabilization in migrating fibroblasts. *Genes Cells.* **12**, 535–546.

Otto, E. A., Trapp, M. L., Schultheiss, U. T., Helou, J., Quarmby, L. M., and Hildebrandt, F. (2008). NEK8 mutations affect ciliary and centrosomal localization and may cause nephronophthisis. *J. Am. Soc. Nephrol.* **19**, 587–592.

Pan, J., and Snell, W. J. (2003). Kinesin II and regulated intraflagellar transport of Chlamydomonas aurora protein kinase. *J. Cell Sci.* **116**, 2179–2186.

Parent, C. A., and Devreotes, P. N. (1999). A cell's sense of direction. *Science* **284**, 765–770.

Park, T. J., Gray, R. S., Sato, A., Habas, R., and Wallingford, J. B. (2005). Subcellular localization and signaling properties of dishevelled in developing vertebrate embryos. *Curr. Biol.* **15**, 1039–1044.

Park, T. J., Haigo, S. L., and Wallingford, J. B. (2006). Ciliogenesis defects in embryos lacking inturned or fuzzy function are associated with failure of planar cell polarity and Hedgehog signaling. *Nat. Genet.* **38**, 303–311.

Park, T. J., Mitchell, B. J., Abitua, P. B., Kintner, C., and Wallingford, J. B. (2008). Dishevelled controls apical docking and planar polarization of basal bodies in ciliated epithelial cells. *Nat. Genet.* **40**, 871–879.

Parsons, J. T. (2003). Focal adhesion kinase: the first ten years. *J. Cell Sci.* **116**, 1409–1416.

Patel, H., and Barber, D. L. (2005). A developmentally regulated Na-H exchanger in Dictyostelium discoideum is necessary for cell polarity during chemotaxis. *J. Cell Biol.* **169**, 321–329.

Pedersen, S. F. (2006). The Na(+)/H (+) exchanger NHE1 in stress-induced signal transduction: implications for cell proliferation and cell death. *Pflugers Arch.* **452**, 249–259.

Pazour, G. J., Dickert, B. L., Vucica, Y., Seeley, E. S., Rosenbaum, J. L., Witman, G. B., and Cole, D. G. (2000). Chlamydomonas IFT88 and its mouse homologue, polycystic kidney disease gene tg737, are required for assembly of cilia and flagella. *J. Cell Biol.* **151**, 709–718.

Pébay, A., Wong, R. C., Pitson, S. M., Wolvetang, E. J., Peh, G. S., Filipczyk, A., Koh, K.L., Tellis, I., Nguyen, L. T., and Pera, M. F. (2005). Essential roles of sphingosine-1-phosphate and platelet-derived growth factor in the maintenance of human embryonic stem cells. *Stem Cells.* **23**, 1541–1548.

Pedersen, L. B., Veland, I. R., Schroder, J. M., and Christensen, S. T. (2008). Assembly of primary cilia. *Dev. Dyn.* **237**, 1993–2006.

Plotnikova, O. V., Golemis, E. A., and Pugacheva, E. N. (2008). Cell cycle-dependent ciliogenesis and cancer. *Cancer Res.* **68**, 2058–2061.

Poole, C. A., Flint, M. H., and Beaumont, B. W. (1985). Analysis of the morphology and function of primary cilia in connective tissues: a cellular cybernetic probe? *Cell Motil.* **5,** 175–193.

Praetorius, H. A., Praetorius, J., Nielsen, S., Frokiaer, J., and Spring, K. R. (2004). Beta1-integrins in the primary cilium of MDCK cells potentiate fibronectin-induced Ca2 + signaling. *Am. J. Physiol Renal Physiol.* **287,** F969–F978.

Privette, L. M., and Petty, E. M. (2008). CHFR: A Novel Mitotic Checkpoint Protein and Regulator of Tumorigenesis. *Transl. Oncol.* **1,** 57–64.

Privette, L. M., Gonzalez, M. E., Ding, L., Kleer, C. G., and Petty, E. M. (2007). Altered expression of the early mitotic checkpoint protein, CHFR, in breast cancers: implications for tumor suppression. *Cancer Res.* **67,** 6064–6074.

Pugacheva, E. N., Jablonski, S. A., Hartman, T. R., Henske, E. P., and Golemis, E. A. (2007). HEF1-dependent Aurora A activation induces disassembly of the primary cilium. *Cell* **129,** 1351–1363.

Putney, L. K., Denker, S. P., and Barber, D. L. (2002). The changing face of the Na + /H + exchanger, NHE1: structure, regulation, and cellular actions. *Annu. Rev. Pharmacol. Toxicol.* **42,** 527–552.

Qin, H., Wang, Z., Diener, D., and Rosenbaum, J. (2007). Intraflagellar transport protein 27 is a small G protein involved in cell-cycle control. *Curr. Biol.* **17,** 193–202.

Raman, D., Baugher, P. J., Thu, Y. M., and Richmond, A. (2007). Role of chemokines in tumor growth. *Cancer Lett.* **256,** 137–165.

Rasmussen, M., Alexander, R. T., Darborg, B. V., Mobjerg, N., Hoffmann, E. K., Kapus, A., and Pedersen, S. F. (2008). Osmotic cell shrinkage activates ezrin/radixin/moesin (ERM) proteins: activation mechanisms and physiological implications. *Am. J. Physiol. Cell Physiol.* **294,** C197–C212.

Reis, R. M., Martins, A., Ribeiro, S. A., Basto, D., Longatto-Filho, A., Schmitt, F. C., and Lopes, J. M. (2005). Molecular characterization of PDGFR-alpha/PDGF-A and c-KIT/SCF in gliosarcomas. *Cell Oncol.* **27,** 319–326.

Reshkin, S. J., Bellizzi, A., Albarani, V., Guerra, L., Tommasino, M., Paradiso, A., and Casavola, V. (2000). Phosphoinositide 3-kinase is involved in the tumor-specific activation of human breast cancer cell Na(+)/H(+) exchange, motility, and invasion induced by serum deprivation. *J. Biol. Chem.* **275,** 5361–5369.

Rohatgi, R., Milenkovic, L., and Scott, M. P. (2007). Patched1 regulates hedgehog signaling at the primary cilium. *Science* **317,** 372–376.

Rosenbaum, J. L., and Witman, G. B. (2002). Intraflagellar transport. *Nat. Rev. Mol. Cell Biol.* **3,** 813–825.

Rosenkranz, S., DeMali, K. A., Gelderloos, J. A., Bazenet, C., and Kazlauskas, A. (1999). Identification of the receptor-associated signaling enzymes that are required for platelet-derived growth factor-AA-dependent chemotaxis and DNA synthesis. *J. Biol. Chem.* **274,** 28335–28343.

Ross, R. (1987). Platelet-derived growth factor. *Annu. Rev. Med.* **38,** 71–79.

Ross, A. J., May-Simera, H., Eichers, E. R., Kai, M., Hill, J., Jagger, D. J., Leitch, C. C., Chapple, J. P., Munro, P. M., Fisher, S., Tan, P. L., Phillips, H. M., *et al.* (2005). Disruption of Bardet–Biedl syndrome ciliary proteins perturbs planar cell polarity in vertebrates. *Nat. Genet.* **37,** 1135–1140.

Saburi, S., Hester, I., Fischer, E., Pontoglio, M., Eremina, V., Gessler, M., Quaggin, S. E., Harrison, R., Mount, R., and McNeill, H. (2008). Loss of Fat4 disrupts PCP signaling and oriented cell division and leads to cystic kidney disease. *Nat. Genet.* **40,** 1010–1015.

Santos, N., and Reiter, J. F. (2008). Building it up and taking it down: The regulation of vertebrate ciliogenesis. *Dev. Dyn.* **237,** 1972–1981.

Satir, P., and Christensen, S. T. (2007). Overview of structure and function of mammalian cilia. *Annu. Rev. Physiol.* **69,** 377–400.

Sato, A., Khadka, D. K., Liu, W., Bharti, R., Runnels, L. W., Dawid, I. B., and Habas, R. (2006). Profilin is an effector for Daam1 in non-canonical Wnt signaling and is required for vertebrate gastrulation. *Development* **133,** 4219–4231.

Schambony, A., Kunz, M., and Gradl, D. (2004). Cross-regulation of Wnt signaling and cell adhesion. *Differentiation* **72,** 307–318.

Schatten, H. (2008). The mammalian centrosome and its functional significance. *Histochem. Cell Biol.* **129,** 667–686.

Schermer, B., Ghenoiu, C., Bartram, M., Muller, R. U., Kotsis, F., Hohne, M., Kuhn, W., Rapka, M., Nitschke, R., Zentgraf, H., Fliegauf, M., Omran, H., et al. (2006). The von Hippel–Lindau tumor suppressor protein controls ciliogenesis by orienting microtubule growth. *J. Cell Biol.* **175,** 547–554.

Schlessinger, K., McManus, E. J., and Hall, A. (2007). Cdc42 and noncanonical Wnt signal transduction pathways cooperate to promote cell polarity. *J. Cell Biol.* **178,** 355–361.

Schneider, L., Clement, C. A., Teilmann, S. C., Pazour, G. J., Hoffmann, E. K., Satir, P., and Christensen, S. T. (2005). PDGFRαα signaling is regulated through the primary cilium in fibroblasts. *Curr. Biol.* **15,** 1861–1866.

Schneider, L., Cammer, M., Lehman, J., Nielsen, S. K., Guerra, C. F., Veland, I. R., Stock, C., Hoffmann, E. K., Yoder, B. K., Schwab, A., Satir, P., and Christensen, S. T. (2008a). Directional cell migration and chemotaxis in wound healing response to PDGF-AA are coordinated by the primary cilium in fibroblasts. *Submitted*.

Schneider, L., Stock, C., Dieterich, P., Satir, P., Schwab, A., Christensen, S. T., and Pedersen, S. P. (2008b). The Na^+/H^+ exchanger NHE1 plays a central role in directional migration stimulated via PDGFRα in the primary cilium. *Submitted*.

Schroder, J. M., Schneider, L., Christensen, S. T., and Pedersen, L. B. (2007). EB1 is required for primary cilia assembly in fibroblasts. *Curr. Biol.* **17,** 1134–1139.

Schwab, A. (2001a). Function and spatial distribution of ion channels and transporters in cell migration. *Am. J. Physiol. Renal. Physiol.* **280,** F739–F747.

Schwab, A. (2001b). Ion channels and transporters on the move. *News Physiol. Sci.* **16,** 29–33.

Schwab, A., Nechyporuk-Zloy, V., Fabian, A., and Stock, C. (2007). Cells move when ions and water flow. *Pflugers Arch.* **453,** 421–432.

Scolnick, D. M., and Halazonetis, T. D. (2000). Chfr defines a mitotic stress checkpoint that delays entry into metaphase. *Nature* **406,** 430–435.

Siegbahn, A., Hammacher, A., Westermark, B., and Heldin, C. H. (1990). Differential effects of the various isoforms of platelet-derived growth factor on chemotaxis of fibroblasts, monocytes, and granulocytes. *J. Clin. Invest.* **85,** 916–920.

Simons, M., Gloy, J., Ganner, A., Bullerkotte, A., Bashkurov, M., Kronig, C., Schermer, B., Benzing, T., Cabello, O. A., Jenny, A., Mlodzik, M., Polok, B., et al. (2005). Inversin, the gene product mutated in nephronophthisis type II, functions as a molecular switch between Wnt signaling pathways. *Nat. Genet.* **37,** 537–543.

Singer, A. J., and Clark, R. A. (1999). Cutaneous wound healing. *N. Engl. J. Med.* **341,** 738–746.

Soma, Y., Mizoguchi, M., Yamane, K., Yazawa, N., Kubo, M., Ihn, H., Kikuchi, K., and Tamaki, K. (2002). Specific inhibition of human skin fibroblast chemotaxis to platelet-derived growth factor A-chain homodimer by transforming growth factor-beta1. *Arch. Dermatol. Res.* **293,** 609–613.

Spektor, A., Tsang, W. Y., Khoo, D., and Dynlacht, B. D. (2007). Cep97 and CP110 suppress a cilia assembly program. *Cell* **130,** 678–690.

Srinivasan, S., Wang, F., Glavas, S., Ott, A., Hofmann, F., Aktories, K., Kalman, D., and Bourne, H. R. (2003). Rac and Cdc42 play distinct roles in regulating PI(3,4,5)P3 and polarity during neutrophil chemotaxis. *J. Cell Biol.* **160,** 375–385.

Srivastava, J., Barber, D. L., and Jacobson, M. P. (2007). Intracellular pH sensors: Design principles and functional significance. *Physiology (Bethesda.).* **22,** 30–39.

Stambolic, V., and Woodgett, J. R. (2006). Functional distinctions of protein kinase B/Akt isoforms defined by their influence on cell migration. *Trends Cell Biol.* **16,** 461–466.

Stock, C., Gassner, B., Hauck, C. R., Arnold, H., Mally, S., Eble, J. A., Dieterich, P., and Schwab, A. (2005). Migration of human melanoma cells depends on extracellular pH and Na + /H + exchange. *J. Physiol.* **567,** 225–238.

Stock, C., Cardone, R. A., Busco, G., Krahling, H., Schwab, A., and Reshkin, S. J. (2008). Protons extruded by NHE1: Digestive or glue? *Eur. J. Cell Biol.*

Stupack, D. G. (2007). The biology of integrins. *Oncology (Williston. Park).* **21,** 6–12.

Takahashi, E., Abe, J., Gallis, B., Aebersold, R., Spring, D. J., Krebs, E. G., and Berk, B. C. (1999). p90(RSK) is a serum-stimulated Na + /H + exchanger isoform-1 kinase. Regulatory phosphorylation of serine 703 of Na + /H + exchanger isoform-1. *J. Biol. Chem.* **274,** 20206–20214.

Thiery, J. P. (2002). Epithelial-mesenchymal transitions in tumour progression. *Nat. Rev. Cancer.* **2,** 442–454.

Thoma, C. R., Frew, I. J., Hoerner, C. R., Montani, M., Moch, H., and Krek, W. (2007). pVHL and GSK3beta are components of a primary cilium-maintenance signalling network. *Nat. Cell Biol.* **9,** 588–595.

Tobin, J. L., and Beales, P. L. (2007). Bardet–Biedl syndrome: beyond the cilium. *Pediatr. Nephrol.* **22,** 926–936.

Tobin, J. L., Di, F. M., Eichers, E., May-Simera, H., Garcia, M., Yan, J., Quinlan, R., Justice, M. J., Hennekam, R. C., Briscoe, J., Tada, M., Mayor, R., et al. (2008). Inhibition of neural crest migration underlies craniofacial dysmorphology and Hirschsprung's disease in Bardet–Biedl syndrome. *Proc. Natl. Acad. Sci. USA* **105,** 6714–6719.

Tominaga, T., and Barber, D. L. (1998). Na-H exchange acts downstream of RhoA to regulate integrin-induced cell adhesion and spreading. *Mol. Biol. Cell* **9,** 2287–2303.

Tucker, R. W., Pardee, A. B., and Fujiwara, K. (1979). Centriole ciliation is related to quiescence and DNA synthesis in 3T3 cells. *Cell* **17,** 527–535.

Uren, A., Yu, J. C., Gholami, N. S., Pierce, J. H., and Heidaran, M. A. (1994). The alpha PDGFR tyrosine kinase mediates locomotion of two different cell types through chemotaxis and chemokinesis. *Biochem. Biophys. Res. Commun.* **204,** 628–634.

van Seventer, G. A., Salmen, H. J., Law, S. F., O'Neill, G. M., Mullen, M. M., Franz, A. M., Kanner, S. B., Golemis, E. A., and van Seventer, J. M. (2001). Focal adhesion kinase regulates beta1 integrin-dependent T cell migration through an HEF1 effector pathway. *Eur. J. Immunol.* **31,** 1417–1427.

van Rossum, A. G., Moolenaar, W. H., and Schuuring, E. (2006). Cortactin affects cell migration by regulating intercellular adhesion and cell spreading. *Exp. Cell Res.* **312,** 1658–1670.

Veeman, M. T., Axelrod, J. D., and Moon, R. T. (2003). A second canon. Functions and mechanisms of beta-catenin-independent Wnt signaling. *Dev. Cell* **5,** 367–377.

Vlad, A., Rohrs, S., Klein-Hitpass, L., and Muller, O. (2008). The first five years of the Wnt targetome. *Cell Signal* **20,** 795–802.

Vu, T. H., and Werb, Z. (2000). Matrix metalloproteinases: effectors of development and normal physiology. *Genes Dev.* **14,** 2123–2133.

Wada, N., Javidan, Y., Nelson, S., Carney, T. J., Kelsh, R. N., and Schilling, T. F. (2005). Hedgehog signaling is required for cranial neural crest morphogenesis and chondrogenesis at the midline in the zebrafish skull. *Development* **132,** 3977–3988.

Wallingford, J. B. (2006). Planar cell polarity, ciliogenesis and neural tube defects. *Hum. Mol. Genet.* 15 Spec No. **2,** R227–R234.

Wang, J., Hamblet, N. S., Mark, S., Dickinson, M. E., Brinkman, B. C., Segil, N., Fraser, S. E., Chen, P., Wallingford, J. B., and Wynshaw-Boris, A. (2006). Dishevelled genes mediate a conserved mammalian PCP pathway to regulate convergent extension during neurulation. *Development* **133,** 1767–1778.

Wang, D., Hu, J., Bobulescu, I. A., Quill, T. A., McLeroy, P., Moe, O. W., and Garbers, D. L. (2007). A sperm-specific Na + /H + exchanger (sNHE) is critical for expression and in vivo bicarbonate regulation of the soluble adenylyl cyclase (sAC). *Proc. Natl. Acad. Sci. USA* **104,** 9325–9330.

Watanabe, D., Saijoh, Y., Nonaka, S., Sasaki, G., Ikawa, Y., Yokoyama, T., and Hamada, H. (2003). The left-right determinant Inversin is a component of node monocilia and other 9 + 0 cilia. *Development* **130,** 1725–1734.

Welch, H. C., Coadwell, W. J., Ellson, C. D., Ferguson, G. J., Andrews, S. R., Erdjument-Bromage, H., Tempst, P., Hawkins, P. T., and Stephens, L. R. (2002). P-Rex1, a PtdIns (3,4,5)P3- and Gbetagamma-regulated guanine-nucleotide exchange factor for Rac. *Cell* **108,** 809–821.

Wheatley, D. N. (1995). Primary cilia in normal and pathological tissues. *Pathobiology* **63,** 222–238.

Wheatley, D. N., Wang, A. M., and Strugnell, G. E. (1996). Expression of primary cilia in mammalian cells. *Cell Biol. Int.* **20,** 73–81.

White, M. C., and Quarmby, L. M. (2008). The NIMA-family kinase, Nek1 affects the stability of centrosomes and ciliogenesis. *BMC. Cell Biol.* **9,** 29.

Wu, K. L., Khan, S., Lakhe-Reddy, S., Jarad, G., Mukherjee, A., Obejero-Paz, C. A., Konieczkowski, M., Sedor, J. R., and Schelling, J. R. (2004). The NHE1 Na + /H + exchanger recruits ezrin/radixin/moesin proteins to regulate Akt-dependent cell survival. *J. Biol. Chem.* **279,** 26280–26286.

Wu, X., Quondamatteo, F., Lefever, T., Czuchra, A., Meyer, H., Chrostek, A., Paus, R., Langbein, L., and Brakebusch, C. (2006). Cdc42 controls progenitor cell differentiation and beta-catenin turnover in skin. *Genes Dev.* **20,** 571–585.

Xie, J. F., Stroumza, J., and Graves, D. T. (1994). IL-1 down-regulates platelet-derived growth factor-alpha receptor gene expression at the transcriptional level in human osteoblastic cells. *J. Immunol.* **153,** 378–383.

Xu, J., and Clark, R. A. (1996). Extracellular matrix alters PDGF regulation of fibroblast integrins. *J. Cell Biol.* **132,** 239–249.

Xue, J., Zhou, D., Yao, H., Gavrialov, O., McConnell, M. J., Gelb, B. D., and Haddad, G. G. (2007). Novel functional interaction between Na + /H + exchanger 1 and tyrosine phosphatase SHP-2. *Am. J. Physiol. Regul. Integr. Comp. Physiol.* **292,** R2406–R2416.

Yamaguchi, H., and Condeelis, J. (2007). Regulation of the actin cytoskeleton in cancer cell migration and invasion. *Biochim. Biophys. Acta.* **1773,** 642–652.

Yan, W., Nehrke, K., Choi, J., and Barber, D. L. (2001). The Nck-interacting kinase (NIK) phosphorylates the Na + -H + exchanger NHE1 and regulates NHE1 activation by platelet-derived growth factor. *J. Biol. Chem.* **276,** 31349–31356.

Yeh, Y. L., Kang, Y. M., Chaibi, M. S., Xie, J. F., and Graves, D. T. (1993). IL-1 and transforming growth factor-beta inhibit platelet-derived growth factor-AA binding to osteoblastic cells by reducing platelet-derived growth factor-alpha receptor expression. *J. Immunol.* **150,** 5625–5632.

Yoder, B. K. (2007). Role of primary cilia in the pathogenesis of polycystic kidney disease. *J. Am. Soc. Nephrol.* **18,** 1381–1388.

Yokote, K., Mori, S., Siegbahn, A., Ronnstrand, L., Wernstedt, C., Heldin, C. H., and Claesson-Welsh, L. (1996). Structural determinants in the platelet-derived growth factor alpha-receptor implicated in modulation of chemotaxis. *J. Biol. Chem.* **271,** 5101–5111.

Yokote, K., Hellman, U., Ekman, S., Saito, Y., Ronnstrand, L., Saito, Y., Heldin, C. H., and Mori, S. (1998). Identification of Tyr-762 in the platelet-derived growth factor alpha-receptor as the binding site for Crk proteins. *Oncogene.* **16,** 1229–1239.

Yoshimura, S., Egerer, J., Fuchs, E., Haas, A. K., and Barr, F. A. (2007). Functional dissection of Rab GTPases involved in primary cilium formation. *J. Cell Biol.* **178,** 363–369.

Yu, J., Moon, A., and Kim, H. R. (2001). Both platelet-derived growth factor receptor (PDGFR)-alpha and PDGFR-beta promote murine fibroblast cell migration. *Biochem. Biophys. Res. Commun.* **282,** 697–700.

Yu, X., Minter-Dykhouse, K., Malureanu, L., Zhao, W. M., Zhang, D., Merkle, C. J., Ward, I. M., Saya, H., Fang, G., van, D. J., and Chen, J. (2005). Chfr is required for tumor suppression and Aurora A regulation. *Nat. Genet.* **37,** 401–406.

Yun, M. S., Kim, S. E., Jeon, S. H., Lee, J. S., and Choi, K. Y. (2005). Both ERK and Wnt/beta-catenin pathways are involved in Wnt3a-induced proliferation. *J. Cell Sci.* **118,** 313–322.

Zhang, Q., Davenport, J. R., Croyle, M. J., Haycraft, C. J., and Yoder, B. K. (2005). Disruption of IFT results in both exocrine and endocrine abnormalities in the pancreas of Tg737(orpk) mutant mice. *Lab Invest.* **85,** 45–64.

Zhang, Q., Murcia, N. S., Chittenden, L. R., Richards, W. G., Michaud, E. J., Woychik, R. P., and Yoder, B. K. (2003). Loss of the Tg737 protein results in skeletal patterning defects. *Dev. Dyn.* **227,** 78–90.

Zhang, X., Zhu, J., Yang, G. Y., Wang, Q. J., Qian, L., Chen, Y. M., Chen, F., Tao, Y., Hu, H. S., Wang, T., and Luo, Z. G. (2007). Dishevelled promotes axon differentiation by regulating atypical protein kinase C. *Nat. Cell Biol.* **9,** 743–754.

Zhou, G. L., Tucker, D. F., Bae, S. S., Bhatheja, K., Birnbaum, M. J., and Field, J. (2006). Opposing roles for Akt1 and Akt2 in Rac/Pak signaling and cell migration. *J. Biol. Chem.* **281,** 36443–36453.

CHAPTER ELEVEN

CILIA INVOLVEMENT IN PATTERNING AND MAINTENANCE OF THE SKELETON

Courtney J. Haycraft* *and* Rosa Serra[†]

Contents

1. Introduction	304
2. Cilia are Required for Anterior–Posterior Limb Patterning	305
3. A Role for IFT/Primary Cilia in Endochondral Bone Formation and Development of the Postnatal Growth Plate	312
4. Primary Cilia in Articular Cartilage	318
5. A Role for Primary Cilia in the Development of the Bone Collar	319
6. Primary Cilia in the Maintenance of Bone	321
7. Primary Cilia in Craniofacial Development	323
8. Primary Cilia in Tooth Development	325
9. Summary	326
Acknowledgments	327
References	327

Abstract

Although the expression of cilia on chondrocytes was described over 40 years ago, the importance of this organelle in skeletal development and maintenance has only recently been recognized. Primary cilia are found on most mammalian cells and have been shown to play a role in chemosensation and mechanosensation. A growing number of human pleiotropic syndromes have been shown to be associated with ciliary or basal body dysfunction. Skeletal phenotypes, including alterations in limb patterning, endochondral bone formation, craniofacial development, and dentition, have been described in several of these syndromes. Additional insights into the potential roles and mechanisms of cilia action in the mammalian skeleton have been provided by research in model organisms including mouse and zebrafish. In this article we describe what is currently known about the localization of cilia in the skeleton as well as the roles and underlying molecular mechanisms of cilia in skeletal development.

---

* Department of Medicine/Division of Nephrology, Medical University of South Carolina, Charleston, SC 29425
[†] Department of Cell Biology, University of Alabama at Birmingham, 1918 University Blvd, Birmingham, AL 35294-0005

1. INTRODUCTION

The primary cilium is an organelle that consists of a microtubule-based axoneme covered by a specialized plasma membrane (Bisgrove and Yost, 2006; Davenport and Yoder, 2005; Pan et al., 2005; Satir and Christensen, 2007). A single, nonmotile primary cilium is present on the surface of most eukaryotic cells. Primary cilia are generated through a process called intraflagellar transport (IFT) (Bisgrove and Yost, 2006; Davenport and Yoder, 2005; Pan et al., 2005; Satir and Christensen, 2007). Protein synthesis does not occur in cilia so the proteins that make up the structure must be transported into the cilium along microtubules. There are two IFT complexes which can be separated biochemically. Complex A mediates retrograde transport of cargoes from the tip to the base of the cilium and complex B mediates anterograde transport of specific cargoes from the base to the tip. Anterograde transport is driven by hetromeric kinesin 2 motors, which are composed of Kif3a and Kif3b motor subunits. Retrograde transport is mediated by dynein 1B. In addition, an assortment of accessory proteins are required for normal ciliary function.

It has been over 40 years since primary cilia were first identified on chondrocytes (Scherft and Daems, 1967). Ultrastructural studies showed that each chondrocyte has one cilium with the 9 + 0 microtubule pattern associated with nonmotile primary cilia (Scherft and Daems, 1967; Wilsman and Fletcher, 1978). It was shown that chondrocyte cilia vary in length from 1 to 4 μm and are about 0.2 μm in width (Poole et al., 2001; Scherft and Daems, 1967). The microtubules of cilia are enriched with detyrosinated and acetylated tubulin, allowing the use of antibodies to these modified forms of tubulin to immunolocalize cilia in a variety of cells including chondrocytes, perichondrial cells, and osteoblasts (Fig. 11.1; Poole et al., 1997, 2001; Song et al. 2007; Xiao et al., 2006). Electron micrograhic studies showed that the chondrocyte cilium projects into the extracellular matrix (Poole et al., 1997). Electron, confocal, and tomographic microscopy indicate that chondrocyte cilia display various bending patterns potentially due to association of the cilium with the surrounding pericellular matrix (Jensen et al., 2004; Poole et al., 2001). Studies using flourescent microscopy suggest that cilia on chondrocytes have a specific orientation. Cilia on articular chondrocytes point away from the articular surface while cilia on columnar chondrocytes protrude from the center of the cell between the Golgi and nucleus (McGlashan et al., 2008, 2007; Song et al., 2007). The significance of this orientation is not clear. Future studies using mathematical models in combination with multiphoton microscopy will help clarify the orientation of cilia in different types of connective tissue (Ascenzi et al., 2007).

Studies over the past decade have revealed an association between defects in the normal structure or function of the primary cilium and various

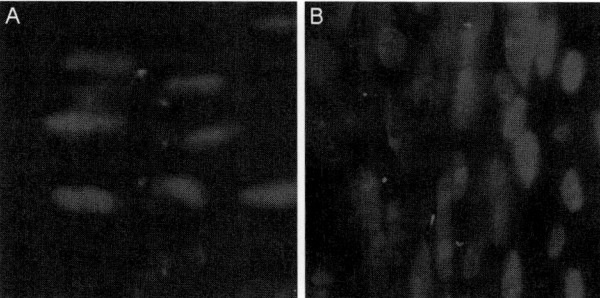

Figure 11.1 *Localization of cilia on chondrocytes and perichondrial cells.* (A) Cilia, visualized by immunostaining with anti-acetylated α-tubulin antibodies (red), are aligned in the center of columns of proliferating chondrocytes in the growth plate of endochondral bones. Nuclei (blue) are located on alternate sides of the cells within a column. (B) In the perichondrium, cilia are present on both elongated and cuboidal cells adjacent to the cartilage anlagen (right side of image). Cilia were visualized by immunostaining with anti-Ift88 antibodies (red). Nuclei are blue. (See Color Insert.)

congenital human conditions including polycystic kidney disease (PKD), *situs inversus*, and retinal degeneration (Bisgrove and Yost, 2006; Davenport and Yoder, 2005; Pan *et al.*, 2005). Pleiotrophic syndromes such as Jeune syndrome, Meckel syndrome, and Bardet–Biedl syndrome, which are associated with mutations in ciliary or basal body genes are characterized by various combinations of pathological changes including those in the skeleton (Table 11.1) (Bisgrove and Yost, 2006; Davenport and Yoder, 2005; Pan *et al.*, 2005). The most common alterations in the skeleton associated with mutations in ciliary genes include polydactyly, shortened bones, and craniofacial dysmorphology. It has been shown that primary cilia are involved in transmitting both chemical and mechanical signals but the molecular mechanisms of these conditions are just now being determined.

To help in this regard, several mouse models have been generated and used to determine the role and mechanism of action for primary cilia in various aspects of skeletal development (Table 11.2). Furthermore, the zebrafish has emerged as another useful model to help uncover the molecular mechanisms of skeletal development. These animal models provide interesting insights into the how the cilium functions in the skeleton.

2. CILIA ARE REQUIRED FOR ANTERIOR–POSTERIOR LIMB PATTERNING

The skeletal elements of the limb must be properly patterned along both the anterior–posterior and proximal–distal axes (for detailed review see Tickle, 2006). During outgrowth of the limb bud, the distal ectoderm forms a specialized epithelium called the apical ectodermal ridge (AER) that

Table 11.1 Human cilopathies affecting the skeleton

Gene affected	OMIM#	Name	References
BBS			
	209900	Bardet–Biedl Syndrome (BBS)	Ansley et al. (2003), Tobin et al. (2008)
CXORF5/Ofd1			
	311200	Oral facial digital syndrome 1 (OFD1)	Ferrante et al. (2006)
	300209	Simpson–Golabi–Behmel syndrome type 2 (SGBS2)	
EVC			
	225500	Ellis–van Creveld syndrome (EvC)	Ruiz-Perez et al. (2007)
IFT80			
	611623	Asphyxiating thoracic dystrophy 2 (ATD2); Juene's asphyxiating thoracic dystorophy	Beales et al. (2007)
MKS1			
	249000	Meckels syndrome type I (MKS1)	Kyttala et al. (2006)
PKD1			
	17390	Autosomal dominant polycyctic kidney disease (ADPKD)	Turco et al. (1993)
RPGRIP1L/Ftm			
	611561	Meckels syndrome type 5 (MKS5)	Delous et al. (2007)
	611560	Joubert syndrome type 7 (JBTS7)	

secretes factors including fibroblast growth factors (Fgfs) essential for outgrowth of the limb. The anterior–posterior patterning of the limb is controlled mainly by a region of cells in the posterior mesenchyme known as the zone of polarizing activity (ZPA). Mice lacking the IFT protein Ift88 develop ectopic digits suggesting a role for cilia in the patterning of the mammalian digits. Indeed, cilia are expressed on both ectodermal and mesenchymal cells of the developing limb; however, only cilia in the mesoderm are required for proper limb patterning (Haycraft et al., 2005; Haycraft et al., 2007).

The major signaling pathway implicated in digit patterning is the hedgehog (Hh) pathway. In addition to an important role in digit patterning, Hh

Table 11.2 Animal models for IFT/cilia in the skeleton

Gene alteration	Strain name	References
Bbs		
Null allele	Bbs4$^{null/null}$; Bbs6$^{null/null}$	Tobin et al. (2008)
Zebrafish Knock down	morpholino	Tayeh et al. (2008), Tobin et al. (2008)
Evc		
Null allele	Evc$^{null/null}$	Ruiz-Perez et al. (2007)
Ftm/RPGRIP1L		
Null allele	Ftm $^{null/null}$	Vierkotten et al. (2007)
Ift80		
Zebrafish Knock down	morpholino	Beales et al. (2007)
Ift88/Tg737/Polaris		
Hypomorphic allele	Ift88^{orpk}	Zhang et al. (2003) Haycraft et al. (2005) McGlashan et al. (2007)
	Ift88^{flexo}	Liu et al. (2005)
Null allele	Ift88^{tm1Rpw}	Zhang et al. (2003) Haycraft et al. (2005) Liu et al. (2005)
Cre-LoxP conditional	Prx1Cre;Ift88$^{LoxP/LoxP}$	Haycraft et al. (2007)
	Col2aCre;Ift88$^{LoxP/LoxP}$	Song et al. (2007)
Kif3a		
Cre-LoxP conditional	Prx1Cre;Kif3a$^{LoxP/LoxP}$	Haycraft et al. (2007) Kolpakova-Hart et al. (2007)
	Col2aCre;Kif3a$^{LoxP/LoxP}$	Song et al. (2007) Koyama et al. (2007)
	Dermo1-Cre;Kif3a$^{LoxP/LoxP}$	Kolpakova-Hart et al. (2007)
	Wnt1-Cre;Kif3a$^{LoxP/LoxP}$	Kolpakova-Hart et al. (2007)

(continued)

Table 11.2 (continued)

Gene alteration	Strain name	References
Ofd1/CXORF5		
Cre-LoxP conditional	pCX-NLS-Cre;Ofd$^{\text{LoxP/LoxP}}$	Ferrante et al. (2006)
Pkd1		
Null allele	Pkd1^{null}	Lu et al. (2001)
Strong hypomorphic/Null	Pkd1$^{\text{m1Bei}}$	Xiao et al. (2006)
Truncation	Pkd1$^{\Delta17-21\beta\text{geo}}$	Boulter et al. (2001)
Thm1		
Null allele	Thm1^{aln}	Tram et al. (2008)

signaling is involved in the development of many tissues in the body (for detailed review see Huangfu and Anderson, 2006). There are three mammalian Hh ligands, Sonic hedgehog (Shh), Indian hedgehog (Ihh), and Desert hedgehog (Dhh). Hh signaling is both positively and negatively regulated. In the absence of Hh ligand, Smothened (Smo) activity is repressed by the receptor Patched1 (Ptc1) or Patched2 (Ptc2). Upon ligand binding, Smo repression is relieved and the Gli transcription factors activate transcription of downstream target genes. There are three mammalian Gli proteins, Gli1, Gli2, and Gli3. Gli3 is processed to generate a potent transcriptional inhibtor (Gli3R) in the absence of Hh while Gli2 is a potent activator of transcription in the presence of ligand. Unlike Gli2 and Gli3, Gli1 is expressed only after pathway activation.

While Hh pathway regulation has been well characterized in *Drosophila*, there are significant differences in the mammalian signaling pathway, such as the requirement for the primary cilium in mammals, which are not fully understood. Recent work has shown that Ptc1 and Smo translocate out of and into the cilium, respectively, upon ligand binding while the Gli proteins localize to the tip of the cilium with SuFu, a regulator of Gli processing and activity (Fig. 11.2; Corbit *et al.*, 2005; Haycraft *et al.*, 2005; Huangfu and Anderson, 2005; Rohatgi *et al.*, 2007). Loss of cilia in *Ift88* or *Kif3a* null mice or in the limb mesoderm only using Prx1-Cre results in extensive polydactyly with up to eight unpatterned digits on each limb (Haycraft *et al.*, 2005, 2007; Liu *et al.*, 2005). Both activation and repression of Hh signaling are affected in *Ift88* and *Kif3a* null embryos and Gli3 processing is disrupted (Haycraft *et al.*, 2005; Liu *et al.*, 2005). Even though localization of Shh mRNA is normal, expression of Ptc1 and Gli1, direct targets of Hh activity are reduced indicating that the activation of Hh signaling is repressed in the limb field (Haycraft *et al.*, 2007). Nevertheless, a block to activation of Shh

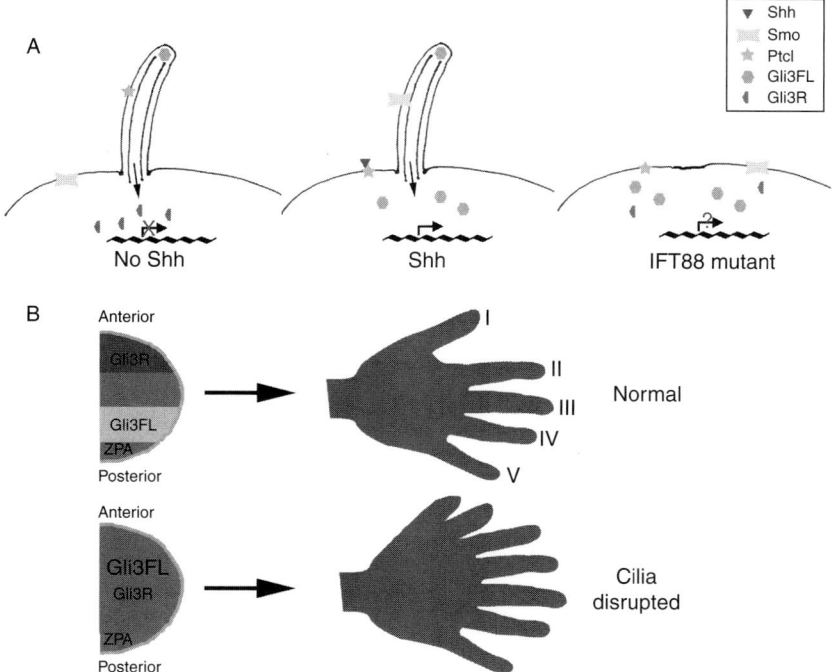

Figure 11.2 *Mammalian Hh signal transduction and limb patterning.* (A) *Hh signal transduction.* In the absence of Hh ligand, Ptc1 localizes to the ciliary membrane. Full length Gli3 (Gli3FL) is found at the distal tip of the cilium and subsequent proteolytic processing results in increased amounts of Gli3 repressor (Gli3R). Gli3R is a potent transcriptional repressor resulting in repression of target gene transcription including Ptc1 and Gli1. Upon ligand binding, Ptc1 translocates out of the cilium and Smo is localized in the cilium membrane. Smo translocation to the cilium inhibits Gli3 processing leading to an increased Gli3FL:Gli3R ratio allowing transcription of target genes. In Ift88 mutants, the cilium is not formed and Gli3 processing is inefficient. As a result, target genes such as Gremlin and Hand2, which are repressed by Gli3R, are expressed while transcription of Gli1 and Ptc1 is not activated likely due to disruption of Gli2 function by an unknown mechanism. (B) *Role of Shh in development of the mammalian limb.* In the developing limb, Shh is expressed by cells in the ZPA (red). Activation of Shh signaling leads to decreased Gli3R levels in the posterior limb bud while the level of Gli3R remains high in the anterior limb bud. In cilia mutants, such as those lacking Ift88 or Kif3a, disruption of the normal gradient of Gli3FL:Gli3R in conjunction with loss of Gli2 activity across the limb bud results in polydactyly. (See Color Insert.)

signaling cannot account for the severe polydactyly observed. *Shh*-null mice have only one digit not polydactyly (Chiang *et al.*, 2001). It was previously shown that mutations in Gli3 result in polydactyly (te Welscher *et al.*, 2002; Wang *et al.*, 2000). In the absence of Hh signaling, Gli3 is normally processed to a repressor form, suppressing transcription of Hh target genes. The expression pattern of two genes, *Gremlin* and *Hand2*, that are normally repressed in the anterior limb bud by Gli3 is expanded in the limbs of IFT/cilia mutants indicating that the loss of the Gli3 repressor and overall misexpression of Shh target genes results in polydactyly (Fig. 11.2; Haycraft et al., 2005; Liu et al., 2005; Haycraft et al., 2007).

Recent work has identified a mouse mutant, $Thm1^{aln}$, containing a mutation in an IFT complex A protein. In contrast to mice with mutations in Ift88, a complex B protein, only Hh pathway repression is disrupted resulting in ectopic expression of Ptc1 and Gli1 in the anterior limb bud (Tran et al., 2008). The reason for the differences in these two models is not fully understood, but anterograde trafficking of IFT proteins remains intact in $Thm1^{aln}$ mutants while both anterograde and retrograde transport are disrupted in $Ift88$ and $Kif3a$ mutants as evidenced by complete loss of the ciliary axoneme.

As described above, cilia play an integral role in Shh signaling and therefore disruption of cilia results in defects in digit patterning. It is not surprising that a number of pleiotropic syndromes that have been shown to be due to mutations in cilia or basal body-localized proteins, for example Bardet–Biedl syndrome (BBS; OMIM #209900), include alterations in the number of digits among their clinical features. BBS features include cystic kidney dysplasia, obesity, hypogonadism and digit patterning alterations including postaxial polydactyly, syndactyly or brachydactyly. Mutations in at least one of 14 distinct genes (BBS1-BBS14) lead to BBS although the same genes have been linked to other pleiotropic syndromes including, McKusick–Kaufman syndrome (MKKS; OMIM #236700) and Meckel syndrome 1 (MKS1; OMIM #249000). Additionally, mutations in more than one BBS gene may lead to alterations in clinical presentation and patients have been identified with mutations in additional BBS genes leading to the hypothesis that BBS can develop through triallelic inheritance (Katsanis et al., 2001, 2002).

While digit patterning defects are seen in a majority of BBS patients, the two BBS mouse models (Bbs4 null and Bbs6 null) do not develop polydactyly or syndactyly (Fath et al., 2005). Pectoral fin development in the zebrafish shares similarities with mammalian limb development including expression of Shh in the posterior mesenchyme. In BBS morpholino-injected zebrafish embryos, the region of Shh expression was increased relative to control-injected embryos although no alterations in pectoral fin size were observed (Tayeh et al., 2008). These results suggest that the polydactyly seen in BBS patients may be due to alterations in Shh signaling. The proposed triallelic inheritance of BBS mentioned above may also play a role in the differences between mouse models and features observed in human patients since the mouse models examined contain mutations in only one BBS gene (Kulaga et al., 2004).

Syndactyly, brachydactyly, or polysyndactyly are also seen in Oral-facial-digital syndrome I (OFD1; OMIM #311200). The gene responsible for OFD1 is X chromosome open reading frame 5 (CXORF5). The protein contains a LIS1 homology domain, which is thought to regulate microtubule dynamics and is localized to the centrosome and basal body of primary cilia (Ferrante et al., 2001; Romio et al., 2004). Studies in mice have shown that cells lacking Ofd1 fail to form a primary cilium. In heterozygous female embryos one allele is randomly inactivated by X inactivation resulting in a chimeric embryo with normal and Ofd1 null cells, whereas males have a single allele (Ferrante et al., 2006). Female mice lacking one copy of the gene mutated in Ofd1 also show severe defects in patterning of the digits including development of seven to nine digits per limb. The neural tube of male embryos shows a reduction in the level of Ptc1 and Gli1 expression indicating a possible requirement for Ofd1 in activating Shh signaling. No change in Ptc1 or Gli1 expression was seen in the limb bud of heterozygous female embryos, however, regions of expanded Hox gene expression were detected in the anterior limb bud (Ferrante et al., 2006). It is unclear how loss of Ofd1 results in ectopic or expanded Hox gene expression and polydactyly, but it is hypothesized to be due to disruption of Gli3 function.

Postaxial polydactyly is also common in Ellis–van Creveld syndrome (EvC; OMIM#225500). EvC is a chondroectodermal dysplasia characterized by numerous skeletal and craniofacial abnormalities. Recently, one of the two proteins mutated in EvC was identified (EVC) and localized to the base of the cilium (Ruiz-Perez et al., 2003, 2007). EVC is not required for the formation or the maintenance of cilia structure but appears to be required for normal Hedghog signaling. In the mouse, Evc expression was restricted to the skeleton and the developing face (Ruiz-Perez et al., 2007). Recently, mice lacking *Evc* were generated but no alterations in digit patterning were seen. The lack of digit defects in *Evc* null mice was attributed to either low levels of Evc expression during early stages of limb patterning or the lack of Gli3 processing defects in Evc mutants despite reduced Gli1 and Ptc1 expression in the developing bones (Ruiz-Perez et al., 2007).

Similarly, polydactyly is prevalent in Meckel–Gruber syndrome (MKS1; OMIM #249000). Meckel–Gruber syndrome is an autosomal lethal disorder due to mutations in one of several distinct loci including MKS1, which was shown to be expressed in the developing digits in mice (Kyttala et al., 2006). Meckels type 5 (OMIM#611561) is a form of Meckel–Gruber syndrome resulting from mutations in Retinitis Pigmentosa GTPase Regulator-interacting-protein 1-like (*RPGRIP1L*) (Delous et al., 2007). *Fantom* (*Ftm*) encodes the mouse homolog of *RPGRIP1L* and mice lacking *Ftm* develop ectopic digits (Vierkotten et al., 2007). Similar to what is seen in *Ift88* mutants, Gli3 processing is disrupted in *Ftm* mutant embryos and excess full length Gli3 is present. Unlike *Ift88* mutants, Ptc1 expression is present although at reduced levels in the developing limbs. Reduced numbers of cilia are seen by SEM on the node and by immunostaining in the developing limb of *Ftm* mutant embryos suggesting that Ftm is required for proper cilia formation or maintenance, although the defects are not as severe as in *Ift88* or *Kif3a* mutants.

3. A ROLE FOR IFT/PRIMARY CILIA IN ENDOCHONDRAL BONE FORMATION AND DEVELOPMENT OF THE POSTNATAL GROWTH PLATE

Most of the bones in the body develop through a process called endochondral bone formation (Cancedda et al., 1995; Colnot, 2005; van der Eerden et al., 2003). Embryonic mesenchymal cells condense at the sites where the skeletal elements will form. Condensed mesenchymal cells then undergo differentiation to become chondrocytes forming the cartilaginous anlagen of the future bone. Proliferation allows for the initial expansion of the skeletal element. Chondrocytes then differentiate toward their terminally differentiated form, the hypertrophic chondrocyte, also allowing for an increase in size of the skeletal element. Hypertrophic chondrocytes do not proliferate but they synthesize a unique extracellular matrix that allows mineralization of the matrix. The hypertrophic cells also secrete factors that attract vasculature to the bone anlagen and promote the differentiation of osteoblasts from cells in the adjacent perichondrium leading to the development of the bone collar. Finally, hypertrophic chondrocytes undergo apoptosis and the mineralized matrix is replaced with osteoblasts forming trabecular bone.

As described above, there are three members of the vertebrate Hh family, of these Ihh is the most important regulator of endochondral bone formation. Ihh is expressed in cells that are committed to become hypertrophic and it regulates proliferation of chondrocytes, the rate of hypertrophic differentiation, as well as the formation of the bone collar (St-Jacques et al., 1999). Regulation of proliferation is thought to be a direct effect of Ihh on chondrocytes while hypertrophic differentiation is mediated indirectly (Long et al., 2001). Ihh acts in a negative feedback loop with Parathyroid Hormone related Protein (PTHrP) and other factors to sense and regulate the rate of chondrocyte differentiation and thus limit hypertrophic differentiation (Lanske et al., 1996; Vortkamp et al., 1996).

Since Ihh is a major regulator of endochondral bone formation and it has been shown that cilia are required to mediate both the activator and repressor functions of Hh signaling (Haycraft et al., 2005; Liu et al., 2005) it makes sense that mice and humans with mutations in ciliary genes often present with defects in the development of the skeleton. Recently, it was demonstrated that at least two human syndromes that include significant defects in endochondral bone formation are associated with mutations in genes that code for ciliary proteins. First, Juene's asphyxiating thoracic dystrophy (ATD2; OMIM #611177) was associated with a missense mutation in one of the WD40 domains of IFT80. IFT80 is localized to the basal body and the ciliary axoneme and is part of IFT complex B (Beales et al., 2007).

The missense mutation resulted in missing or shortened cilia. Although a connection to Hedghog signaling could not be examined in human tissues, knock down of IFT80 in zebrafish using morpholino technology resulted in skeletal defects as well as renal cysts (Beales et al., 2007). Some of the skeletal defects resembled those seen in Shh depleted fish. Furthermore, Ptc1 expression was down-regulated in the IFT80 morphants suggesting alterations in Hedgehog signaling (Beales et al., 2007).

As mentioned above, EvC is characterized by numerous skeletal and craniofacial abnormalities. One of the proteins mutated in EvC is localized to the base of the cilium and is required for normal Hedgehog signaling (Ruiz-Perez et al., 2007). In the growth plate, Evc was expressed in the perichondrium as well as in resting and proliferating chondrocytes (Ruiz-Perez et al., 2007). Expression was not detected in late prehypertophic or hypertrophic cells. Disruption of *Evc* in mice resulted in a variety of skeletal and craniofacial abnormalities as well as alterations in the teeth and nails. The phenotypes were similar to that seen in EvC patients (Ruiz-Perez et al., 2007). Alterations in chondrocyte proliferation or early differentiation were not observed; however, accelerated hypertrophic differentiation was observed in the *Evc*-null growth plates. Defects in the growth plate were further associated with diminished Ihh signaling as measured by a reduction in the expression of Ptc1 and Gli1; however, processing of Gli3 to the repressor form was normal. The results suggest that Evc is specifically involved in Hh-dependent activation of gene expression. This is in contrast to studies in which cilia themselves are absent and both activator and repressor functions of Hh are affected (Haycraft et al., 2005, 2007; Liu et al., 2005; Song et al., 2007). Phenotypic differences in *Evc*-null mice and mice with conditional mutations in *Kif3a* or *Ift88* (see below) can in part be accounted for by intact Gli3 function in the *Evc* mutant mice.

Mutations in *Pkd1* are associated with autosomal dominant polycystic kidney disease (ADPKD; OMIN#173900). Skeletal malformations have been detected in unusual patients with this form of PKD (Turco et al., 1993). Mice with mutations in *Pkd1* demonstrate defects in embryonic endochondral bone formation as well as osteopenia in adults (Boulter et al., 2001; Lu et al., 2001; Xiao et al., 2006). Pkd1 mRNA is expressed in early condensing mesenchyme and later in prechondrogenic tissue (Guillaume et al., 1999). As skeletal development progresses, it becomes enriched in the perichondrium and invertervertebral disc. *Pkd1* mutant mice demonstrate shortened long bones as well as defects in the formation of the vertebrae (spina bifida occulta) (Lu et al., 2001). A delay in hypertrophic differentiation, vascularization, and skeletal mineralization was also noted (Boulter et al., 2001; Lu et al., 2001; Xiao et al., 2006). The molecular mechanism accounting for the defects seen in endochondral bone formation in *Pkd1* mutant mice is not known.

Mice with mutations in other ciliary genes also demonstrate alterations in endochondral bone formation resulting in shortening of the bones in the limbs (Haycraft et al., 2007; McGlashan et al., 2007; Song et al., 2007; Zhang et al., 2003). Mice with conditional deletion of *Ift88* or *Kif3a* in early limb mesenchyme (Prx1-Cre;Ift88$^{loxP/loxP}$ or Prx1-Cre;Kif3a$^{loxP/loxP}$) have defects in embryonic endonchondral bone formation including accelerated hypertrophic differentiation as well as a delay in vascularization of the primary ossification center resulting in the persistence of hypertrophic cells in the growth plate (Haycraft et al., 2007). Defects were observed as early as 15.5 days of gestation. Mice in which *Kif3a* was deleted using Dermo1-Cre had a similar, although milder, phenotype to the Prx1-Cre mutants (Kolpakova-Hart et al., 2007). The phenotypes resembled those seen in mice with germline mutations in *Ihh* (St-Jacques et al., 1999). Ptc1 and Gli1 mRNA were dramatically reduced in the long bones of Prx1-Cre; Ift88$^{loxP/loxP}$ and Prx1-Cre;Kif3a$^{loxP/loxP}$ mice relative to controls, supporting of a role for IFT/cilia in mediating Ihh signaling during embryonic endochondral bone formation (Fig. 11.3A). Expression of Ptc1 and Gli1 was reduced in both chondrocytes and perichondral cells suggesting Ihh signaling was disrupted in both compartments. Somewhat surprisingly, PTHrP expression was maintained; perhaps due to the loss of Gli3 repressor function (Hilton et al., 2005; Koziel et al., 2005). In contrast to the Prx1-Cre and Dermo1-Cre mutant mice, mice in which *Ift88* or *Kif3a* was deleted using Col2a-Cre (Col2a-Cre;Ift88$^{loxP/loxP}$ or Col2a-Cre;Kif3a$^{loxP/loxP}$) did not demonstrate alterations in embryonic endochondral bone formation (Song et al., 2007). Immunoflourescent staining confirmed that cilia were missing in chondrocytes by E15.5 days. Furthermore, Ptc1 expression was dramatically reduced in embryonic chondrocytes but expression was maintained in the perichondrium (Fig. 11.3A). The differences in the phenotype between Prx1-Cre and Col2a-Cre mutants were likely due to the expression pattern of Col2a-Cre, which targets chondrocytes but does not efficiently target cells in the perichondrium (Ovchinnikov et al., 2000). Prx1-Cre and Dermo1-Cre target the perichondrium and cartilage (Logan et al., 2002; Yu et al., 2003). The combined results suggest an important role for perichondral IFT/cilia in regulating embryonic endochondral bone formation possibly by mediating signaling by Ihh. There is prior evidence indicating the importance of perichondral Ihh signaling in the regulation of hypertrophic differentiation that supports this model (Alvarez et al., 2002; Colnot, 2005; Long et al., 2001; Vortkamp et al., 1996).

Although Col2a-Cre;Ift88$^{loxP/loxP}$ and Col2a-Cre;Kif3a$^{loxP/loxP}$ mice did not demonstrate defects in embryonic endochondral bone formation, the mice demonstrated postnatal dwarfism due to a progressive loss of the growth plate (Song et al., 2007). After birth, the growth plate allows for continued longitudinal growth of the bone through continued proliferation, differentiation, and replacement by bone. The cells in the growth plate

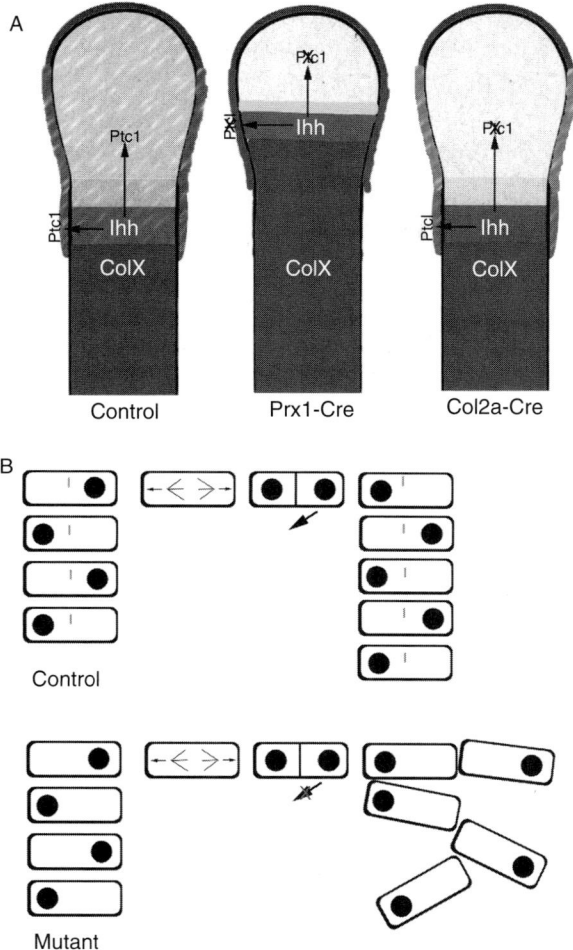

Figure 11.3 *A role for primary cilia in endochondral bone formation and the postnatal growth plate.* (A) *Primary cilia regulate Ihh signaling during embryonic endochondral bone formation.* Ihh is expressed in chondrocytes that are committed to hypertrophic differentiation (red area). Ihh normally acts on both prehypertrophic chondrocytes (gray and yellow areas) and the perichondrium (green). Ptc1, a downstream target of Ihh signaling, was used to determine how depletion of cilia affects Ihh signaling during endochondral bone formation. In normal control mice, cilia are present on chondrocytes as well as cells in the perichondrium and Ptc1 expression is detected in both cell types. In mice in which ciliogenesis is disrupted in Prx1-Cre expressing cells, cilia are absent from both chondrocytes and cells in the perichondrium. Disruption of ciliogenesis in the cartilage and in the perichondrium resulted in accelerated hypertrophic differentiation (blue area) similar to that seen in the Ihh-null mice. Ptc1 expression was dramatically reduced in both chondrocytes and cells of the perichondrium confirming that Ihh signaling was disrupted. In contrast, when ciliogenesis was disrupted using Col2a-Cre, which only targets the chondrocytes, hypertrophic differentiation was normal. Ptc1 expression

are aligned into columns representing a continuum of differentiation. The cells are maintained in columns through a process that has recently been referred to as rotation. During cell division, chondrocytes divide perpendicular to the long axis of the bone. This is followed by a series of cell shape changes and movements that result in the positioning of flat cells one on top of the other (Fig. 11.3B; Dodds, 1930). Very little is known about this process. The first alterations seen in the growth plate of Col2a-Cre; Kif3a$^{loxP/loxP}$ mice were at postnatal day 7 (P7) and included a decrease in the length of the proliferative zone associated with reduced cell proliferation. By P10 days, the columnar organization of the growth plate was severely altered. In addition to accelerated hypertrophy, alterations in the localization of activated focal adhesion kinase (FAK) and disorganization of the actin cytoskeleton suggested defects in the process of chondrocyte rotation (Fig. 11.3B). The phenotype of Col2a-Cre;Ift88$^{loxP/loxP}$ and Col2a-Cre;Kif3a$^{loxP/loxP}$ mice had some similarities to mice with conditional deletion of Ihh induced in postnatal cartilage (Col2a-CreER;Ihh$^{loxP/loxP}$) (Maeda *et al.*, 2007). Nevertheless, alterations in Ptc1 expression were not detected by RT-PCR or *in situ* hybridization suggesting the phenotype was independent of Ihh signaling. It was proposed that defects in rotation could be the result of alterations in cell adhesion and shape mediated by mechanical stimulation (Aszodi *et al.*, 2003; Jensen *et al.*, 2004; Park *et al.*, 2006).

It was previously suggested that chondrocyte cilia transmit mechanical forces through their interaction with the surrounding ECM (Poole *et al.*, 1985; Jensen *et al.*, 2004). It is known in other cell types that cilia are involved in transmitting both chemical and mechanical signals. The role of cilia in transmitting mechanical signals is most well characterized in the kidney where the protein products for genes associated with polycycstic kidney disease (PKD) are localized to cilia (Nauli and Zhou, 2004; Yoder *et al.*, 2002). Autosomal dominant PKD results from mutations in *Pkd1* or *Pkd2*, which encode polycystin 1 (Pc1) and polycystin 2 (Pc2). Pc1 and Pc2 are integral membrane proteins that form a regulatory protein and a Ca^{2+} channel in the primary cilium of renal epithelial cells. Cilia on renal cells are sensitive to fluid shear stress and respond with changes in the concentration of cytosolic Ca^{2+} (Praetorius and Spring, 2001). Cells without cilia or cells

---

was reduced in chondrocytes but maintained in the perichondrium. (B) *Primary cilia mediate chondrocyte rotation in the postnatal growth plate.* In normal control mice, flat cells in the growth plate have polarity and are aligned in columns parallel to the long axis of the bone. Cell division occurs perpendicular to the long axis. One cell then migrates under the other to form the columns. This process is called chondrocyte rotation. Cilia are present on chondrocytes in the columns and are required for normal chondrocyte rotation. In cilia depleted growth plates, chondrocytes still divide perpendicular to the long axis of the bone and polarity is maintained; however, the orientation of the cells one to another is altered and the cells are not maintained in columns. (See Color Insert.)

in which the function of Pc1 or Pc2 has been blocked do not respond to mechanical stimulation (Praetorius and Spring, 2003a,b). It is possible that mechanical forces are transmitted through the cartilage extracellular matrix to chondrocytes by bending of the cilium (Jensen et al., 2004). Using electron microscopy is has been shown that the chondrocyte cilium projects into the extracellular matrix and is tightly associated with the Golgi apparatus (Poole et al., 1997). A combination of additional imaging techniques including confocal and tomographic microscopy indicate that chondrocyte cilia display various bending patterns in association with the surrounding pericellular matrix (Jensen et al., 2004; Poole et al., 2001). Some of the bending patterns fit with a model of shear stress while others suggested deflection by the matrix. Furthermore, the close anatomical association of the cilium with the microtubule cytoskeleton, Golgi apparatus, and microtubule organizing centers within the cell support a model of direct signaling between the matrix, the cilium, and the inside of the cell (Poole et al., 1997). Cilia were dramatically depleted from the cartilage of Col2a-Cre; Kif3a$^{loxP/loxP}$ mice by E15.5 days, nevertheless, alterations in the organization of the growth plate were not observed until 7–10 days after birth, a time when the young mice are starting to become subjected to significant mechanical load. Alterations in the perception of mechanical load as a result of the loss of cilia could lead to alterations in the growth plate and articular cartilage. Previously, it was shown that when young rats are subjected to unloading, alterations in the growth plate are observed including a reduction in the height of the growth plate due to reduced number of cells in the proliferative zone. Older rats were less responsive suggesting that there is a window of development that requires appropriate loading (Yu et al., 2003; Zerath et al., 1997).

The molecular mechanism whereby cilia potentially mediate mechanical signals in cartilage is not known. It has been shown that integrins are present on the chondrocyte cilium and integrin dependent signaling cascades have been described in chondrocyte mechanotransduction (McGlashan et al., 2006; Millward-Sadler and Salter, 2004; Praetorius et al., 2004). Furthermore, deletion of chondrocyte $\beta1$ integrin results in defects in rotation that include alterations in cell shape and orientation with some similarities to that seen in postnatal mice with disrupted cilia (Aszodi et al., 2003). In support of this model, deletion of *Pkhd1*, a ciliary protein mutated in autosomal recessive PKD, in cultured renal epithelial cells resulted in alterations in cell adhesion and activation of FAK supporting a link between primary cilia and integrin mediated cell-ECM interactions in the kidney (Mai et al., 2005).

Growth plates with some similarities to those in the long bones are present at the base of skull. Synchondroses consist of two mirror image growth plates that are important for development and growth of the cranial base. The synchondroses in Col2a-Cre;Kif3a$^{loxP/loxP}$ mice have been characterized at postnatal stages and compared to mice with inducible and conditional deletion of Ihh (Col2a-CreER;Ihh$^{loxP/loxP}$) (Koyama et al., 2007). Similar to what was seen in the postnatal long bone, the synchondroses were disorganized with significant reduction in chondrocyte proliferation. Histologically hypertrophic cells were present but expression of Col10a mRNA was dramatically reduced indicating a delay in complete hypertrophic differentiation. Ptc1 and Gli1 expression were reduced in chondrocytes suggesting Hh signaling was disrupted in the cranial base growth plates. Nevertheless, the cranial base in mice deficient in Ihh only minimally resembled the cranial base of mice in which cilia were disrupted suggesting IFT/cilia also have unique Ihh-independent roles in synchondroisis development (Koyama et al., 2007).

Mice harboring the *Ift88^{orpk}* mutation also demonstrate short limbs (McGlashan et al., 2007). Minor alterations in the shape and organization of growth plate chondrocytes were detected by P4. In contrast to Col2a-Cre;Ift88$^{loxP/loxP}$ and Col2a-Cre;Kif3a$^{loxP/loxP}$ mice, *Ift88^{orpk}* mutants demonstrated a delay in hypertrophic differentiation as measured by a reduction in the expression domain of Type X collagen protein with continued expression of Type II collagen. The differences in the *Ift88^{orpk}* and Col2a-Cre models may reflect differences between systemic effects of germline mutation versus conditional deletion or the effect of a hypomorphic versus a null allele.

Recently, it was shown that another IFT protein, IFT46, is localized to cilia and preferentially expressed in early hypertrophic chondrocytes (Gouttenoire et al., 2007). IFT46 was initially identified as a BMP-responsive gene. BMP is a known mediator of chondrocyte differentiation. Knock down of IFT46 using siRNA technology in cultured chondrocytes stimulated the expression of genes that are primarily expressed by prehypertrophic columnar chondrocytes. The results suggest that IFT46 regulates specific aspects of growth plate biology. Since BMP regulated expression of IFT46, it is possible that BMP mediates the function of cilia in early hypertrophic cells.

4. Primary Cilia in Articular Cartilage

Articular cartilage differs from the growth plate in that it is maintained as a mature resting cartilage. It is required for normal joint function but has limited capacity for repair. Osteoarthritis is a leading cause of disability in the world so understanding how articular cartilage is maintained is of primary importance. Mechanical load is a critical factor in maintaining articular cartilage but how the load is sensed is not known. Since it is likely that primary cilia function as mechanosensors, understanding the role of primary cilia in articular cartilage is of interest. Recently, the fate of primary cilia on articular chondrocytes during the progression of bovine osteoarthritis was investigated (McGlashan et al., 2008). Cilia were present on a subset of cells within all the layers, superficial, middle and deep, of normal articular cartilage. The incidence of cilia was greater in the deep zone cartilage. Cilia in normal cartilage also had a distinct orientation. As seen by others using mouse tissue, cilia in the superficial layer were oriented away from the articular surface (McGlashan et al., 2007; Poole et al., 1985; Wilsman et al., 1980). The majority of cilia in the middle and deep zones also projected away from the surface. Primary cilia were present during all stages of osteoarthritis examined; however, the proportion of ciliated cells within each zone increased except for that in the deep zone, which was not significantly different from the normal controls. In contrast to normal cartilage, cilia within cell clusters at the eroding osteoarthritic surface projected toward the center of the cluster. The majority of cilia in middle and deep zone osteoarthritic cartilage projected away from the articular surface similar to normal cartilage. The significance of this orientation remains unclear.

A biological role for primary cilia in maintaining articular chondrocyte function is suggested in mice with mutations in ciliary genes. Alterations in the shape of cells in the superficial layers of the articular cartilage were described in $Ift88^{orpk}$ mice (McGlashan et al., 2007). In addition, the cartilage in these mice demonstrated reduced toluidine blue and Type II collagen staining suggesting defects in the maintenance of the cartilage phenotype. Col2a-Cre;Kif3a$^{loxP/loxP}$ mice had alterations in the shape of the joints and alterations in the organization of the articular cartilage but expression of aggrecan was maintained up to P30 days (Song et al., 2007). Dermo-1Cre;Kif3a$^{loxP/loxP}$ mice had severely misshapen joint surfaces that were thought to be due to early patterning defects in the articular cartilage (Kolpakova-Hart et al., 2007). Severe systemic defects in the $Ift88^{orpk}$ mouse and perinatal lethality in Dermo-1Cre;Kif3a$^{loxP/loxP}$ mice preclude comprehensive studies on osteoarthritis. Progression of osteoarthritis in the Col2a-Cre;Kif3a$^{loxP/loxP}$ line has not yet been addressed.

5. A ROLE FOR PRIMARY CILIA IN THE DEVELOPMENT OF THE BONE COLLAR

As the chondrocytes in the developing skeleton begin to undergo hypertrophy, mesenchymal cells in the perichondrium begin to differentiate into osteoblasts and secrete osteoid, which will mineralize to form the mature bone (Day and Yang, 2008). Ihh signaling induces expression of Wnt ligands in the perichondrium and canonical Wnt signaling which is essential to drive cells to differentiate along the osteoblast lineage. In the absence of Wnt signaling, perichondrial cells differentiate into chondrocytes or adipocytes.

Wnt signaling is essential for development of many tissues (Clevers, 2006; Veeman et al., 2003). In the mammalian genome there are at least 19 Wnt ligands which bind to a subset of seven pass membrane receptors, the Frizzled proteins. Wnt signaling is further subdivided into two general categories, canonical and noncanonical. In canonical Wnt signaling, the absence of ligand results in β-catenin phosphorylation by glycogen synthase kinase-3β (GSK3β) and targetting for proteosome-mediated degradation. Upon ligand binding, Disheveled inhibits GSK3β activity leading to stabilization of β-catenin and its subsequent translocation to the nucleus. In the nucleus, β-catenin interacts with members of the LEF/TCF family to activate transcription of target genes. In contrast to canonical Wnt signaling, noncanonical signaling does not require β-catenin. Noncanonical signaling is not as well characterized at the molecular level as canonical Wnt signaling, but has been shown to be required for morphological movements such as convergent extension in the inner ear in mice and the body axis in *Xenopus* and zebrafish. Convergent extension movements during embryonic development require the planar cell polarity (PCP) signaling pathway, which leads to cytoskeletal alterations through RhoA and JNK activation.

Cilia have been implicated in regulating both canonical and noncanonical Wnt signaling (Beales et al., 2007; Corbit et al., 2008; Gerdes et al., 2007; Simons et al., 2005). Overexpression of Inversin, which localizes to cilia and is required for left–right axis formation in mice, leads to disruption of canonical Wnt signaling (Simons et al., 2005). In contrast, cells lacking Kif3a show enhanced canonical Wnt signaling due to constitutive phosphorylation of disheveled (Corbit et al., 2008). These data have led to the hypothesis that the cilium and basal body may regulate the switch between canonical and noncanonical Wnt signaling. Defects in PCP have also been shown in BBS morpholino-injected zebrafish embryos resulting in shortening of the body along the anterior–posterior axis (Gerdes et al., 2007; Tayeh et al., 2008; Tobin et al., 2008).

The bone collar fails to develop in Prx1-Cre;Ift88/Kif3a$^{loxP/loxP}$ mutant mice, but it is not affected in the limbs of Col2a-Cre;Kif3a$^{loxP/loxP}$ mice suggesting cilia on the perichondrial cells are essential for bone collar formation and the defects are not due to loss of cilia on the chondrocytes (Haycraft et al., 2007; Song et al., 2007). In these mice and Dermo1-Cre;Kif3a$^{loxP/loxP}$ mutant mice, ectopic cartilage is seen along the diaphysis of the forming bone and Ihh signaling is lost as indicated by lack of Ptc1 or Gli1 expression (Haycraft et al., 2007; Kolpakova-Hart et al., 2007). While loss of Ihh signaling has been shown to result in defects in bone collar formation, it does not result in ectopic cartilage (St-Jacques et al., 1999). In contrast, disruption of β-catenin in the perichondrium results in a similar phenotype as seen in Prx1-Cre;Ift88/Kif3a$^{loxP/loxP}$ mutant mice and suggests that canonical Wnt signaling is disrupted in the perichondrium (Hilton et al., 2005; Hu et al., 2005). One function of Ihh is to induce expression of Wnt ligands. It is not known whether the similarity to conditional deletion of β-catenin is due to a requirement for cilia in canonical Wnt signaling in the perichondrial cells or loss of Wnt ligand expression downstream of Ihh signaling.

While the bone collar surrounding the long bones of Col2a-Cre; Kif3a$^{loxP/loxP}$ mutant mice develop normally, there are defects in bone formation in the synchondroses of the skull including excessive intramembraneous ossification and ectopic cartilage formation (Koyama et al., 2007). In these mutants, Gli1 and Ptc1 expression is present in the perichondrium and may be expanded where cilia formation is not disrupted. In agreement with expanded Ihh signaling, syndecan 3 expression, which has been proposed to restrict Ihh signaling in the chondrocytes, is absent providing a mechanism for increased Ptc1 and Gli1 expression in the adjacent perichondrium (Koyama et al., 2007).

6. Primary Cilia in the Maintenance of Bone

Mature osteoblasts secrete osteoid that is subsequently mineralized to form bone. As bone formation progresses, a subset of osteoblasts become trapped in the bone and differentiate into osteocytes. Osteocyte cell bodies reside within the lacunae of the bone and they extend processes through the canaliculi where they communicate through gap junctions. Canonical Wnt signaling has been implicated in the control of postnatal bone formation since activating mutations in the Frizzled co-receptor Lrp5 lead to increased bone mass in adults (reviewed in Bonewald and Johnson, 2008; Macsai et al., 2008).

Throughout life, the bones of the mature skeleton are remodeled in response to many stimuli including mechanical loading. Loss of mechanical stimulation leads to loss of bone mass while increased loading leads to formation of bone. Intriguingly, both Pc1 and Pc2 are localized to the primary cilium of cultured osteoblasts and osteocytes suggesting a role for the cilium in mechanosensation in the bone as well as the kidney as described above (Nauli and Zhou, 2004; Xiao et al., 2008). In mice, loss of Pc1 results in delayed vascular invasion and a thin bone collar. Specifically, *Pkd1* null samples show decreased Runx2-II expression as well as decreases in expression of other osteoblast/osteocyte specific genes such as osteocalcin and osterix (Xiao et al., 2006). *Pkd1* null osteoblasts cultured *in vitro* also had lower basal intracellular calcium than wild type controls. Despite the decrease in Runx2-II expression, no change in Runx2-I expression was detected (Xiao et al., 2006). Runx2-II is required for terminal osteoblast differentiation while Runx2-I is essential to early stages of osteoblast differentiation (Ducy et al., 1997; Xiao et al., 2004, 2005). This selective decrease in Runx2-II expression may explain the Pkd1 bone phenotype. Indeed, *Pkd1* null bones resemble those in *Runx2-II* mutant mice.

In addition to the embryonic defects in *Pkd1* null mice, heterozygous *Pkd1* adult mice show a 9% reduction in bone mineral density at 12 weeks of age (Xiao et al., 2006). Both trabecular bone volume and cortical bone thickness were affected suggesting a general requirement for Pc1 in postnatal bone maintenance. More recent work has shown that mice lacking one allele of *Pkd1* and one allele of *Runx2-II* have a more pronounced phenotype than either heterozygote alone (Xiao et al., 2008). While Pkd2 expression has been shown both in cultured osteoblasts and osteocytes by *in situ* hybridization, the role of Pc2 in the skeleton remains unknown (Markowitz et al., 1999; Xiao et al., 2006). However, loss of the Pc2-coupling domain of Pc1 abolished Runx2-II expression *in vitro* suggesting that Pc2 also plays a role in development and maintenance of the mammalian skeleton. Pc1 activation of Runx2-II transcription was also shown to require Pc1-mediated intracellular calcium increases resulting in increased binding of the transcription factor NFI to the Runx2-II promoter region (Xiao et al., 2008). How these defects are related to the proposed mechanosensation by Pc1, and Pc2 awaits further investigation.

In contrast to the requirement for intracellular calcium signaling via Pc1 shown for Runx2-II expression, other work has demonstrated that mechanosensation in bone cells occurs through a calcium-independent mechanism (Malone et al., 2007). MC3T3-E1 osteoblasts were subjected to dynamic flow and the cilium was observed to deflect in response to flow suggesting that it is positioned to receive and respond to mechanical stimuli *in vivo*. Mechanical stimulation has been shown to upregulate expression of osteopontin and synthesis of prostaglandin E_2 (Malone et al., 2007). In MC3T3-E1 or MLO-Y4 cells treated with Ift88-directed siRNA, no change in prostaglandin E_2 release was detected. Similarly, osteopontin mRNA levels did not increase in Ift88-siRNA treated cells. Calcium influx in response to flow has been shown and requires Pc2 in kidney epithelia (Praetorius and Spring, 2001). In contrast, no correlation between calcium influx in response to fluid flow, and expression of a cilium was observed in MC3T3-E1 or MLO-Y4 cells suggesting that cilia-mediated mechanosensation is calcium independent (Malone et al., 2007). Treatment of cells with gadolinium chloride to disrupt Pc2 function had no effect on calcium influx in response to flow indicating that calcium influx in response to flow in bone cells occurs through a different mechanism than in kidney epithelia.

The apparent discrepancies between the two proposed roles for cilia and Pc1 in the skeleton have yet to be investigated although it the results are not mutually exclusive. The work performed *in vivo* and with cultured cells did not examine the response of the cells to dynamic flow while the experiments performed under flow conditions did not investigate the potential role of Pc1 in the responses observed nor did the authors examine Runx2-II expression. It should be noted that alterations in bone mineral density have not been reported in patients with ADPKD. Additional experiments both *in vivo* and *in vitro* are necessary to address the role of cilia and Pc1 and Pc2 in osteoblast function and mechanosensation.

7. Primary Cilia in Craniofacial Development

Many human ciliopathies are characterized by defects in craniofacial development. The face develops from the branchial arches through a series of very complicated morphogenetic events (Chai and Maxson, 2006). The bones of the face are derived from cranial neural crest and cranial mesoderm. In addition to malformations in the digits as described above, OFD1 is characterized by malformations in the craniofacial region including clefts in the jaw, tongue, and upper lip. Female mice in which *CXORF5/Ofd1* was deleted using a ubiquitously expressed Cre demonstrated craniofacial malformations including a severe cleft palate and short snout (Ferrante *et al.*, 2006). Males died early in gestation with defects in left–right patterning. The molecular mechanism of the craniofacial defects was not examined; however, alterations in Hh signaling were noted in the neural tube, and alterations in Hox gene expression were detected in early limb buds (Ferrante *et al.*, 2006).

One of the characteristics of Meckels type 5 is cleft lip and palate. Meckels type 5 and Joubert type 7 (OMIN#611560) are allelic and, as described above, result from mutations in RPGRIP1L (Delous, 2007). A mouse homolog of this protein is called Fantom (Ftm). Null mutations of *Ftm* in mouse result in, among many defects, hypoplastic lower jaw and cleft lip with similarities to the human condition (Delous, 2007, Vierkotten *et al.*, 2007). RPGRIP1L and Ftm are both localized to the basal body but do not appear to disrupt the formation or maintenance of cilia structure (Delous *et al.*, 2007; Vierkotten *et al.*, 2007). Molecular mechanisms underlying the craniofacial defects have not been examined but alterations in Hh signaling were detected in mouse embryonic fibroblast cultured from mutant mice suggesting Hh signaling could be involved.

Craniofacial dymorphology has also been observed in some patients with BBS. Characteristic facial dysmorphology includes displacement of the nose, mid face hypoplasia, and mild mandibular retrognatia. *Bbs4*-null and *Bbs6*-null mice demonstrated a short snout due to premaxillary and mandibular hypoplasia (Tobin *et al.*, 2008). Using morpholino technology, fish in which various BBS proteins were depleted demonstrated defects in craniofacial development that were considered similar to those observed in humans and mice. Specifically, the mandibles were reduced and the branchial arch derivates were hypoplastic. Bbs8 depleted fish had the most severe phenotype including partial cyclopia, and in some cases the mandibles were completely missing (Tobin et al., 2008). Many of the bones in the face are derived from neural crest cells and it has been shown that migrating neural crest cells bear primary cilia (Tobin et al., 2008). Using the zebrafish model is was suggested that the craniofacial defects observed in BBS mutant fish were due to defects in migration of neural crest cells. It was shown that the neural crest proliferated and was specified and maintained. Further experiments with the zebrafish suggested that noncanonical Wnt, PCP, signaling was involved. Induction of noncanonical Wnt signaling using either a truncated form of Dishevelled (Dvl) or a membrane targeted form of Dvl partially rescued the phenotype in Bbs8 depleted fish (Tobin et al., 2008). Although Hh signaling was also disrupted, it was concluded that the defect in neural crest migration was primarily due to a defect in PCP signaling since a drug that blocks Hh signaling, cyclopamine, did not have any effect on migration of cranial neural crest (Tobin et al., 2008).

Craniofacial abnormalities including cleft palate and supernumerary teeth have been described in $Ift88^{orpk}$ mice (Zhang et al., 2003). A more severe craniofacial phenotype including severe frontonasal dysplasia and shortening of the lower jaw as well as profound cleft secondary palate was documented in mice with conditional deletion of Kif3a in Wnt1 expressing neural crest cells (Kolpakova-Hart et al., 2007). Patterning defects in the midline of the face including mid-facial clefting, missing tongue and missing incisors were also observed. Both endochondral and intramembranous bones of the head and face, which are primarily derived from neural crest, were affected. Since Wnt1-Cre expression was restricted to the neural crest, as expected structures derived from cephalic mesoderm were not affected. Similarities in Wnt1-Cre;Kif3a$^{loxP/loxP}$ mice to the craniofacial phenotypes seen in mice with conditional deletion of Smo in Wnt1 expressing cells suggested Hh signaling was altered in the absence of Kif3a (Jeong et al., 2004). In support of this model, Gli1 mRNA was limited to the epithelial structures and was absent from the mesenchyme in mutant mice whereas it was expressed in both the epithelial and mesenchymal components of the midface region in control mice (Kolpakova-Hart et al., 2007). The results suggested that Hh signaling was disrupted in the neural crest-derived mesenchyme leading to some of the defects observed. However, mid-facial clefting, like that seen in the Wnt1-Cre;Kif3a$^{loxP/loxP}$ mutants, was not observed in Smo mutants suggesting defects in midline patterning were due to disruptions in the Hh-independent repressor function of Gli3 although alterations in other signaling pathways, for example, PCP, could not be excluded (Kolpakova-Hart et al., 2007). Potential effects on neural crest migration were not specifically examined. Together, craniofacial defects in humans and mice with alterations in cilia-related genes indicate an important role for primary cilia and IFT in craniofacial development.

8. Primary Cilia in Tooth Development

Like the skeleton, the teeth are mineralized tissues and therefore share some similarities with the bones. Although they do not erupt through the gingiva until after birth in both mice and humans, teeth begin to develop during embryogenesis from the oral ectoderm and neural crest-derived mesenchyme (reviewed in Miletich and Sharpe, 2003). The oral ectoderm first thickens, then invaginates as tooth development begins. Underlying neural crest cells condense around the invaginating ectoderm and reciprocal signaling interactions elaborate the structure through several stages as development progresses. As in many tissues, both Hh and Wnt signaling play important roles at several stages in tooth development including patterning and morphogenesis. The mature tooth is composed of enamel secreted by epithelial-derived ameloblasts overlaying mineralized dentin secreted by neural crest-derived mesenchyme cells. While the odontoblasts are retained within the pulp cavity following eruption, the ameloblasts are lost. The dental pulp is the inner cavity of the tooth and contains many cell types including vasculature, mesenchymal stem cells and nerves. There are significant differences in tooth development and patterning between mice and humans. While mice develop a single set of permanent dentition consisting of incisors separated from molars by a toothless diastema, humans develop two sets of dentition, deciduous and permanent. The deciduous teeth are shed during childhood and replaced by the permanent dentition that is retained. Humans also develop additional tooth types between the incisors and molars including the canines and premolars. Despite these differences, the mouse has proved to be an invaluable research model to study tooth development.

Cilia have been observed on both ameloblasts and odontoblasts *in vivo* using electron microscopy (Magloire *et al.*, 2004; Sasano, 1986). On odontoblasts, cilia are hypothesized to play a role in mechanotransduction although this has yet to be investigated (Magloire *et al.*, 2004). Ameloblasts showed dynamic expression of cilia throughout their differentiation in rats and were predicted to be present on all postmitotic amelobalsts based on their prevelance in serial sections (Sasano, 1986).

In agreement with a role for cilia in tooth patterning, $Ift88^{orpk}$ mutants develop an ectopic molar mesial to the first molar although the mechanism for this has yet to be reported (Zhang *et al.*, 2003). As mentioned above, Wnt1-Cre;Kif3a$^{loxP/loxP}$ mutant mice lack incisors but this may be due to midline defects and may not reflect a requirement for cilia in the initiation of incisor development (Kolpakova-Hart *et al.*, 2007).

Defects in tooth number and morphology have also been reported in several ciliopathies but the molecular mechanism underlying the defects has not been investigated. Natal teeth, the eruption of teeth through the gingiva prior to birth, are a common feature of EvC. Additionally, EvC patients often present with missing permanent teeth (hypodontia or oligodontia) and malformed, often pointed, cusps. In agreement with a role for Evc in tooth morphogenesis, the first molar of *Evc* null mice is often smaller than that found in controls (Ruiz-Perez *et al.*, 2007). Significant tooth defects are also observed in patients with OFD1. The maxillary canines are often malpositioned and supernumary deciduous teeth are frequent including the canines and premolars. In addition to defects in the number of teeth, canines are often misshapen. Alterations in the dimensions of the pulp cavity have also been reported in a small number of BBS patients but there has been no investigation of the underlying cause and no defects in tooth morphogenesis have been reported in BBS mutant mice. Overall the role of the cilium in the formation of the mammalian dentition has not been examined in detail and awaits further investigation, but defects in Shh and Wnt signaling may be involved based on the phenotype observed in other tissues.

9. Summary

Defects in the primary cilium were first shown to be involved in the development of cysts in the kidney less than 10 years ago (reviewed in Bisgrove and Yost, 2006; Davenport and Yoder, 2005; Pan *et al.*, 2005; Satir and Christensen, 2007). Since that time, the expression and function of cilia has become an area of intense research. Mutations in proteins localized to the primary cilium or basal body lead to a diverse group of pleiotropic syndromes including BBS, MKS, EvC, and OFD1, which show defects in the skeleton including patterning of the digits, endochondral bone formation, craniofacial development and the dentition.

Cilia have been shown to play a role in signaling pathways important in the development of the skeleton including Hh and Wnt (Bisgrove and Yost, 2006; Pan *et al.*, 2005). Many of the defects in digit patterning are attributed to loss of activation or repression of Shh signaling. In the developing bones, Ihh signaling is also disrupted although the potential role for cilia in Wnt signaling in the perichondrium either through induction of Wnt ligand expression or canonical signaling distinct from that described in other tissues is still under investigation (Haycraft *et al.*, 2007; Koyama *et al.*, 2007).

Despite the fact the both bone cells and kidney epithelia respond to fluid flow and express primary cilia as well as the mechanosensitive Pc1 and Pc2 proteins, there are distinct differences in the function of this organelle in the two cell types (Malone *et al.*, 2007; Nauli and Zhou, 2004; Praetorius *et al.*, 2003). Fluid flow in canine kidney epithelia results in Pc2 mediated calcium influx while no link to calcium influx and cilia is seen in bone cells. In cartilage, cilia are in direct contact with the extracellular matrix but the potential function of Pc1 and Pc2 in the cartilage has not been determined. This highlights the importance of studying the function of the cilium and cilia-localized proteins in multiple cells types throughout development and postnatally since cell types may respond differently.

The various roles of cilia in the formation and maintenance of the mammalian skeleton are an emerging area of research and will provide important insight into the growing number of roles cilia play throughout the body.

ACKNOWLEDGMENTS

We would like to thank the members of our labs and S. R. McGlashan for helpful discussions during preparation of this manuscript.

REFERENCES

Alvarez, J., Sohn, P., Zeng, X., Doetschman, T., Robbins, D. J., and Serra, R. (2002). TGFbeta2 mediates the effects of hedgehog on hypertrophic differentiation and PTHrP expression. *Development* **129,** 1913–1924.

Ascenzi, M. G., Lenox, M., and Farnum, C. (2007). Analysis of the orientation of primary cilia in growth plate cartilage: A mathematical method based on multiphoton microscopical images. *J. Struct. Biol.* **158,** 293–306.

Aszodi, A., Hunziker, E. B., Brakebusch, C., and Fassler, R. (2003). Beta1 integrins regulate chondrocyte rotation, G1 progression, and cytokinesis. *Genes Dev.* **17,** 2465–2479.

Beales, P. L., Bland, E., Tobin, J. L., Bacchelli, C., Tuysuz, B., Hill, J., Rix, S., Pearson, C. G., Kai, M., Hartley, J., Johnson, C., Irving, M., et al. (2007). IFT80, which encodes a conserved intraflagellar transport protein, is mutated in Jeune asphyxiating thoracic dystrophy. *Nat. Genet.* **39,** 727–729.

Bisgrove, B. W., and Yost, H. J. (2006). The roles of cilia in developmental disorders and disease. *Development* **133,** 4131–4143.

Bonewald, L. F., and Johnson, M. L. (2008). Osteocytes, mechanosensing and Wnt signaling. *Bone* **42,** 606–615.

Boulter, C., Mulroy, S., Webb, S., Fleming, S., Brindle, K., and Sandford, R. (2001). Cardiovascular, skeletal, and renal defects in mice with a targeted disruption of the Pkd1 gene. *Proc. Natl. Acad. Sci. USA* **98,** 12174–12179.

Cancedda, R., Cancedda, F. D., and Castagnola, P. (1995). Chondrocyte Differentiation. *International Review of Cytology* **159,** 265–358.

Chai, Y., and Maxson, R. E., Jr. (2006). Recent advances in craniofacial morphogenesis. *Dev. Dyn.* **235,** 2353–2375.

Chiang, C., Litingtung, Y., Harris, M. P., Simandl, B. K., Li, Y., Beachy, P. A., and Fallon, J. F. (2001). Manifestation of the limb prepattern: Limb development in the absence of sonic hedgehog function. *Dev. Biol.* **236,** 421–435.

Clevers, H. (2006). Wnt/beta-catenin signaling in development and disease. *Cell* **127,** 469–480.

Colnot, C. (2005). Cellular and molecular interactions regulating skeletogenesis. *J. Cell Biochem.* **95,** 688–697.

Corbit, K. C., Aanstad, P., Singla, V., Norman, A. R., Stainier, D. Y., and Reiter, J. F. (2005). Vertebrate smoothened functions at the primary cilium. *Nature* **437,** 1018–1021.

Corbit, K. C., Shyer, A. E., Dowdle, W. E., Gaulden, J., Singla, V., Chen, M. H., Chuang, P. T., and Reiter, J. F. (2008). Kif3a constrains beta-catenin-dependent Wnt signalling through dual ciliary and non-ciliary mechanisms. *Nat. Cell Biol.* **10,** 70–76.

Davenport, J. R., and Yoder, B. K. (2005). An incredible decade for the primary cilium: A look at a once-forgotten organelle. *Am. J. Physiol. Renal Physiol.* **289,** F1159–F1169.

Day, T. F., and Yang, Y. (2008). Wnt and hedgehog signaling pathways in bone development. *J. Bone Joint. Surg. Am.* **90**(Suppl 1), 19–24.

Delous, M., Baala, L., Salomon, R., Laclef, C., Vierkotten, J., Tory, K., Golzio, C., Lacoste, T., Besse, L., Ozilou, C., Moutkine, I., Hellman, N. E., et al. (2007). The ciliary gene RPGRIP1L is mutated in cerebello-oculo-renal syndrome (Joubert syndrome type B) and Meckel syndrome. *Nat. Genet.* **39,** 875–881.

Dodds, G. S. (1930). Row formation and other types of arrangement of cartilage cells in endochondral ossification. *Anatomical Record* **46,** 385–399.

Ducy, P., Zhang, R., Geoffroy, V., Ridall, A. L., and Karsenty, G. (1997). Osf2/Cbfa1: A transcriptional activator of osteoblast differentiation. *Cell* **89,** 747–754.

Fath, M. A., Mullins, R. F., Searby, C., Nishimura, D. Y., Wei, J., Rahmouni, K., Davis, R. E., Tayeh, M. K., Andrews, M., Yang, B., Sigmund, C. D., Stone, E. M., et al. (2005). Mkks-null mice have a phenotype resembling Bardet-Biedl syndrome. *Hum. Mol. Genet.* **14,** 1109–1118.

Ferrante, M. I., Giorgio, G., Feather, S. A., Bulfone, A., Wright, V., Ghiani, M., Selicorni, A., Gammaro, L., Scolari, F., Woolf, A. S., Sylvie, O., Bernard, L., et al. (2001). Identification of the gene for oral-facial-digital type I syndrome. *Am. J. Hum. Genet.* **68,** 569–576.

Ferrante, M. I., Zullo, A., Barra, A., Bimonte, S., Messaddeq, N., Studer, M., Dolle, P., and Franco, B. (2006). Oral-facial-digital type I protein is required for primary cilia formation and left-right axis specification. *Nat. Genet.* **38,** 112–117.

Gerdes, J. M., Liu, Y., Zaghloul, N. A., Leitch, C. C., Lawson, S. S., Kato, M., Beachy, P. A., Beales, P. L., DeMartino, G. N., Fisher, S., Badano, J. L., and Katsanis, N. (2007). Disruption of the basal body compromises proteasomal function and perturbs intracellular Wnt response. *Nat. Genet.* **39,** 1350–1360.

Gouttenoire, J., Valcourt, U., Bougault, C., Aubert-Foucher, E., Arnaud, E., Giraud, L., and Mallein-Gerin, F. (2007). Knockdown of the intraflagellar transport protein IFT46 stimulates selective gene expression in mouse chondrocytes and affects early development in zebrafish. *J. Biol. Chem.* **282,** 30960–30973.

Guillaume, R., D'Agati, V., Daoust, M., and Trudel, M. (1999). Murine Pkd1 is a developmentally regulated gene from morula to adulthood: Role in tissue condensation and patterning. *Dev. Dyn.* **214,** 337–348.

Haycraft, C. J., Banizs, B., Aydin-Son, Y., Zhang, Q., Michaud, E. J., and Yoder, B. K. (2005). Gli2 and Gli3 localize to cilia and require the intraflagellar transport protein polaris for processing and function. *PLoS Genet.* **1,** e53.

Haycraft, C. J., Zhang, Q., Song, B., Jackson, W. S., Detloff, P. J., Serra, R., and Yoder, B. K. (2007). Intraflagellar transport is essential for endochondral bone formation. *Development* **134,** 307–316.

Hilton, M. J., Tu, X., Cook, J., Hu, H., and Long, F. (2005). Ihh controls cartilage development by antagonizing Gli3, but requires additional effectors to regulate osteoblast and vascular development. *Development* **132,** 4339–4351.

Hu, H., Hilton, M. J., Tu, X., Yu, K., Ornitz, D. M., and Long, F. (2005). Sequential roles of Hedgehog and Wnt signaling in osteoblast development. *Development* **132,** 49–60.

Huangfu, D., and Anderson, K. V. (2005). Cilia and Hedgehog responsiveness in the mouse. *Proc. Natl. Acad. Sci. USA* **102,** 11325–11330.

Huangfu, D., and Anderson, K. V. (2006). Signaling from Smo to Ci/Gli: Conservation and divergence of Hedgehog pathways from *Drosophila* to vertebrates. *Development* **133**, 3–14.

Jensen, C. G., Poole, C. A., McGlashan, S. R., Marko, M., Issa, Z. I., Vujcich, K. V., and Bowser, S. S. (2004). Ultrastructural, tomographic and confocal imaging of the chondrocyte primary cilium *in situ*. *Cell Biol. Int.* **28**, 101–110.

Jeong, J., Mao, J., Tenzen, T., Kottmann, A. H., and McMahon, A. P. (2004). Hedgehog signaling in the neural crest cells regulates the patterning and growth of facial primordia. *Genes Dev.* **18**, 937–951.

Katsanis, N., Ansley, S. J., Badano, J. L., Eichers, E. R., Lewis, R. A., Hoskins, B. E., Scambler, P. J., Davidson, W. S., Beales, P. L., and Lupski, J. R. (2001). Triallelic inheritance in Bardet-Biedl syndrome, a Mendelian recessive disorder. *Science* **293**, 2256–2259.

Katsanis, N., Eichers, E. R., Ansley, S. J., Lewis, R. A., Kayserili, H., Hoskins, B. E., Scambler, P. J., Beales, P. L., and Lupski, J. R. (2002). BBS4 is a minor contributor to Bardet-Biedl syndrome and may also participate in triallelic inheritance. *Am. J. Hum. Genet.* **71**, 22–29.

Kolpakova-Hart, E., Jinnin, M., Hou, B., Fukai, N., and Olsen, B. R. (2007). Kinesin-2 controls development and patterning of the vertebrate skeleton by Hedgehog- and Gli3- dependent mechanisms. *Dev. Biol.* **309**, 273–284.

Koyama, E., Young, B., Nagayama, M., Shibukawa, Y., Enomoto-Iwamoto, M., Iwamoto, M., Maeda, Y., Lanske, B., Song, B., Serra, R., and Pacifici, M. (2007). Conditional Kif3a ablation causes abnormal hedgehog signaling topography, growth plate dysfunction, and excessive bone and cartilage formation during mouse skeletogenesis. *Development* **134**, 2159–2169.

Koziel, L., Wuelling, M., Schneider, S., and Vortkamp, A. (2005). Gli3 acts as a repressor downstream of Ihh in regulating two distinct steps of chondrocyte differentiation. *Development* **132**, 5249–5260.

Kulaga, H. M., Leitch, C. C., Eichers, E. R., Badano, J. L., Lesemann, A., Hoskins, B. E., Lupski, J. R., Beales, P. L., Reed, R. R., and Katsanis, N. (2004). Loss of BBS proteins causes anosmia in humans and defects in olfactory cilia structure and function in the mouse. *Nat. Genet.* **36**, 994–998.

Kyttala, M., Tallila, J., Salonen, R., Kopra, O., Kohlschmidt, N., Paavola-Sakki, P., Peltonen, L., and Kestila, M. (2006). MKS1, encoding a component of the flagellar apparatus basal body proteome, is mutated in Meckel syndrome. *Nat. Genet.* **38**, 155–157.

Lanske, B., Karapalis, A. C., Lee, K., Luz, A., Vortkamp, A., Pirro, A., Karperien, M., Defize, L. H. K., Ho, C., Mulligan, R. C., Abou-Samra, A.-B., Juppner, H., *et al.* (1996). PTH/PTHrP receptor in early development and indian hedgehog-regulated bone growth. *Science* **273**, 663–666.

Liu, A., Wang, B., and Niswander, L. A. (2005). Mouse intraflagellar transport proteins regulate both the activator and repressor functions of Gli transcription factors. *Development* **132**, 3103–3111.

Logan, M., Martin, J. F., Nagy, A., Lobe, C., Olson, E. N., and Tabin, C. J. (2002). Expression of Cre recombinase in the developing mouse limb bud driven by a Prxl enhancer. *Genesis* **33**, 77–80.

Long, F., Zhang, X. M., Karp, S., Yang, Y., and McMahon, A. P. (2001). Genetic manipulation of hedgehog signaling in the endochondral skeleton reveals a direct role in the regulation of chondrocyte proliferation. *Development* **128**, 5099–5108.

Lu, W., Shen, X., Pavlova, A., Lakkis, M., Ward, C. J., Pritchard, L., Harris, P. C., Genest, D. R., Perez-Atayde, A. R., and Zhou, J. (2001). Comparison of Pkd1-targeted mutants reveals that loss of polycystin-1 causes cystogenesis and bone defects. *Hum. Mol. Genet.* **10**, 2385–2396.

Macsai, C. E., Foster, B. K., and Xian, C. J. (2008). Roles of Wnt signalling in bone growth, remodelling, skeletal disorders and fracture repair. *J. Cell Physiol.* **215,** 578–587.

Maeda, Y., Nakamura, E., Nguyen, M. T., Suva, L. J., Swain, F. L., Razzaque, M. S., Mackem, S., and Lanske, B. (2007). Indian Hedgehog produced by postnatal chondrocytes is essential for maintaining a growth plate and trabecular bone. *Proc. Natl. Acad. Sci. USA* **104,** 6382–6387.

Magloire, H., Couble, M. L., Romeas, A., and Bleicher, F. (2004). Odontoblast primary cilia: Facts and hypotheses. *Cell Biol. Int.* **28,** 93–99.

Mai, W., Chen, D., Ding, T., Kim, I., Park, S., Cho, S. Y., Chu, J. S., Liang, D., Wang, N., Wu, D., Li, S., Zhao, P., Zent, R., and Wu, G. (2005). Inhibition of Pkhd1 impairs tubulomorphogenesis of cultured IMCD cells. *Mol. Biol. Cell* **16,** 4398–4409.

Malone, A. M., Anderson, C. T., Tummala, P., Kwon, R. Y., Johnston, T. R., Stearns, T., and Jacobs, C. R. (2007). Primary cilia mediate mechanosensing in bone cells by a calcium-independent mechanism. *Proc. Natl. Acad. Sci. USA* **104,** 13325–13330.

Markowitz, G. S., Cai, Y., Li, L., Wu, G., Ward, L. C., Somlo, S., and D'Agati, V. D. (1999). Polycystin-2 expression is developmentally regulated. *Am. J. Physiol.* **277,** F17–F25.

McGlashan, S. R., Cluett, E. C., Jensen, C. G., and Poole, C. A. (2008). Primary cilia in osteoarthritic chondrocytes: From chondrons to clusters. *Dev. Dyn.* **237,** 2013–2020.

McGlashan, S. R., Haycraft, C. J., Jensen, C. G., Yoder, B. K., and Poole, C. A. (2007). Articular cartilage and growth plate defects are associated with chondrocyte cytoskeletal abnormalities in Tg737(orpk) mice lacking the primary cilia protein polaris. *Matrix Biol.* **26,** 234–246.

McGlashan, S. R., Jensen, C. G., and Poole, C. A. (2006). Localization of extracellular matrix receptors on the chondrocyte primary cilium. *J. Histochem. Cytochem.* **54,** 1005–1014.

Miletich, I., and Sharpe, P. T. (2003). Normal and abnormal dental development. *Hum. Mol. Genet.* **12**(Spec No 1), R69–R73.

Millward-Sadler, S. J., and Salter, D. M. (2004). Integrin-dependent signal cascades in chondrocyte mechanotransduction. *Ann. Biomed. Eng.* **32,** 435–446.

Nauli, S. M., and Zhou, J. (2004). Polycystins and mechanosensation in renal and nodal cilia. *Bioessays* **26,** 844–856.

Ovchinnikov, D. A., Deng, J. M., Ogunrinu, G., and Behringer, R. R. (2000). Col2a1-directed expression of Cre recombinase in differentiating chondrocytes in transgenic mice. *Genesis* **26,** 145–146.

Pan, J., Wang, Q., and Snell, W. J. (2005). Cilium-generated signaling and cilia-related disorders. *Lab. Invest.* **85,** 452–463.

Park, T. J., Haigo, S. L., and Wallingford, J. B. (2006). Ciliogenesis defects in embryos lacking inturned or fuzzy function are associated with failure of planar cell polarity and Hedgehog signaling. *Nat. Genet.* **38,** 303–311.

Poole, C. A., Flint, M. H., and Beaumont, B. W. (1985). Analysis of the morphology and function of primary cilia in connective tissues: A cellular cybernetic probe? *Cell Motil.* **5,** 175–193.

Poole, C. A., Jensen, C. G., Snyder, J. A., Gray, C. G., Hermanutz, V. L., and Wheatley, D. N. (1997). Confocal analysis of primary cilia structure and colocalization with the Golgi apparatus in chondrocytes and aortic smooth muscle cells. *Cell Biol. Int.* **21,** 483–494.

Poole, C. A., Zhang, Z. J., and Ross, J. M. (2001). The differential distribution of acetylated and detyrosinated alpha-tubulin in the microtubular cytoskeleton and primary cilia of hyaline cartilage chondrocytes. *J. Anat.* **199,** 393–405.

Praetorius, H. A., Frokiaer, J., Nielsen, S., and Spring, K. R. (2003). Bending the primary cilium opens Ca2+-sensitive intermediate-conductance K+ channels in MDCK cells. *J. Membr. Biol.* **191,** 193–200.

Praetorius, H. A., Praetorius, J., Nielsen, S., Frokiaer, J., and Spring, K. R. (2004). Beta1-integrins in the primary cilium of MDCK cells potentiate fibronectin-induced Ca2+ signaling. *Am. J. Physiol. Renal. Physiol.* **287,** F969–F978.

Praetorius, H. A., and Spring, K. R. (2001). Bending the MDCK cell primary cilium increases intracellular calcium. *J. Membr. Biol.* **184,** 71–79.

Praetorius, H. A., and Spring, K. R. (2003a). Removal of the MDCK cell primary cilium abolishes flow sensing. *J. Membr. Biol.* **191,** 69–76.

Praetorius, H. A., and Spring, K. R. (2003b). The renal cell primary cilium functions as a flow sensor. *Curr. Opin. Nephrol. Hypertens.* **12,** 517–520.

Rohatgi, R., Milenkovic, L., and Scott, M. P. (2007). Patched1 regulates hedgehog signaling at the primary cilium. *Science* **317,** 372–376.

Romio, L., Fry, A. M., Winyard, P. J., Malcolm, S., Woolf, A. S., and Feather, S. A. (2004). OFD1 is a centrosomal/basal body protein expressed during mesenchymal-epithelial transition in human nephrogenesis. *J. Am. Soc. Nephrol.* **15,** 2556–2568.

Ruiz-Perez, V. L., Blair, H. J., Rodriguez-Andres, M. E., Blanco, M. J., Wilson, A., Liu, Y. N., Miles, C., Peters, H., and Goodship, J. A. (2007). Evc is a positive mediator of Ihh-regulated bone growth that localises at the base of chondrocyte cilia. *Development* **134,** 2903–2912.

Ruiz-Perez, V. L., Tompson, S. W., Blair, H. J., Espinoza-Valdez, C., Lapunzina, P., Silva, E. O., Hamel, B., Gibbs, J. L., Young, I. D., Wright, M. J., and Goodship, J. A. (2003). Mutations in two nonhomologous genes in a head-to-head configuration cause Ellis-van Creveld syndrome. *Am. J. Hum. Genet.* **72,** 728–732.

Sasano, Y. (1986). Dynamic behavior of ciliated centrioles in rat incisor ameloblasts during cell differentiation. *Arch. Histol. Jpn.* **49,** 437–448.

Satir, P., and Christensen, S. T. (2007). Overview of structure and function of mammalian cilia. *Annu. Rev. Physiol.* **69,** 377–400.

Scherft, J. P., and Daems, W. T. (1967). Single cilia in chondrocytes. *J. Ultrastruct. Res.* **19,** 546–555.

Simons, M., Gloy, J., Ganner, A., Bullerkotte, A., Bashkurov, M., Kronig, C., Schermer, B., Benzing, T., Cabello, O. A., Jenny, A., Mlodzik, M., Polok, B., et al. (2005). Inversin, the gene product mutated in nephronophthisis type II, functions as a molecular switch between Wnt signaling pathways. *Nat. Genet.* **37,** 537–543.

Song, B., Haycraft, C. J., Seo, H. S., Yoder, B. K., and Serra, R. (2007). Development of the post-natal growth plate requires intraflagellar transport proteins. *Dev. Biol.* **305,** 202–216.

St-Jacques, B., Hammerschmidt, M., and McMahon, A. P. (1999). Indian hedgehog signaling regulates proliferation and differentiation of chondrocytes and is essential for bone formation. *Genes Dev.* **13,** 2072–2086.

Tayeh, M. K., Yen, H. J., Beck, J. S., Searby, C. C., Westfall, T. A., Griesbach, H., Sheffield, V. C., and Slusarski, D. C. (2008). Genetic interaction between Bardet-Biedl syndrome genes and implications for limb patterning. *Hum. Mol. Genet.* **17,** 1956–1967.

te Welscher, P., Zuniga, A., Kuijper, S., Drenth, T., Goedemans, H. J., Meijlink, F., and Zeller, R. (2002). Progression of vertebrate limb development through SHH-mediated counteraction of GLI3. *Science* **298,** 827–830.

Tickle, C. (2006). Making digit patterns in the vertebrate limb. *Nat. Rev. Mol. Cell Biol.* **7,** 45–53.

Tobin, J. L., Di Franco, M., Eichers, E., May-Simera, H., Garcia, M., Yan, J., Quinlan, R., Justice, M. J., Hennekam, R. C., Briscoe, J., Tada, M., Mayor, R., et al. (2008). Inhibition of neural crest migration underlies craniofacial dysmorphology and Hirschsprung's disease in Bardet-Biedl syndrome. *Proc. Natl. Acad. Sci. USA* **105,** 6714–6719.

Tran, P. V., Haycraft, C. J., Besschetnova, T. Y., Turbe-Doan, A., Stottmann, R. W., Herron, B. J., Chesebro, A. L., Qiu, H., Scherz, P. J., Shah, J. V., Yoder, B. K., and Beier, D. R. (2008). THM1 negatively modulates mouse sonic hedgehog signal transduction and affects retrograde intraflagellar transport in cilia. *Nat. Genet.* **40,** 403–410.

Turco, A. E., Padovani, E. M., Chiaffoni, G. P., Peissel, B., Rossetti, S., Marcolongo, A., Gammaro, L., Maschio, G., and Pignatti, P. F. (1993). Molecular genetic diagnosis of autosomal dominant polycystic kidney disease in a newborn with bilateral cystic kidneys detected prenatally and multiple skeletal malformations. *J. Med. Genet.* **30,** 419–422.

van der Eerden, B. C., Karperien, M., and Wit, J. M. (2003). Systemic and local regulation of the growth plate. *Endocr. Rev.* **24,** 782–801.

Veeman, M. T., Axelrod, J. D., and Moon, R. T. (2003). A second canon. Functions and mechanisms of beta-catenin-independent Wnt signaling. *Dev. Cell* **5,** 367–377.

Vierkotten, J., Dildrop, R., Peters, T., Wang, B., and Ruther, U. (2007). Ftm is a novel basal body protein of cilia involved in Shh signalling. *Development* **134,** 2569–2577.

Vortkamp, A., Lee, K., Lanske, B., Segre, G. V., Kroneberg, H. M., and Tabin, C. J. (1996). Regulation of rate of chondrocyte differentiation by indian hedgehog and PTH-related protein. *Science* **273,** 613–621.

Wang, B., Fallon, J. F., and Beachy, P. A. (2000). Hedgehog-regulated processing of Gli3 produces an anterior/posterior repressor gradient in the developing vertebrate limb. *Cell* **100,** 423–434.

Wilsman, N. J., Farnum, C. E., and Reed-Aksamit, D. K. (1980). Incidence and morphology of equine and murine chondrocytic cilia. *Anat. Rec.* **197,** 355–361.

Wilsman, N. J., and Fletcher, T. F. (1978). Cilia of neonatal articular cartilage: Incident and morphology. *Anatomical Record* **190,** 871–889.

Xiao, Z., Awad, H. A., Liu, S., Mahlios, J., Zhang, S., Guilak, F., Mayo, M. S., and Quarles, L. D. (2005). Selective Runx2-II deficiency leads to low-turnover osteopenia in adult mice. *Dev. Biol.* **283,** 345–356.

Xiao, Z., Zhang, S., Magenheimer, B. S., Luo, J., and Quarles, L. D. (2008). Polycystin-1 regulates skeletogenesis through stimulation of the osteoblast-specific transcription factor RUNX2-II. *J. Biol. Chem.* **283,** 12624–12634.

Xiao, Z., Zhang, S., Mahlios, J., Zhou, G., Magenheimer, B. S., Guo, D., Dallas, S. L., Maser, R., Calvet, J. P., Bonewald, L., and Quarles, L. D. (2006). Cilia-like structures and polycystin-1 in osteoblasts/osteocytes and associated abnormalities in skeletogenesis and Runx2 expression. *J. Biol. Chem.* **281,** 30884–30895.

Xiao, Z. S., Hjelmeland, A. B., and Quarles, L. D. (2004). Selective deficiency of the "bone-related" Runx2-II unexpectedly preserves osteoblast-mediated skeletogenesis. *J. Biol. Chem.* **279,** 20307–20313.

Yoder, B. K., Hou, X., and Guay-Woodford, L. M. (2002). The polycystic kidney disease proteins, polycystin-1, polycystin-2, polaris, and cystin, are co-localized in renal cilia. *J. Am. Soc. Nephrol.* **13,** 2508–2516.

Yu, K., Xu, J., Liu, Z., Sosic, D., Shao, J., Olson, E. N., Towler, D. A., and Ornitz, D. M. (2003). Conditional inactivation of FGF receptor 2 reveals an essential role for FGF signaling in the regulation of osteoblast function and bone growth. *Development* **130,** 3063–3074.

Zerath, E., Holy, X., Mouillon, J. M., Farbos, B., Machwate, M., Andre, C., Renault, S., and Marie, P. J. (1997). TGF-beta2 prevents the impaired chondrocyte proliferation induced by unloading in growth plates of young rats. *Life Sci.* **61,** 2397–2406.

Zhang, Q., Murcia, N. S., Chittenden, L. R., Richards, W. G., Michaud, E. J., Woychik, R. P., and Yoder, B. K. (2003). Loss of the Tg737 protein results in skeletal patterning defects. *Dev. Dyn.* **227,** 78–90.

CHAPTER TWELVE

Olfactory Cilia: Our Direct Neuronal Connection to the External World

Dyke P. McEwen, Paul M. Jenkins, *and* Jeffrey R. Martens

Contents

1. Olfaction as a Sensory Modality	334
2. Anatomy and Organization of the Olfactory Epithelium	335
2.1. Gross anatomy	335
2.2. Cell types and ultrastructure	337
2.3. Regeneration	338
3. Structure of Olfactory Cilia	338
3.1. Cilia axoneme	339
3.2. Lipid composition	340
3.3. Ciliary necklace/transition zone	342
3.4. Basal body	343
3.5. Ciliary rootlet	343
4. Formation of Olfactory Cilia	344
5. Intraflagellar Transport	347
6. Ciliary Proteome	349
6.1. Olfactory signaling through the canonical G protein-coupled pathway	349
6.2. Signaling through noncanonical pathways	351
6.3. Bioinformatics and expression profiling	352
6.4. Ciliary proteomics	353
6.5. Regulation of ciliary protein entry	354
7. Olfactory Cilia and Disease	355
7.1. Olfactory ciliopathies	355
7.2. Head trauma	356
7.3. Chronic rhinosinusitus	357
7.4. The olfactory system as a pathogenic target	358
8. Summary	358
References	359

Department of Pharmacology, University of Michigan, Ann Arbor, MI 48109–5632

Current Topics in Developmental Biology, Volume 85   © 2008 Elsevier Inc.
ISSN 0070-2153, DOI: 10.1016/S0070-2153(08)00812-0   All rights reserved.

Abstract

An organism's awareness of its surroundings is dependent on sensory function. As antennas to our external environment, cilia are involved in fundamental biological processes such as olfaction, photoreception, and touch. The olfactory system has adapted this organelle for its unique sensory function and optimized it for detection of external stimuli. The elongated and tapering structure of olfactory cilia and their organization into an overlapping meshwork bathed by the nasal mucosa is optimized to enhance odor absorption and detection. As many as 15–30 nonmotile, sensory cilia on dendritic endings of single olfactory sensory neurons (OSNs) compartmentalize signaling molecules necessary for odor detection allowing for efficient and spatially confined responses to sensory stimuli. Although the loss of olfactory cilia or deletion of selected components of the olfactory signaling cascade leads to anosmia, the mechanisms of ciliogenesis and the selected enrichment of signaling molecules remain poorly understood. Much of our current knowledge is the result of elegant electron microscopy studies describing the structure and organization of the olfactory epithelium and cilia. New genetic and cell biological approaches, which compliment these early studies, show promise in elucidating the mechanisms of olfactory cilia assembly, maintenance, and compartmentalization. Importantly, emerging evidence suggests that olfactory dysfunction represents a previously unrecognized clinical manifestation of multiple ciliary disorders. Future work investigating the mechanisms of olfactory dysfunction combining both clinical studies with basic science research will provide us important new information regarding the pathogenesis of human sensory perception diseases.

1. Olfaction as a Sensory Modality

Chemosensory systems, such as the olfactory and gustatory systems, give us an awareness of our immediate environment by allowing us to detect airborne and fluidborne stimuli. The gustatory and olfactory systems are thought to have evolved from the chemical sensing apparatuses, which can be found on almost all creatures in the animal kingdom (Hildebrand and Shepherd, 1997). The olfactory system is responsible for the detection of volatile chemicals dissolved in the air around us. While the olfactory system is necessary for detecting odors and crucial for our sense of taste, it also plays important roles in our quality of life, health, and safety. Dysosmia (impaired sense of smell) or anosmia (loss of ability to smell) can prevent us from detecting signs of danger such as smoke or spoiled food, and also can lead to medical problems such as weight gain and poor nutrition (Toller, 1999). Impaired olfactory function is thought to affect over 2 million Americans and over 50% of those over the age of 65 (Murphy et al., 2002), however, this may be a gross underestimate given that olfactory dysfunction frequently goes unreported (Nguyen-Khoa *et al.*, 2007). While in some cases we understand the cause of olfactory dysfunction, in at least 20% of cases the underlying etiology remains unknown (Jafek, 2000).

Inhalation of odorants across the surface of the olfactory epithelium (OE) initiates the olfactory signaling cascade, which involves the binding of odorants to receptors localized on the cilia of olfactory sensory neurons (OSNs). In the canonical pathway, activated odorant receptors (ORs) act through a stimulatory G protein-coupled mechanism to activate adenylyl cyclase type III (ACIII) and increase the ciliary concentration of cAMP. Olfactory cyclic nucleotide-gated (CNG) channels open in response to cAMP binding and allow the depolarization of the OSN that is further amplified by the Ca^{2+}-activated Cl^- channel. All of the components necessary for odorant detection are enriched in olfactory cilia, and any perturbation in the localization of these components or in cilia themselves causes impaired olfactory function. In this chapter, we will focus on olfactory cilia including structure and function, developmental formation and relation to human disease.

2. ANATOMY AND ORGANIZATION OF THE OLFACTORY EPITHELIUM

2.1. Gross anatomy

In humans, the olfactory and respiratory epithelia line the three bony turbinates and cartilaginous septum of the nasal passage. In other mammals such as rodents there is a larger number of turbinates (I, II, IIb, III, IV), which presumably either fused or were lost during evolution to bipedalism (Harkema *et al.*, 2006). In the human, the respiratory epithelium, which is responsible for warming, cleaning, and humidifying inspired air, lines most of the inferior and middle turbinates as well as a portion of the superior turbinate. The OE, however, is segregated from the respiratory epithelium and is responsible for the detection of volatile chemicals dissolved in the air. It lines part of the nasal septum, the remainder of the superior turbinate and potentially part of the middle turbinate with a combined surface area of 1–2 cm^2 (Bucher, 1973; Leopold *et al.*, 2000). As odorant-containing air passes across the turbinates it is exposed to the OE, which contains the sensory element of the olfactory system, the OSN.

Inspired odorants bind to ORs localized on the ciliary membrane of OSNs to stimulate the olfactory sensory cascade (Fig. 12.1E). Bundled axons from OSNs form the olfactory nerve and serve to transmit their information to synapses on mitral cells and tufted cells in the glomeruli of the main olfactory bulb (MOB) (Hinds and Hinds, 1976a,b). From the

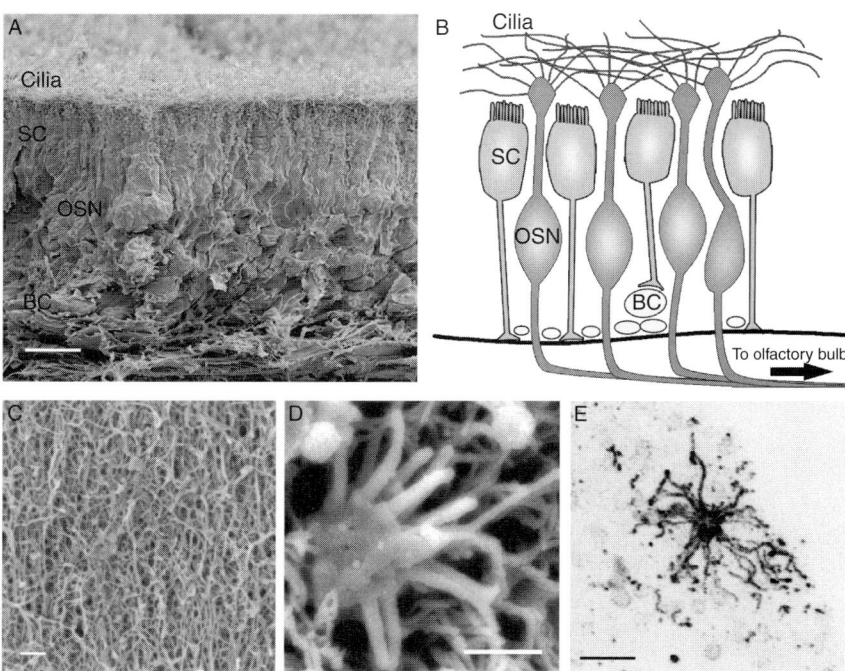

Figure 12.1 *Anatomy of the olfactory epithelium.* Mouse olfactory epithelium was dissected, fixed in glutaraldehyde, and processed for scanning electron microscopy as previously described (McEwen *et al.*, 2007). Scanning electron micrographs were captured using an Amray (Drogheda, Ireland) 1910FE field emission scanning electron microscope at 5 kV. Images were recorded digitally with Semicaps software. (A) Side view of a scanning electron micrograph at 1030× zoom from mouse olfactory epithelium (image courtesy of Wanda Layman and Donna Martin, Department of Human Genetics, University of Michigan). The layers of the olfactory epithelium are labeled on the left of the image. For (A) and (B), OSN = olfactory sensory neuron, SC = sustentacular cell, BC = basal cell. Scale bar represents 10 μm. (B) Cartoon representation of the organization of the various cell types in the olfactory epithelium. (C) Scanning electron micrograph of a surface view of the olfactory epithelium shown at 7400 × zoom. The dense meshwork of overlapping cilia across the surface of the epithelium is visible. Scale bar represents 1 μm. (D) Scanning electron micrograph of a single dendritic knob showing multiple cilia extending from the surface of the knob. Image is shown at 16,000× zoom. Scale bar is 1 μm. (E) Immunocytochemistry of mouse olfactory epithelium stained for an odorant receptor, mOR28 (antibody courtesy of Dr. Richard Axel). A 14-μm thick slice of mouse olfactory epithelium was stained as described previously (McEwen *et al.*, 2007). Numerous cilia can be observed expressing mOR28 and extending from a single dendritic knob. Scale bar is 5 μm.

MOB, olfactory information is sent to higher order centers of the brain for processing, including the amygdala, anterior olfactory nucleus, olfactory tubercle, piriform cortex, and entorhinal cortex (Haberly, 2001; Lledo *et al.*, 2005).

2.2. Cell types and ultrastructure

The main OE is a stratified epithelium composed of several cell types (Fig. 12.1A and B). The OSN is the main sensory cell, which houses the elements of the olfactory sensory cascade. OSNs are bipolar neurons with long axons projecting through the bony cribiform plate into the olfactory bulb, and relatively short dendrites terminating in a specialized ending termed a dendritic knob. The dendritic knob contains multiple basal bodies from which the olfactory cilia project into the mucous of the OE (Fig. 12.1D) (Cuschieri and Bannister, 1975a,b).

Surrounding the OSNs is a layer of supporting cells, termed sustentacular cells (Fig. 12.1A and B). The sustentacular cells contain many microvilli on their apical surface that underlie the cilia layer of the OE. Sustentacular cells have been shown to play a role in water balance, regulation of mucous ion composition (along with the Bowman's glands), drug metabolism, and purinergic modulation of odor sensitivity (Carr et al., 2001; Hegg et al., 2003; Kern and Pitovski, 1997; Menco et al., 1998), however, their precise function remains unknown.

In addition to the OSN and sustentacular cell, there is also a population of stem cells, termed basal cells, capable of replenishing the OSN and sustentacular cell population (Fig. 12.1A and B). The basal cell layer is composed of two types of cells: the globose basal cell (GBC) and the horizontal basal cell (HBC). Although it is a matter of continued debate, it is thought that HBCs divide slowly and replenish the GBCs, which in turn allow the regeneration of new OSNs (Caggiano et al., 1994; Iwai et al., 2008; for review see Murdoch and Roskams, 2007).

While sustentacular cells possess microvilli, there are also five other distinct microvillous cells types in the OE that are found in much lower abundance than OSNs, sustentacular cells, and basal cells. These cells, while sharing the common feature of microvilli, are distinct in their morphological characteristics and distribution (Lin et al., 2007; Moran et al., 1982a,b; Rowley et al., 1989).

Together, these cell types combine to form the main OE, lying on top of the basal lamina along the dorsal roof of the nasal turbinates and along the nasal septum. The basal cell layer, consisting of both GBCs and HBCs, lies immediately superficial to the basal lamina. The OSN cell bodies are arranged in a layer superficial to the basal cell layer. The OSN dendrites project through the layer of sustentacular cells where the dendritic knobs terminate near the apical surface of the sustentacular cells. Olfactory cilia project from these dendritic knobs into the mucous layer of the OE as an intermingled web (Fig. 12.1C), which lies on top of the microvilli of the sustentacular cells.

In mammals, the mucous layer bathing the olfactory cilia is a mixture of water secretions of approximately 5 μm in thickness (Menco, 1980c). The mucous is secreted primarily by the Bowman's gland, although there may be some regulation of the mucous composition by the sustentacular cells (Menco et al., 1998). In general, the nasal mucosa contains mucopolysaccharides, immune factors, metabolizing enzymes, and odorant-binding proteins (OBPs). These components serve to protect the epithelium from damage by physical stresses, exposure to toxicants, or infection.

One main functional component of the olfactory mucosa is the family of OBPs. OBPs are lipocalin family members secreted from the lateral nasal gland into the olfactory mucosa that serve to passively shuttle hydrophobic ligands through the aqueous mucous to the olfactory cilia. There are multiple homologous forms of the lipophilic OBP (OBPI-OBPIV in mice), which bind odorants in the micromolar range with distinct ligand specificity (Lobel et al., 1998; Pes and Pelosi, 1995; Pevsner et al., 1990). In addition to the potential chaperone role for OBPs, there have been other physiological roles suggested, such as protection from toxicants (Marinari et al., 1984), prevention of saturation of the ORs (Burchell, 1991; Schofield, 1988), and even acting as a required cofactor for odorant binding to the OR (Pelosi, 1994).

2.3. Regeneration

The OSN is one of only a few types of neurons that continually regenerate throughout adult life. OSNs die through apoptotic processes and are replaced by neurons derived from the progenitor basal cells approximately every 30–90 days (Farbman, 1990; Graziadei and Graziadei, 1979b; Mackay-Sim and Kittel, 1991). This regeneration is often accelerated following insult and allows for the repair of the OE after loss/damage of OSNs through sickness, exposure to environmental toxicants, or following invasion from pathogens (Graziadei and DeHan, 1973; Graziadei and Graziadei, 1979a,b; Graziadei and Metcalf, 1971; Graziadei et al., 1978; Harding et al., 1977). Interestingly, even noninhalatory routes of toxicant exposure can lead to damage and regeneration in the OE through the formation of systemic reactive intermediates (Bergman et al., 2002). In contrast, at least one report suggests that OSNs not exposed to toxicants may survive for over a year (Hinds et al., 1984). Importantly for our topic, evidence suggests that deciliation may be an early step, or perhaps a trigger, for increased neuronal damage/death and basal cell regeneration following toxicant exposure (Calderon-Garciduenas et al., 1998).

3. STRUCTURE OF OLFACTORY CILIA

Almost any cell in the human body is capable of forming a cilium (see http://www.bowserlab.org/primarycilia/cilialist.html for a comprehensive list). Cilia are typically divided into classes based on their axonemal structure and motility. The ciliary axoneme is most often composed of nine doublets of microtubules arranged symmetrically around a central core that either contains ((9 + 2) configuration) or lacks ((9 + 0) configuration) a central doublet of microtubules. Traditionally, cilia of the (9 + 2) configuration have been termed motile, whereas cilia of the (9 + 0) configuration have been termed nonmotile or primary cilia. Motile (9 + 2) cilia and flagella, which utilize structures called dynein arms along with the energy from ATP hydrolysis to generate movement, play important roles in fluid flow, sexual reproduction, and airway clearance. Nonmotile (9 + 0) cilia are commonly found as single primary cilia that help regulate cell-cycle progression, oncogenesis, and renal function. However, these classifications are not steadfast. For example, rare motile (9 + 0) cilia can be found in the embryonic node and allow the development of proper left–right asymmetry in the body (Okada et al., 2005). Nonmotile (9 + 2) cilia can be found in sensory organs such as the inner ear and OE (Dabdoub and Kelley, 2005; Menco, 1984). Although olfactory cilia have the (9 + 2) microtubule configuration normally found in motile cilia, they lack the dynein arms and are thus rendered immotile (Menco, 1984). Interestingly, some non-mammalian vertebrates, such as goldfish and frogs (Lidow and Menco, 1984; Reese, 1965), display motile olfactory cilia, which have an axoneme resembling that of respiratory cilia in their proximal segments and are suggested to play a role in odorant clearance (Bronshtein and Minor, 1973; Mair et al., 1982).

3.1. Cilia axoneme

Much of what we currently know about the structure of olfactory cilia derives from early electron microscopy studies (Cuschieri and Bannister, 1975a,b; Menco, 1980a,c; Menco and Morrison, 2003; Reese, 1965). These reports showed that the mammalian olfactory cilium is approximately 50–60 μm in length and can be divided into two distinct sections termed the proximal and distal segments. The thicker proximal segment in a (9 + 2) configuration projects 2–3 μm from the basal body with a thickness of around 0.3 μm (Menco, 1997). The thinner distal segment projects the remaining $\sim$50 μm with a distinct axonemal configuration of 1–4 singlet microtubules, most commonly consisting of a pair of singlet microtubules (Menco, 1997).

Interestingly, the proximal and distal cilia segments may represent distinct sub-cellular compartments. Throughout development, signaling proteins display a differential distribution between these two regions. In nascent cilia, signaling proteins are more uniformly distributed between the proximal and distal segments. In mature cilia, the signaling proteins, such as Gα_{olf}, ACIII, and CNG channel, appear to preferentially localize to the long distal segment where the odorant first makes contact with the OSN (Fig. 12.1E)

(Flannery et al., 2006; Matsuzaki et al., 1999; Menco, 1997). This clustering of signaling molecules at the site of odorant exposure may increase the efficiency of odorant-stimulated signaling. The distal segments of the olfactory cilia are oriented parallel to the epithelial surface. Because there are many cilia (10–30) per cell and because they project 50–60 μm from the dendritic knob there is substantial overlap of cilia from different OSNs (Fig. 12.1C and D)(Menco, 1997). This intertwined mat of cilia increases the sensory surface of the OE by over 40 times thus increasing our ability to detect odorants (Fig. 12.1C) (Menco, 1992).

The ciliary axoneme is composed of polymers of α and β tubulin, which form the structural backbone for the cilium (reviewed in Rosenbaum and Witman, 2002; Scholey, 2003). These microtubules provide the roadway for molecular motors to move their cargo into and out of the cilium. Olfactory ciliary axonemes are oriented with the plus end located in the distal tip of the cilium, which means that plus end-directed motors carry cargo to the tip of the cilium, while minus end-directed motors are responsible for the return of cargo (reviewed in Rosenbaum and Witman, 2002; Scholey, 2003).

Posttranslational modifications of tubulin have been discovered to play functional roles in the regulation of cargo transport (reviewed in Hammond et al., 2008). Many modifications to tubulin of the ciliary axoneme have been found, including acetylation (α), polyglutamylation ($\alpha + \beta$), polyglycylation ($\alpha + \beta$), and detyrosination (α). While all of these modifications have been detected in olfactory cilia, their precise functional relevance is poorly understood (Pathak et al., 2007; Schwarzenbacher et al., 2005). However, a recent study found that loss of an enzyme responsible for polyglutamylation in zebrafish caused a loss of olfactory cilia (Pathak et al., 2007), indicating a role for posttranslational tubulin modifications in assembly or maintenance of olfactory cilia.

3.2. Lipid composition

The ciliary axoneme is encased in a membrane sheath formed by the lipid bilayer, which most certainly plays an integral role in olfactory signaling. There is a historic interest in the role of these lipids in the regulation of olfaction. Prior to the discovery of olfactory receptors, numerous reports hypothesized on the potential role of membrane lipids in the modulation of odorant transduction and odor recognition (Cherry et al., 1970; Kashiwayanagi et al., 1987, 1997, 1990; Nomura and Kurihara, 1987a,b; Russell et al., 1989). The lipophilicity of odorants and the ability of various odorants to induce changes in membrane fluidity suggested that membrane lipids might play an important role in olfactory signaling (Cherry et al., 1970; Kashiwayanagi et al., 1987, 1990, 1997; Nomura and Kurihara, 1987a,b; Russell et al., 1989). While it is now well established that odor recognition is mediated by G protein-coupled receptors (GPCRs), this does not exclude the possibility for a role of ciliary membrane lipids in the modulation of odorant transduction. In fact, there may exist a dynamic reciprocity between odorant signaling proteins and membrane lipids in olfactory cilia such that perturbation of membrane lipids can affect olfactory signaling.

Recently, there is growing evidence for the role of lipid rafts in the organization of olfactory signaling proteins that are highly concentrated in the cilia (Brady et al., 2004; Kobayakawa et al., 2002; Schreiber et al., 2000). In OE, Schreiber and colleagues (Schreiber et al., 2000) demonstrated that the G protein and adenylyl cyclase isoforms involved in odorant signaling associate with lipid rafts. They also reported that G_{olf} and ACIII interact with the cholesterol binding protein, caveolin, and that disruption of the caveolin interaction inhibits odorant-induced cAMP production in OSNs. Additionally, the recently identified stomatin-related olfactory (SRO) protein (Goldstein et al., 2003; Kobayakawa et al., 2002) has been shown to associate with lipid rafts in olfactory cilia and bind both caveolin and ACIII. Importantly, anti-SRO antibodies stimulated cAMP production in fractionated cilia membranes suggesting that rafts and/or a caveolin/lipid/protein complex regulate odorant signaling (Kobayakawa et al., 2002). In further support of this, early ultrastructural data from the Menco laboratory comparing olfactory cilia membranes to that of respiratory cilia led them to conclude that that the outer leaflet membranes of olfactory cilia are thicker than inner leaflets (Lidow and Menco, 1984). This is consistent with a potential enrichment of sphingolipids that are localized almost exclusively to the outer leaflet. Interestingly, the bilayer thickness of lipid raft domains will be greater than surrounding membrane due to sterol packing and the fact that raft hydrocarbon chains are longer and straighter than the acyl chains in the surrounding phospholipid regions (Tillman and Cascio, 2003). The enrichment of certain lipids is supported by work in invertebrates that has shown that the ciliary membrane of *Paramecium* is highly enriched with sphingolipids (Andrews and Nelson, 1979). Furthermore, these investigators later showed that ciliary membrane excitability in the same invertebrate model was sensitive to sterol composition (Hennessey et al., 1983). Others have reported that there is an enrichment of cholesterol in the ciliary shaft, but not the necklace region, of epithelial cilia that extends during ciliogenesis (Chailley et al., 1983). Perhaps another interesting feature of mammalian cilia is the distinct lipid composition at the base of the cilium. For example, Madin–Darby canine kidney cells demonstrate an annulus of condensed lipids at the base of their primary cilia (Vieira et al., 2006). This peripheral evidence supports the hypothesis that the lipid composition of olfactory cilia may be specialized to support olfactory signal transduction. Surprisingly, however, there is virtually no information regarding the precise lipid composition of this important membrane structure in the olfactory system.

Nevertheless, in addition to the potential role for lipids in the nucleation of olfactory signaling complexes, the lipid composition of olfactory ciliary membranes may be important for several reasons.

Ciliary membrane lipids may be an important pharmacological consideration for the intranasal route of drug delivery. One recognized limitation of intranasal drug delivery, an approach often used to bypass the blood–brain barrier, is the temporary or permanent loss of olfactory function (Agarwal and Mishra, 1999; Illum, 2003). To improve transport across the nasal membrane, cyclodextrins (CD) are often used in drug formulations. These molecules are cyclic oligosaccharides that contain a hydrophobic binding cavity capable of incorporating a drug (Challa et al., 2005). CDs exchange the drug contents of their cavity with the lipids in a plasma membrane. Importantly, these molecules are often used to deplete membranes of lipids and have been shown to disrupt lipid raft/caveolae formation. Using similar logic, it is not surprising that lipid-lowering drugs (i.e., statins) can cause anosmia and dysosmia in patients (Doty et al., 2003; Weber et al., 1992); this is listed as one of the manufacturer's recognized side effects on the FDA-approved product labeling for atorvastatin (Lipitor^{TM}).

Finally, one of the more clearly defined roles for lipid microdomains is as a portal of entry for viral pathogens. As discussed in detail later in this chapter, several viruses enter the body through the nasal cavity, gain access to the brain, and spread transneuronally to other parts of the central nervous system (CNS) using the olfactory system as a gateway. This suggests that the content and organization of ciliary lipids may be important not only in maintaining the integrity of olfactory signaling but also as a permissive entry site for invading pathogens.

3.3. Ciliary necklace/transition zone

At the extreme proximal end of the olfactory cilium, where the lipid membrane sheath meets the dendritic knob, there exists a region termed the "ciliary necklace." This highly ordered domain is marked by a spiraling array of membrane particles (Andres, 1969; Gilula and Satir, 1972; Menco, 1980d), which connect to the basal body just below the ciliary axoneme (Satir and Christensen, 2007). While most cilia types possess a ciliary necklace, olfactory cilia typically have more strands per cilium than their respiratory counterparts (Menco, 1980d). The formation of the ciliary necklace precedes ciliogenesis as a patch of membrane, and in malformed cilia there are still necklace-like structures (Carson et al., 1981; Menco, 1980d). Interestingly, ciliary transport proteins have been found to be localized at the ciliary necklace indicating that it may serve as a cargo docking site connecting the ciliary shaft to the protein complexes at the base of the cilium (Deane et al., 2001).

3.4. Basal body

The protein complex at the base of the cilium is formed by the basal body, a modified centriole that migrates to the plasma membrane prior to ciliogenesis. The basal bodies are duplicated *en masse* in the cell body of the OSN before they migrate to the dendritic knob (Cuschieri and Bannister, 1975a, b; Dirksen, 1974; Hagiwara *et al.*, 2004; Schwarzenbacher *et al.*, 2005). Basal bodies, like the ciliary axoneme, are composed of nine sets of microtubules arranged in a radial symmetry. However, unlike the axoneme, basal bodies are composed of polymers of triplet microtubules of γ tubulin rather than doublet microtubules of α and β tubulin. The basal body serves as the microtubule organizing center (MTOC) in the dendritic knob with the axonemal tubules projecting from the basal body, such that the plus ends orient toward the distal tip of the cilium (Burton, 1992).

In addition to serving as MTOCs for the ciliary axoneme, the basal bodies are associated with electron-dense satellite particles that appear to also be MTOCs (Burton, 1992). These organizing centers serve as nucleation sites for microtubules that project from the dendritic knob back through the dendrite toward the cell body (Burton and Laveri, 1985). Some of the MTOCs are connected to the basal body through a sheath of material that surrounds the basal body and thickens at its proximal end. The basal bodies and sheath are connected to the plasma membrane through nine struts which correspond to the electron-dense endings of the ciliary rootlet (Menco, 1980d).

3.5. Ciliary rootlet

The ciliary rootlet, first described over a century ago, is a cytoskeletal feature found projecting from the basal body in ciliated cells (Engelmann, 1880). Although the structural components of the ciliary rootlet are beginning to be elucidated (Yang *et al.*, 2002), still very little is known about its function. It has been proposed that the ciliary rootlet is important for the stability of sensory cilia. In these studies, photoreceptor connecting cilia from mice lacking rootletin, a component of the ciliary rootlet, displayed fragility at their basal body (Yang *et al.*, 2005). Although olfactory cilia were not examined in the rootletin-null mice, OSNs have been shown to express components of the ciliary rootlet in a localization consistent with the dendritic knob/basal body region (McClintock *et al.*, 2008; Yamamoto, 1976). Given that rootletin has been shown to affect cilia stability and that OSNs have been shown to express components of the rootlet, the prediction would be that olfactory cilia lacking a rootlet may also show fragility and become detached from the OSN more easily upon physical stresses such as sneezing.

4. Formation of Olfactory Cilia

Much of what we currently know about ciliogenesis derives from the study of lower vertebrates. In the mouse, the olfactory placode is first visible at embryonic day 9 (E9) postfertilization (Cuschieri and Bannister, 1975a,b; Menco, 1980a,b; Menco and Morrison, 2003; Schwarzenbacher et al., 2005). At E10, an invagination of the olfactory placode leads to the development of the olfactory pit. At this point, two different cell types are visible: a population that is electron dense (proliferative basal cells) and those that appear light (differentiated OSNs) (Cuschieri and Bannister, 1975a,b; Menco, 1980a,b; Menco and Morrison, 2003). It is not until E11, however, that the dendrites begin to form and extend toward the apical surface. Further, the olfactory pit deepens and secondary recesses, which will eventually develop into turbinates, appear (Cuschieri and Bannister, 1975a,b; Menco, 1980a,b; Menco and Morrison, 2003). During this period of extensive proliferation, OSN growth and maturation is most pronounced in the deep recesses of the olfactory pit (Cuschieri and Bannister, 1975a,b; Menco, 1980a,b; Menco and Morrison, 2003).

The earliest detectable signs of neuronal differentiation are observed at E10, where, using electron microscopy, a small population of OSNs extending dendrites toward the apical surface can be detected (Cuschieri and Bannister, 1975a). By E11, several morphological changes occur in these OSNs, suggesting the initiation of ciliogenesis. First, in the perinuclear region of these neurons, numerous microtubules, and microfilaments have formed and extend vertically toward the apical surface (Cuschieri and Bannister, 1975a; Menco and Farbman, 1985; Menco and Morrison, 2003). Second, the terminal portion of the dendrite now extends past the apical surface and into the lumen of the nasal cavity, where it begins to swell indicating the formation of the dendritic knob (Cuschieri and Bannister, 1975a; Menco and Farbman, 1985; Menco and Morrison, 2003). Finally, and perhaps most importantly, centriole duplication has occurred and groups of centrioles are amassed in the perinuclear region of the neuron (Fig. 12.2A) (Cuschieri and Bannister, 1975a; Menco and Farbman, 1985; Menco and Morrison, 2003).

By E12, the number of OSNs with well-formed dendrites and dendritic knobs has increased markedly. The dendritic knobs in these neurons are filled with mitochondria, small coated vesicles, and numerous microtubules (Cuschieri and Bannister, 1975a; Menco and Farbman, 1985; Menco and Morrison, 2003). These microtubules are arranged in two distinct populations; one is arranged concentrically around the periphery of the knob while the other is arranged longitudinally and extends deep into the dendrite (Cuschieri and Bannister, 1975a; Menco and Farbman, 1985; Menco and Morrison, 2003). Further, the groups of centrioles observed at E11 have

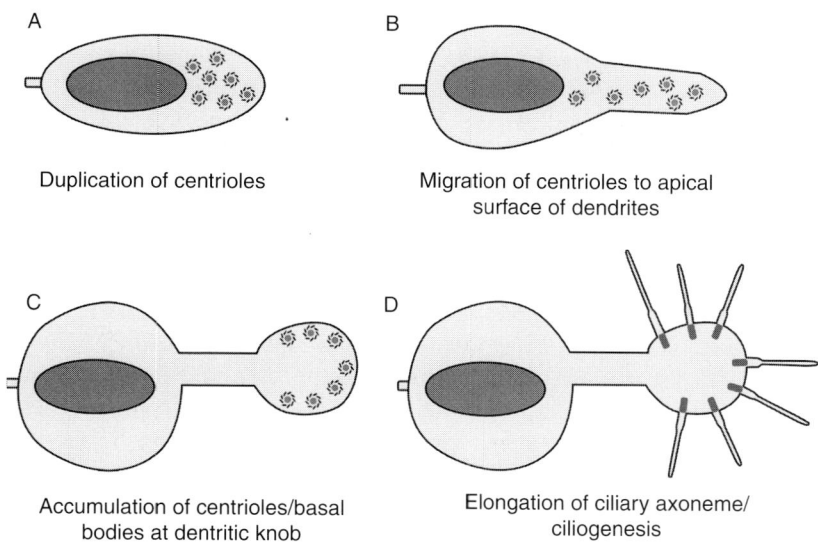

Figure 12.2 *The major steps of OSN ciliogenesis.* Cartoon representations are shown for the major phases of ciliogenesis in the olfactory epithelium. (A) In the first step, centrioles duplicate and accumulate in the perinuclear space prior to dendrite formation. (B) The OSN dendrite extends and the duplicated centrioles migrate toward the apical surface of the olfactory epithelium. (C) The terminal ending of the dendrite extends past the apical surface of the olfactory epithelium and begins to swell, forming the dendritic knob. The centrioles accumulate at the dendritic knob and begin to arrange around the periphery of the knob, where they will become the ciliary basal bodies. (D) The axonemes of the newly formed cilia grow and extend into the lumen of the nasal cavity. Also, the peripheral structures of the basal bodies form, anchoring the cilia into the plasma membrane.

begun to migrate out to the dendritic knob (Fig. 12.2B) and appear either as rosette-like clusters at the center of the knob or are dispersed singly around the knob periphery (Fig. 12.2C). In the OSNs with the centrioles arranged around the periphery, the first hint of ciliogenesis is occurring when a single, primary cilium of approximately 1 μm is observed extending into the nasal cavity (Cuschieri and Bannister, 1975a; Menco and Farbman, 1985; Menco and Morrison, 2003; Schwarzenbacher et al., 2005). In the next stage of development, the microtubule-based axoneme of these newly formed cilia begins to elongate and the basal body, formed by the migrating centrioles, matures and is anchored at the plasma membrane (Fig. 12.2D) (Cuschieri and Bannister, 1975a,b; Dirksen, 1974; Hagiwara et al., 2004; Schwarzenbacher et al., 2005). By E13 or E14, multiple cilia up to 2 μm in length can be seen extending from a single dendritic knob (Fig. 12.2D). Over the next several days, olfactory cilia elongate and can reach up to 60 μm prior to birth. Intraflagellar transport (IFT), which will be discussed in more detail below, plays a key role in the growth and maintenance of cilia

(Rosenbaum and Witman, 2002; Scholey, 2003). Postnatally, the cilia will continue to grow and can reach up to 200 μm in length in some vertebrate populations (Menco and Morrison, 2003; Reese, 1965; Seifert, 1971). Due to this length, OSN cilia create a meshwork across the surface of the OE, thus increasing the surface area of the OE up to 40 times and enhancing the likelihood of odorant detection (Lidow and Menco, 1984; Menco and Morrison, 2003).

For the process of olfaction, it is important to consider not only how cilia are formed, but also when the odorant signaling molecules are expressed and localized to cilia. The majority of studies investigating the developmental expression of olfactory signaling molecules have probed for mRNA expression using either RT-PCR, northern blot, or *in situ* hybridization analysis (Margalit and Lancet, 1993; Saito *et al.*, 1998; Schwarzenbacher *et al.*, 2004, 2005; Strotmann *et al.*, 1995). Interestingly, not all of the components necessary for odor detection are expressed at the same point during development. The first proteins to be expressed are the ORs, of which a subset begin to be expressed at E11, as determined by both mRNA and protein expression (Saito *et al.*, 1998; Schwarzenbacher *et al.*, 2004, 2005). This expression occurs prior to ciliogenesis, resulting in the accumulation of the OR protein in high density at the dendritic knob (Schwarzenbacher *et al.*, 2005). Surprisingly, not all ORs are expressed at the same time in development. Two of the earliest expressing ORs identified, mOR256-17 and V1, begin to be expressed at E11 (Saito *et al.*, 1998; Schwarzenbacher *et al.*, 2004, 2005). By E12, several more receptors, mOR5, mOR14, mOR18–2, mOR37, mOR111–5, mOR124, and mOR171–24 exhibit increased expression in OSNs (Schwarzenbacher *et al.*, 2004; Strotmann *et al.*, 1995). The diversity of OR expression continues to increase over the next several days of embryonic and postnatal development (Margalit and Lancet, 1993; Saito *et al.*, 1998; Schwarzenbacher *et al.*, 2004). As for the other components of the olfactory signaling cascade, they are expressed later in embryonic development. ACIII is first detected around E15, while G_{olf} and the CNG channel are expressed at E16 and E19, respectively (Margalit and Lancet, 1993). Due to this range in developmental expression patterns, it is assumed that odor detection cannot occur until all proteins are present in olfactory cilia. However, this has yet to be determined.

The protein expression of one specific OR, mOR256–17, has been used to track ciliogenesis (Schwarzenbacher *et al.*, 2004). Specifically, as mentioned previously, the earliest expressing ORs are present prior to the initiation of ciliogenesis. mOR256–17 accumulates at the dendritic knob in high density at this stage. As cilia begin to form and elongate, the pattern of OR localization changes. At E11, when a primary cilium around 1 μm can be observed on a subset of neurons, the OR remains localized to the knob and at the very proximal portions of the cilia. As early as E12, when the cilia reach 2 μm or longer, mOR256–17 migrates almost exclusively to

OSN cilia, and, by E13, this localization is complete (Schwarzenbacher et al., 2004). The physiologic relevance of OR expression prior to ciliogenesis, however, remains to be determined.

5. Intraflagellar Transport

The formation of cilia occurs through an evolutionarily conserved process termed IFT, which was first discovered in the laboratory of Joel Rosenbaum in *Chlamydomonas* (Kozminski et al., 1993). Since cilia lack the necessary components for protein synthesis, cargo must be synthesized in the cell and carried into the cilia through IFT, which involves movement along microtubules by molecular motors in complex with transport molecules, called IFT particles (reviewed in Rosenbaum and Witman, 2002; Scholey, 2003, 2008). Given that the basic mechanisms of IFT are widely conserved not only between cilia types, but also often between species, we presume that these mechanisms studied in invertebrates are also acting in mammalian olfactory cilia.

Much of what we know about IFT in chemosensory cilia is from work in *Caenorhabditis elegans* where IFT can be visualized in real time using GFP-tagged motors (Orozco et al., 1999). Transport in the anterograde direction, towards the distal, plus end of the cilium microtubules, has been shown to involve kinesin motors (Cole et al., 1998), whereas retrograde transport back into the cell is accomplished with the cytoplasmic dynein motor (Pazour et al., 1998). These microtubule-based motors utilize the energy from ATP hydrolysis to move cargo processively along the microtubules to their destination.

Work in *C. elegans* has shown that the formation and maintenance of the chemosensory ciliary axoneme and the delivery of cargo is accomplished through coordination of two kinesin motors: the heterotrimeric kinesin-II motor and the homodimeric OSM-3 (Snow et al., 2004). The conservation between IFT mechanisms was shown in the mammalian kidney, where the heterotrimeric kinesin-II motor, consisting of the two motor subunits KIF3a, KIF3b, and the accessory protein, KAP3, is necessary for ciliogenesis (Lin et al., 2003). However, differences are beginning to be recognized between specialized cilia types in invertebrates and mammals (Jenkins et al., 2006; Ou et al., 2005). Although loss of function of the OSM-3 homolog, KIF17, impaired ciliary trafficking of the olfactory CNG channel, it had no effect on cilia length as predicted by work in *C. elegans* (Jenkins et al., 2006; Ou et al., 2005). Future studies, however, are necessary to determine if these mechanisms are functioning directly in mammalian olfactory cilia. Interestingly, OSM-3 operates on singlet microtubules of the distal segments of *C. elegans* cilia (Ou et al., 2005), and since olfactory cilia have such prominent distal segments it is likely that KIF17 is also functioning on distal segments in the mammalian olfactory cilium. Nevertheless, the differences in kinesin-2 regulation of cilia length between the cilia of C. *elegans* and mammalian cilia highlight the need to further explore the mechanisms of IFT in mammalian olfactory cilia.

Early work using electron microscopy of frog olfactory cilia found cargo in complex with IFT particles seen as electron-dense regions along the ciliary axoneme (Reese, 1965). It is known that IFT motors associate with two distinct complexes of transport proteins called IFT proteins, named for their molecular weight. These two complexes comprise 17 highly conserved proteins, termed complex A and complex B (Cole, 2003). Complex A consists of IFT144, 140, 139, 122, and possibly 43, while complex B consists of IFT 172, 88, 81, 80, 74/72, 57/55, 52, 46, 27, and 20. Defects in either complex can impair IFT and cause a host of human diseases (reviewed in Blacque *et al.*, 2008). The precise role of the IFT complexes in mammalian olfactory cilia transport remains undefined. IFT proteins have been shown to share significant homology with Golgi-localized clathrin trafficking machinery (Avidor-Reiss *et al.*, 2004). Interestingly, the clathrin AP-1 μ adaptor, UNC-101, has been shown to be responsible for the localization of ORs to the cilia of C. *elegans* (Dwyer *et al.*, 2001).

Another complex of proteins that have been shown to be involved in cilia assembly and maintenance are the Bardet–Biedl syndrome (BBS) proteins. BBS is a pleiotropic ciliopathy that includes phenotypes such as retinal degeneration, polydactyly, obesity, anosmia, and others (discussed in more detail below). There are 12 known BBS proteins (BBS1–12), which encode proteins involved in different stages of cilia transport. While there are a variety of ciliary phenotypes associated with defects in BBS proteins, loss of function of BBS1 and BBS4 caused impaired olfactory function (Iannaccone *et al.*, 2005; Kulaga *et al.*, 2004). Interestingly, mice null for BBS1 or BBS4 may exhibit defects in olfactory cilia maintenance or assembly, although the mechanism for this defect remains unknown (Iannaccone *et al.*, 2005; Kulaga *et al.*, 2004).

A recent report suggests that there is a dynamic reciprocity between the signaling function of cilia and its structural maintenance through IFT. Work in C. *elegans* by Mukhopadhyay *et al.* has shown that the loss of activation of the sensory signaling cascade modulates the structure of the AWB neuron modified sensory cilia (Mukhopadhyay *et al.*, 2008). This sensory signaling-dependent remodeling was shown to be dependent on kinesin-II as well as BBS proteins (Mukhopadhyay *et al.*, 2008). This is similar to a previous study showing that structure of AWC neuron cilia is also linked to sensory function (Roayaie *et al.*, 1998). While structural changes have been reported in mice deprived of odorant stimulation by naris occlusion (Farbman *et al.*, 1988), it would be interesting to examine changes in cilia architecture due to loss of olfactory cues. Regardless, this suggests a feedback interaction between the ciliary proteins involved in assembly and maintenance and those participating in signaling.

6. Ciliary Proteome

Cilia contain a subset of cellular proteins which comprise a population distinct from the extraciliary compartment (Inglis *et al.*, 2006). Emerging techniques, such as bioinformatic screens or proteomic analyses, are yielding new insights into cilia-related genes novel proteins that may be involved in olfactory signaling or ciliary structure and maintenance.

6.1. Olfactory signaling through the canonical G protein-coupled pathway

Perhaps the most recognized subset of proteins enriched in the olfactory cilia is that of the canonical signaling pathway. This cascade begins when odorants dissolve in the nasal mucosa where they gain access to the OSN. The odorant binds to the OR on the cilia of OSNs to initiate the odorant detection pathway (Fig. 12.1E). The OR activates the olfactory G protein, which stimulates ACIII to increase the local concentration of cAMP. The CNG channel is activated by the increased cAMP and opens to allow Ca^{2+} influx into the OSN. This signal is amplified by the Ca^{2+}-activated Cl^- channel which binds Ca^{2+} and opens allowing Cl^- efflux from the OSN leading to further depolarization (reviewed in Ronnett and Moon, 2002). The function of the olfactory system is dependent on the ciliary localization of each of these proteins described below.

6.1.1. Odorant receptors

The discovery of the family of ORs was described by Linda Buck and Richard Axel in a groundbreaking publication in 1991 (Buck and Axel, 1991), which led to the awarding of the Nobel Prize in Physiology or Medicine in 2004. These receptors represent the most numerous member of the family of 7-transmembrane GPCRs, with humans having ~400 functional genes and ~600 pseudogenes (Gilad and Lancet, 2003). Interestingly, only one OR is expressed in any single OSN. Given the tremendous diversity of odorants and the finite number of OR genes each OR must have the ability to bind multiple odorants (Buck, 2004; Malnic *et al.*, 1999; Mombaerts, 2004, 2006). While some specific ligands have been identified (reviewed in Mombaerts, 2004), difficulties with functional expression in nonciliated heterologous systems have slowed OR characterization perhaps suggesting a requirement for cilia in OR expression.

6.1.2. Olfactory G proteins

The main olfactory G protein is a heterotrimeric stimulatory G protein comprising $G\alpha_{olf}$, $\beta 1$, and $\gamma 13$ (Jones and Reed, 1989; Kerr *et al.*, 2008). This heterotrimer provides the link between the ORs and ACIII in mature

OSNs. Interestingly, Gα_s is also expressed in OSNs and appears precede the developmental expression of Gα_{olf} (Belluscio et al., 1998; Menco et al., 1994). For reasons unknown, there is a phenotypic switch in olfactory cilia from Gα_s to Gα_{olf} later in the maturation of the OE.

6.1.3. Adenylyl cyclase

The olfactory heterotrimeric G protein is a stimulatory G protein that mediates downstream signaling through the stimulation of adenylyl cyclase. The major form of adenylyl cyclase in the OSN is ACIII, which was cloned by Reed and colleagues in 1990 (Bakalyar and Reed, 1990). ACIII, like most components of the olfactory signaling cascade, is critical for odorant detection as genetic deletion leads to anosmia (Wong et al., 2000). ACIII is a Ca^{2+}-calmodulin stimulated isoform of adenylyl cyclase that is responsible for the elevation of intracellular cAMP in the cilia of OSNs (Bakalyar and Reed, 1990; Choi et al., 1992).

6.1.4. CNG channel

CNG channels were first discovered in retinal photoreceptors and olfactory neurons, where they modulate the membrane potential in response to stimulus-induced changes in the intracellular concentrations of cyclic nucleotides (Fesenko et al., 1985; Firestein and Werblin, 1989; Nakamura and Gold, 1987). Although CNG channels have now been found in many other neuronal and nonneuronal cells, their physiological roles in nonsensory tissues remain obscure (Finn et al., 1996). The functional role of CNG channels in the olfactory system is firmly established. The olfactory CNG channel comprises three distinct subunits, CNGA2, CNGA4, and CNGB1b, in a 2:1:1 stoichiometry (Bonigk et al., 1999; Zheng and Zagotta, 2004). CNGA2 is the only subunit capable of forming functional homotetramers, while CNGA4 and CNGB1b serve to modulate properties of the channel such as ion selectivity, nucleotide sensitivity, and Ca^{2+}/Calmodulin regulation (Kaupp and Seifert, 2002). Recently, the CNGB1b subunit has been found to be necessary for delivery of the CNG channel to cilia (Jenkins et al., 2006; Michalakis et al., 2006), and this trafficking was shown to be dependent on a the kinesin motor protein, KIF17, and a C-terminal "RVxP" trafficking motif on CNGB1b (Jenkins et al., 2006).

6.1.5. Ca^{+2}-activated Cl^- channel

Ca^{2+} entry through the CNG channel directly stimulates the Ca^{2+}-activated Cl^- channel. Although chloride conductance is often considered an inhibitory process, in olfactory cilia there is a reverse chloride gradient resulting in an efflux of chloride with the opening of the channel. This current further depolarizes the neuron and amplifies the signal from the CNG channel causing the generation of an action potential. While functional studies have firmly established a role for the olfactory Ca^{2+}-activated Cl^- channel (Boccaccio and Menini, 2007; Kaneko *et al.*, 2006; Kleene and Gesteland, 1991; Kurahashi and Yau, 1993; Pifferi *et al.*, 2006; Qu *et al.*, 2003; Reisert *et al.*, 2003, 2005; Reuter *et al.*, 1998), its molecular identity has remained elusive.

6.2. Signaling through noncanonical pathways

While the canonical olfactory signaling cascade has long been thought to be the main pathway involved in odor detection, recent evidence suggests that at least three other pathways exist that are uniquely involved in detecting changes in the external environment. In the OE, a subset of neurons has been shown to express the guanylyl cyclase type D (GC-D) receptor, while lacking the canonical signaling components (Juilfs *et al.*, 1997; Ma, 2007; Meyer *et al.*, 2000). GC-D expression seems to be specific to the olfactory system, where its localization is widely dispersed over multiple turbinates of the epithelium (Breer *et al.*, 2006; Fulle *et al.*, 1995). As its name implies, upon stimulation, the GC-D receptor generates cGMP, which then signals to a cGMP-responsive CNG channel subunit, CNGA3 (Leinders-Zufall *et al.*, 2007). Importantly, all of the components of this cascade are enriched in olfactory cilia, suggesting that they are involved in chemosensation (Juilfs *et al.*, 1997; Leinders-Zufall *et al.*, 2007; Meyer *et al.*, 2000). While the physiologic relevance of this system has been controversial, recent evidence suggests that, in rodents, these neurons are responsible for the detection of hormone peptides as well as natural urine stimuli (Leinders-Zufall *et al.*, 2007). These data are supported in both the GC-D and CNGA3 knockout mice, which are unresponsive to these stimuli, while canonical odor detection remains intact (Leinders-Zufall *et al.*, 2007). Perhaps more controversial is the hypothesis that these neurons act as the CO_2 sensor in the OE, which may be involved in detecting signs of danger in the external environment (Hu *et al.*, 2007).

More recently, a third subset of OSNs has been shown to express a family of receptors termed trace amino acid receptors (TAARs) (Borowsky *et al.*, 2001; Gloriam *et al.*, 2005; Lewin, 2006; Liberles and Buck, 2006). In addition to their expression in OSNs, TAARs are also expressed outside of the main OE, in the Gruenberg ganglion (Fleischer *et al.*, 2007). The topology of these receptors includes seven transmembrane-spanning segments, linking them with the GPCR family of proteins. However, sequence analysis of the TAARs shows the most similarities with dopamine and serotonin, but not odorant, receptors (Liberles and Buck, 2006). Nine subtypes of TAARs have been cloned, with eight of them being specifically expressed in the OE (Ma, 2007). Despite co-localization with G_{olf}, it remains to be determined if TAARs couple to the canonical downstream signaling mechanisms involved in odor detection (Liberles and Buck, 2006). As with the odorant and GC-D receptors, the TAARs localize to OSN cilia, again indicating a chemosensory function (Liberles and Buck, 2006). In support of this function, three TAARs expressed in the OE have been shown to respond *in vitro* to volatile amino acids found in mouse urine (Liberles and Buck, 2006), suggesting that the TAARs are involved in detecting social cues from neighboring animals.

Finally, in most mammals and perhaps even primates, a separate organ, termed the vomeronasal organ (VNO), is involved in pheromone detection. Recent evidence suggests that the VNO can also detect small, volatile odorants (Breer *et al.*, 2006; Dulac and Torello, 2003; Ma, 2007; Sam *et al.*, 2001). The mammalian VNO possesses both ciliated and microvillar sensory neurons whose projections synapse in the accessory olfactory bulb located on the dorsal-posterior side of the MOB (Cuschieri and Bannister, 1975b). While no distinct VNO has been shown to exist in humans beyond early development (Boehm and Gasser, 1993), both of the vomeronasal receptors, V1R and V2R, are expressed in the OE (Giorgi *et al.*, 2000; Rodriguez, 2004; Rodriguez and Mombaerts, 2002; Witt and Hummel, 2006). In the rodent, both the V1R and the V2R localize to cilia of VNO neurons where they couple to either the G_i (V1R) or G_o (V2R) family of G proteins (Dulac and Torello, 2003). Unlike the canonical odorant signaling pathway, the V1Rs and V2Rs couple to a PLC-mediated pathway, producing the downstream signaling molecules diacyl glycerol and phosphatidylinositol-3-phosphate. Stimulation of this pathway leads to arachadonic acid production and, ultimately, to TRPC2 activation, converting the chemical signal into an electrical response (Liman *et al.*, 1999; Lucas *et al.*, 2003).

Taken together, these data indicate that the mammalian nose is a complex organ with multiple systems designed to detect everything from simple odors to pheromones and social cues. Despite differences between these systems, the underlying commonality remains that chemical detection, whether in the OE or the VNO, occurs at the level of the cilium. Consequently, while mutation or deletion of a single signaling protein will only affect a single pathway, disrupting proteins involved in ciliary formation or maintenance should render all of these systems ineffective.

6.3. Bioinformatics and expression profiling

While we are beginning to more fully understand the proteins involved in the signaling and maintenance of olfactory cilia, the full scope of genes involved in the various aspects of cilia function has yet to be identified (McClintock *et al.*, 2008). Recent advances in technology have vastly improved our ability to use bioinformatics as a tool to identify novel genes involved in various cellular processes, such as cilia formation and function. Hundreds of genes present in numerous ciliated species have recently been identified to be important in cilia-related functions (Avidor-Reiss et al., 2004; Blacque et al., 2006; Li et al., 2004; McClintock et al., 2008; Pazour et al., 2005; Smith et al., 2005; Stolc et al., 2005). While these studies are informative in identifying cilia-related genes, it is not clear that the mechanisms or gene products regulating ciliary function are entirely conserved between invertebrate and mammalian species. Only a handful of studies have concentrated on identifying cilia-related genes in mammals, with only two focusing on olfactory cilia (Klimmeck et al., 2008; McClintock et al., 2008; Ostrowski et al., 2002; Sammeta et al., 2007; Su et al., 2004). Using complementary approaches, comparing olfactory cilia to other ciliated cell types and calcium calmodulin-affinity column purification from isolated cilia, these two studies have identified over 100 cilia-related genes of known and unknown function in OSNs (Klimmeck et al., 2008; McClintock et al., 2008). While these approaches have proved useful in identifying new gene products, perhaps the most intriguing use will be to identify changes in gene expression in known ciliopathies, which will yield insights into the underlying mechanisms of these cilia disorders.

6.4. Ciliary proteomics

While it is interesting to identify novel gene products involved in cilia function, perhaps more important is to characterize the function of these newly identified proteins. In the study using calmodulin-affinity purification from isolated cilia, not only were proteins identified from the olfactory signaling cascade, but they also identified proteins involved in cytoprotection, cytoskeletal proteins, and proteins involved in modification of ciliary proteins (Klimmeck et al., 2008). In a similar study, isolated cilia preparations were subject to mass spectrometric analysis, from which 268 proteins were identified (Mayer et al., 2008). Of these proteins, 49% were transmembrane proteins, 41% were cytosolic, and 10% were cytoskeletal proteins (Mayer et al., 2008). Within the membrane fraction, the traditional signaling components were identified, along with ER and Golgi proteins, including proteins involved in metabolism, protein biosynthesis, and signal transduction. One such protein was PDE1C, which had been previously shown to localize to olfactory cilia and is thought to aid in odorant signal termination (Mayer et al., 2008). The cytosolic fraction included such proteins as calmodulin, calreticulin and two 14–3–3 isoforms, which are known to interact with and be modulated by calmodulin. The cytoskeletal fractions included tubulin, various keratin isoforms, and actin (Mayer et al., 2008). The presence of actin indicates that this preparation contains not only olfactory cilia, but also extraciliary proteins, since actin is not found in olfactory cilia. Thus, the isolation of pure populations of cilia, free of contamination from other membranes or organelles, appears to be a limiting factor (Klimmeck et al., 2008; Mayer et al., 2008). Given the large number of newly identified targets that may be expressed throughout the entire OSN, the question remains if they are enriched in the cilium. Nevertheless, these proteins were detected as part of the ciliary proteome, and the challenge remains to demonstrate their function and physiological relevance. This may not prove simple as illustrated by the case of olfactory marker protein (OMP). While this olfactory-specific protein, which is expressed in the cilia and throughout the OSN, was identified over 30 years ago, its precise function remains elusive(Youngentob and Margolis, 1999; Youngentob et al., 2001, 2003, 2004). Regardless, use of OMP as an identifier of mature OSNs and its OSN-selective promoter have proved invaluable for cell biological and genetic studies of olfactory function. The potential for the discovery of other such markers of olfactory cilia or function alone justifies this proteomic approach.

6.5. Regulation of ciliary protein entry

Although our knowledge of proteins localized to olfactory cilia is expanding, from a mechanistic perspective it is interesting to consider the fact that only a subset of cellular proteins is able to gain access to the cilium. Since the cilium contains a protein population distinct from the extraciliary compartment (Inglis et al., 2006), there must be a barrier to diffusion that restricts entry into the cilium. This selective gate is thought to occur at the basal body through interactions with a large complex of proteins (Rosenbaum and Witman, 2002; Scholey, 2003). Recently, mutation in the cilia/centrosomal protein CEP290 has been implicated in the specific mislocalization of olfactory G proteins (McEwen et al., 2007). Importantly, mutation in CEP290 did not globally alter cilia structure and all other olfactory signaling molecules tested were localized normally, indicating that in olfactory cilia, regulation of cargo entry is distinct for different proteins.

Growing interest in ciliary protein trafficking has led to the identification of amino acid sequences necessary for entry of cargo into cilia. For example, the "RVxP" motif originally identified in polycystin-2 (Geng et al., 2006), was found to be necessary for the ciliary delivery of the olfactory CNG channel (Jenkins et al., 2006). Additionally, several ORs were recently found to contain a ciliary targeting motif consisting of (AX[S/A]XQ) which was sufficient to drive ciliary localization of nonciliary receptors (Berbari et al., 2008). The precise mechanisms by which these motifs control ciliary localization remain unknown. Interestingly, only a subset of ciliary proteins express these motifs indicating that there are multiple potential ciliary targeting motifs that most likely act through distinct ciliary entry mechanisms.

Defects in these processes or the potential loss of OSN cilia following trauma, inflammation, or pathogen entry may all contribute to the etiology of olfactory disorders and, together, highlight the need for more understanding of the mechanisms and molecular machinery necessary for ciliary transport in OSNs.

REFERENCES

Afzelius, B. A. (2004). Cilia-related diseases. *J. Pathol.* **204,** 470–477.
Agarwal, V., and Mishra, B. (1999). Recent trends in drug delivery systems: Intranasal drug delivery. *Indian J. Exp. Biol.* **37,** 6–16.
Andres, K. H. (1969). Der olfaktorische Saum der Katze. *Z Zellforsch Mikrosk Anat.* **96,** 140–154.
Andrews, D., and Nelson, D. L. (1979). Biochemical studies of the excitable membrane of Paramecium tetraurelia. II. Phospholipids of ciliary and other membranes. *Biochim. Biophys. Acta* **550,** 174–187.
Avidor-Reiss, T., Maer, A. M., Koundakjian, E., Polyanovsky, A., Keil, T., Subramaniam, S., and Zuker, C. S. (2004). Decoding cilia function: Defining specialized genes required for compartmentalized cilia biogenesis. *Cell* **117,** 527–539.
Bakalyar, H. A., and Reed, R. R. (1990). Identification of a specialized adenylyl cyclase that may mediate odorant detection. *Science* **250,** 1403–1406.
Baker, H., and Genter, M. B. (2003). The olfactory system and the nasal mucosa as portals of entry of viruses, drugs, and other exogenous agents into the brain. *In* "Handbook on Olfaction and Gustation" (R. L. Doty, ed.), pp. 909–950. Informa Health Care, New York.
Bardet, G. (1995). On congenital obesity syndrome with polydactyly and retinitis pigmentosa (a contribution to the study of clinical forms of hypophyseal obesity). 1920. *Obes Res.* **3,** 387–399.
Beales, P. L., Elcioglu, N., Woolf, A. S., Parker, D., and Flinter, F. A. (1999). New criteria for improved diagnosis of Bardet–Biedl syndrome: Results of a population survey. *J. Med. Genet.* **36,** 437–446.
Beales, P. L., Badano, J. L., Ross, A. J., Ansley, S. J., Hoskins, B. E., Kirsten, B., Mein, C. A., Froguel, P., Scambler, P. J., Lewis, R. A., Lupski, J. R., and Katsanis, N. (2003). Genetic interaction of BBS1 mutations with alleles at other BBS loci can result in non-Mendelian Bardet–Biedl syndrome. *Am. J. Hum. Genet.* **72,** 1187–1199.
Belluscio, L., Gold, G. H., Nemes, A., and Axel, R. (1998). Mice deficient in G(olf) are anosmic. *Neuron* **20,** 69–81.
Benninger, M. S., Ferguson, B. J., Hadley, J. A., Hamilos, D. L., Jacobs, M., Kennedy, D. W., Lanza, D. C., Marple, B. F., Osguthorpe, J. D., Stankiewicz, J. A., Anon, J., Denneny, J., *et al.* (2003). Adult chronic rhinosinusitis: Definitions, diagnosis, epidemiology, and pathophysiology. *Otolaryngol. Head Neck Surg.* **129,** S1–32.
Berbari, N. F., Johnson, A. D., Lewis, J. S., Askwith, C. C., and Mykytyn, K. (2008). Identification of ciliary localization sequences within the third intracellular loop of G protein-coupled receptors. *Mol. Biol. Cell* **19,** 1540–1547.
Bergman, U., Ostergren, A., Gustafson, A. L., and Brittebo, B. (2002). Differential effects of olfactory toxicants on olfactory regeneration. *Arch Toxicol.* **76,** 104–112.
Biedl, A. (1995). A pair of siblings with adiposo-genital dystrophy. 1922. *Obes. Res.* **3,** 404.
Blacque, O. E., Li, C., Inglis, P. N., Esmail, M. A., Ou, G., Mah, A. K., Baillie, D. L., Scholey, J. M., and Leroux, M. R. (2006). The WD repeat-containing protein IFTA-1 is required for retrograde intraflagellar transport. *Mol. Biol. Cell* **17,** 5053–5062.

Blacque, O. E., Cevik, S., and Kaplan, O. I. (2008). Intraflagellar transport: From molecular characterisation to mechanism. *Front Biosci.* **13**, 2633–2652.

Boccaccio, A., and Menini, A. (2007). Temporal development of cyclic nucleotide-gated and Ca2+-activated Cl- currents in isolated mouse olfactory sensory neurons. *J. Neurophysiol.* **98**, 153–160.

Boehm, N., and Gasser, B. (1993). Sensory receptor-like cells in the human foetal vomeronasal organ. *Neuroreport* **4**, 867–870.

Bonigk, W., Bradley, J., Muller, F., Sesti, F., Boekhoff, I., Ronnett, G. V., Kaupp, U. B., and Frings, S. (1999). The native rat olfactory cyclic nucleotide-gated channel is composed of three distinct subunits. *J. Neurosci.* **19**, 5332–5347.

Borowsky, B., Adham, N., Jones, K. A., Raddatz, R., Artymyshyn, R., Ogozalek, K. L., Durkin, M. M., Lakhlani, P. P., Bonini, J. A., Pathirana, S., Boyle, N., Pu, X., *et al.* (2001). Trace amines: Identification of a family of mammalian G protein-coupled receptors. *Proc. Natl. Acad. Sci. USA* **98**, 8966–8971.

Brady, J. D., Rich, T. C., Le, X., Stafford, K., Fowler, C. J., Lynch, L., Karpen, J. W., Brown, R. L., and Martens, J. R. (2004). Functional role of lipid raft microdomains in cyclic nucleotide-gated channel activation. *Mol. Pharmacol.* **65**, 503–511.

Breer, H., Fleischer, J., and Strotmann, J. (2006). The sense of smell: Multiple olfactory subsystems. *Cell Mol. Life Sci.* **63**, 1465–1475.

Brodie, M., and Elvidge, A. R. (1934). The portal of entry and transmission of the virus of poliomyelitis. *Science* **79**, 235–236.

Bronshtein, A. A., and Minor, A. V. (1973). Significance of flagellae and their mobility for olfactory receptor function. *Dokl Akad Nauk SSSR.* **213**, 987–989.

Bucher, O. (1973). Cytologie, Histologie und Mikroskopische Anatomie des Menschen Bern: Huber, Bern.

Buck, L. B. (2004). Olfactory receptors and odor coding in mammals. *Nutr. Rev.* **62**, S184–188; discussion S224–S241.

Buck, L., and Axel, R. (1991). A novel multigene family may encode odorant receptors: A molecular basis for odor recognition. *Cell* **65**, 175–187.

Burchell, B. (1991). Turning on and turning off the sense of smell. *Nature* **350**, 16–17.

Burton, P. R. (1992). Ultrastructural studies of microtubules and microtubule organizing centers of the vertebrate olfactory neuron. *Microsc. Res. Tech.* **23**, 142–156.

Burton, P. R., and Laveri, L. A. (1985). The distribution, relationships to other organelles, and calcium-sequestering ability of smooth endoplasmic reticulum in frog olfactory axons. *J. Neurosci.* **5**, 3047–3060.

Caggiano, M., Kauer, J. S., and Hunter, D. D. (1994). Globose basal cells are neuronal progenitors in the olfactory epithelium: A lineage analysis using a replication-incompetent retrovirus. *Neuron* **13**, 339–352.

Calderon-Garciduenas, L., Rodriguez-Alcaraz, A., Villarreal-Calderon, A., Lyght, O., Janszen, D., and Morgan, K. T. (1998). Nasal epithelium as a sentinel for airborne environmental pollution. *Toxicol. Sci.* **46**, 352–364.

Carr, V. M., Menco, B. P., Yankova, M. P., Morimoto, R. I., and Farbman, A. I. (2001). Odorants as cell-type specific activators of a heat shock response in the rat olfactory mucosa. *J. Comp. Neurol.* **432**, 425–439.

Carson, J. L., Collier, A. M., Knowles, M. R., Boucher, R. C., and Rose, J. G. (1981). Morphometric aspects of ciliary distribution and ciliogenesis in human nasal epithelium. *Proc. Natl. Acad. Sci. USA* **78**, 6996–6999.

Chailley, B., Boisvieux-Ulrich, E., and Sandoz, D. (1983). Evolution of filipin-sterol complexes and intramembrane particle distribution during ciliogenesis. *J. Submicrosc. Cytol.* **15**, 275–280.

Challa, R., Ahuja, A., Ali, J., and Khar, R. K. (2005). Cyclodextrins in drug delivery: An updated review. *AAPS PharmSciTech.* **6**, E329–357.

Cherry, R. J., Dodd, G. H., and Chapman, D. (1970). Small molecule-lipid membrane interactions and the puncturing theory of olfaction. *Biochim. Biophys. Acta* **211**, 409–416.

Choi, E. J., Xia, Z., and Storm, D. R. (1992). Stimulation of the type III olfactory adenylyl cyclase by calcium and calmodulin. *Biochemistry* **31**, 6492–6498.

Cideciyan, A. V., Aleman, T. S., Jacobson, S. G., Khanna, H., Sumaroka, A., Aguirre, G. K., Schwartz, S. B., Windsor, E. A., He, S., Chang, B., Stone, E. M., and Swaroop, A. (2007). Centrosomal-ciliary gene CEP290/NPHP6 mutations result in blindness with unexpected sparing of photoreceptors and visual brain: Implications for therapy of Leber congenital amaurosis. *Hum. Mutat.* **28**, 1074–1083.

Cole, D. G. (2003). The intraflagellar transport machinery of Chlamydomonas reinhardtii. *Traffic* **4**, 435–442.

Cole, D. G., Diener, D. R., Himelblau, A. L., Beech, P. L., Fuster, J. C., and Rosenbaum, J. L. (1998). Chlamydomonas kinesin-II-dependent intraflagellar transport (IFT): IFT particles contain proteins required for ciliary assembly in Caenorhabditis elegans sensory neurons. *J. Cell Biol.* **141**, 993–1008.

Costanzo, R. M., and Becker, D. P. (1986). Smell and taste disorders in head injury and neurosurgery patients New York: MacMillian Publishing Company, New York.

Costanzo, R. M., and Miwa, T. (2006). Posttraumatic olfactory loss. *Adv. Otorhinolaryngol.* **63**, 99–107.

Cullen, M. M., and Leopold, D. A. (1999). Disorders of smell and taste. *Med. Clin. North Am.* **83**, 57–74.

Cuschieri, A., and Bannister, L. H. (1975a). The development of the olfactory mucosa in the mouse: Electron microscopy. *J. Anat.* **119**, 471–498.

Cuschieri, A., and Bannister, L. H. (1975b). The development of the olfactory mucosa in the mouse: Light microscopy. *J. Anat.* **119**, 277–286.

Dabdoub, A., and Kelley, M. W. (2005). Planar cell polarity and a potential role for a Wnt morphogen gradient in stereociliary bundle orientation in the mammalian inner ear. *J. Neurobiol.* **64**, 446–457.

Deane, J. A., Cole, D. G., Seeley, E. S., Diener, D. R., and Rosenbaum, J. L. (2001). Localization of intraflagellar transport protein IFT52 identifies basal body transitional fibers as the docking site for IFT particles. *Curr. Biol.* **11**, 1586–1590.

den Hollander, A. I., Koenekoop, R. K., Yzer, S., Lopez, I., Arends, M. L., Voesenek, K. E., Zonneveld, M. N., Strom, T. M., Meitinger, T., Brunner, H. G., Hoyng, C. B., van den Born, L. I., *et al.* (2006). Mutations in the CEP290 (NPHP6) gene are a frequent cause of Leber congenital amaurosis. *Am. J. Hum. Genet.* **79**, 556–561.

Ding, X., and Dahl, A. R. (2003). Olfactory Mucosa: Composition, enzymatic localization, and metabolism. *In* "Handbook on Olfaction and Gustation" (R. L. Doty, ed.), pp. 98–135. Informa Health Care, New York.

Dirksen, E. R. (1974). Ciliogenesis in the mouse oviduct. A scanning electron microscope study. *J. Cell Biol.* **62**, 899–904.

Doty, R. L. (2008). The olfactory vector hypothesis of neurodegenerative disease: Is it viable? *Ann. Neurol.* **63**, 7–15.

Doty, R. L., Perl, D. P., Steele, J. C., Chen, K. M., Pierce, J. D., Jr., Reyes, P., and Kurland, L. T. (1991). Odor identification deficit of the parkinsonism-dementia complex of Guam: Equivalence to that of Alzheimer's and idiopathic Parkinson's disease. *Neurology* **41**, 77–80; discussion 80–81.

Doty, R. L., Yousem, D. M., Pham, L. T., Kreshak, A. A., Geckle, R., and Lee, W. W. (1997). Olfactory dysfunction in patients with head trauma. *Arch Neurol.* **54**, 1131–1140.

Doty, R. L., Philip, S., Reddy, K., and Kerr, K. L. (2003). Influences of antihypertensive and antihyperlipidemic drugs on the senses of taste and smell: A review. *J. Hypertens.* **21**, 1805–1813.

Douek, E., Bannister, L. H., and Dodson, H. C. (1975). Recent advances in the pathology of olfaction. *Proc. R. Soc. Med.* **68,** 467–470.

Dulac, C., and Torello, A. T. (2003). Molecular detection of pheromone signals in mammals: From genes to behaviour. *Nat. Rev. Neurosci.* **4,** 551–562.

Dwyer, N. D., Adler, C. E., Crump, J. G., L'Etoile, N. D., and Bargmann, C. I. (2001). Polarized dendritic transport and the AP-1 mu1 clathrin adaptor UNC-101 localize odorant receptors to olfactory cilia. *Neuron* **31,** 277–287.

Engelmann, T. W. (1880). Zur anatomie und physiologie der flimmerzellen. *Arch Gel Physiol.* **23,** 505–535.

Farbman, A. I. (1990). Olfactory neurogenesis: Genetic or environmental controls? *Trends Neurosci.* **13,** 362–365.

Farbman, A. I., Brunjes, P. C., Rentfro, L., Michas, J., and Ritz, S. (1988). The effect of unilateral naris occlusion on cell dynamics in the developing rat olfactory epithelium. *J. Neurosci.* **8,** 3290–3295.

Fesenko, E. E., Kolesnikov, S. S., and Lyubarsky, A. L. (1985). Induction by cyclic GMP of cationic conductance in plasma membrane of retinal rod outer segment. *Nature* **313,** 310–313.

Finn, J. T., Grunwald, M. E., and Yau, K. W. (1996). Cyclic nucleotide-gated ion channels: An extended family with diverse functions. *Annu. Rev. Physiol.* **58,** 395–426.

Firestein, S., and Werblin, F. (1989). Odor-induced membrane currents in vertebrate-olfactory receptor neurons. *Science* **244,** 79–82.

Flannery, R. J., French, D. A., and Kleene, S. J. (2006). Clustering of cyclic-nucleotide-gated channels in olfactory cilia. *Biophys. J.* **91,** 179–188.

Fleischer, J., Schwarzenbacher, K., and Breer, H. (2007). Expression of trace amine-associated receptors in the Grueneberg ganglion. *Chem. Senses* **32,** 623–631.

Flexner, S. (1917). Mechanisms that defend the body from poliomyelitic infection, (a) external or extra-nervous, (b) internal or nervous. *Proc. Natl. Acad. Sci. USA* **3,** 416–418.

Fulle, H. J., Vassar, R., Foster, D. C., Yang, R. B., Axel, R., and Garbers, D. L. (1995). A receptor guanylyl cyclase expressed specifically in olfactory sensory neurons. *Proc. Natl. Acad. Sci. USA* **92,** 3571–3575.

Geng, L., Okuhara, D., Yu, Z., Tian, X., Cai, Y., Shibazaki, S., and Somlo, S. (2006). Polycystin-2 traffics to cilia independently of polycystin-1 by using an N-terminal rvxp motif. *J. Cell Sci.* **119,** 1383–1395.

Gilad, Y., and Lancet, D. (2003). Population differences in the human functional olfactory repertoire. *Mol. Biol. Evol.* **20,** 307–314.

Gilula, N. B., and Satir, P. (1972). The ciliary necklace. A ciliary membrane specialization. *J. Cell Biol.* **53,** 494–509.

Giorgi, D., Friedman, C., Trask, B. J., and Rouquier, S. (2000). Characterization of nonfunctional V1R-like pheromone receptor sequences in human. *Genome Res.* **10,** 1979–1985.

Gloriam, D. E., Bjarnadottir, T. K., Yan, Y. L., Postlethwait, J. H., Schioth, H. B., and Fredriksson, R. (2005). The repertoire of trace amine G-protein-coupled receptors: Large expansion in zebrafish. *Mol. Phylogenet. Evol.* **35,** 470–482.

Goldstein, B. J., Kulaga, H. M., and Reed, R. R. (2003). Cloning and characterization of SLP3: A novel member of the stomatin family expressed by olfactory receptor neurons. *J. Assoc. Res. Otolaryngol.* **4,** 74–82.

Graziadei, P. P., and DeHan, R. S. (1973). Neuronal regeneration in frog olfactory system. *J. Cell. Biol.* **59,** 525–530.

Graziadei, G. A., and Graziadei, P. P. (1979a). Neurogenesis and neuron regeneration in the olfactory system of mammals. II. Degeneration and reconstitution of the olfactory sensory neurons after axotomy. *J. Neurocytol.* **8,** 197–213.

Graziadei, P. P., and Graziadei, G. A. (1979b). Neurogenesis and neuron regeneration in the olfactory system of mammals. I. Morphological aspects of differentiation and structural organization of the olfactory sensory neurons. *J. Neurocytol.* **8,** 1–18.

Graziadei, P. P., and Metcalf, J. F. (1971). Autoradiographic and ultrastructural observations on the frog's olfactory mucosa. *Z Zellforsch Mikrosk Anat.* **116,** 305–318.

Graziadei, P. P., Levine, R. R., and Graziadei, G. A. (1978). Regeneration of olfactory axons and synapse formation in the forebrain after bulbectomy in neonatal mice. *Proc. Natl. Acad. Sci. USA.* **75,** 5230–5234.

Haberly, L. B. (2001). Parallel-distributed processing in olfactory cortex: New insights from morphological and physiological analysis of neuronal circuitry. *Chem. Senses* **26,** 551–576.

Hagiwara, H., Ohwada, N., and Takata, K. (2004). Cell biology of normal and abnormal ciliogenesis in the ciliated epithelium. *Int. Rev. Cytol.* **234,** 101–141.

Hammond, J. W., Cai, D., and Verhey, K. J. (2008). Tubulin modifications and their cellular functions. *Curr. Opin. Cell. Biol.* **20,** 71–76.

Harding, J., Graziadei, P. P., Monti Graziadei, G. A., and Margolis, F. L. (1977). Denervation in the primary olfactory pathway of mice. IV. Biochemical and morphological evidence for neuronal replacement following nerve section. *Brain Res.* **132,** 11–28.

Harkema, J. R., Carey, S. A., and Wagner, J. G. (2006). The nose revisited: A brief review of the comparative structure, function, and toxicologic pathology of the nasal epithelium. *Toxicol. Pathol.* **34,** 252–269.

Hasegawa, S., Yamagishi, M., and Nakano, Y. (1986). Microscopic studies of human olfactory epithelia following traumatic anosmia. *Arch Otorhinolaryngol.* **243,** 112–116.

Hegg, C. C., Greenwood, D., Huang, W., Han, P., and Lucero, M. T. (2003). Activation of purinergic receptor subtypes modulates odor sensitivity. *J. Neurosci.* **23,** 8291–8301.

Hennessey, T. M., Andrews, D., and Nelson, D. L. (1983). Biochemical studies of the excitable membrane of Paramecium tetraurelia. VII. Sterols and other neutral lipids of cells and cilia. *J. Lipid Res.* **24,** 575–587.

Hichri, H., Stoetzel, C., Laurier, V., Caron, S., Sigaudy, S., Sarda, P., Hamel, C., Martin-Coignard, D., Gilles, M., Leheup, B., Holder, M., Kaplan, J., et al. (2005). Testing for triallelism: Analysis of six BBS genes in a Bardet–Biedl syndrome family cohort. *Eur. J. Hum. Genet.* **13,** 607–616.

Hildebrand, J. G., and Shepherd, G. M. (1997). Mechanisms of olfactory discrimination: converging evidence for common principles across phyla. *Annu. Rev. Neurosci.* **20,** 595–631.

Hinds, J. W., and Hinds, P. L. (1976a). Synapse formation in the mouse olfactory bulb. I. Quantitative studies. *J. Comp. Neurol.* **169,** 15–40.

Hinds, J. W., and Hinds, P. L. (1976b). Synapse formation in the mouse olfactory bulb. II. Morphogenesis. *J Comp Neurol.* **169,** 41–61.

Hinds, J. W., Hinds, P. L., and McNelly, N. A. (1984). An autoradiographic study of the mouse olfactory epithelium: Evidence for long-lived receptors. *Anat. Rec.* **210,** 375–383.

Hu, J., Zhong, C., Ding, C., Chi, Q., Walz, A., Mombaerts, P., Matsunami, H., and Luo, M. (2007). Detection of near-atmospheric concentrations of CO_2 by an olfactory subsystem in the mouse. *Science* **317,** 953–957.

Iannaccone, A., Mykytyn, K., Persico, A. M., Searby, C. C., Baldi, A., Jablonski, M. M., and Sheffield, V. C. (2005). Clinical evidence of decreased olfaction in Bardet–Biedl syndrome caused by a deletion in the BBS4 gene. *Am. J. Med. Genet. A.* **132,** 343–346.

Illum, L. (2003). Nasal drug delivery – possibilities, problems and solutions. *J. Control Release* **87,** 187–198.

Inglis, P. N., Boroevich, K. A., and Leroux, M. R. (2006). Piecing together a ciliome. *Trends Genet.* **22,** 491–500.

Iwai, N., Zhou, Z., Roop, D. R., and Behringer, R. R. (2008). Horizontal basal cells are multipotent progenitors in normal and injured adult olfactory epithelium. *Stem Cells* **26,** 1298–1306.

Jafek, B. W. (2000). Evaluation and treatment of anosmia. *Curr. Opin. Otolaryngol. Head Neck Surg.* **8,** 63–67.

Jafek, B. W., Eller, P. M., Esses, B. A., and Moran, D. T. (1989). Post-traumatic anosmia. Ultrastructural correlates. *Arch Neurol.* **46,** 300–304.

Jafek, B. W., Murrow, B., Michaels, R., Restrepo, D., and Linschoten, M. (2002). Biopsies of human olfactory epithelium. *Chem. Senses* **27,** 623–628.

Jenkins, P. M., Hurd, T. W., Zhang, L., McEwen, D. P., Brown, R. L., Margolis, B., Verhey, K. J., and Martens, J. R. (2006). Ciliary targeting of olfactory CNG channels requires the CNGB1b subunit and the kinesin-2 motor protein, KIF17. *Curr. Biol.* **16,** 1211–1216.

Jones, D. T., and Reed, R. R. (1989). Golf: An olfactory neuron specific-G protein involved in odorant signal transduction. *Science* **244,** 790–795.

Juilfs, D. M., Fulle, H. J., Zhao, A. Z., Houslay, M. D., Garbers, D. L., and Beavo, J. A. (1997). A subset of olfactory neurons that selectively express cgmp-stimulated phosphodiesterase (PDE2) and guanylyl cyclase-D define a unique olfactory signal transduction pathway. *Proc. Natl. Acad. Sci. USA* **94,** 3388–3395.

Kaneko, H., Mohrlen, F., and Frings, S. (2006). Calmodulin contributes to gating control in olfactory calcium-activated chloride channels. *J. Gen. Physiol.* **127,** 737–748.

Kashiwayanagi, M., Sai, K., and Kurihara, K. (1987). Cell suspensions from porcine olfactory mucosa. Changes in membrane potential and membrane fluidity in response to various odorants. *J. Gen. Physiol.* **89,** 443–457.

Kashiwayanagi, M., Suenaga, A., Enomoto, S., and Kurihara, K. (1990). Membrane fluidity changes of liposomes in response to various odorants. Complexity of membrane composition and variety of adsorption sites for odorants. *Biophys J.* **58,** 887–895.

Kashiwayanagi, M., Sasaki, K., Iida, A., Saito, H., and Kurihara, K. (1997). Concentration and membrane fluidity dependence of odor discrimination in the turtle olfactory system. *Chem. Senses* **22,** 553–563.

Kaupp, U. B., and Seifert, R. (2002). Cyclic nucleotide-gated ion channels. *Physiol. Rev.* **82,** 769–824.

Kern, R. C., and Pitovski, D. Z. (1997). Localization of 11 beta-hydroxysteroid dehydrogenase: Specific protector of the mineralocorticoid receptor in mammalian olfactory mucosa. *Acta Otolaryngol.* **117,** 738–743.

Kerr, D. S., Von Dannecker, L. E., Davalos, M., Michaloski, J. S., and Malnic, B. (2008). Ric-8B interacts with Galphaolf and Ggamma13 and co-localizes with Galphaolf, Gbeta1 and Ggamma13 in the cilia of olfactory sensory neurons. *Mol. Cell Neurosci.* **38,** 341–348.

Kleene, S. J., and Gesteland, R. C. (1991). Calcium-activated chloride conductance in frog olfactory cilia. *J. Neurosci.* **11,** 3624–3629.

Klimmeck, D., Mayer, U., Ungerer, N., Warnken, U., Schnolzer, M., Frings, S., and Mohrlen, F. (2008). Calcium-signaling networks in olfactory receptor neurons. *Neuroscience* **151,** 901–912.

Klysik, M. (2008). Ciliary syndromes and treatment. *Pathol. Res. Pract.* **204,** 77–88.

Kobayakawa, K., Hayashi, R., Morita, K., Miyamichi, K., Oka, Y., Tsuboi, A., and Sakano, H. (2002). Stomatin-related olfactory protein, SRO, specifically expressed in the murine olfactory sensory neurons. *J. Neurosci.* **22,** 5931–5937.

Koenekoop, R. K. (2004). An overview of Leber congenital amaurosis: A model to understand human retinal development. *Surv. Ophthalmol.* **49,** 379–398.

Kozminski, K. G., Johnson, K. A., Forscher, P., and Rosenbaum, J. L. (1993). A motility in the eukaryotic flagellum unrelated to flagellar beating. *Proc. Natl. Acad. Sci. USA* **90,** 5519–5523.

Kulaga, H. M., Leitch, C. C., Eichers, E. R., Badano, J. L., Lesemann, A., Hoskins, B. E., Lupski, J. R., Beales, P. L., Reed, R. R., and Katsanis, N. (2004). Loss of BBS proteins causes anosmia in humans and defects in olfactory cilia structure and function in the mouse. *Nat. Genet.* **36,** 994–998.

Kurahashi, T., and Yau, K. W. (1993). Co-existence of cationic and chloride components in odorant-induced current of vertebrate olfactory receptor cells. *Nature* **363,** 71–74.

Leber, T. (1869). Uber retinitis pigmentosa und angeborene amaurose. *Graefes Arch Clin. Exp. Ophthalmol.* **15,** 1–25.

Leinders-Zufall, T., Cockerham, R. E., Michalakis, S., Biel, M., Garbers, D. L., Reed, R. R., Zufall, F., and Munger, S. D. (2007). Contribution of the receptor guanylyl cyclase GC-D to chemosensory function in the olfactory epithelium. *Proc. Natl. Acad. Sci. USA* **104,** 14507–14512.

Leopold, D. A., Hummel, T., Schwob, J. E., Hong, S. C., Knecht, M., and Kobal, G. (2000). Anterior distribution of human olfactory epithelium. *Laryngoscope* **110,** 417–421.

Lewin, A. H. (2006). Receptors of mammalian trace amines. *AAPS J.* **8,** E138–145.

Li, J. B., Gerdes, J. M., Haycraft, C. J., Fan, Y., Teslovich, T. M., May-Simera, H., Li, H., Blacque, O. E., Li, L., Leitch, C. C., Lewis, R. A., Green, J. S., Parfrey, P. S., *et al.* (2004). Comparative genomics identifies a flagellar and basal body proteome that includes the BBS5 human disease gene. *Cell* **117,** 541–552.

Liberles, S. D., and Buck, L. B. (2006). A second class of chemosensory receptors in the olfactory epithelium. *Nature* **442,** 645–650.

Lidow, M. S., and Menco, B. P. (1984). Observations on axonemes and membranes of olfactory and respiratory cilia in frogs and rats using tannic acid-supplemented fixation and photographic rotation. *J. Ultrastruct. Res.* **86,** 18–30.

Liman, E. R., Corey, D. P., and Dulac, C. (1999). TRP2: A candidate transduction channel for mammalian pheromone sensory signaling. *Proc. Natl. Acad. Sci. USA* **96,** 5791–5796.

Lin, F., Hiesberger, T., Cordes, K., Sinclair, A. M., Goldstein, L. S., Somlo, S., and Igarashi, P. (2003). Kidney-specific inactivation of the KIF3A subunit of kinesin-II inhibits renal ciliogenesis and produces polycystic kidney disease. *Proc. Natl. Acad. Sci. USA* **100,** 5286–5291.

Lin, W., Margolskee, R., Donnert, G., Hell, S. W., and Restrepo, D. (2007). Olfactory neurons expressing transient receptor potential channel M5 (TRPM5) are involved in sensing semiochemicals. *Proc. Natl. Acad. Sci. USA* **104,** 2471–2476.

Lledo, P. M., Gheusi, G., and Vincent, J. D. (2005). Information processing in the mammalian olfactory system. *Physiol. Rev.* **85,** 281–317.

Lobel, D., Marchese, S., Krieger, J., Pelosi, P., and Breer, H. (1998). Subtypes of odorant-binding proteins – heterologous expression and ligand binding. *Eur. J. Biochem.* **254,** 318–324.

Lucas, P., Ukhanov, K., Leinders-Zufall, T., and Zufall, F. (2003). A diacylglycerol-gated cation channel in vomeronasal neuron dendrites is impaired in TRPC2 mutant mice: mechanism of pheromone transduction. *Neuron* **40,** 551–561.

Ma, M. (2007). Encoding olfactory signals via multiple chemosensory systems. *Crit. Rev. Biochem. Mol. Biol.* **42,** 463–480.

Mackay-Sim, A., and Kittel, P. W. (1991). On the life span of olfactory receptor neurons. *Eur. J. Neurosci.* **3,** 209–215.

Mair, R. G., Gesteland, R. C., and Blank, D. L. (1982). Changes in morphology and physiology of olfactory receptor cilia during development. *Neuroscience* **7,** 3091–3103.

Malnic, B., Hirono, J., Sato, T., and Buck, L. B. (1999). Combinatorial receptor codes for odors. *Cell* **96,** 713–723.

Margalit, T., and Lancet, D. (1993). Expression of olfactory receptor and transduction genes during rat development. *Brain Res. Dev. Brain Res.* **73,** 7–16.

Marinari, U. M., Ferro, M., Sciaba, L., Finollo, R., Bassi, A. M., and Brambilla, G. (1984). DNA-damaging activity of biotic and xenobiotic aldehydes in Chinese hamster ovary cells. *Cell Biochem. Funct.* **2**, 243–248.

Matsuzaki, O., Bakin, R. E., Cai, X., Menco, B. P., and Ronnett, G. V. (1999). Localization of the olfactory cyclic nucleotide-gated channel subunit 1 in normal, embryonic and regenerating olfactory epithelium. *Neuroscience* **94**, 131–140.

Mayer, U., Ungerer, N., Klimmeck, D., Warnken, U., Schnolzer, M., Frings, S., and Mohrlen, F. (2008). Proteomic analysis of a membrane preparation from rat olfactory sensory cilia. *Chem. Senses* **33**, 145–162.

McClintock, T. S., Glasser, C. E., Bose, S. C., and Bergman, D. A. (2008). Tissue expression patterns identify mouse cilia genes. *Physiol. Genomics* **32**, 198–206.

McEwen, D. P., Koenekoop, R. K., Khanna, H., Jenkins, P. M., Lopez, I., Swaroop, A., and Martens, J. R. (2007). Hypomorphic CEP290/NPHP6 mutations result in anosmia caused by the selective loss of G proteins in cilia of olfactory sensory neurons. *Proc. Natl. Acad. Sci. USA* **104**, 15917–15922.

Menco, B. (1992). Ultrastructural studies on membrane, cytoskeletal, mucous, and protective compartments in olfaction. *Microsc. Res. Tech.* **22**, 215–224.

Menco, B. P. (1980a). Qualitative and quantitative freeze-fracture studies on olfactory and nasal respiratory epithelial surfaces of frog, ox, rat, and dog. II. Cell apices, cilia, and microvilli. *Cell Tissue Res.* **211**, 5–29.

Menco, B. P. (1980b). Qualitative and quantitative freeze-fracture studies on olfactory and nasal respiratory epithelial surfaces of frog, ox, rat, and dog. III. Tight-junctions. *Cell Tissue Res.* **211**, 361–373.

Menco, B. P. (1980c). Qualitative and quantitative freeze-fracture studies on olfactory and nasal respiratory structures of frog, ox, rat, and dog. I. A general survey. *Cell Tissue Res.* **207**, 183–209.

Menco, M. (1980d). Qualitative and quantitative freeze-fracture studies on olfactory and respiratory epithelial surfaces of frog, ox, rat, and dog. IV. Ciliogenesis and ciliary necklaces (including high-voltage observations). *Cell Tissue Res.* **212**, 1–16.

Menco, B. P. (1984). Ciliated and microvillous structures of rat olfactory and nasal respiratory epithelia. A study using ultra-rapid cryo-fixation followed by freeze-substitution or freeze-etching. *Cell Tissue Res.* **235**, 225–241.

Menco, B. P. (1997). Ultrastructural aspects of olfactory signaling. *Chem. Senses* **22**, 295–311.

Menco, B. P., and Farbman, A. I. (1985). Genesis of cilia and microvilli of rat nasal epithelia during pre-natal development. II. Olfactory epithelium, a morphometric analysis. *J. Cell Sci.* **78**, 311–336.

Menco, B. P., and Morrison, E. E. (2003). Morphology of the mammalian olfactory epithelium: form, fine structure, function, and pathology. *In* "Handbook on Olfaction and Gustation" (R. L. Doty, ed.), pp. 17–49. Informa Health Care, New York.

Menco, B. P., Tekula, F. D., Farbman, A. I., and Danho, W. (1994). Developmental expression of G-proteins and adenylyl cyclase in peripheral olfactory systems. Light microscopic and freeze-substitution electron microscopic immunocytochemistry. *J. Neurocytol.* **23**, 708–727.

Menco, B. P., Birrell, G. B., Fuller, C. M., Ezeh, P. I., Keeton, D. A., and Benos, D. J. (1998). Ultrastructural localization of amiloride-sensitive sodium channels and Na+,K(+)-atpase in the rat's olfactory epithelial surface. *Chem. Senses* **23**, 137–149.

Meyer, M. R., Angele, A., Kremmer, E., Kaupp, U. B., and Muller, F. (2000). A cgmp-signaling pathway in a subset of olfactory sensory neurons. *Proc. Natl. Acad. Sci. USA* **97**, 10595–10600.

Michalakis, S., Reisert, J., Geiger, H., Wetzel, C., Zong, X., Bradley, J., Spehr, M., Huttl, S., Gerstner, A., Pfeifer, A., Hatt, H., Yau, K. W., *et al.* (2006). Loss of

CNGB1 protein leads to olfactory dysfunction and subciliary cyclic nucleotide-gated channel trapping. *J. Biol. Chem.* **281**, 35156–35166.

Mombaerts, P. (2004). Genes and ligands for odorant, vomeronasal and taste receptors. *Nat. Rev. Neurosci.* **5**, 263–278.

Mombaerts, P. (2006). Axonal wiring in the mouse olfactory system. *Annu. Rev. Cell Dev. Biol.* **22**, 713–737.

Moran, D. T., Rowley, J. C., III, and Jafek, B. W. (1982a). Electron microscopy of human olfactory epithelium reveals a new cell type: the microvillar cell. *Brain Res.* **253**, 39–46.

Moran, D. T., Rowley, J. C., 3rd, Jafek, B. W., and Lovell, M. A. (1982b). The fine structure of the olfactory mucosa in man. *J. Neurocytol.* **11**, 721–746.

Mukhopadhyay, S., Lu, Y., Shaham, S., and Sengupta, P. (2008). Sensory signaling-dependent remodeling of olfactory cilia architecture in C. Elegans. *Dev. Cell* **14**, 762–774.

Murdoch, B., and Roskams, A. J. (2007). Olfactory epithelium progenitors: insights from transgenic mice and in vitro biology. *J. Mol. Histol.* **38**, 581–599.

Murphy, C., Schubert, C. R., Cruickshanks, K. J., Klein, B. E., Klein, R., and Nondahl, D. M. (2002). Prevalence of olfactory impairment in older adults. *JAMA* **288**, 2307–2312.

Mykytyn, K., Mullins, R. F., Andrews, M., Chiang, A. P., Swiderski, R. E., Yang, B., Braun, T., Casavant, T., Stone, E. M., and Sheffield, V. C. (2004). Bardet–Biedl syndrome type 4 (BBS4)-null mice implicate Bbs4 in flagella formation but not global cilia assembly. *Proc. Natl. Acad. Sci. USA* **101**, 8664–8669.

Mykytyn, K., Nishimura, D. Y., Searby, C. C., Beck, G., Bugge, K., Haines, H. L., Cornier, A. S., Cox, G. F., Fulton, A. B., Carmi, R., Iannaccone, A., Jacobson, S. G., et al. (2003). Evaluation of complex inheritance involving the most common Bardet–Biedl syndrome locus (BBS1). *Am. J. Hum. Genet.* **72**, 429–437.

Nakamura, T., and Gold, G. H. (1987). A cyclic nucleotide-gated conductance in olfactory receptor cilia. *Nature* **325**, 442–444.

Nguyen-Khoa, B. A., Goehring, E. L., Jr., Vendiola, R. M., Pezzullo, J. C., and Jones, J. K. (2007). Epidemiologic study of smell disturbance in 2 medical insurance claims populations. *Arch Otolaryngol. Head Neck Surg.* **133**, 748–757.

Nishimura, D. Y., Fath, M., Mullins, R. F., Searby, C., Andrews, M., Davis, R., Andorf, J. L., Mykytyn, K., Swiderski, R. E., Yang, B., Carmi, R., Stone, E. M., et al. (2004). Bbs2-null mice have neurosensory deficits, a defect in social dominance, and retinopathy associated with mislocalization of rhodopsin. *Proc. Natl. Acad. Sci. USA* **101**, 16588–16593.

Nomura, T., and Kurihara, K. (1987a). Effects of changed lipid composition on responses of liposomes to various odorants: Possible mechanism of odor discrimination. *Biochemistry* **26**, 6141–6145.

Nomura, T., and Kurihara, K. (1987b). Liposomes as a model for olfactory cells: changes in membrane potential in response to various odorants. *Biochemistry* **26**, 6135–6140.

Okada, Y., Takeda, S., Tanaka, Y., Belmonte, J. C., and Hirokawa, N. (2005). Mechanism of nodal flow: a conserved symmetry breaking event in left–right axis determination. *Cell* **121**, 633–644.

Orozco, J. T., Wedaman, K. P., Signor, D., Brown, H., Rose, L., and Scholey, J. M. (1999). Movement of motor and cargo along cilia. *Nature* **398**, 674.

Ostrowski, L. E., Blackburn, K., Radde, K. M., Moyer, M. B., Schlatzer, D. M., Moseley, A., and Boucher, R. C. (2002). A proteomic analysis of human cilia: identification of novel components. *Mol. Cell. Proteomics* **1**, 451–465.

Ou, G., Blacque, O. E., Snow, J. J., Leroux, M. R., and Scholey, J. M. (2005). Functional coordination of intraflagellar transport motors. *Nature* **436**, 583–587.

Pathak, N., Obara, T., Mangos, S., Liu, Y., and Drummond, I. A. (2007). The zebrafish fleer gene encodes an essential regulator of cilia tubulin polyglutamylation. *Mol. Biol. Cell* **18,** 4353–4364.

Pazour, G. J., Agrin, N., Leszyk, J., and Witman, G. B. (2005). Proteomic analysis of a eukaryotic cilium. *J. Cell Biol.* **170,** 103–113.

Pazour, G. J., Wilkerson, C. G., and Witman, G. B. (1998). A dynein light chain is essential for the retrograde particle movement of intraflagellar transport (IFT). *J. Cell Biol.* **141,** 979–992.

Pelosi, P. (1994). Odorant-binding proteins. *Crit. Rev. Biochem. Mol. Biol.* **29,** 199–228.

Pes, D., and Pelosi, P. (1995). Odorant-binding proteins of the mouse. *Comp. Biochem. Physiol. B Biochem. Mol. Biol.* **112,** 471–479.

Pevsner, J., Hou, V., Snowman, A. M., and Snyder, S. H. (1990). Odorant-binding protein. Characterization of ligand binding. *J. Biol. Chem.* **265,** 6118–6125.

Pifferi, S., Pascarella, G., Boccaccio, A., Mazzatenta, A., Gustincich, S., Menini, A., and Zucchelli, S. (2006). Bestrophin-2 is a candidate calcium-activated chloride channel involved in olfactory transduction. *Proc. Natl. Acad. Sci. USA* **103,** 12929–12934.

Qu, Z., Wei, R. W., Mann, W., and Hartzell, H. C. (2003). Two bestrophins cloned from Xenopus laevis oocytes express Ca(2+)-activated Cl(−) currents. *J. Biol. Chem.* **278,** 49563–49572.

Raviv, J. R., and Kern, R. C. (2004). Chronic sinusitis and olfactory dysfunction. *Otolaryngol. Clin. North Am.* **37,** 1143–1157, v–vi.

Reese, T. S. (1965). Olfactory cilia in the frog. *J. Cell Biol.* **25,** 209–230.

Reisert, J., Bauer, P. J., Yau, K. W., and Frings, S. (2003). The Ca-activated Cl channel and its control in rat olfactory receptor neurons. *J. Gen. Physiol.* **122,** 349–363.

Reisert, J., Lai, J., Yau, K. W., and Bradley, J. (2005). Mechanism of the excitatory Cl- response in mouse olfactory receptor neurons. *Neuron* **45,** 553–561.

Reiter, E. R., DiNardo, L. J., and Costanzo, R. M. (2004). Effects of head injury on olfaction and taste. *Otolaryngol. Clin. North Am.* **37,** 1167–1184.

Reuter, D., Zierold, K., Schroder, W. H., and Frings, S. (1998). A depolarizing chloride current contributes to chemoelectrical transduction in olfactory sensory neurons *in situ*. *J. Neurosci.* **18,** 6623–6630.

Roayaie, K., Crump, J. G., Sagasti, A., and Bargmann, C. I. (1998). The G alpha protein ODR-3 mediates olfactory and nociceptive function and controls cilium morphogenesis in C. Elegans olfactory neurons. *Neuron* **20,** 55–67.

Rodriguez, I. (2004). Pheromone receptors in mammals. *Horm. Behav.* **46,** 219–230.

Rodriguez, I., and Mombaerts, P. (2002). Novel human vomeronasal receptor-like genes reveal species-specific families. *Curr. Biol.* **12,** R409–411.

Ronnett, G. V., and Moon, C. (2002). G proteins and olfactory signal transduction. *Annu. Rev. Physiol.* **64,** 189–222.

Rosenbaum, J. L., and Witman, G. B. (2002). Intraflagellar transport. *Nat. Rev. Mol. Cell Biol.* **3,** 813–825.

Rowley, J. C., 3rd, Moran, D. T., and Jafek, B. W. (1989). Peroxidase backfills suggest the mammalian olfactory epithelium contains a second morphologically distinct class of bipolar sensory neuron: the microvillar cell. *Brain Res.* **502,** 387–400.

Russell, Y., Evans, P., and Dodd, G. H. (1989). Characterization of the total lipid and fatty acid composition of rat olfactory mucosa. *J. Lipid Res.* **30,** 877–884.

Saito, H., Mimmack, M., Kishimoto, J., Keverne, E. B., and Emson, P. C. (1998). Expression of olfactory receptors, G-proteins and axcams during the development and maturation of olfactory sensory neurons in the mouse. *Brain Res. Dev. Brain Res.* **110,** 69–81.

Sam, M., Vora, S., Malnic, B., Ma, W., Novotny, M. V., and Buck, L. B. (2001). Neuropharmacology. Odorants may arouse instinctive behaviours. *Nature* **412,** 142.

Sammeta, N., Yu, T. T., Bose, S. C., and McClintock, T. S. (2007). Mouse olfactory sensory neurons express 10,000 genes. *J. Comp. Neurol.* **502,** 1138–1156.

Satir, P., and Christensen, S. T. (2007). Overview of structure and function of mammalian cilia. *Annu. Rev. Physiol.* **69,** 377–400.

Schofield, P. R. (1988). Carrier-bound odorant delivery to olfactory receptors. *Trends Neurosci.* **11,** 471–472.

Scholey, J. M. (2003). Intraflagellar transport. *Annu. Rev. Cell Dev. Biol.* **19,** 423–443.

Scholey, J. M. (2008). Intraflagellar transport motors in cilia: moving along the cell's antenna. *J. Cell Biol.* **180,** 23–29.

Schreiber, S., Fleischer, J., Breer, H., and Boekhoff, I. (2000). A possible role for caveolin as a signaling organizer in olfactory sensory membranes. *J. Biol. Chem.* **275,** 24115–24123.

Schultz, E. W., and Gebhardt, L. P. (1936). Chemoprophylaxis of Poliomyelitis: A Progress Report. *Cal. West Med.* **45,** 138–140.

Schwarzenbacher, K., Fleischer, J., Breer, H., and Conzelmann, S. (2004). Expression of olfactory receptors in the cribriform mesenchyme during prenatal development. *Gene Exp. Patterns* **4,** 543–552.

Schwarzenbacher, K., Fleischer, J., and Breer, H. (2005). Formation and maturation of olfactory cilia monitored by odorant receptor-specific antibodies. *Histochem. Cell Biol.* **123,** 419–428.

Seiden, A. M., and Duncan, H. J. (2001). The diagnosis of a conductive olfactory loss. *Laryngoscope* **111,** 9–14.

Seifert, K. (1971). Light and electron microscopic studies on the organ of Jacobson (vomeronasal organ) of cats. *Arch Klin Exp Ohren Nasen Kehlkopfheilkd.* **200,** 223–251.

Smith, J. C., Northey, J. G., Garg, J., Pearlman, R. E., and Siu, K. W. (2005). Robust method for proteome analysis by MS/MS using an entire translated genome: Demonstration on the ciliome of Tetrahymena thermophila. *J. Proteome Res.* **4,** 909–919.

Snow, J. J., Ou, G., Gunnarson, A. L., Walker, M. R., Zhou, H. M., Brust-Mascher, I., and Scholey, J. M. (2004). Two anterograde intraflagellar transport motors cooperate to build sensory cilia on C. Elegans neurons. *Nat. Cell Biol.* **6,** 1109–1113.

Stoetzel, C., Laurier, V., Davis, E. E., Muller, J., Rix, S., Badano, J. L., Leitch, C. C., Salem, N., Chouery, E., Corbani, S., Jalk, N., Vicaire, S., *et al.* (2006). BBS10 encodes a vertebrate-specific chaperonin-like protein and is a major BBS locus. *Nat. Genet.* **38,** 521–524.

Stolc, V., Samanta, M. P., Tongprasit, W., and Marshall, W. F. (2005). Genome-wide transcriptional analysis of flagellar regeneration in Chlamydomonas reinhardtii identifies orthologs of ciliary disease genes. *Proc. Natl. Acad. Sci. USA* **102,** 3703–3707.

Strotmann, J., Wanner, I., Helfrich, T., and Breer, H. (1995). Receptor expression in olfactory neurons during rat development: in situ hybridization studies. *Eur. J. Neurosci.* **7,** 492–500.

Su, A. I., Wiltshire, T., Batalov, S., Lapp, H., Ching, K. A., Block, D., Zhang, J., Soden, R., Hayakawa, M., Kreiman, G., Cooke, M. P., Walker, J. R., *et al.* (2004). A gene atlas of the mouse and human protein-encoding transcriptomes. *Proc. Natl. Acad. Sci. USA* **101,** 6062–6067.

Sumner, D. (1964). Post-traumatic anosmia. *Brain* **87,** 107–120.

Tabaton, M., Monaco, S., Cordone, M. P., Colucci, M., Giaccone, G., Tagliavini, F., and Zanusso, G. (2004). Prion deposition in olfactory biopsy of sporadic Creutzfeldt–Jakob disease. *Ann. Neurol.* **55,** 294–296.

Tillman, T. S., and Cascio, M. (2003). Effects of membrane lipids on ion channel structure and function. *Cell Biochem. Biophys.* **38,** 161–190.

Toller, S. V. (1999). Assessing the impact of anosmia: Review of a questionnaire's findings. *Chem. Senses* **24,** 705–712.

Vieira, O. V., Gaus, K., Verkade, P., Fullekrug, J., Vaz, W. L., and Simons, K. (2006). FAPP2, cilium formation, and compartmentalization of the apical membrane in polarized Madin–Darby canine kidney (MDCK) cells. *Proc. Natl. Acad. Sci. USA* **103,** 18556–18561.

Weber, R., Raschka, C., and Bonzel, T. (1992). Toxic drug-induced hyposmia with lovastatin. *Laryngorhinootologie.* **71,** 483–484.

Witt, M., and Hummel, T. (2006). Vomeronasal versus olfactory epithelium: is there a cellular basis for human vomeronasal perception? *Int. Rev. Cytol.* **248,** 209–259.

Wong, S. T., Trinh, K., Hacker, B., Chan, G. C., Lowe, G., Gaggar, A., Xia, Z., Gold, G. H., and Storm, D. R. (2000). Disruption of the type III adenylyl cyclase gene leads to peripheral and behavioral anosmia in transgenic mice. *Neuron* **27,** 487–497.

Yamamoto, M. (1976). An electron microscopic study of the olfactory mucosa in the bat and rabbit. *Arch Histol. Jpn.* **38,** 359–412.

Yang, J., Gao, J., Adamian, M., Wen, X. H., Pawlyk, B., Zhang, L., Sanderson, M. J., Zuo, J., Makino, C. L., and Li, T. (2005). The ciliary rootlet maintains long-term stability of sensory cilia. *Mol. Cell Biol.* **25,** 4129–4137.

Yang, J., Liu, X., Yue, G., Adamian, M., Bulgakov, O., and Li, T. (2002). Rootletin, a novel coiled-coil protein, is a structural component of the ciliary rootlet. *J. Cell Biol.* **159,** 431–440.

Youngentob, S. L., and Margolis, F. L. (1999). OMP gene deletion causes an elevation in behavioral threshold sensitivity. *Neuroreport* **10,** 15–19.

Youngentob, S. L., Margolis, F. L., and Youngentob, L. M. (2001). OMP gene deletion results in an alteration in odorant quality perception. *Behav. Neurosci.* **115,** 626–631.

Youngentob, S. L., Kent, P. F., and Margolis, F. L. (2003). OMP gene deletion results in an alteration in odorant-induced mucosal activity patterns. *J. Neurophysiol.* **90,** 3864–3873.

Youngentob, S. L., Pyrski, M. M., and Margolis, F. L. (2004). Adenoviral vector-mediated rescue of the OMP-null behavioral phenotype: enhancement of odorant threshold sensitivity. *Behav. Neurosci.* **118,** 636–642.

Zheng, J., and Zagotta, W. N. (2004). Stoichiometry and assembly of olfactory cyclic nucleotide-gated channels. *Neuron* **42,** 411–421.

CHAPTER THIRTEEN

Ciliary Dysfunction in Developmental Abnormalities and Diseases

Neeraj Sharma, Nicolas F. Berbari, *and* Bradley K. Yoder

Contents

1. Introduction	372
2. The Oak Ridge Polycystic Kidney (*Orpk*) Mouse: A Model For Human Ciliopathies	374
3. Functions and Phenotypes Associated with Abnormal Motile Cilia	376
3.1. Primary Cilia Dyskinesia	377
3.2. Identifying candidate PCD proteins	378
3.3. Cilia and left–right asymmetry defects	379
3.4. Role of cilia and flagella in reproductive system	381
3.5. Functions of motile cilia in brain	381
3.6. Motile cilia and neurodevelopment	383
4. Functions and Diseases Associated with Immotile Cilium	383
4.1. Primary cilia and cell cycle	384
4.2. Sensory abnormalities associated with cilia mutants	386
4.3. Neuronal cilia and obesity phenotypes	387
4.4. Ciliary connection to PKD related phenotypes in the kidney, liver, and pancreas	390
4.5. NPHP (omim 256100)	396
4.6. MKS (omim 249000)	398
4.7. BBS (omim 209900)	401
4.8. JBTS (omim 213300)	405
4.9. SLSN (omim 266900)	406
4.10. OFD (omim 311200)	406
4.11. ALMS (omim 203800)	408
5. Oligogenic Inheritance and Clinical Variability in the Ciliopathies	409
6. Cilia Proteome and Genome Databases and the Identification of Human Ciliopathy Genes	410
Acknowledgments	412
References	413

Department of Cell Biology, University of Alabama at Birmingham, School of Medicine, Birmingham, Alabama

Abstract

Cilia are small microtubule-based cellular appendages that are broadly classified as being either motile or immotile (primary cilia). Since their initial discovery several centuries ago, motile cilia have been of general interest to basic scientists and others who study the dynamics and physiological relevance of their motility. More recent discoveries have found that motile and immotile cilia, the later of which are present on nearly all cells in the mammalian body, also have major roles during development and in postnatal life. Dysfunction of the cilium is the basis for multiple human genetic disorders that have collectively been called the ciliopathies. The phenotypes associated with cilia dysfunction in mammals are diverse and include randomization of the left–right body axis, abnormalities in neural tube closure and patterning, skeletal defects such as polydactyly, cystic kidney, liver, and pancreatic diseases, blindness and anosmia, behavioral and cognitive defects, and obesity. The connection between disease and developmental defects due to the loss of ciliary function has brought the efforts of the biomedical research establishment to bear on this underappreciated and long overlooked organelle. Several groups have applied *en silico*, genetic, and biochemical approaches to identify the components of the cilia proteome. The resulting datasets have contributed to a remarkable increase in the rate at which human ciliopathy disease loci are being identified. This intense basic and clinical research interest has revealed that the cilium is a very complex sensory machine involved in transducing extracellular stimuli involved in many different signaling pathways into cellular responses. Although major advances have been made in understanding the importance of the cilium, it remains enigmatic how the cilium functions to coordinate signaling pathways and how loss of this organelle results in the severe defects observed in human ciliopathies.

1. INTRODUCTION

Although cilia and flagella and the processes responsible for their formation have been a great curiosity in the realm of cell biology for over a century, only recently has there been an insurgence of interest in the cilium as a clinically relevant organelle. In large part, the renaissance began with the discovery that cilia mutant mice exhibit random left–right body axis specification and that mutations in cilia localized proteins cause various forms of human renal cystic diseases. From these seminal studies, a new paradigm rapidly emerged in which the cilium serves a complex range of functions important during development and for maintaining normal tissue physiology. The dysfunction of this small microtubule-based organelle, which was once considered to be a vestigial remnant of our evolutionary past, is now known to be a major contributor to a remarkable spectrum of diseases and developmental disorders referred to as the ciliopathies. Although the cilium is currently appreciated as an essential organelle, the challenge to elucidate its normal functions during development as well as in adult tissue homeostasis remains. Concomitant with this challenge is assessing how dysfunction of the cilium disrupts signaling events that contribute to the severe disease phenotypes associated with ciliopathies.

Cilia and flagella are highly complex structures that require more than 500 proteins for normal maintenance and function (Ostrowski et al., 2002; Pazour, 2004; Pazour et al., 2005). Since the cilium is devoid of ribosomes, cells must transport all cilia subcomponents from the cell body to the cilium. This is achieved through a process known as intraflagellar transport (IFT) which was first described in the laboratory of Dr. Rosenbaum (Kozminski et al., 1993, 1995, 1998). Briefly, IFT mediates the bidirectional transport of axonemal subunits and signaling machinery between the base and tip of the cilium (Cole, 2003; Pazour et al., 2002a; Scholey, 2003). A detailed description of IFT and its importance for cilia assembly and signaling is provided in Chapter 2 of this volume.

Cilia are found on most cells in the mammalian body (Olsen, 2005; Satir and Christensen, 2008). The ubiquitous nature of the cilium is in agreement with the wide spectrum of disease phenotypes associated with ciliary mutants (Afzelius, 2004; Eley et al., 2005). They extend from the surface of cells into the local environment and are classified as either motile or nonmotile. The primary cilium ($9 + 0$ microtubule architecture) is immotile and is present as a solitary structure on most cells including epithelia, fibroblasts, and neurons. However, there are also examples where a solitary cilium on a cell can be motile. This is observed on cells of the embryonic node where the rotational motility of the primary cilium is required for normal left–right body axis specification (for more detail on cilia and left–right axis formation, see Chapter 6, this volume) (Essner et al., 2002) and the flagella on sperm which are essential for cell movement (Inaba, 2003; Mitchell, 2007). In contrast, epithelial cells lining the ventricles of the brain (Worthington and Cathcart, 1963), the respiratory (Lansley et al., 1992) and reproductive tracts assemble hundreds of motile cilia ($9 + 2$ microtubule architecture) (Lyons et al., 2006) that beat in a synchronized manner to direct fluid movement. Disruption of normal cilia motility can result in a number of disease states in humans, such as hydrocephalus, chronic bronchiectasis, sinusitis, *situs inversus*, and infertility. Both motile and nonmotile cilia are thought to have important sensory functions that allow cells to receive and respond to extracellular stimuli (Huang et al., 2004; Josef et al., 2005; Pan and Snell, 2002).

Much of our current understanding of ciliary function and cilia mediated signaling activities have come from studies in model organisms such as *Chlamydomonas, Trypanosome, Tetrahymena, Caenorhabditis elegans,* zebrafish, and mice (Brown et al., 1999; Dentler, 2005; Evans et al., 2006; Hou et al., 2007; Krock and Perkins, 2008; Pan and Snell, 2005; Pan et al., 2006; Pedersen et al., 2005). In lower eukaryotes, disruption of cilia does not have a major impact on the development or survival of the organism, thus, these simpler model systems have had major utility. Their viability has allowed the identification of ciliary components, analysis of cilia assembly processes, and the study of cilia mediated regulation of signaling pathways. In contrast, disruption of cilia assembly in vertebrates, such as mice, is lethal at around embryonic day 9.5 (Murcia et al., 2000; Nonaka et al., 1998). This is due to severe developmental defects involving abnormal neural tube closure, neural tube patterning, and body axis specification. In humans, mutations that completely disrupt cilia formation (such as IFT null mutations) have yet to be identified which is likely due to lethality. However, there are now multiple examples where mutations in cilia proteins cause human diseases (see Tables 13.1–13.6) such as Primary Cilia Dyskinesia (PCD [OMIM 242650]), Polycystic Kidney Disease (PKD [OMIM 173900]), Nephronophthisis (NPHP [OMIM 256100]), Joubert syndrome (JBTS [OMIM 21330]), Meckel–Gruber syndrome (MKS [OMIM 249000]), Leber Congenital Amaurosis (LCA [OMIM 611755]), Senior–Loken syndrome (SLSN [OMIM 266900]), Kartagener syndrome (KS [OMIM 244400]), Alstrom syndrome (ALMS [OMIM 203800]), and Orofaciodigital type 1 syndrome (OFD1[OMIM 311200]) Asphyxiating thoracic Dystrophy (ATD [OMIM 208500]), Bardet–Biedl syndrome (BBS [OMIM 209900]) to mention a few. In this chapter, we will describe the functions attributed to the various forms of cilia and discuss the current understanding of cilia dysfunction and the associated disease and developmental phenotypes.

2. THE OAK RIDGE POLYCYSTIC KIDNEY (ORPK) MOUSE: A MODEL FOR HUMAN CILIOPATHIES

The best example of the spectrum of phenotypes associated with cilia dysfunction in mammals is seen in the oak ridge polycystic kidney mouse model (ORPK, $ift88^{Tg737Rpw}$). The ORPK mouse was originally described as a model for human autosomal recessive polycystic kidney disease (ARPKD). ORPK mice have a hypomorphic mutation in a gene initially called *Tg737* and now referred to as *ift88*. This model originated in the colony at the Oak Ridge National Laboratory though a large-scale transgene insertional mutagenesis project. The transgene integrated into an intron of *ift88* near the 3' end of the gene, partially disrupting its expression (Moyer et al., 1994; Yoder et al., 1995). The ORPK mutant mouse was a subject of a recent review and thus will not be described in detail here (Lehman et al., 2008). Briefly, the hypomorphic *ift88* allele in ORPK mutants results in short and malformed primary and motile cilia. As a consequence of these abnormal cilia, the ORPK mutants develop cysts in the kidney, liver, and pancreas and have hyperplasia and dysplasia of the pancreatic and biliary ducts (Fig. 13.1). In addition, these mutants exhibit hydrocephaly, retinal degeneration, anosmia, cerebellar hypoplasia, ataxia, abnormal neuroblast migration, sterility, skin and hair follicle defects, and skeletal abnormalities.

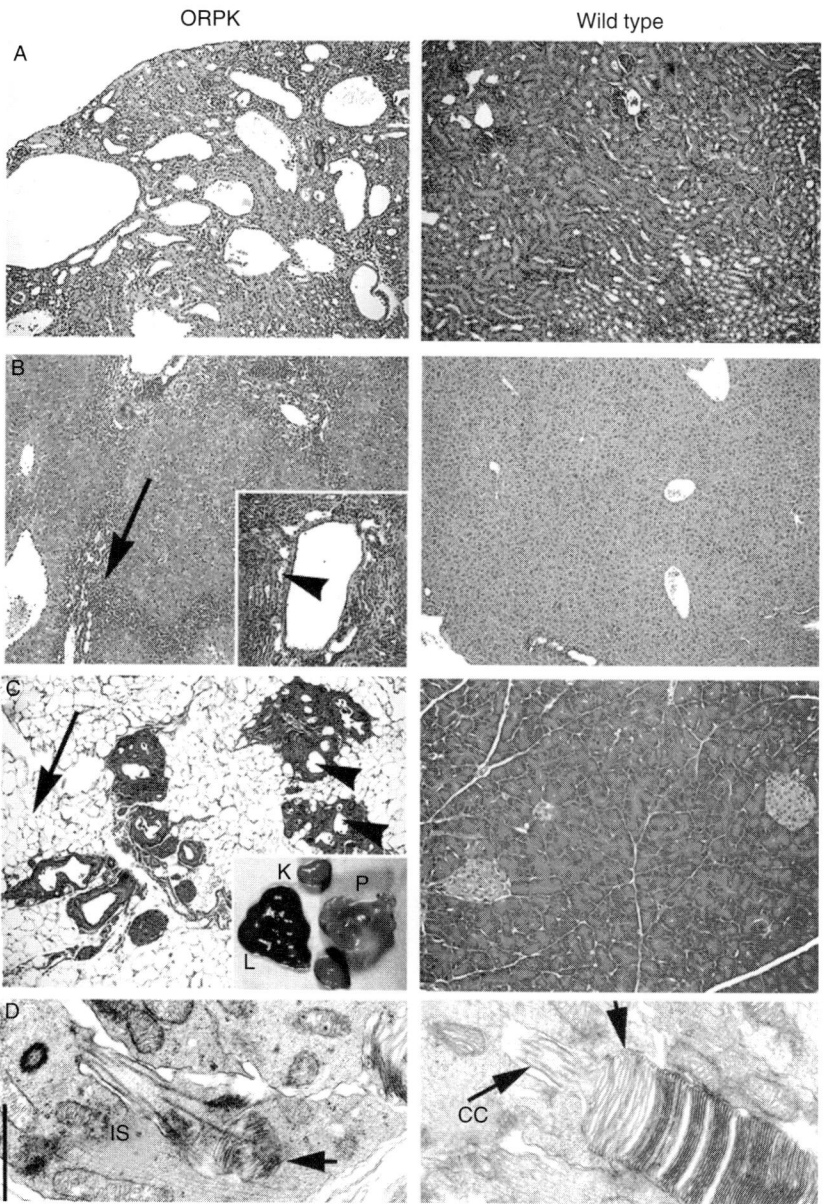

Figure 13.1 Characteristic phenotypes in the ORPK mouse. Left panels show various disease phenotypes observed in different tissue types in ORPK (FVB/N background) mutants and right panels show wild-type controls. (A) Enlarged cysts present in the kidney. (B) Biliary (arrow) and bile duct (arrow head) hyperplasia developed in the liver. Inset shows higher magnification view of a central vein displaying multiple dysplastic bile ductules. (C) Dilated ducts (arrowhead) and acini atrophy (arrow) in the OPRK mutant pancreas. Inset show the individual affected organs (K, kidney;

In most cases, the phenotypes in ORPK mutants have been associated with defects in ciliogenesis, cilia motility, or cilia signaling activities. The ORPK mutants serve as an excellent model to study the multiple phenotypes associated with cilia dysfunction and to demonstrate the critical importance of the cilium in normal tissue physiology.

3. Functions and Phenotypes Associated with Abnormal Motile Cilia

Arguably, the best known cilia are the motile forms exemplified by those found on cells of the respiratory tract and ependymal cells lining the ventricles in the brain. In the respiratory tract, these cilia play important roles in mucus clearance; while the ependymal cilia mediate movement of cerebral spinal fluid. Motile cilia are also present in the female reproductive system facilitating the movement of the ovum. These motile cilia with $9+2$ axoneme architecture, are normally present in large numbers on the cell surface, and beat in an orchestrated wavelike manner.

The motile cilium is a very complex machine. This machine consists of a central pair of single microtubules that are surrounded by nine pairs ($9+2$) of microtubules doublets consisting of an A and B tubule (Inaba, 2003). Dynein arms extend from the A tubule toward the B tubule. An array of radial spoke proteins connect dynein motors and the membrane sheath surrounding the central microtubule doublets (Fig. 13.2). Waveform movement of epithelial cilia is achieved by sliding of the microtubule doublets relative to one another in an ATP dependent manner. ATP hydrolysis in the outer dynein arms regulates the formation and breaking of the cross bridges causing movement of an A tubule of one doublet along the B tubule of an adjacent doublet. As the microtubules are sliding on one side of the axoneme, the doublets on the other side are moving in the opposite direction. This sliding motion leads to the bending of the cilium by the selective activation and deactivation of the radial spoke cross bridges in the axoneme. Cross sections through motile cilia indicate there is structural asymmetry in the axoneme. An outer dynein arm is not present on the number one microtubule doublet along most of the axoneme which reflect asymmetry of dynein function. These data suggest there are structural components within the axoneme that determine the direction of ciliary beat.

L, liver; P, pancreas). (D) P10. The ORPK mutants develop disrupted discs and misshaped OSs filled with amorphous material. (IS, inner segment; CC, connecting cilium). Panel D images are reproduced from Pazour *et al.* (2002) *J. Cell Biol.* (Copyright © 2002, The Rockefeller University Press). All panels were reproduced from Lehman *et al.* (2008), *Dev. Dyn.* (Copyright © 2008, Wiley-Liss). (See Color Insert.)

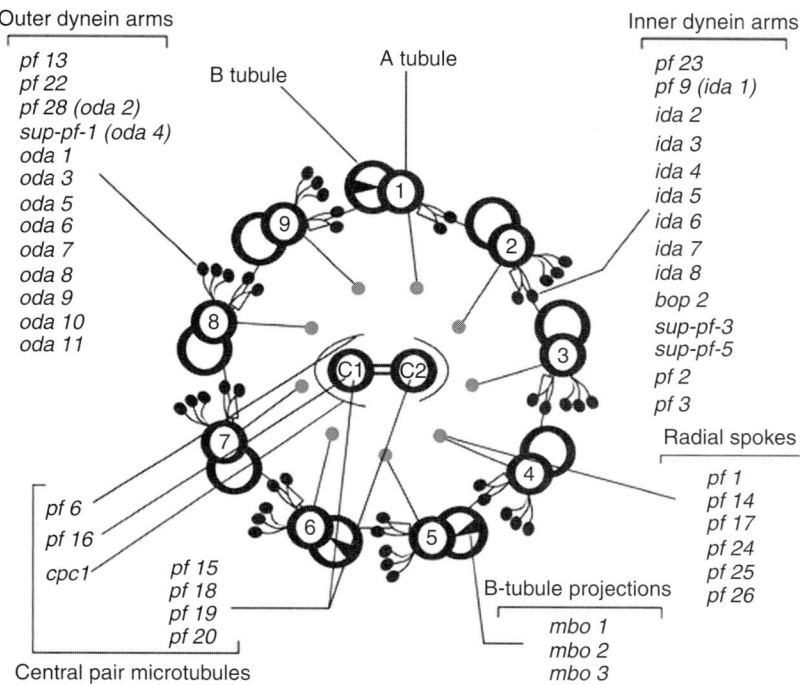

Figure 13.2 Schematic representation of the flagellar axoneme. This 9 + 2 arrangement of microtubules is seen when the axoneme is viewed in cross section with the transmission electron microscope. Each doublet microtubule consists of A and B tubules: the A tubule is a complete microtubule with 13 protofilaments (protofilament not shown), while the B tubule contains 10 protofilaments (protofilaments not shown). Inner and outer dynein arms are attached to the A tubule. Radial spokes are projected towards the center of the axoneme and may interact with central pair projections. Mutations affecting the structure or motility of flagellar are mentioned below each structural component of the axoneme (Image adapted from Porter, M. E., and Sale, W. S. (2000). *J. Cell Biol.*) (Copyright © 2000, The Rockefeller University Press).

Mutations that affect proper axoneme assembly, dynein motor activity or positioning can disrupt normal cilia beat and lead to disease phenotypes in mammals. This is normally evident through four phenotypic manifestations including abnormal embryonic turning and left–right body axis formation, chronic respiratory infections and rhinitis, sterility, and hydrocephalus.

3.1. Primary Cilia Dyskinesia

PCD (formerly immotile cilia syndrome) is an autosomal recessive disorder resulting from abnormalities in the structure and motility of cilia. The majority of PCD patients in the perinatal period exhibit significant respiratory distress; suggesting cilia are crucial for effective clearance of fetal lung fluid. The PCD phenotype is characterized by chronic upper and lower respiratory tract infection and abnormal left–right asymmetry (mirror image reversal or *situs inversus*) in nearly 50% of affected patients (Bush *et al.*, 1998; Lie and Ferkol, 2007). The triad of sinusitis, bronchiectasis, and *situs inversus* is known as Kartagener's syndrome (KS) while the PCD is used to represent congenital abnormalities of ciliary motility.

The clinical features in PCD reflect the distribution of motile cilia. Mucociliary clearance requires the coordination of multiple cell types in several anatomical locations. The pseudostratified columnar respiratory epithelium consists of ciliated, nonciliated, mucus, and basal cells. Cilia line the nasal cavity, paranasal sinuses, eustachian tubes, and middle ear of the upper respiratory tract. Ciliated cells are also found lining the trachea to the terminal bronchioles in the lower respiratory epithelium. Interestingly, the density and size of the motile cilia are not uniform, and generally both decrease as airway size decreases. Each ciliated cell is decorated with approximately 100–200 cilia on its surface which beat in a synchronous manner (Boysen, 1982). The lung and nasal passages require proper mucociliary clearance as they are constantly exposed to air-borne pathogens, toxins, and particulate matter. Mucociliary clearance requires appropriate interactions between ciliated epithelium and mucus (Wanner *et al.*, 1996). The mucus traps inhaled particles and pathogens and the synchronous ciliary movement moves the mucus layer. This clearance can be disturbed by viral and bacterial infections and also by inherited diseases such as PCD or cystic fibrosis.

Early ultrastructural studies conducted on sperm and respiratory epithelium by Bjorn Afzelius and others revealed that cilia of KS patients often lack dynein arms on the outer doublet microtubules (Afzelius, 1976; Afzelius *et al.*, 1975; Pedersen and Mygind, 1976; Pedersen and Rebbe, 1975). The loss of cilia and sperm motility and subsequent decreased mucociliary clearance in these patients were attributed to these ultrastructural abnormalities. Since that time, multiple structural defects in cilia have been demonstrated in KS patients including absence of radial spokes (Sturgess *et al.*, 1979), central pair microtubules (Sturgess *et al.*, 1980), and central sheath (Afzelius and Eliasson, 1979). More recently, Rutland *et al.* demonstrated that PCD can also occur in patients where axonemal ultrastructure is not overtly altered. These cilia beat in a disorganized fashion due to misorientation of the central pair which is normally aligned perpendicular to the direction of the beat. This misorientation of the central pair results in each cilia beating in an independent direction and generates ineffective propulsion of mucus along the mucosal surface causing chronic bronchiectasis (Rutland and de Iongh, 1990).

3.2. Identifying candidate PCD proteins

The ultrastructural architecture of cilia is highly conserved from protists to humans. Detailed examination of the *Chlamydomonas* flagella axoneme, which shares a high degree of homology with the mammalian motile cilium and spermatozoa tail, has provided extensive insights into the structure, function, and genetics of the human cilium and also identified candidate genes important for cilia motility (Blair and Dutcher, 1992; Li et al., 2004; Pazour et al., 2005). Thus, the study of Chlamydomonas has helped to elucidate several PCD candidate genes (Pennarun et al., 1999; Wilkerson et al., 1994). Based on the data obtained from Chlamydomonas, Tetrahymena, and other lower organisms, outer dynein arms are known to have significant roles in generating ciliary beat frequency whereas the inner dynein arms mainly regulate the ciliary beat waveform (Hilfinger and Julicher, 2008; Kamiya, 2002; Taylor et al., 1999; Wood et al., 2007).

Genome-wide linkage analysis revealed extensive locus heterogeneity and failed to identify a single major gene responsible for most cases of PCD (Blouin et al., 2000), possibly a reflection of the complex nature of the cilia axoneme. However studies into the genetic basis of PCD in humans have primarily identified mutations in genes encoding outer dynein chains. Pennarun et al. demonstrated that mutations in *DNAI1* are present in some PCD patients and that the cilia axoneme from these patients lacked outer dynein arms (Pennarun et al., 1999). Similar observations were made in *Chlamydomonas* with mutations in the *DNAI1* homolog IC78 which caused outer dynein arm assembly defects and reduced flagella motility (Mitchell and Kang, 1991). *DNAH5* is a second locus associated with PCD and is the homolog of *Chlamydomonas* γ heavy chain (Olbrich et al., 2002). However, surprisingly, a recent study indicated that 53% of PCD patients with known outer dynein arm ultrastructural defects were due to mutations in *DNAH5* (Hornef et al., 2006). Two other dynein protein genes *DNAH11* (Bartoloni et al., 2002) and *DNAH7* (Zhang et al., 2002) have also been implicated in PCD; however, their direct connection to PCD is not well established.

3.3. Cilia and left–right asymmetry defects

The high incidence of *situs inversus* observed in PCD patients (KS) suggested that motile cilia have a role in left–right asymmetry specification (Fig. 13.3A). The cilia connection was clarified when a motile form of cilia was discovered on an embryonic organizing structure called the node (Fig. 13.3B–B′), an important signaling center involved in establishing the body axis. These cilia are present as solitary structures on node cells and it remains a matter of debate as to whether they have $9+0$ or $9+2$ architecture. In contrast to the normal wave-like motion of the motile cilia in the lung, node cilia were found to rotate with the cilia axoneme being slightly angled relative to the location of the basal body. This tilted rotational beating of the cilium causes directional movement of fluid across the node surface that is thought to regulate axis specification. A more detailed description of motile node cilia and their role in left–right axis formation can be found in Chapter 6, this volume.

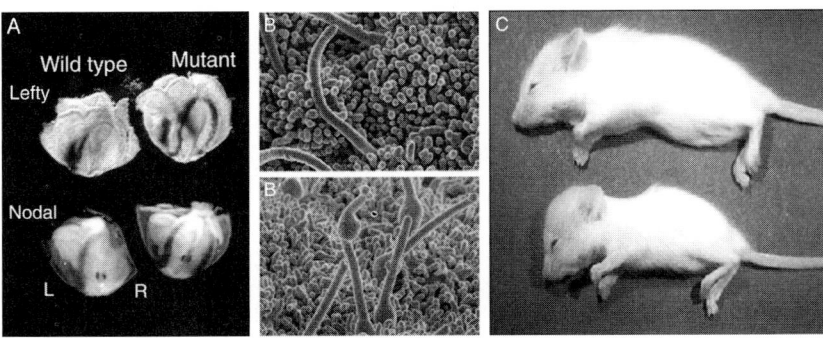

Figure 13.3 Motile cilia defects. Left–right asymmetry defects in day 7.5 mouse embryos. *In situ* hybridization for two genes expressed asymmetrically in the embryonic node (Lefty and Nodal). In Wild-type embryos, both are expressed on the left while in cilia mutant embryos, this expression pattern is lost (Panel A, reproduced with permission of the Company of Biologists, adapted from Murcia *et al.* (2000) Copyright © 2000, The company of Biologists). Scanning electron microscopy of the embryonic node showing wild-type primary cilia (Panel B) and Dync2h1 mutant cilia with atypical bulges (Panel B' Images courtesy of Danwei Huangfu and Kathryn Anderson). Mice mutant for the gene *Hydin* (*Hyd*rocephalus *In*ducing) develop perinatal hydrocephalus due to ependymal cell motile cilia defects (Panel C image adapted from Robinson *et al.* (2002), Copyright © 2002, Springer, New York). (See Color Insert.)

Some of the initial work connecting cilia, flow, and left–right axis specification came from the Hirokawa lab on mice lacking KIF3A and KIF3B, two subunits of the heterotrimeric kinesin-II motor protein required for IFT and cilia formation (Marszalek *et al.*, 1999; Nonaka *et al.*, 1998). Identical phenotypes were observed in the *ift88* mutants (Fig. 13.3A) (Murcia *et al.*, 2000). In the inversus viscerum (*iv/iv*) mutant mouse where the left–right dynein (*lrd*) is disrupted, cilia do form but they are immotile (Layton, 1976; Supp *et al.*, 1997). As with the *kif3a* and *kif3b* mutants and human PCD patients, roughly half of the embryos display normal asymmetry whereas other half exhibited *situs inversus*. The immotility of the cilia in the *iv* mutants and lack of monocilia in the kinesin mutants (*kif3a* or *kif3b*) strongly suggested that rotational movement of cilia is required to induce breaking of body asymmetry via generation of leftward fluid flow. Nonaka *et al.* further demonstrated this by applying an artificially generated nodal flow to embryos which was sufficient to determine laterality (Nonaka *et al.*, 2002). Interestingly, this artificial flow was also able to rescue the situs of mutant mouse embryos with immotile cilia.

Two basic models have been proposed to explain how cilium-generated fluid might establish left–right body axis. In the first model, flow results in deflection of nonmotile cilia located at the edge of the node activating a calcium signaling pathway. Intriguingly, this calcium signal requires polycystin-2, a TRP channel protein that localizes to the cilium and is mutated in human PKD (see Section 4.4 below (McGrath et al., 2003)). The second model proposes that left–right asymmetry is generated by leftward movement of membrane-sheathed particles (Tanaka et al., 2005). These "nodal vesicular parcels" carry morphogens such as sonic hedgehog and retinoic acid that are released at the left side of the node to establish the body axis. Unfortunately neither of these models adequately rationalizes the complex laterality defects (e.g., *situs inversus totalis*, left isomerism, *situs inversus thoracalis*, *situs inversus abdominalis*, and left and right isomerism) that are present in humans and mice with congenital ciliary motility defects. Although the precise mechanism by which cilia and flow dictate axis specification remains controversial, the involvement of nodal cilia motility and axis determination is well established.

3.4. Role of cilia and flagella in reproductive system

Flagella and motile cilia are analogous structures with flagella generally being larger than cilia and having a different form of motility. They are both constructed through IFT. Ultrastructurally, the axonemes of the sperm tail and motile cilia are very similar. Thus, it is not surprising that male PCD patients can be infertile due to sperm immotility. In addition to PCD, sperm motility defects are also a feature of other human ciliopathies such as BBS. Missense mutations in BBS2 have been reported with defects in the axoneme of spermatozoa (Fath et al., 2005; Mykytyn et al., 2004; Nishimura et al., 2004) and mice with mutations in BBS2, BBS4, and BBS6 are infertile due to failed flagella assembly.

Although less common, infertility is also associated with ciliary dysfunction in females. Motile cilia are found on cells of the fallopian tubes which, along with muscular contractions and flow of tubule secretions, are responsible for oocyte transit. In the female reproductive system, hormones regulate the cyclical morphological changes in the fallopian tube epithelium that affect the ciliated cells and cilia beat frequency (Bush et al., 1998; Donnez et al., 1985; Verhage et al., 1979).

3.5. Functions of motile cilia in brain

Hydrocephalus is caused by excessive accumulation of cerebrospinal fluid (CSF) as a result of blockage of aqueducts connecting the brain ventricles, impaired CSF movement, lack of CSF absorption, or increased CSF production. Ependymal cells lining the brain ventricles possess 9 + 2 cilia (Dalen et al., 1971). As in the airways, these cilia beat synchronously to push CSF through the ventricular system. Thus, it is somewhat surprising that hydrocephalus is not a frequent characteristic of PCD (Bush et al., 1998) nor have ciliary defects been associated with other forms of hydrocephaly in humans, although it is commonly seen among the cilia mutants in mice and zebrafish (Banizs et al., 2005; Kramer-Zucker et al., 2005). For example, mice with mutations in *mdnah5* (mouse axonemal dynein heavy chain gene) have impaired motility of cilia on ependymal cells of the brain ventricles that disrupts CSF movement and these mutants present with hydrocephalus (Ibanez-Tallon et al., 2002, 2004). Another example of hydrocephalus associated with cilia abnormalities in mice is the hy3 autosomal recessive mouse model (Fig. 13.3, panel C). The hy3 mouse has a mutation in the protein Hydin. Hydin is highly expressed in the ciliated ependymal cell layer lining the ventricles, the ciliated epithelial cells in the respiratory tract, and the oviduct and spermatocytes in the testis (Davy and Robinson, 2003). Hydin appears to be present in organisms possessing motile cilia. It localizes to the C2 tubule of the central apparatus (Lechtreck and Witman, 2007) and interacts with central pair proteins CPC1 and kinesin-like protein 1 (Klp1). In *Chlamydomonas*, knockdown of hydin leads to short flagella lacking the C2b projection of the C2 tubule. Lechtreck et al. confirmed these *Chlamydomonas* observations in the hydin mutant mouse. Their studies show that the length and density of cilia on ependymal cells in the brains of hydin mutant animals are normal and the only detectable structural defect present in ciliary axoneme was the absence of a projection from one of the two central microtubules (Lechtreck et al., 2008). The phenotypic consequence of the hydin mutation is that the cilia arrested either in the beginning of the effective stroke or the recovery stroke. Overall, the hydin mutant cilia are unable to bend normally, ciliary beat frequency is reduced, and the cilia paused frequently, all suggesting that the loss of cilia motility is responsible for the hydrocephalic phenotype. Hydrocephaly has also been observed in mice with mutations in several other cilia or basal body localized proteins including the *Pcdp1* (primary ciliary dyskinesia protein 1) (Lee et al., 2008), B9 domain protein Stumpy (Town et al., 2008), *SPAG6* (Sperm Associated Antigen 6 or PF16) (Sapiro et al., 2002), the transcription factor *Hfh4* (*Foxj1*) (Kosaki et al., 2004), and *Tg737* (IFT88, Polaris) (Banizs et al., 2005).

The exact mechanism by which loss of cilia function results in hydrocephaly is not known. Based on mouse models of hydrocephalus, two cilia-related pathogenic mechanisms have been suggested. The first and most widely expected model is that impaired cilia motility disrupts CSF flow which initiates hydrocephaly. Expansion of the ventricles then contributes to obstruction of the cerebral aqueduct that further exacerbates the phenotype. However, it remains uncertain whether CSF accumulation and ventricle expansion associated with loss of cilia directly results in duct closure or whether beating of cilia on cells lining the duct itself has a role in preventing stenosis.

The second mechanism originated from studies in the ORPK ($IFT88^{Tg737RPW}$) mice mentioned above. In ORPK mice, the ependymal cells lining the brain ventricles have aberrantly formed cilia that cause disorganized beating and impaired CSF movement (Banizs et al., 2005).

However, the loss of cilia beat and abnormal CSF flow appears not to be the initiating factor in these mutants since the phenotype is present prior to the development of motile cilia on the ependymal cells and aqueduct remains patent at early stages of the disease. An interesting observation made in these mutants was that loss of cilia did result in altered function of the choroid plexus epithelium which is responsible for production of CSF. Choroid plexus epithelial cells have either tufts of motile cilia or a single immotile cilium. The functions of these cilia are largely unknown. However, in ORPK mice defects in the cilia of the choroid plexus result in elevated intracellular cAMP levels, altered Na (+)/HCO3(−) transporter activity with abnormal intracellular pH regulation, and increased chloride secretion into the CSF (Banizs et al., 2005, 2007). These findings indicate that one initiating factor in cilia mutants, at least in ORPK mice, may be excess CSF production. At later stages, the subsequent defects in CSF movement caused by impaired motility of cilia along with duct stenosis may cause progression to the severe phenotype seen in the older mutants.

3.6. Motile cilia and neurodevelopment

Based on studies conducted in the ORPK mice, another function of motile cilia on ependymal cells in the brain is to direct long-range neuroblast migration from the stem cell region of the subventricular zone (SVZ) towards their final destination in the olfactory bulb. In ORPK mutants, the neuroblasts do not align or migrate properly, a property that was not intrinsic to the migratory neuroblasts, but rather, is dependent on the CSF environment and on the flow of CSF. Data from Sawamoto et al. indicate that this likely involves the release of a chemorepellent by the choroid plexus that forms a gradient generated by ependymal cilia beating and movement of the CSF (Sawamoto et al., 2006). It will be interesting to see whether similar defects in neuroblast migration occur in human ciliopathies which might explain phenotypes such as mental retardation or sensory deficits (e.g., nociception or anosmia).

4. Functions and Diseases Associated with Immotile Cilium

The immotile primary cilium is present on most cells in the mammalian body with few exceptions such as bone marrow-derived cells (Wheatley et al., 1996). For a recent listing of cell types known to have primary cilia, see the Primary Cilia Resource Site (http://www.primary-cilium.co.uk/). Until recently, primary cilia were thought to be of relative unimportance; however, this perception has changed. The organelle is now known to be critical for normal development as well as for tissue homeostasis in adults. The functions of primary cilia are rapidly expanding and include mechanosensation, chemosensation, photoreception, and olfaction as well as new roles as a mediator of signaling activities that allow cells to receive and respond normally to extracellular stimuli. As discussed below, primary cilia dysfunction in mammals can cause a wide range of disease phenotypes in diverse tissues and as well as severe developmental abnormalities. Despite the rather impressive advances made over the past decade demonstrating the importance of primary cilia, there remains a great deal to be learned about this organelle. Strikingly, the functional role of the cilium on most cell types remains unknown. Further, determining how loss of the primary cilia results in aberrant regulation of signaling pathways leading to disease and developmental abnormalities will be of great interest. The wide spectrum of defects now associated with cilia dysfunction accentuates the importance of elucidating the mechanism controlling cilia assembly and maintenance and how this organelle functions to integrate complex signaling activities. Below we discuss several major diseases and developmental abnormalities that are linked with primary cilia dysfunction.

4.1. Primary cilia and cell cycle

The primary cilium is anchored to the cell through the basal bodies which are derived from the centrosome in a manner that is coordinately regulated with the cell cycle (Fig. 13.4) (Plotnikova et al., 2008). Centrosomes are composed of two centrioles that are surrounded by pericentriolar material. The centrosome/centriole plays an important role in organizing cytoplasmic interphase microtubules and mitotic spindles during cell division (Mikule et al., 2007). In nondividing cells, the cilium is present on the cell surface during the G1-S phase of the cell cycle (Ho and Tucker, 1989; Plotnikova et al., 2008; Quarmby and Parker, 2005). As cells enter the cell cycle, the cilium and basal body disassemble freeing the centrioles to function as the microtubule organizing center for the mitotic spindles. Whether this reflects a role for cilia in cell cycle control or indicates that cell cycle arrest is a necessity for ciliogenesis to occur remains a matter of debate.

In support of a direct role for cilia in cell cycle control are data from Robert et al. They demonstrated that siRNA knockdown of IFT88 expression promotes cell-cycle progression to S and G2/M phases (Robert et al., 2007). Their data identified a direct interaction between IFT88 and CHE-1 a protein that functions to inhibit retinoblastoma's (Rb) growth suppressing activity (note CHE-1 is not an IFT or cilia related protein). An additional connection between cilia/basal body and cell cycle is supported by data indicating that several centriolar/centrosomal proteins which are essential for cell cycle progression are also required for primary cilium formation (Graser et al., 2007; Mikule et al., 2007). Furthermore, the activity of AuroraA kinase, a known regulator of mitosis, has a role in regulating cilia

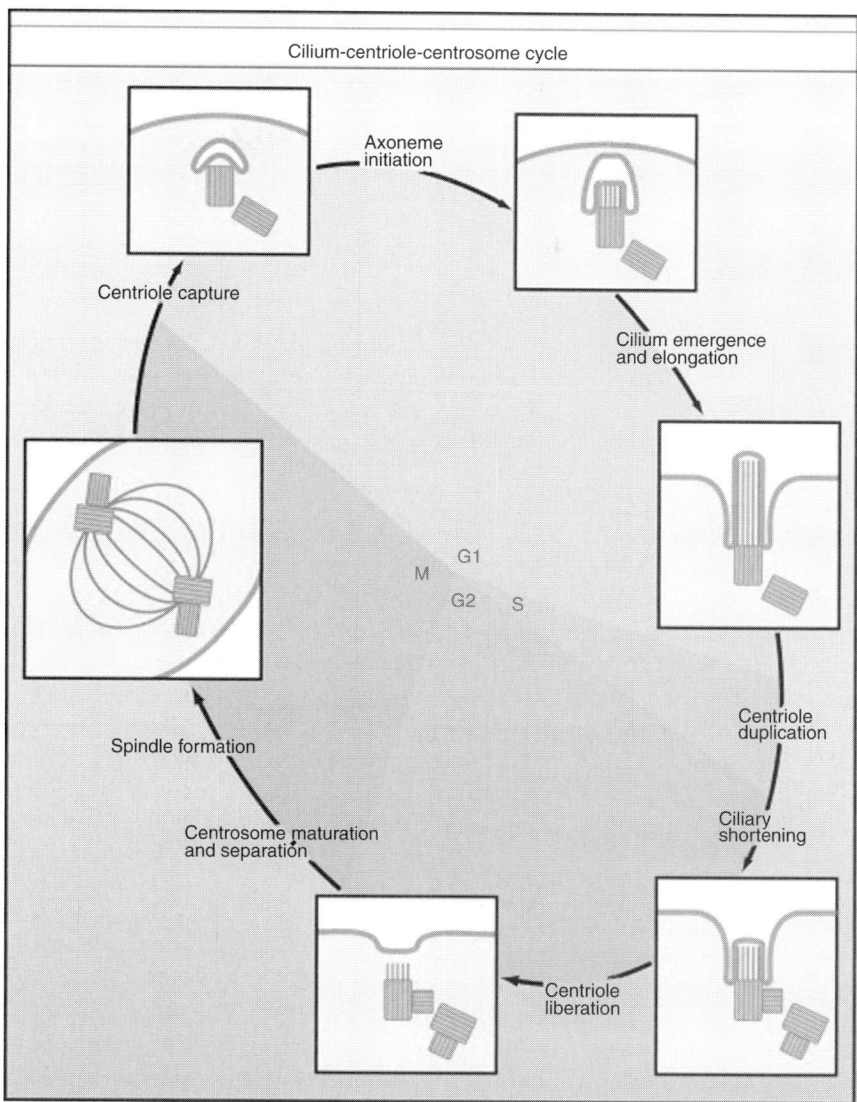

Figure 13.4 The primary cilium and the cell cycle. Formation of the primary cilium occurs in fully differentiated cells in the G0 phase of the cell cycle. A cilium axoneme protrudes from the basal body (centrosome) which consists of two centrioles (mother centriole: vertical yellow cylinder, daughter centriole: diagonally placed yellow cylinder). As cells enter the S phase of the cell cycle, the cilium starts to resorb. The centrioles duplicate, elongate, and attain the mature stage during late G2/M phase of the cycle. The mature centrosomes separate and migrate to the poles of the nucleus in position to nucleate mitotic spindles. (Image reproduced from Junmin Pan and William Snell (2007), *Cell*, copyright © 2007, Cell Press).

stability. Inhibition of AuroraA activity blocks ciliary disassembly while increased activity initiates cilia loss (Pugacheva et al., 2007). Together these findings raise the intriguing possibility that cilia may function as a tumor suppressor by inhibiting entry into the cell cycle; however, presence of cilia on tumor cells in the skin would argue against this possibility (Schafer, 1969). On the other hand, cilia are involved in regulation of key signaling pathways (e.g., sonic hedgehog) that play roles in the formation of tumors in humans such as basal cell carcinoma (Oro et al., 1997). Thus, the connection between cilia, cancer, and cell cycle regulation still need to be more fully evaluated.

A confounding issue with regards to cilia and cell cycle control is that loss of *ift88 in vivo* does not appear to have a consistent effect on proliferation. In pancreatic ducts of ORPK mutants there is an increase in proliferation compared to controls (Cano et al., 2004; Zhang et al., 2005) while in other tissues, such as the cerebellum, the phenotype is characterized by a failure of proliferation of progenitor cells (Chizhikov et al., 2007). In addition, induction of cilia loss in adult kidneys with *kif3a* or *ift88* conditional mutants does not produce a marked increase in proliferation until late stages of the cystic disease (Davenport et al., 2007; Patel et al., 2008). Together these data argue against a general link between loss of cilia and induction of cell proliferation at least *in vivo*; however, it is possible that there will be tissue or cell type specific effects on proliferation that must be further analyzed.

4.2. Sensory abnormalities associated with cilia mutants

The primary cilium plays important roles in many sensory activities including olfaction, photoreception, nociception, and possibly in somatosensory responses involved in touch. Examples of ciliary function in sensory reception are evident from studies in model organisms where cilia have been disrupted. For example, in *C. elegans* cilia are present on sensory neurons in the head and tail where they are necessary for chemotaxis toward attractants and avoidance responses to anoxic environments and noxious compounds (Bargmann, 2006). A number of odorant receptors involved in these responses have been identified and have been shown to localize to cilia on these neurons. Similarly, in mice, olfactory receptors concentrate in nonmotile cilia that extend from the dendritic knobs of neurons in the olfactory bulb. Odorant binding to the receptors in the cilium leads to activation of cyclic nucleotide gated ion channels (CNG) and ultimately results in action potentials and olfaction (Menini, 1999). Anosmic behavior has been reported in mice and human patients with mutations in BBS genes (Fath et al., 2005; Nishimura et al., 2004; Ross et al., 2005). For more detailed discussion on cilia and olfaction, see Chapter 12, this volume.

In the retina, rod and cone photoreceptors possess highly modified cilia that sense and transduce light signals. As such, degeneration of the retinal rod and cone photoreceptors is a common phenotype associated with cilia related disorders in humans including NPHP/SLS, ALMS, and BBS. The rod and cone photoreceptors extend axons to make contact with higher order bipolar and horizontal cell neurons, while the dendrites extend toward the retinal surface. The dendrites end in a structure called the outer segment (OS). The OS consists of elaborate discs which are derived from the plasma membrane. The axoneme of the cilium functions as the backbone of the OS structure. The region joining the OS to the cell body is referred to as the connecting cilium, which extends from the basal body. Since all OS proteins must pass through the narrow connecting cilium, it is thought that this region functions as a key regulator of protein entry into cilia. The OS is a highly dynamic structure that is undergoing continual turnover (Besharse and Hollyfield, 1979; Hollyfield *et al.*, 1977) and its construction and maintenance is dependant on IFT (Pazour *et al.*, 2002a). Thus, in the absence of IFT, the OS collapses resulting in impaired vision.

Detection of light by the photoreceptors is dependent on rhodopsin, a photoreactive protein that localizes to the discs in the OS. Under dark conditions, elevated levels of cGMP keep the CNG open. In the presence of light, cGMP levels are reduced through rhodopsin mediated activation of the heterotrimeric G protein transducin and cGMP phosophodiesterase (PDE). In addition, rhodopsin itself is subject to regulation by phosphorylation and binding of arrestin, which moves between the different segments of the photoreceptor in light dependant manner. The mechanism responsible for rhodopsin and arrestin transport into the OS has been a matter of debate with models proposing simple diffusion or a requirement for IFT. The importance of IFT in OS assembly and protein movement is evidenced by the phenotypes resulting from mutations in the *kif3a*, *ift88* and in the ORPK mutants. In these lines, the photoreceptors are lost after a failure in OS morphogenesis and rhodopsin is mistargeted. Defects in photoreceptor maintenance and formation are also present in several of the BBS mutant mice. In *bbs4* mutants, Abd-El-Barr *et al.* have shown that this phenotype is associated with the accumulation of vesicles at the base of the connecting cilium (Abd-El-Barr *et al.*, 2007). These data are in agreement with the proposed role for the BBSome (see below) in regulating cilia membrane formation and for trafficking into the OS. For more detailed discussion of photoreceptor cilia see the recent review by Insinna and Besharse (Insinna and Besharse, 2008).

4.3. Neuronal cilia and obesity phenotypes

One of the most intriguing and difficult phenotypes to explain in the ciliopathies is the connection between cilia dysfunction and obesity. This is a characteristic phenotype seen in human patients with BBS and ALMS.

The gene products involved in these syndromes encode proteins that localize to the cilium or the basal bodies. Their exact roles in ciliogenesis or signaling remain a matter of debate.

Insights into the connection between cilia and obesity came from analysis of conditional mutations that disrupt IFT in mice. Using the Cre/LoxP system, Davenport *et al.* demonstrated that the obesity phenotype associated with *ift88* mutants is due to disruption of IFT in pro-opiomelanocortin (POMC) neurons in the hypothalamus (Fig. 13.5, panel A) (Davenport *et al.*, 2007). Disruption of the cilium/IFT on these neurons in the hypothalamus results in hyperphagia, indicating that functional cilia are needed for normal satiation responses.

POMC neurons are activated by leptin, an anorexigenic peptide derived from adipose. Binding of leptin to its receptor ObRb activates the Jak-Stat signaling cascade inducing the production and release of the anorectic peptide α-MSH. α-MSH binds to the melanocortin 4 receptor (MC4R) located on higher order neurons to inhibit feeding. Current data suggest that the POMC neurons lacking the cilium/IFT are unable to effectively respond to leptin. The mechanism responsible for this resistance is unknown; however, one possibility would be a requirement for ciliary localization of ObRb to bind leptin for signaling activity. A similar mechanism has been shown for the proteins patched and smoothened in the hedgehog signaling pathway. However, to date, there are no data to support leptin receptor localization in the cilium of POMC neurons.

Another connection between neuronal cilia and obesity came from studies using BBS mice by Berbari *et al.* Their data indicate that BBS proteins are needed for normal localization of G-protein coupled receptors (GPCRs) to

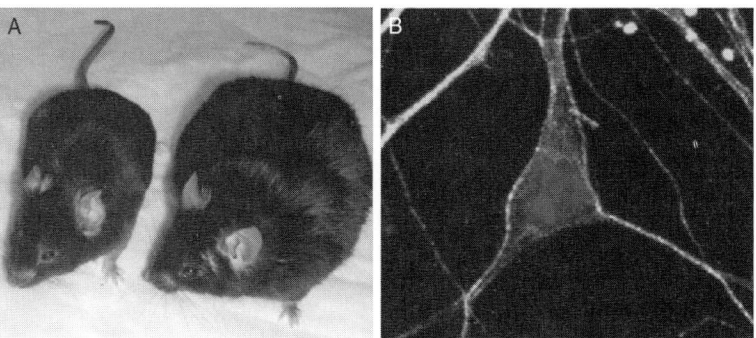

Figure 13.5 Neuronal cilia have a role in feeding behavior. The loss of neuronal cilia results in hyperphagia induced obesity in mice. Panel A shows an obese mouse that has the conditional allele of IFT88 and a neuronal specific Cre transgene (SynI-Cre), next to its Cre negative littermate (Image courtesy of Mandy Croyle). Panel B shows a primary hypothalamic cultured neuron (neuronal marker β-Tubulin III in green) with a Mchr1 positive cilium in red (Image Courtesy of Kirk Mykytyn). (See Color Insert.)

the cilium (Berbari et al., 2008b). Intriguingly, one of these receptors was melanin-concentrating hormone receptor 1 (MCHR1) (Fig. 13.5, panel B), which they show fails to enter the cilium of hypothalamic neurons in mice lacking either BBS2 or BBS4. MCHR1 antagonists are known to decrease feeding activity, thus, if obesity in the ciliopathies is due to effects on MCHR1 signaling, it would suggest that cilia have an inhibitory role on the MCHR1 pathway and that ciliary defects would then abrogate this inhibitory effect. A possible connection between the leptin data from Davenport et al. and the MCHR1 results of Berbari et al. is that MCHR1 is also a central target of leptin signaling (Kokkotou et al., 2001). In addition to activating POMC neurons and inducing α-MSH release, leptin inhibits production of several orexigenic neuropeptides, one of which is MCH (Tritos et al., 2001). Thus, loss of cilia and the subsequent impact on the MCH signaling pathway may lead to leptin resistance through a secondary route.

Further evidence for a dysfunction of the cilium in obesity has come from two recent genome-wide association studies that identified a variant in the fat mass and obesity associated (*fto*) locus in obese humans (Dina et al., 2007; Frayling, 2007). *Fto* was originally identified in the mouse *Fused toes* mutants that have a 1.6 Mb deletion removing several genes including fantom (*ftm*) and the *Iroquois B* cluster of genes (*irx3,5, 6*), and *fto*. The *Fused toes* mutants are not viable and present with phenotypes due to altered Shh signaling including polydactyly, neural tube, and left–right axis defects. The variant identified in these obese patients occurs in an intron of *fto*. This was found to disrupt a binding site for Cut-like 1, CCAAT displacement protein (CutL1, also called Cux1) (Stratigopoulos et al., 2008). In addition to regulating *fto* expression, this site regulates expression of *ftm*, which is also called RPGRIP1L. Although it is currently unknown which of the two genes may be involved in the obesity phenotype, the fact that RPGRIPL1 is a cilia protein and mutations in this gene are known to cause both MKS and JBTS, makes it interesting to speculate that the obesity in these human patients may be due to effects on RPGRIPL1 expression. This would establish another connection between cilia and obesity in humans.

Despite the importance of cilia on sensory neurons in chemosensation and mechanosensation in lower eukaryotes such as *C. elegans* and *Drosophila*, the function of this organelle on mammalian neurons has been relatively ignored. Along with defects in satiation responses in cilia mutants, data from Tan et al. also support the importance of cilia for neuronal function in mammals. Their data show that BBS patients have a significant diminution in their ability to detect vibration and that *bbs1* or *bbs4* mice have defects in nociceptive responses to heat (Tan et al., 2007). This is thought to involve cilia on peripheral sensory neurons. Intriguingly, *C. elegans* with mutations in the BBS proteins also have thermosensory deficits suggesting an evolutionarily conserved signaling pathway involving the neuronal cilium (Tan et al., 2007). Further evidence suggesting the importance of neuronal cilia is the fact that a number of GPCRs such as a serotonin and a somatostatin receptor localize to this organelle (Brailov *et al.*, 2000; Hamon *et al.*, 1999; Handel *et al.*, 1999). In addition, MCHR1 localizes to neuronal cilia in many places in the brain and MCH signaling is known to participate in functions other than orexigenic responses (Berbari *et al.*, 2008a; Pissios *et al.*, 2006). Cilia are found on most if not all neurons (Fig. 13.5B). Thus it will be interesting to evaluate whether the general loss of cilia on CNS neurons has effects on behaviors such as addiction, stress responses, learning, memory, anxiety, activity, and aggression.

4.4. Ciliary connection to PKD related phenotypes in the kidney, liver, and pancreas

Primary cilia ranging in size from 2–5 μm in length extend off most cells of the epithelium into the lumen of the nephron with the possible exception of intercalated cells in the collecting duct. Cilia length varies both with age and nephron segment and reports have ranged from 5–15 μm and sometimes as long as 55 μm (Roth *et al.*, 1988; Schwartz *et al.*, 1997). The position of the renal cilium protruding into the lumen led to the hypothesis that one function of this organelle may be to serve as a mechanosensor, detecting fluid movement through the nephron. In 1997, Bowser and coworkers were able to demonstrate that the primary cilium of cultured kidney cells is able to bend in response to the fluid flow (Schwartz *et al.*, 1997). Later, seminal work by Praetorius and Spring demonstrated that primary cilium on dog kidney cells (MDCK) do indeed function as a mechanosenstive sensors that respond to flow by elevating intracellular calcium (Praetorius and Spring, 2001, 2003). The calcium response induced by flow was ablated in kidney cells that are unable to assemble a primary cilium. The functional significance of the increase in calcium induced by flow remains unknown.

Renal cysts are a clinical feature shared by numerous human syndromes such as PKD (Fig. 13.6A), OFD, MKS (Fig. 13.6C), BBS (Fig. 13.6B),

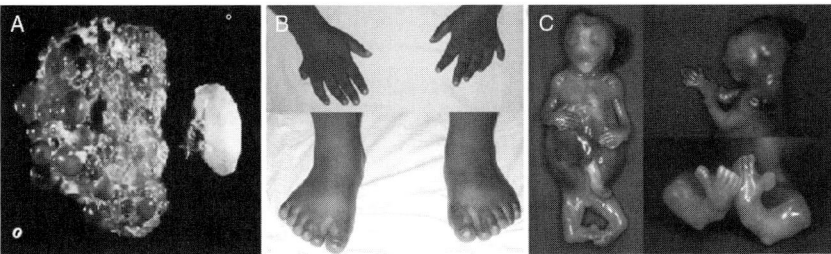

Figure 13.6 Examples of clinical features of ciliopathies. A polycystic kidney in shown next to a normal kidney in panel A. Polydactyly is a cardinal feature of BBS, panel B. Embryonic lethality with exencephaly and polydactyly is observed in MKS, panel C. (See Color Insert.)

NPHP, and JBTS. The most common hereditary form of cystic kidney disease in humans is PKD. PKD can be inherited as an autosomal dominant (ADPKD) or an autosomal recessive (ARPKD) trait affecting 1/1000 and 1/20,000 in live births, respectively. In ARPKD, cysts are largely present in the collecting ducts, whereas in autosomal dominant PKD (ADPKD) cysts develop from any region of the nephron. In both conditions, epithelial cells lining the cysts appear less differentiated and are characterized by increased rates of proliferation and apoptosis, aberrant protein trafficking, and abnormal fluid and ion transport (Torres and Harris, 2006). ARPKD is mainly an infantile disease with cystic kidneys and hepatic fibrosis. Mutations in *Pkhd1*, which encodes a large single transmembrane protein polyductin or fibrocystin (Onuchic et al., 2002; Ward et al., 2002, 2003) have been identified in ARPKD patients. In ADPKD, mutations have been detected in *Pkd1* and *Pkd2* genes, which encode polycystin 1 (PC1) and polycystin 2 (PC2), respectively (Hughes et al., 1995; Mochizuki et al., 1996). Although ADPKD is inherited as a dominant trait, it is recessive at the molecular level requiring a somatic mutation in PC1 or PC2. PC1 and PC2 are both transmembrane proteins that function as a complex by associating with one another through their C-terminal tails (Qian et al., 1997). PC2 is believed to be a transient receptor potential (TRP) like cation channel. Despite being identified over a decade ago, the function of PC1 has not yet been fully defined. Recently it was found that both PC1 and PC2 undergo posttranslational cleavage events and that the C-terminal region of PC1 translocates to the nucleus (Huang et al., 2007; Qian et al., 2002). The fate and activity of the N-terminal extracellular domain is not known. In the case of PC1, this cleavage event has functional importance since mice expressing a noncleavable form of PC1 present with a cystic phenotype (Yu et al., 2007). These data raise the possibility that proteolytic release of PC1 and PC2 peptides into the lumen of the nephron may have roles in communication between cells in distant parts of the nephron.

The initial connection between PKD and cilia came from an unexpected study by Barr et al. who were conducting genetic screens for mutations affecting male mating responses in *C. elegans*. A gene called *lov-1* was identified as a result of this screen and found to encode the homolog of *pkd-1* (Barr and Sternberg, 1999; Barr et al., 2001). Through homology they also identified the homolog of *pkd-2* which functions in the same pathway as *lov-1*. Expression of LOV-1 and PKD-2 as GFP fusions in transgenic worms demonstrated that these proteins localize in the cilium of the sensory neurons of the male. Subsequently, the mammalian PC1 and PC2 proteins were also found to reside in the cilium, lateral membrane, and in the endoplasmic reticulum of multiple different cell types (Geng et al., 1997; Guillaume and Trudel, 2000; Ong et al., 1999; Pazour et al., 2002b; Yoder et al., 2002). Importantly, renal epithelia cells lacking either PC1 or PC2 are unable to elicit a calcium response despite the presence of a morphologically normal cilium (Nauli et al., 2003, 2006). This suggests that one function of the polycystin complex is in the mechanosensory response of the cilium to fluid movement. Similar findings have now been reported for cilia and the polycystins in various tissues such as the endothelium, biliary duct epithelium, and as indicated above in the node during left–right axis formation (Housset, 2005; Karcher et al., 2005; Kim et al., 2000).

The second major association between renal cysts and cilia again came from studies in *C. elegans* as well as in *Chlamydomonas* and mice. This occurred when the homolog of the *Tg737* (*ift88*) gene was found to be mutated in the ORPK mouse model of PKD and to be *osm-5* in *C. elegans* and *ift88* in *Chlamydomonas* (Haycraft et al., 2001; Pazour et al., 2000; Qin et al., 2001). These studies revealed that the gene associated with the cystic kidney disease in the ORPK mutants encoded a component of the IFT raft needed for cilia or flagella assembly. Analyses of renal cell lines or perfused kidney tubules isolated from the ORPK mutants indicate that loss of the cilium impairs normal mechanosensory responses induced by flow (Liu et al., 2005b).

Once these connections between cilia dysfunction and cyst formation were established there was a surge in the number of human and mouse cystic kidney disease loci identified that were found to encode ciliary localized proteins (Tables 13.1–13.7). This now includes genes involved in MKS, BBS, JBTS, ARPKD, OFD, and NPHP, as well as several novel genes, such as stumpy, in mice for which human disease associations have not yet been found. In a large part, the rapid rate at which cystic kidney disease loci were being identified resulted from characterization of the cilia or flagella proteome in humans, mice, *C. elegans,Chlamydomonas*, and *Drosophila*. These proteomic datasets were generated through a combination of computational, genetic, and classical biochemical approaches. The reason for this was that the ciliome databases allowed genes within the candidate disease interval to be prioritized based on the prediction that cystic kidney disease genes were going to have an association with the cilium.

Table 13.1 PKD genes

Disease locus	Gene	Protein sequence characterization	Subcellular localization
PKD1 (16p13)	Polycystin-1	Coiled-coil domain	Cilium
PKD2 (4q21)	Polycystin-2	TRP-like cation channel	Cilium
PKHD1 (6p21)	Fibrocystin	Single pass type I membrane protein	Membrane/ cilium/ nucleus

Initial data suggested that the cystic kidney disease phenotypes resulting from ciliary dysfunction in mice and humans were caused by the loss of the calcium mechanosensory response normally generated through cilia deflection and fluid flow. However, this mechanism of cystogenesis has recently been brought into question. Several studies using conditional null alleles of *ift88, kif3a,* and *pkd1* to disrupt cilia formation or function in the early developing kidney result in a severe and rapid onset cystic phenotype (Davenport et al., 2007; Lantinga-van Leeuwen et al., 2007; Piontek et al., 2007). However, if cilia are ablated in the kidney of adult mice, cyst development is delayed until 6 months to a year after cilia were initially lost. Further analyses in conditional mutants indicate that there is a critical switch point near postnatal days 12–14 where loss of cilia after that point no longer results in a rapid cystogenic program. Interestingly, the timing of the switch point correlates with a large-scale change in gene expression profiles, completion of nephrogenic differentiation, and a reduction in proliferation rates (Piontek et al., 2007). If loss of mechanosensation was the driving mechanism involved in cyst formation, it would be difficult to reconcile with the critical switch point and the long delay required for cyst formation in the adult induced cilia mutants. Thus, there appear to be additional factors that are needed to promote cyst formation in the absence of cilia. These findings could have important implications for human PKD, if therapeutic approaches could be devised to maintain PKD-1, PKD-2, or ciliary functions at early stages of the disease, since this should greatly retard cyst progression and disease severity into adulthood.

The dramatic reduction in proliferation rates that occur after the critical switch point suggested that proliferation may be a variable that influences the rate of cyst formation and expansion. This was assessed in the adult-induced cilia mutants by ischemic reperfusion (IR). After IR injury, the repair process results in a transient increase in proliferation. Intriguingly, this was found to induce rapid cyst formation in a cilia mutant background in contrast to what is seen in the contralateral cilia mutant kidney that was not injured (Patel et al., 2008). Similar results have been reported in the cilia mutants after repair and proliferation were induced by renal injury with nephrotoxin dichloro [14C] vinyl cysteine (DCVC) (Dr. Dorien Peters, personal communication). However, previous studies have demonstrated that increased proliferation in itself is not sufficient to induce renal cyst formation indicating that additional factors are needed (Ledford et al., 2002). Further insight into the possible mechanism connecting cilia, cysts, and proliferation came from Fischer *et al.* They demonstrated in the *Hnf1a* mutant mouse and *pck* rat (mutation in *pkdh1*) that renal cyst formation was associated with abnormal orientation of the mitotic spindles relative to the axis of tubule (Fischer et al., 2006). This was subsequently observed in several other cystic kidney disease models, including *kif3a* and *ift88*, with defects in ciliogenesis (Patel et al., 2008) (Sharma and Yoder, unpublished data). According to this model, misalignment of the spindle would cause tubules to increase diameter rather than elongate (Fischer and Pontoglio, 2006; Fischer et al., 2006).

The mechanism responsible for the spindle orientation defects in cystic kidneys is not yet well understood, but the current model proposes there are defects in the planar cell polarity (PCP) pathway (Fanto and McNeill, 2004). PCP controls how cells organize themselves within a tissue for example hair or bristle orientation in the wing and body of *Drosophila*. Additionally, in mammals the patterning of the stereocilia in the cochlea of the inner ear also requires the PCP pathway (see chapter 8 of this issue or (Simons and Mlodzik, 2008) for a recent review.). In addition, the PCP pathway is important for the convergent–extension (CE) process during gastrulation that results in narrowing of the embryo along the left–right axis and elongation along the anterior–posterior axis. Importantly, mutations in ciliary proteins are known to have CE like phenotypes in zebrafish and to cause inner ear patterning defects in mice.

Mechanisms controlling PCP appear to be conserved from flies to humans and are best characterized in *Drosophila* (Simons and Mlodzik, 2008). Atypical cadherins Fat4 (Ft) and Dashsous (Dsch) along with the transmembrane protein four-jointed (Fjx) regulate the distribution of the "core" PCP proteins such as frizzled (fz), Van Gogh (Vang), Prickel (Pk), disheveled (Dsh), flamingo (Fmi) in specific subcellular domains (Saburi et al., 2008). The distribution of these proteins function to organize the cell surface and cytoskeleton through the activity of the small GTPase proteins, RhoA and Rac or other PCP effectors such as Inturned (In) and Fuzzy (Fy).

In support of the involvement of PCP in cystic kidney disease, Saburi et al. recently demonstrated that disruption of Fat4 in the kidney cause renal cysts (Saburi et al., 2008). This phenotype in the *Fat4* mice was further acerbated by mutations in another PCP gene *fjx* and the cysts in these mice were associated with altered mitotic spindle orientation. Together these findings support the involvement of the PCP pathway in cyst formation; however, additional studies are needed to determine how the PCP is connecting cilia/IFT and the spindle pole. It is feasible that PCP and IFT affect spindle orientation by regulating the anchoring or migration of the centrioles/basal bodies to the correct position in the cell (see Chapter 8, this volume), defects in which could alter the mitotic axis. An additional possibility that has not yet been addressed is that the tubules in the kidney elongates by intercalation of cells through a CE-like process that requires PCP and activity of IFT/cilia. Interestingly, there are several migration related phenotypes already reported in mice and zebrafish with mutations in ciliary proteins (see Chapter 10, this volume) (Jones et al., 2008; Ross et al., 2005).

In addition to the kidney, the liver and pancreas are also tissues commonly affected in PKD patients. In the liver, primary cilia are present on the biliary epithelium (cholangiocytes) but not on hepatocytes. In the pancreas,

Table 13.3 MKS genes

Disease locus	Gene	Protein sequence characterization	Subcellular localization
MKS1 (17q23)	MKS1	B9 Domain	Centrosome
MKS2 (11q13)	MKS2	Unknown	Unknown
MKS3 (8q21)	TMEM67	Putative 7 transmembrane protein	Basal body/cilium
MKS4 (12q21)	CEP290	Coiled-coil domains/bipartite nuclear localization sequence	Basal body/centrosome
MKS5 (16q)	RPGRIP1L	Coiled-coil domains/bipartite nuclear localization sequence	Basal body/centrosome
MKS6 (4p15)	CC2D2A	C2 Calcium binding domain/coiled coil domain	Unknown

possibility that the B9 proteins are functionally related and may contribute in human ciliopathies.

MKS3 encodes a seven-transmembrane receptor protein, Meckelin. The only existing model for MKS3 is the missense mutation identified in the *wpk* rat which develops PKD and neural tube defects along with a wide spectrum of phenotypes in other tissues (Smith *et al.*, 2006). Recently, Dawe *et al.* demonstrated that both MKS1 and Meckelin (MKS3) are broadly expressed in epithelial cells, including proximal renal tubules and biliary epithelial cells, and that they are required for primary cilium formation based on expression knockdown studies (Dawe *et al.*, 2007). They also showed that MKS1 preferentially localize to basal bodies, while Meckelin was present in both primary cilium and the plasma membrane in ciliated cells. Their study also revealed that MKS1 and Meckelin physically associate and that centrioles fail to migrate to the apical membrane in MKS1 and MKS3 silenced cells. Failure of centriole migration is thought to be the cause of the defects in cilia formation.

Mutations in RPGRIP1L and cep290 cause multiorgan abnormalities that are found in MKS patients. Mutations in both of these genes have also been identified in other human syndromes involving the cilium (Table 13.3 and 13.7). They are being discussed later in this chapter.

CC2D2A is the newest addition to the MKS loci that was identified in MKS6 patients (Tallila *et al.*, 2008). CC2D2A is a novel protein that contains a C2 domain. The C2 domain is a ~120 amino acid sequence that functions as a Ca^{2+} calcium dependant phospholipid binding module, related to the B9 domain that targets proteins to membranes. Although the function of CC2D2A is unknown, it is required for ciliogenesis, since cells isolated from MKS6 patients do not form cilia. The proposed function of CC2D2A is that of a calcium sensor activated by flow mechanosensory signals possibly mediated by the polycystins.

Importantly, several of the MKS related proteins are conserved and are present in lower eukaryotes. This will play an important role in elucidating the possible functions of these proteins. In this regard, the B9 domain containing protein *MKS1* in *C. elegans* is called *xbx-7* and is expressed specifically in ciliated sensory neurons (Efimenko *et al.*, 2005). Furthermore, the XBX-7 protein localizes to the base of the cilium but is not present in the cilium (Williams *et al.*, 2008). Intriguingly, the homologs of the two other B9 domain proteins TZA-1 (also called MKSL-2) and TZA-2 (also called MKSL-1, homolog of stumpy), colocalize with XBX-7 (MKS1) at the base of the cilium. Recent data indicating that XBX-7 (MKS1), TZA-1, and TZA-2 all colocalize at the transition zones at the base of the cilium in worms strongly support that these two additional B9 proteins are good candidates for novel MKS loci. Furthermore XBX-7 (MKS1) was found to require both TZA-1 and TZA-2 for proper localization while TZA-2 requires only TZA-1 (Williams *et al.*, 2008). TZA-1 appeared to be independent of the other B9 domain proteins. These data support the idea that the B9 proteins form a complex at the base of the cilium; however, the role that these proteins play at the transition zone is uncertain.

In contrast to studies with RNA knock downs in mammalian systems, *C. elegans* with mutations in all three B9 proteins simultaneously do not exhibit overt cilia morphology defects, nor do any of the single mutants. These data indicate that these proteins are not required for ciliogenesis, at least in this model system. A very intriguing finding from additional studies in *C. elegans* was that severe cilia morphology defects and abnormal orientation and positioning of cilia were generated when any of the B9 protein gene mutations were on the *nph1* or *nph4* mutant background (Williams *et al.*, 2008). Since there is no requirement for the NPH proteins for B9 protein localization or the B9 proteins for NPH localization at the transition zone, these data argue that there are two independent complexes involved in the ciliopathies that localize to the base of the cilium in *C. elegans*. One consists of the Nephrocystin (NPHP) proteins and the other the MKS/B9 domain proteins. These two complexes appear to function redundantly in determining the orientation and positioning of the cilium. Whether similar complexes are present and function in mammalian systems needs to be further addressed. This possibility is further supported by the fact that disruption of *nph1* in mice does not exhibit a renal cystic phenotype (Jiang et al., 2008). In addition, whether the transmembrane protein MKS3 or the other MKS-related proteins CC2D2A, CEP290, and RPGRIPL1 fit into this complex and will have functional interactions with the NPH complex has not yet been analyzed; however, this possibility seems likely, based on the fact that mutations in *cep290* and *rpgripl1* have already been identified as both NPH and MKS loci.

4.7. BBS (omim 209900)

BBS is a rare genetically heterogeneous disorder with multiorgan involvement that displays clinical variability. Phenotypes observed in BBS patients include obesity, polydactyly, mental retardation, kidney abnormalities, hypogenitalism, and diabetes, (Fig. 13.6). (Blacque and Leroux, 2006; Fath et al., 2005; Kulaga et al., 2004; Li et al., 2004; Mykytyn et al., 2004; Nishimura et al., 2004; Tobin and Beales, 2007). The Bardet–Biedl phenotype also includes a large number of sensory deficits such as vision loss associated with retinal degeneration, as well as anosmia, and defective hearing (Tan et al., 2007). In contrast to MKS, BBS mutations are not neonatal lethal and result in more mild phenotypes. Fourteen BBS genes have been identified, to date, (BBS1–14) and in most cases the corresponding proteins localize to the basal body or the cilium (Blacque and Leroux, 2006; Tobin and Beales, 2007) (Table 13.4).

As with MKS and NPHP proteins, the function of the BBS proteins remains obscure; however, recent data suggest they are involved in cilia function and intracellular transport (Nachury et al., 2007). Proteins involved in BBS have domains that are suggestive of being chaperones (BBS6, 10, and 12) involved in protein folding (Kim et al., 2005), E3 ubiquitin-ligase (BBS11) involved in proteosomal degradation (Chiang et al., 2006; Saccone et al., 2008), ARF family protein (BBS3) involved in intracellular vesicle trafficking and microtubule assembly (Blacque and Leroux, 2006; Chiang et al., 2006), DM16 repeat motifs of unknown function (BBS5) (Li et al., 2004), TPR motifs (BBS4 and BBS8) involved in protein interactions (Ansley et al., 2003), and β-propeller domain (BBS1, 2, and 7) proteins (Badano et al., 2003). The other BBS proteins have no homologies that would be suggestive of their function (Blacque and Leroux, 2006; Tobin and Beales, 2007).

Several studies in diverse model systems now support a role for BBS proteins in functions associated with cilia. The BBS proteins are present in most ciliated organisms with the exception of BBS6 and BBS10, which appear to have arisen later in vertebrates. Most of the BBS proteins identified to date localize to the cilium or at the basal body (Table 13.4). A clear

Table 13.4 BBS genes

Disease locus	Gene	Protein sequence characterization	Subcellular localization
BBS1[a] (11q13)	BBS1	Novel	Basal body/cilium
BBS2[a] (16q22)	BBS2	Novel	Basal body/cilium
BBS3 (3p12)	ARL6	ADP-ribosylation factor-like protein 6	Basal body/cilium
BBS4[a] (15q22)	BBS4	Novel, TPR repeats	Basal body/cilium
BBS5[a] (2q31)	BBS5	Novel	Basal body/cilium
BBS6 (20p12)	MKKS	Chaperonin-like	Basal body
BBS7[a,b] (4q27)	FLJ10715	Novel	Basal body/cilium
BBS8[a,b] (14q31)	TTC8	Novel, TPR repeats	Basal body/cilium
BBS9[a] (7p14)	PTHB1	Novel	Basal body/cilium
BBS10 (12q21)	C12ORF58	Chaperonin-like	Unknown
BBS11 (9q31)	TRIM32	E3 Ubiquitin ligase	Unknown
BBS12 (4q27)	FLJ35630	Chaperonin-like	Unknown
BBS13 (17q23)	MKS1	B9 Domain	Centrosome
BBS14 (12q21)	CEP290	Coiled-coil domains/bipartite nuclear localization sequence	Basal body/centrosome

[a] Indicates part of the BBsome complex.
[b] Are thought to be IFT particle and cargo adaptors in *C. elegans*.

example of BBS proteins functioning in cilia is seen for BBS7 and BBS8 in C. elegans (Blacque et al., 2004). The BBS7 and BBS8 proteins form a connection between complex A and complex B of the IFT particle (see Chapter 2, this volume for more details on IFT) (Ansley et al., 2003; Blacque et al., 2004). In the absence of BBS7 or BBS8 the two IFT subcomplexes move at distinct rates along the cilia axoneme. How this may affect ciliary function is not known. Additional support is seen with the knockdown bbs5 in Chlamydomonas which results in the loss of flagella (Li et al., 2004).

In mammalian systems, the role of the BBS proteins in cilia is less well defined and appears to depend on which type of cilia or flagella are analyzed and which BBS protein is being affected. Male BBS2, BBS4, and BBS6 mutant mice are sterile due to the absence of flagella on sperm (Fath et al., 2005; Mykytyn et al., 2004; Nishimura et al., 2004) and BBS6 mutants have abnormal cilia on photoreceptors and display abnormal ciliary structure and function on olfactory neurons and in a fraction of the respiratory cilia (Shah et al., 2008). In the kidney, initial observations found that the primary cilium on renal epithelium of BBS mutant mice were normal in number, length and morphology despite the formation of cysts (Mykytyn et al., 2004). In contrast to these data, in cultured renal epithelial cells from BBS4 null mice cilia were found to be abnormally long (Mokrzan et al., 2007). The reason for these divergent observations is not known but could reflect analysis in different cell types or in vitro and in vivo conditions.

Work by Nachury et al. has recently shown that BBS proteins function as part of a large complex that they have called the BBSome (Nachury et al., 2007). BBS9 appears to be critical as an organizing subunit as it interacts with multiple BBS proteins including BBS1, 2, 4, 5, and 7, and 8 based on tandem affinity purification analysis (Nachury, 2008). In contrast, BBS3, BBS6, BBS10, BBS11, and BBS12 do not appear to be part of the BBSome. The BBSome is thought to localize to centriolar satellites around the basal body in the cytoplasm and to the membrane of the cilium. Through BBS1, the BBSome is able to interact with Rabin8 to promote cilia membrane extension. Rabin8 is a guanine exchange factor (GEF) for RAB8 which functions in docking of Golgi derived vesicles to the base of the cilium. Furthermore, preventing Rab8 (GTP) production in zebrafish blocks ciliation and results in classical BBS phenotypes (Nachury et al., 2007). These data indicate that one role for the BBSome may be to regulate entry of components into the cilium or onto the IFT particles. This is further supported by recent data indicating that the G protein-coupled receptors somatostatin receptor type 3 (Sstr3) and melanin-concentrating hormone receptor 1 (Mchr1) that normally reside in the primary cilium of central neurons of wild-type mice fail to enter the cilium in BBS mutant cells (Berbari et al., 2008b).

The cause of the renal cysts in BBS is not certain. However, it is interesting that BBS1, BBS4, and BBS6 mutants exhibit misoriented cochlear stereocilia bundles in the inner ear indicative of PCP defects (Ross et al., 2005). In addition, BBS mutants share other PCP related phenotypes including exencephaly and incomplete neural tube closure (Ross et al., 2005) and PCP related phenotypes associated with convergent–extension have been reported in BBS knockdown mutants in zebrafish (Tayeh et al., 2008). Finally, cystic phenotypes are observed in mice that are double heterozygotes for BBS6 and the PCP gene Vangl2. Together these data argue that renal cysts in the BBS mutants may be related to defects in regulation of the PCP pathway as has been proposed for PKD.

Knockdown studies of BBS proteins in zebrafish have indicated another function of the BBS proteins in regulating Wnt signaling activity (See Chapter 7 of this issue for a discussion of cilia in Wnt signaling.). Data from Gerdes et al. revealed that knockdown of the BBS proteins results in a large enhancement in the response to canonical Wnt3a compared to controls cells (Gerdes et al., 2007). Interestingly, this also impaired the ability of the noncanonical Wnt5a from suppressing the canonical pathway induced by Wnt3a. These data further demonstrated that suppression of BBS4 impaired proteosomal targeting and degradation of β-catenin leading to its accumulation. In addition to PCP defects, this may be another connection to cystic kidney disease since expression of a stabilized form of β-catenin in the mouse kidney is known to cause renal cyst formation and in most models of PKD β-catenin levels are markedly elevated (Benzing et al., 2007; Calvet and Grantham, 2001; Lin et al., 2003; Qian et al., 2005; Saadi-Kheddouci et al., 2001). In part, this may involve altered phosphorylation and activation of Disheveled, a protein known to regulate canonical versus noncanonical signaling activities (Simons et al., 2005). Similar observations were made with mutant alleles of *kif3a*, the kinesin involved in IFT, in cell culture and in mice suggesting that these effects are associated with a ciliary dysfunction (Corbit et al., 2008).

Polydactyly is a cardinal feature of BBS patients as well as in several mice with mutations in IFT genes, but interestingly has not been reported in mouse models of BBS. Polydactyly in mice, chicks, and humans is often associated with alterations in the sonic hedgehog (Shh) signaling pathway due to an increase in Shh expression on the anterior side of the limb or due to loss of Gli3, a transcriptional repressor of the Shh pathway (Babbs et al., 2008; Tanaka et al., 2000; Tickle, 2003). Analysis of IFT mutant mice has demonstrated that cilia are required for cells to respond to Shh and that in the absence of the cilium the pathway becomes deregulated with loss of both Gli activator and repressor function (Haycraft et al., 2005, 2007; Huangfu et al., 2003; Liu et al., 2005a; May et al., 2005) (for more detailed discussion of cilia and hedgehog signaling and skeletal defects see Chapters 9 and 11, this volume).

To explore whether polydactyly in BBS is associated with the Shh pathway, Tayeh et al. used knockdown approaches with combinations of

BBS genes in zebrafish (Tayeh et al., 2008). Their data demonstrated a strong genetic interaction between the BBS genes that caused an anterior expansion of Shh in the pectoral fin and changes in patterning of skeletal elements indicative of polydactyly. These findings are in contrast to what is observed in most complex B anterograde IFT mutant mice where Shh expression does not change (Liu et al., 2005a; Ocbina and Anderson, 2008). Rather, the pathway becomes ectopically activated due to loss of Gli3 repressor activity. However, it is in agreement with other mouse mutations that affect complex A retrograde IFT, such as alien (IFT139), where the Shh pathway is ectopically activated on the anterior of the limb (Tran et al., 2008). This observation is intriguing in light of additional data indicating that knockdown of zebrafish *bbs2, bbs4–bbs8,* or *bbs11* resulted in a delay of retrograde transport.

4.8. JBTS (omim 213300)

JBTS is a group of recessive genetically heterogeneous disorders with phenotype variability that involves renal cysts, ataxia, mental retardation, obesity, oculomotor apraxia, polydactyly, retinal degeneration, hypotonia, and neonatal breathing dysfunction (Baala et al., 2007; Braddock et al., 2007; Cantagrel et al., 2008; Parisi et al., 2007; Valente et al., 2008). JBTS patients also exhibit what is called the molar tooth malformation (MTM). MTM is a complex brainstem malformation caused by aplasia and marked hypoplasia of the cerebellar vermis. None of these features by themselves are diagnostic of JBTS and it is now believed that JBTS is a spectrum of disorders involving vermis hypoplasia.

One of the most studied features of JBTS has been the cerebellar hypoplasia. Development of the cerebellum requires synchronized regulation of progenitor cell proliferation, neuronal differentiation, and migration (Sillitoe and Joyner, 2007). Progenitor cells originate from the rhombic region. Cells from the rhombic lip migrate over the surface of the developing cerebellum to form the external granular layer (EGL). Progenitor cells located in the EGL undergo a massive proliferation in response to sonic hedgehog (Shh) that is secreted from the Purkinje cells. As EGL cells differentiate they migrate into the cerebellar cortex to form the internal granule layer (IGL). Formation of the folia in the cerebellum is associated with these ingrowths of the cells and proliferation of the EGL progenitors which occurs during postnatal development.

Clues to what is occurring in JBTS patients were obtained from analysis of mutants with conditional alleles of *ift88* or *kif3a* that were crossed with GFAP-Cre or Nestin-Cre deletor strains (Chizhikov et al., 2007). Both of these deletor lines express Cre recombinase early in the progenitors of the EGL. These mice appeared normal at birth but have cerebellar hypoplasia with foliation abnormalities by postnatal day 4 and become severely ataxic.

The cause of the cerebellar phenotype was a failure in the expansion of the progenitor population in the EGL. This appeared to be due to the inability of these cells to respond to Shh, as was also shown in the limb buds and neural tubes of the *ift88* or *kif3a* mutants (Haycraft et al., 2007; Huangfu and Anderson, 2005). These findings suggest that granule cell proliferation defects are central to the cerebellar phenotype observed in JBTS.

Seven loci (AHI1, NPHP1, RPGRIP1L, CEP290/NPHP6, ARL13B, and TMEM67) have now been linked to JBTS in humans (Table 13.5) (Coll et al., 1997; Connacher et al., 1987; Doege et al., 1964; Feather et al., 1997; Ferrante et al., 2006; Thauvin-Robinet et al., 2006). Intriguingly, three of these genes are also associated with the MKS, and two with NPHP/SLSN (Table 13.7). AHI1 (Abelson helper integration 1 gene) encodes a coiled–coil domain protein with seven WD40 repeats, a SH3 domain, and SH3 binding sites leading to the speculation that AHI1 functions in signal transduction as an adaptor molecule. NPHP1 is a centrosome/cilia protein with an SH3 and coiled-coil domain that is known to interact with AH1. Arl13b belongs to the Ras GTPase family of proteins. It is mutated in the lethal *hennin* mutant mouse that displays Shh- like phenotypes (Caspary et al., 2007). TMEM67 is a seven transmembrane protein in cilia that is related to the frizzled receptors with no known function. CEP290 localizes to cilia, centrosomes, and nucleus and is thought to be involved in chromosome segregation (Frank et al., 2008; Sayer et al., 2006). Interestingly, RPGRIP1L localizes at the basal body and centrosomes and is associated with CEP290 and binds directly to NPHP4 (Chang et al., 2006; Mollet et al., 2002; Roepman et al., 2005).

4.9. SLSN (omim 266900)

SLSN is an autosomal recessive disease with the main clinical features of both NPHP and Leber's Congenital Amaurosis (LCA OMIM 204000). There have been 6 loci associated with SLSN (Table 13.6). In fact all of the genes mutated in SLSN to date have been observed in other ciliopathies, such as the nephrocystins, CEP290, and IQCB1 (Table 13.7).

4.10. OFD (omim 311200)

Oral-facial-digital type I (OFD1) syndrome is a heterogeneous developmental disorder characterized by craniofacial abnormalities, postaxial polydactyly, central nervous system disorders, and in 15% of cases cystic kidney disease (Coll et al., 1997; Connacher et al., 1987). The protein product of OFD1 localizes to the centriole/basal body and has also recently been reported in the primary cilium and the nucleus. The function of the OFD1 protein remains unknown. In murine *ofd1* mutants cilia are absent from the luminal surface of kidney glomerular and tubular cells (Ferrante

Table 13.5 JBTS genes

Disease locus	Gene	Protein sequence characterization	Subcellular localization
JBTS1 (9q34)	Unknown	Unknown	Unknown
JBTS2 (11p12)	Unknown	Unknown	Unknown
JBTS3 (6q23)	AHI1/Jouberin	SH3 Domain/WD40 repeat domains/coiled-coil domain	Centrosome/cilium/adherens Junctions
JBTS4 (2q13)	Nephrocystin-1	SH3 Domain/coiled-coil domain	
JBTS5 (12q21)	CEP290	Coiled-coil domains/bipartite nuclear localization sequence	Basal body/centrosome
JBTS6 (8q21)	TMEM67	Putative 7 transmembrane protein	Basal body/cilium
JBTS7 (16q12)	RPGRIP1L	Coiled-coil domains/bipartite nuclear localization sequence	Basal body/centrosome/cilium
JBTS8 (3p12.3-q12.3)	Arl13b	Small GTPase	

Table 13.6 SLSN genes

Disease locus	Gene	Protein sequence characterization	Subcellular localization
SLSN1 (2q13)	Nephrocystin-1	SH3 Domain/coiled-coil domain	Centrosome/cilium/adherens junctions
SLSN2	Unknown	Unknown	Unknown
SLSN3 (3q22)	Nephrocystin-3	IQ calmodulin binding domain	Cilium
SLSN4 (1p36)	Nephrocystin-4	Novel	Basal body/centrosome/cilium
SLSN5 (3q21)	IQCB1	IQ calmodulin binding domain	Cilium
SLSN6 (12q21)	CEP290	Coiled-coil domains/bipartite nuclear localization sequence	Basal body/centrosome

et al., 2006). Furthermore, *ofd1* null mutants were found to lack node cilia, and display left–right body axis patterning defects and also possess morphological defects involving the Shh pathway in the neural tube and limb bud similar to those observed in *ift172,kif3a*, and *ift88* mutants (Ferrante *et al.*, 2006).

4.11. ALMS (omim 203800)

ALMS is a rare disorder with sensory impairment, cone-rod retinal dystrophy, hearing defects, renal failure, cardiomyopathy, and hepatic dysfunction, and morbid obesity (Hearn *et al.*, 2005; Ozgul *et al.*, 2007). ALMS1 localizes to the centrosome and basal body and the spectrum of phenotypes resembles that seen in BBS. Most ALMS1 mutations result in expression of a C-terminal truncated protein. Analysis of the human ALMS1 fibroblasts or kidney collecting duct epithelial cells or ALMS1 gene trap mouse revealed no overt defects in cilia formation; however, using knockdown approaches in renal cell lines Li *et al.* were able to show that cilia assembly was disrupted (Li *et al.*, 2007). Intriguingly, the cilia assembly defects could be restored by expression of an N-terminal truncated human ALMS1 mutant protein in the knockdown cells.

Table 13.7 Genes associated with multiple ciliopathies

Gene	Locus	Associated ciliopathies
MKS1	17q23	MKS1, BBS13
TMEM67	8q21	MKS3, JBTS6
CEP290	12q21	MKS4, NPHP6, BBS14, JBTS5, SLSN6
Nephrocystin-1	2q13	NPHP1, JBTS4, SLSN1
RPGRIP1L	16q	MKS5, NPHP8, JBTS7
IQCB1	3q21	NPHP5, SLSN5
Nephrocystin-3	3q22	NPHP3, SLSN3
Nephrocystin-4	1p36	NPHP4, SLSN4

5. OLIGOGENIC INHERITANCE AND CLINICAL VARIABILITY IN THE CILIOPATHIES

One fascinating aspect of the ciliopathies is their range of clinical variability. For example, the embryonic lethality observed in MKS as compared to the less severe multisystem disorders, such as BBS, or the more tissue specific disorder PKD. One can reconcile such a range of clinical outcomes when one thinks of the previously unappreciated complexity of the cilium.

In mouse models, it has become apparent that complete ablation of cilia formation and function leads to early embryonic lethality with severe neural tube and body axis defects. Based on these data, it is not too surprising that only one syndrome (Jeune asphyxiating thoracic dystrophy (JATD; OMIM 208500) in humans has been associated with mutations affecting an IFT gene (*Ift80*). Furthermore, in JATD patients, the mutation in *ift80* is believed to be hypomorphic and cilia likely retain some function (Beales *et al.*, 2007). In contrast, other ciliopathies with multisystem involvement such as BBS likely result from defects in specific signaling processes that require cilia function (Blacque and Leroux, 2006; Kulaga *et al.*, 2004). In fact, it is thought that BBS mutations are not true "null" mutations with regards to cilia function (Nachury *et al.*, 2007). Rather, the data support a role for the BBS proteins in regulating entry of specific signaling proteins into the cilium. Interestingly, BBS patients suffer from epistatic interactions that may contribute to the phenotypic variability. Indeed, genetic modifiers of BBS have been identified (Leitch *et al.*, 2008). Furthermore, a non-Mendelian phenomenon termed triallellic inheritance has been reported in a small number of BBS patients (Katsanis *et al.*, 2001). In triallelic inheritance patients with monogenic recessive mutations for one of the BBS mutations also possessed a heterozygous mutation at another BBS locus. Thus, it appears that genetic mutation load in the cilia proteome as well as genetic modifiers contribute to the variability seen in the cilia related diseases.

On the molecular level, the identification of the BBsome has supported such a hypothesis. Nachury *et al.* propose that missing one of the peripheral BBSome components acts as a BBSome hypomorph thus resulting in the array of phenotypes observed in BBS mice models and patients (Nachury *et al.*, 2007). However, they predict that if the core BBsome component (BBS 9) was lacking this would result in embryonic lethality. Further testing of such a hypothesis is needed and will shed light on the mechanisms behind the clinical features observed in BBS.

Not only does the overall genetic load play a role in the clinical presentation in the ciliopathies, but the phenotype also appears to be determined by the type of mutation that occurs within a gene. Clear examples of this are observed with mutations in NPHP1 and NPHP6 (Table 13.7), which have been found in patients diagnosed with NPHP, JBTS, SLSN, and in the case of NPHP6 also MKS. Intriguingly, these syndromes display significantly different clinical features. MKS is a very severe developmental disorder which is nonviable. On the other hand, NPHP is a form of pediatric cystic kidney disease where patients can survive into the perinatal period and young adulthood. However, when there is concomitant NPHP along with retinopathy it is clinically known as SLSN while JBTS also presents with cerebellar vermis aplasia and ataxia. While in the clinic these are identified as distinct disorders, from a basic science standpoint it is interesting that mutations in the same gene can cause such diverse phenotypic features. This ultimately argues, at least in the case of MKS, NPHP, SLS, and JBTS that the ciliopathies may represent a clinical spectrum of disease presentations that are determined by the nature of the mutation that occurs within a specific gene. This has lead to the proposal that the lethality in MKS is associated with mutations that are more likely to be functional null, or combined with additional mutations in other ciliopathy genes, while the less severe phenotypes observed in NPH or JBTS patients result from milder missense type of mutations that occur in the same gene.

Finally, in a disease such as PKD it is the kidney, liver, and pancreas that are largely the affected organs. In this case, the null mutations occur in a protein involved specifically in signaling rather than in a protein involved in regulating multiple facets of cilia function or cilia assembly, such as the IFT genes. The later class would be expected to have more significant clinical presentation.

6. Cilia Proteome and Genome Databases and the Identification of Human Ciliopathy Genes

An important resource that has helped propel the field of cilia biology from basic science into the realm of biomedical science has been the compilation of cilia/flagella proteomes from multiple organisms. Several different strategies have been utilized to define the protein makeup of cilia and each has its benefits and drawbacks. One of the initial approaches was based on the insightful observation of Swoboda *et al.* that multiple IFT and cilia genes in *C. elegans* are coregulated by the transcription factor DAF-19 through an X-box sequence motif located in the promoter of these genes (Swoboda *et al.*, 2000). Based on this observation, *en silico* analyses were conducted in the worm to identify candidate cilia genes based on the presence of a correctly positioned X-box regulator domain. The X-box is a small imperfect 14 nucleotide sequence and as such is found by random chance throughout the genome. Thus, one of the drawbacks of this method was its low stringency and high rate of false positives. Chen *et al.* further refined this X-box analysis by comparing genomes between several related nematode species which greatly reduced the number of potential targets (Chen *et al.*, 2006). Another limitation of the X-box screen is that expression of most cilia genes is not regulated by DAF-19, thus, there is also a high false negative rate. Other innovative *en silico* approaches subtracted the genomes of ciliated and nonciliated organisms to identify potential cilia related genes (Avidor-Reiss *et al.*, 2004; Li *et al.*, 2004). Additional studies used microarray and serial analysis of gene expression (SAGE) to compare expression profiles in ciliated and nonciliated neurons in *C. elegans* (Blacque *et al.*, 2005). More direct proteomic methods have also been employed. Several groups have utilized mass spectrometry to identify proteins in purified motile and nonmotile cilia preparations isolated from human trachea and photoreceptor cells and from motile flagella of *Chlamydomonas* (Ostrowski *et al.*, 2002; Pazour *et al.*, 2005). Table 13.8 gives some examples of these studies that have aimed at defining the protein complement of cilia.

Together, these combinations of diverse approaches resulted in the identification of several hundred to thousands of candidate cilia genes. Subsequent analyses of these proteomic datasets have uncovered the true complexity of this organelle. Among the proteins are a wide spectrum of signal transduction molecules such as receptors and kinases that are likely involved in signaling roles for the cilium (Inglis *et al.*, 2006). These finding suggest that the phenotypes observed in murine and human cilia related disease may result from effects on numerous pathways. Another feature that will likely emerge when proteomes are completed for primary cilia in other tissues is that cilia may not all be functionally equivalent. Thus, some signaling pathways may be present in cilia of some tissues but absent from others. It is becoming evident that the function of a particular cilium within certain cell types will depend on the specific signaling proteins that are enriched in its ciliary compartment.

The ongoing analyses of the proteins identified through these various approaches has demonstrated the utility of such screens as many of the candidate genes were subsequently found to encode proteins involved in ciliogenesis or cilia signaling. In addition, several of these genes were also

Table 13.8 Cilia proteome and genome databases

Type of analysis	General results	Studies
X-box promoter analysis in C. elegans	Revealed novel cilia genes, including signaling proteins	Swoboda et al. (2000), Efimenko et al. (2005), Blacque et al. (2005)
En Silico comparative genomics	Revealed novel cilia genes, including human disease genes, also able to identify genes necessary for cilia function, but not located in the cilium itself	Li et al. (2004), Avidor-Reiss et al. (2004)
Transcriptome analysis	Revealed disease genes and proteins outside of the cilia that play a role in cilia biology (i.e., chaperonin CCT, DNA helicases, qilin)	Stolc et al., Blacque et al. (2005)
Proteomics of motile cilia	Revealed several proteins that have roles in signal transduction (i.e., receptors, small GTPases)	Ostrowski et al. (2002), Smith et al. (2006), Pazour et al. (2005), Broadhead et al. (2006)

associated with human ciliopathies such as PKD, JBTS, NPHP, MKS, and BBS. Another important outcome of the proteomic studies and the subsequent mining and metaanalyses of the resulting databases (http://www.sfu.ca/~leroux/ciliome_database.htm and http://www.ciliaproteome.org) is that they have greatly expedited the identification of novel human disease genes. With positional mapping of disease loci in humans and mice and the presence of ciliopathy-like phenotype, several groups have utilized the proteomes to prioritize and subsequently identify disease genes in multiple disorders (Adams et al., 2008; Chen et al., 2006).

ACKNOWLEDGMENTS

This work was supported in part by the National Institutes of Health RO1 (NIDDK 65655, NIDDK 075996, HD056030) to BKY, National Kidney Foundation Postdoctoral award to NS, and a T32 award postdoctoral award (5T32HL007553, Dr. Janet Yother, UAB) to NFB. We thank Dr. Kathryn Anderson, Dr. Kirk Mykytyn for unpublished images. We also thank Mandy J. Croyle for technical assistance and Amber O'Connor and Sarah Mollo for critical reading of the manuscript.

REFERENCES

Abd-El-Barr, M. M., Sykoudis, K., Andrabi, S., Eichers, E. R., Pennesi, M. E., Tan, P. L., Wilson, J. H., Katsanis, N., Lupski, J. R., and Wu, S. M. (2007). Impaired photoreceptor protein transport and synaptic transmission in a mouse model of Bardet–Biedl syndrome. *Vision Res.* **47,** 3394–3407.

Adams, M., Smith, U. M., Logan, C. V., and Johnson, C. A. (2008). Recent advances in the molecular pathology, cell biology and genetics of ciliopathies. *J. Med. Genet.* **45,** 257–267.

Afzelius, B. A. (1976). A human syndrome caused by immotile cilia. *Science* **193,** 317–319.

Afzelius, B. A. (2004). Cilia-related diseases. *J. Pathol.* **204,** 470–477.

Afzelius, B. A., and Eliasson, R. (1979). Flagellar mutants in man: On the heterogeneity of the immotile-cilia syndrome. *J. Ultrastruct. Res.* **69,** 43–52.

Afzelius, B. A., Eliasson, R., Johnsen, O., and Lindholmer, C. (1975). Lack of dynein arms in immotile human spermatozoa. *J. Cell Biol.* **66,** 225–232.

Alexiev, B. A., Lin, X., Sun, C. C., and Brenner, D. S. (2006). Meckel–Gruber syndrome: Pathologic manifestations, minimal diagnostic criteria, and differential diagnosis. *Arch. Pathol. Lab. Med.* **130,** 1236–1238.

Ansley, S. J., Badano, J. L., Blacque, O. E., Hill, J., Hoskins, B. E., Leitch, C. C., Kim, J. C., Ross, A. J., Eichers, E. R., Teslovich, T. M., Mah, A. K., Johnsen, R. C., et al. (2003). Basal body dysfunction is a likely cause of pleiotropic Bardet–Biedl syndrome. *Nature* **425,** 628–633.

Avidor-Reiss, T., Maer, A. M., Koundakjian, E., Polyanovsky, A., Keil, T., Subramaniam, S., and Zuker, C. S. (2004). Decoding cilia function: Defining specialized genes required for compartmentalized cilia biogenesis. *Cell* **117,** 527–539.

Baala, L., Romano, S., Khaddour, R., Saunier, S., Smith, U. M., Audollent, S., Ozilou, C., Faivre, L., Laurent, N., Foliguet, B., Munnich, A., Lyonnet, S., et al. (2007). The Meckel–Gruber syndrome gene, MKS3, is mutated in Joubert syndrome. *Am. J. Hum. Genet.* **80,** 186–194.

Babbs, C., Furniss, D., Morriss-Kay, G. M., and Wilkie, A. O. (2008). Polydactyly in the mouse mutant Doublefoot involves altered Gli3 processing and is caused by a large deletion in cis to Indian hedgehog. *Mech. Dev.* **125,** 517–526.

Badano, J. L., Ansley, S. J., Leitch, C. C., Lewis, R. A., Lupski, J. R., and Katsanis, N. (2003). Identification of a novel Bardet–Biedl syndrome protein, BBS7, that shares structural features with BBS1 and BBS2. *Am. J. Hum. Genet.* **72,** 650–658.

Banizs, B., Pike, M. M., Millican, C. L., Ferguson, W. B., Komlosi, P., Sheetz, J., Bell, P. D., Schwiebert, E. M., and Yoder, B. K. (2005). Dysfunctional cilia lead to altered ependyma and choroid plexus function, and result in the formation of hydrocephalus. *Development* **132,** 5329–5339.

Banizs, B., Komlosi, P., Bevensee, M. O., Schwiebert, E. M., Bell, P. D., and Yoder, B. K. (2007). Altered pH(i) regulation and Na(+)/HCO3(−) transporter activity in choroid plexus of cilia-defective Tg737(orpk) mutant mouse. *Am. J. Physiol. Cell Physiol.* **292,** C1409–C1416.

Bargmann, C. I. (2006). Chemosensation in C. elegans. *WormBook* 1–29.

Barr, M. M., and Sternberg, P. W. (1999). A polycystic kidney-disease gene homologue required for male mating behaviour in C. elegans. *Nature* **401,** 386–389.

Barr, M. M., DeModena, J., Braun, D., Nguyen, C. Q., Hall, D. H., and Sternberg, P. W. (2001). The *Caenorhabditis elegans* autosomal dominant polycystic kidney disease gene homologs lov-1 and pkd-2 act in the same pathway. *Curr. Biol.* **11,** 1341–1346.

Bartoloni, L., Blouin, J. L., Pan, Y., Gehrig, C., Maiti, A. K., Scamuffa, N., Rossier, C., Jorissen, M., Armengot, M., Meeks, M., Mitchison, H. M., Chung, E. M., et al. (2002). Mutations in the DNAH11 (axonemal heavy chain dynein type 11) gene cause one form of *situs inversus* totalis and most likely primary ciliary dyskinesia. *Proc. Natl. Acad. Sci. USA* **99,** 10282–10286.

Beales, P. L., Bland, E., Tobin, J. L., Bacchelli, C., Tuysuz, B., Hill, J., Rix, S., Pearson, C. G., Kai, M., Hartley, J., Johnson, C., Irving, M., et al. (2007). IFT80, which encodes a conserved intraflagellar transport protein, is mutated in Jeune asphyxiating thoracic dystrophy. *Nat. Genet.* **39,** 727–729.

Benzing, T., Gerke, P., Hopker, K., Hildebrandt, F., Kim, E., and Walz, G. (2001). Nephrocystin interacts with Pyk2, p130(Cas), and tensin and triggers phosphorylation of Pyk2. *Proc. Natl. Acad. Sci. USA* **98,** 9784–9789.

Benzing, T., Simons, M., and Walz, G. (2007). Wnt signaling in polycystic kidney disease. *J. Am. Soc. Nephrol.* **18,** 1389–1398.

Berbari, N. F., Johnson, A. D., Lewis, J. S., Askwith, C. C., and Mykytyn, K. (2008a). Identification of ciliary localization sequences within the third intracellular loop of G protein-coupled receptors. *Mol. Biol. Cell* **19,** 1540–1547.

Berbari, N. F., Lewis, J. S., Bishop, G. A., Askwith, C. C., and Mykytyn, K. (2008b). Bardet–Biedl syndrome proteins are required for the localization of G protein-coupled receptors to primary cilia. *Proc. Natl. Acad. Sci. USA* **105,** 4242–4246.

Besharse, J. C., and Hollyfield, J. G. (1979). Turnover of mouse photoreceptor outer segments in constant light and darkness. *Invest. Ophthalmol. Vis. Sci.* **18,** 1019–1024.

Blacque, O. E., and Leroux, M. R. (2006). Bardet–Biedl syndrome: An emerging pathomechanism of intracellular transport. *Cell. Mol. Life Sci.* **63,** 2145–2161.

Blacque, O. E., Reardon, M. J., Li, C., McCarthy, J., Mahjoub, M. R., Ansley, S. J., Badano, J. L., Mah, A. K., Beales, P. L., Davidson, W. S., Johnsen, R. C., Audeh, M., et al. (2004). Loss of *C. elegans* BBS-7 and BBS-8 protein function results in cilia defects and compromised intraflagellar transport. *Genes Dev.* **18,** 1630–1642.

Blacque, O. E., Perens, E. A., Boroevich, K. A., Inglis, P. N., Li, C., Warner, A., Khattra, J., Holt, R. A., Ou, G., Mah, A. K., McKay, S. J., Huang, P., et al. (2005). Functional genomics of the cilium, a sensory organelle. *Curr. Biol.* **15,** 935–941.

Blair, D. F., and Dutcher, S. K. (1992). Flagella in prokaryotes and lower eukaryotes. *Curr. Opin. Genet. Dev.* **2,** 756–767.

Blouin, J. L., Meeks, M., Radhakrishna, U., Sainsbury, A., Gehring, C., Sail, G. D., Bartoloni, L., Dombi, V., O'Rawe, A., Walne, A., Chung, E., Afzelius, B. A., et al. (2000). Primary ciliary dyskinesia: A genome-wide linkage analysis reveals extensive locus heterogeneity. *Eur. J. Hum. Genet.* **8,** 109–118.

Blowey, D. L., Querfeld, U., Geary, D., Warady, B. A., and Alon, U. (1996). Ultrasound findings in juvenile nephronophthisis. *Pediatr. Nephrol.* **10,** 22–24.

Boysen, M. (1982). The surface structure of the human nasal mucosa. I. Ciliated and metaplastic epithelium in normal individuals. A correlated study by scanning/transmission electron and light microscopy. *Virchows Arch. B Cell Pathol. Incl. Mol. Pathol.* **40,** 279–294.

Braddock, S. R., Henley, K. M., and Maria, B. L. (2007). The face of Joubert syndrome: A study of dysmorphology and anthropometry. *Am. J. Med. Genet. A* **143A,** 3235–3242.

Brailov, I., Bancila, M., Brisorgueil, M. J., Miquel, M. C., Hamon, M., and Verge, D. (2000). Localization of 5-HT(6) receptors at the plasma membrane of neuronal cilia in the rat brain. *Brain Res.* **872,** 271–275.

Brown, J. M., Marsala, C., Kosoy, R., and Gaertig, J. (1999). Kinesin-II is preferentially targeted to assembling cilia and is required for ciliogenesis and normal cytokinesis in *Tetrahymena*. *Mol. Biol. Cell* **10,** 3081–3096.

Broadhead, R., Dawe, H. R., Farr, H., Griffiths, S., Hart, S. R., Portman, N., Shaw, M. K., Ginger, M. L., Gaskell, S. J., McKean, P. G., and Gull, K. (2006). Flagellar motility is required for the viability of the bloodstream trypanosome. *Nature* **440,** (7081), 224–227.

Bush, A., Cole, P., Hariri, M., Mackay, I., Phillips, G., O'Callaghan, C., Wilson, R., and Warner, J. O. (1998). Primary ciliary dyskinesia: Diagnosis and standards of care. *Eur. Respir. J.* **12,** 982–988.

Calvet, J. P., and Grantham, J. J. (2001). The genetics and physiology of polycystic kidney disease. *Semin. Nephrol.* **21,** 107–123.

Cano, D. A., Murcia, N. S., Pazour, G. J., and Hebrok, M. (2004). Orpk mouse model of polycystic kidney disease reveals essential role of primary cilia in pancreatic tissue organization. *Development* **131,** 3457–3467.

Cantagrel, V., Silhavy, J. L., Bielas, S. L., Swistun, D., Marsh, S. E., Bertrand, J. Y., Audollent, S., Attie-Bitach, T., Holden, K. R., Dobyns, W. B., Traver, D., Al-Gazali, L., *et al.* (2008). Mutations in the cilia gene ARL13B lead to the classical form of Joubert syndrome. *Am. J. Hum. Genet.* **83,** 170–179.

Caspary, T., Larkins, C. E., and Anderson, K. V. (2007). The graded response to sonic hedgehog depends on cilia architecture. *Dev. Cell* **12,** 767–778.

Chakrabarti, A., Schatten, H., Mitchell, K. D., Crosser, M., and Taylor, M. (1998). Chloral hydrate alters the organization of the ciliary basal apparatus and cell organelles in sea urchin embryos. *Cell Tissue Res.* **293,** 453–462.

Chang, B., Khanna, H., Hawes, N., Jimeno, D., He, S., Lillo, C., Parapuram, S. K., Cheng, H., Scott, A., Hurd, R. E., Sayer, J. A., Otto, E. A., *et al.* (2006). In-frame deletion in a novel centrosomal/ciliary protein CEP290/NPHP6 perturbs its interaction with RPGR and results in early-onset retinal degeneration in the rd16 mouse. *Hum. Mol. Genet.* **15,** 1847–1857.

Chen, N., Mah, A., Blacque, O. E., Chu, J., Phgora, K., Bakhoum, M. W., Newbury, C. R., Khattra, J., Chan, S., Go, A., Efimenko, E., Johnsen, R., *et al.* (2006). Identification of ciliary and ciliopathy genes in *Caenorhabditis elegans* through comparative genomics. *Genome Biol.* **7,** R126.

Chiang, A. P., Beck, J. S., Yen, H. J., Tayeh, M. K., Scheetz, T. E., Swiderski, R. E., Nishimura, D. Y., Braun, T. A., Kim, K. Y., Huang, J., Elbedour, K., Carmi, R., *et al.* (2006). Homozygosity mapping with SNP arrays identifies TRIM32, an E3 ubiquitin ligase, as a Bardet–Biedl syndrome gene (BBS11). *Proc. Natl. Acad. Sci. USA* **103,** 6287–6292.

Chizhikov, V. V., Davenport, J., Zhang, Q., Shih, E. K., Cabello, O. A., Fuchs, J. L., Yoder, B. K., and Millen, K. J. (2007). Cilia proteins control cerebellar morphogenesis by promoting expansion of the granule progenitor pool. *J. Neurosci.* **27,** 9780–9789.

Cole, D. G. (2003). The intraflagellar transport machinery of *Chlamydomonas reinhardtii*. *Traffic* **4,** 435–442.

Coll, E., Torra, R., Pascual, J., Botey, A., Ara, J., Perez, L., Ballesta, F., and Darnell, A. (1997). Sporadic orofaciodigital syndrome type I presenting as end-stage renal disease. *Nephrol. Dial. Transplant.* **12,** 1040–1042.

Connacher, A. A., Forsyth, C. C., and Stewart, W. K. (1987). Orofaciodigital syndrome type I associated with polycystic kidneys and agenesis of the corpus callosum. *J. Med. Genet.* **24,** 116–118.

Corbit, K. C., Shyer, A. E., Dowdle, W. E., Gaulden, J., Singla, V., Chen, M. H., Chuang, P. T., and Reiter, J. F. (2008). Kif3a constrains beta-catenin-dependent Wnt signalling through dual ciliary and non-ciliary mechanisms. *Nat. Cell Biol.* **10,** 70–76.

Dalen, H., Schlapfer, W. T., and Mamoon, A. (1971). Cilia on cultured ependymal cells examined by scanning electron microscopy. *Exp. Cell Res.* **67,** 375–379.

Davenport, J. R., Watts, A. J., Roper, V. C., Croyle, M. J., van Groen, T., Wyss, J. M., Nagy, T. R., Kesterson, R. A., and Yoder, B. K. (2007). Disruption of intraflagellar transport in adult mice leads to obesity and slow-onset cystic kidney disease. *Curr. Biol.* **17,** 1586–1594.

Davy, B. E., and Robinson, M. L. (2003). Congenital hydrocephalus in hy3 mice is caused by a frameshift mutation in Hydin, a large novel gene. *Hum. Mol. Genet.* **12,** 1163–1170.

Dawe, H. R., Smith, U. M., Cullinane, A. R., Gerrelli, D., Cox, P., Badano, J. L., Blair-Reid, S., Sriram, N., Katsanis, N., Attie-Bitach, T., Afford, S. C., Copp, A. J., *et al.* (2007). The Meckel–Gruber Syndrome proteins MKS1 and meckelin interact and are required for primary cilium formation. *Hum. Mol. Genet.* **16,** 173–186.

Dentler, W. (2005). Intraflagellar transport (IFT) during assembly and disassembly of *Chlamydomonas* flagella. *J. Cell Biol.* **170,** 649–659.

Dina, C., Meyre, D., Gallina, S., Durand, E., Korner, A., Jacobson, P., Carlsson, L. M., Kiess, W., Vatin, V., Lecoeur, C., Delplanque, J., Vaillant, E., *et al.* (2007). Variation in FTO contributes to childhood obesity and severe adult obesity. *Nat. Genet.* **39,** 724–726.

Doege, T. C., Thuline, H. C., Priest, J. H., Norby, D. E., and Bryant, J. S. (1964). Studies of a family with the oral-facial-digital syndrome. *N. Engl. J. Med.* **271,** 1073–1078.

Donnez, J., Casanas-Roux, F., Caprasse, J., Ferin, J., and Thomas, K. (1985). Cyclic changes in ciliation, cell height, and mitotic activity in human tubal epithelium during reproductive life. *Fertil. Steril.* **43,** 554–559.

Efimenko, E., Bubb, K., Mak, H. Y., Holzman, T., Leroux, M. R., Ruvkun, G., Thomas, J. H., and Swoboda, P. (2005). Analysis of xbx genes in *C. elegans*. *Development* **132,** 1923–1934.

Eley, L., Yates, L. M., and Goodship, J. A. (2005). Cilia and disease. *Curr. Opin. Genet. Dev.* **15,** 308–314.

Essner, J. J., Vogan, K. J., Wagner, M. K., Tabin, C. J., Yost, H. J., and Brueckner, M. (2002). Conserved function for embryonic nodal cilia. *Nature* **418,** 37–38.

Evans, J. E., Snow, J. J., Gunnarson, A. L., Ou, G., Stahlberg, H., McDonald, K. L., and Scholey, J. M. (2006). Functional modulation of IFT kinesins extends the sensory repertoire of ciliated neurons in *Caenorhabditis elegans*. *J. Cell Biol.* **172,** 663–669.

Fanto, M., and McNeill, H. (2004). Planar polarity from flies to vertebrates. *J. Cell Sci.* **117,** 527–533.

Fath, M. A., Mullins, R. F., Searby, C., Nishimura, D. Y., Wei, J., Rahmouni, K., Davis, R. E., Tayeh, M. K., Andrews, M., Yang, B., Sigmund, C. D., Stone, E. M., *et al.* (2005). Mkks-null mice have a phenotype resembling Bardet–Biedl syndrome. *Hum. Mol. Genet.* **14,** 1109–1118.

Feather, S. A., Woolf, A. S., Donnai, D., Malcolm, S., and Winter, R. M. (1997). The oral-facial-digital syndrome type 1 (OFD1), a cause of polycystic kidney disease and associated malformations, maps to Xp22.2–Xp22.3. *Hum. Mol. Genet.* **6,** 1163–1167.

Ferrante, M. I., Zullo, A., Barra, A., Bimonte, S., Messaddeq, N., Studer, M., Dolle, P., and Franco, B. (2006). Oral-facial-digital type I protein is required for primary cilia formation and left–right axis specification. *Nat. Genet.* **38,** 112–117.

Fick, G. M., Johnson, A. M., Hammond, W. S., and Gabow, P. A. (1995). Causes of death in autosomal dominant polycystic kidney disease. *J. Am. Soc. Nephrol.* **5,** 2048–2056.

Fischer, E., and Pontoglio, M. (2006). Planar cell polarity and polycystic kidney disease. *Med. Sci. (Paris)* **22,** 576–578.

Fischer, E., Legue, E., Doyen, A., Nato, F., Nicolas, J. F., Torres, V., Yaniv, M., and Pontoglio, M. (2006). Defective planar cell polarity in polycystic kidney disease. *Nat. Genet.* **38,** 21–23.

Frank, V., den Hollander, A. I., Bruchle, N. O., Zonneveld, M. N., Nurnberg, G., Becker, C., Du Bois, G., Kendziorra, H., Roosing, S., Senderek, J., Nurnberg, P., Cremers, F. P., *et al.* (2008). Mutations of the CEP290 gene encoding a centrosomal protein cause Meckel–Gruber syndrome. *Hum. Mutat.* **29,** 45–52.

Frayling, T. M. (2007). Genome-wide association studies provide new insights into type 2 diabetes aetiology. *Nat. Rev. Genet.* **8,** 657–662.

Geng, L., Segal, Y., Pavlova, A., Barros, E. J., Lohning, C., Lu, W., Nigam, S. K., Frischauf, A. M., Reeders, S. T., and Zhou, J. (1997). Distribution and developmentally regulated expression of murine polycystin. *Am. J. Physiol.* **272,** F451–F459.

Gerdes, J. M., Liu, Y., Zaghloul, N. A., Leitch, C. C., Lawson, S. S., Kato, M., Beachy, P. A., Beales, P. L., DeMartino, G. N., Fisher, S., Badano, J. L., and Katsanis, N. (2007). Disruption of the basal body compromises proteasomal function and perturbs intracellular Wnt response. *Nat. Genet.* **39,** 1350–1360.

Gradilone, S. A., Masyuk, A. I., Splinter, P. L., Banales, J. M., Huang, B. Q., Tietz, P. S., Masyuk, T. V., and Larusso, N. F. (2007). Cholangiocyte cilia express TRPV4 and detect changes in luminal tonicity inducing bicarbonate secretion. *Proc. Natl. Acad. Sci. USA* **104,** 19138–19143.

Graser, S., Stierhof, Y. D., Lavoie, S. B., Gassner, O. S., Lamla, S., Le Clech, M., and Nigg, E. A. (2007). Cep164, a novel centriole appendage protein required for primary cilium formation. *J. Cell Biol.* **179,** 321–330.

Guillaume, R., and Trudel, M. (2000). Distinct and common developmental expression patterns of the murine Pkd2 and Pkd1 genes. *Mech. Dev.* **93,** 179–183.

Hamon, M., Doucet, E., Lefevre, K., Miquel, M. C., Lanfumey, L., Insausti, R., Frechilla, D., Del Rio, J., and Verge, D. (1999). Antibodies and antisense oligonucleotide for probing the distribution and putative functions of central 5-HT6 receptors. *Neuropsychopharmacology* **21,** 68S–76S.

Handel, M., Schulz, S., Stanarius, A., Schreff, M., Erdtmann-Vourliotis, M., Schmidt, H., Wolf, G., and Hollt, V. (1999). Selective targeting of somatostatin receptor 3 to neuronal cilia. *Neuroscience* **89,** 909–926.

Haycraft, C. J., Swoboda, P., Taulman, P. D., Thomas, J. H., and Yoder, B. K. (2001). The C. elegans homolog of the murine cystic kidney disease gene Tg737 functions in a ciliogenic pathway and is disrupted in osm-5 mutant worms. *Development* **128,** 1493–1505.

Haycraft, C. J., Banizs, B., Aydin-Son, Y., Zhang, Q., Michaud, E. J., and Yoder, B. K. (2005). Gli2 and Gli3 localize to cilia and require the intraflagellar transport protein polaris for processing and function. *PLoS Genet.* **1,** e53.

Haycraft, C. J., Zhang, Q., Song, B., Jackson, W. S., Detloff, P. J., Serra, R., and Yoder, B. K. (2007). Intraflagellar transport is essential for endochondral bone formation. *Development* **134,** 307–316.

Hearn, T., Spalluto, C., Phillips, V. J., Renforth, G. L., Copin, N., Hanley, N. A., and Wilson, D. I. (2005). Subcellular localization of ALMS1 supports involvement of centrosome and basal body dysfunction in the pathogenesis of obesity, insulin resistance, and type 2 diabetes. *Diabetes* **54,** 1581–1587.

Hildebrandt, F., and Zhou, W. (2007). Nephronophthisis-associated ciliopathies. *J. Am. Soc. Nephrol.* **18,** 1855–1871.

Hilfinger, A., and Julicher, F. (2008). The chirality of ciliary beats. *Phys. Biol.* **5,** 16003.

Ho, P. T., and Tucker, R. W. (1989). Centriole ciliation and cell cycle variability during G1 phase of BALB/c 3T3 cells. *J. Cell Physiol.* **139,** 398–406.

Hollyfield, J. G., Besharse, J. C., and Rayborn, M. E. (1977). Turnover of rod photoreceptor outer segments. I. Membrane addition and loss in relationship to temperature. *J. Cell Biol.* **75,** 490–506.

Hornef, N., Olbrich, H., Horvath, J., Zariwala, M. A., Fliegauf, M., Loges, N. T., Wildhaber, J., Noone, P. G., Kennedy, M., Antonarakis, S. E., Blouin, J. L., Bartoloni, L., et al. (2006). DNAH5 mutations are a common cause of primary ciliary dyskinesia with outer dynein arm defects. *Am. J. Respir. Crit. Care Med.* **174,** 120–126.

Hou, X., Mrug, M., Yoder, B. K., Lefkowitz, E. J., Kremmidiotis, G., D'Eustachio, P., Beier, D. R., and Guay-Woodford, L. M. (2002). Cystin, a novel cilia-associated protein, is disrupted in the cpk mouse model of polycystic kidney disease. *J. Clin. Invest.* **109,** 533–540.

Hou, Y., Qin, H., Follit, J. A., Pazour, G. J., Rosenbaum, J. L., and Witman, G. B. (2007). Functional analysis of an individual IFT protein: IFT46 is required for transport of outer dynein arms into flagella. *J. Cell Biol.* **176**, 653–665.

Housset, C. (2005). Cystic liver diseases. Genetics and cell biology. *Gastroenterol. Clin. Biol.* **29**, 861–869.

Huang, K., Kunkel, T., and Beck, C. F. (2004). Localization of the blue-light receptor phototropin to the flagella of the green alga Chlamydomonas reinhardtii. *Mol. Biol. Cell* **15**, 3605–3614.

Huang, K., Diener, D. R., Mitchell, A., Pazour, G. J., Witman, G. B., and Rosenbaum, J. L. (2007). Function and dynamics of PKD2 in *Chlamydomonas reinhardtii* flagella. *J. Cell Biol.* **179**, 501–514.

Huangfu, D., and Anderson, K. V. (2005). Cilia and Hedgehog responsiveness in the mouse. *Proc. Natl. Acad. Sci. USA* **102**, 11325–11330.

Huangfu, D., Liu, A., Rakeman, A. S., Murcia, N. S., Niswander, L., and Anderson, K. V. (2003). Hedgehog signalling in the mouse requires intraflagellar transport proteins. *Nature* **426**, 83–87.

Hughes, J., Ward, C. J., Peral, B., Aspinwall, R., Clark, K., San Millan, J. L., Gamble, V., and Harris, P. C. (1995). The polycystic kidney disease 1 (PKD1) gene encodes a novel protein with multiple cell recognition domains. *Nat. Genet.* **10**, 151–160.

Ibanez-Tallon, I., Gorokhova, S., and Heintz, N. (2002). Loss of function of axonemal dynein Mdnah5 causes primary ciliary dyskinesia and hydrocephalus. *Hum. Mol. Genet.* **11**, 715–721.

Ibanez-Tallon, I., Pagenstecher, A., Fliegauf, M., Olbrich, H., Kispert, A., Ketelsen, U. P., North, A., Heintz, N., and Omran, H. (2004). Dysfunction of axonemal dynein heavy chain Mdnah5 inhibits ependymal flow and reveals a novel mechanism for hydrocephalus formation. *Hum. Mol. Genet.* **13**, 2133–2141.

Inaba, K. (2003). Molecular architecture of the sperm flagella: Molecules for motility and signaling. *Zool. Sci.* **20**, 1043–1056.

Inglis, P. N., Boroevich, K. A., and Leroux, M. R. (2006). Piecing together a ciliome. *Trends Genet.* **22**, 491–500.

Insinna, C., and Besharse, J. C. (2008). Intraflagellar transport and the sensory outer segment of vertebrate photoreceptors. *Dev. Dyn.* **237**, 1982–1992.

Jauregui, A. R., and Barr, M. M. (2005). Functional characterization of the *C. elegans* nephrocystins NPHP-1 and NPHP-4 and their role in cilia and male sensory behaviors. *Exp. Cell Res.* **305**, 333–342.

Jiang, S. T., Chiou, Y. Y., Wang, E., Lin, H. K., Lee, S. P., Lu, H. Y., Wang, C. K., Tang, M. J., and Li, H. (2008). Targeted disruption of Nphp1 causes male infertility due to defects in the later steps of sperm morphogenesis in mice. *Hum. Mol. Genet.* **17**, 3368–3379.

Jones, C., Roper, V. C., Foucher, I., Qian, D., Banizs, B., Petit, C., Yoder, B. K., and Chen, P. (2008). Ciliary proteins link basal body polarization to planar cell polarity regulation. *Nat. Genet.* **40**, 69–77.

Josef, K., Saranak, J., and Foster, K. W. (2005). An electro-optic monitor of the behavior of *Chlamydomonas reinhardtii* cilia. *Cell Motil. Cytoskeleton* **61**, 83–96.

Kamiya, R. (2002). Functional diversity of axonemal dyneins as studied in *Chlamydomonas* mutants. *Int. Rev. Cytol.* **219**, 115–155.

Kaplan, B. S., Fay, J., Shah, V., Dillon, M. J., and Barratt, T. M. (1989). Autosomal recessive polycystic kidney disease. *Pediatr. Nephrol.* **3**, 43–49.

Karcher, C., Fischer, A., Schweickert, A., Bitzer, E., Horie, S., Witzgall, R., and Blum, M. (2005). Lack of a laterality phenotype in Pkd1 knock-out embryos correlates with absence of polycystin-1 in nodal cilia. *Differentiation* **73**, 425–432.

Katsanis, N., Ansley, S. J., Badano, J. L., Eichers, E. R., Lewis, R. A., Hoskins, B. E., Scambler, P. J., Davidson, W. S., Beales, P. L., and Lupski, J. R. (2001). Triallelic inheritance in Bardet–Biedl syndrome, a Mendelian recessive disorder. *Science* **293**, 2256–2259.

Keller, L. C., Romijn, E. P., Zamora, I., Yates, J. R., III, and Marshall, W. F. (2005). Proteomic analysis of isolated *Chlamydomonas* centrioles reveals orthologs of ciliary-disease genes. *Curr. Biol.* **15**, 1090–1098.

Kennedy, J. R., Jr., and Brittingham, E. (1968). Fine structure changes during chloral hydrate deciliation of *Paramecium caudatum*. *J. Ultrastruct. Res.* **22**, 530–545.

Kim, K., Drummond, I., Ibraghimov-Beskrovnaya, O., Klinger, K., and Arnaout, M. A. (2000). Polycystin 1 is required for the structural integrity of blood vessels. *Proc. Natl. Acad. Sci. USA* **97**, 1731–1736.

Kim, J. C., Ou, Y. Y., Badano, J. L., Esmail, M. A., Leitch, C. C., Fiedrich, E., Beales, P. L., Archibald, J. M., Katsanis, N., Rattner, J. B., and Leroux, M. R. (2005). MKKS/BBS6, a divergent chaperonin-like protein linked to the obesity disorder Bardet–Biedl syndrome, is a novel centrosomal component required for cytokinesis. *J. Cell Sci.* **118**, 1007–1020.

Kokkotou, E. G., Tritos, N. A., Mastaitis, J. W., Slieker, L., and Maratos-Flier, E. (2001). Melanin-concentrating hormone receptor is a target of leptin action in the mouse brain. *Endocrinology* **142**, 680–686.

Kosaki, K., Ikeda, K., Miyakoshi, K., Ueno, M., Kosaki, R., Takahashi, D., Tanaka, M., Torikata, C., Yoshimura, Y., and Takahashi, T. (2004). Absent inner dynein arms in a fetus with familial hydrocephalus-situs abnormality. *Am. J. Med. Genet. A* **129A**, 308–311.

Kozminski, K. G., Johnson, K. A., Forscher, P., and Rosenbaum, J. L. (1993). A motility in the eukaryotic flagellum unrelated to flagellar beating. *Proc. Natl. Acad. Sci. USA* **90**, 5519–5523.

Kozminski, K. G., Beech, P. L., and Rosenbaum, J. L. (1995). The *Chlamydomonas* kinesin-like protein FLA10 is involved in motility associated with the flagellar membrane. *J. Cell Biol.* **131**, 1517–1527.

Kozminski, K. G., Forscher, P., and Rosenbaum, J. L. (1998). Three flagellar motilities in *Chlamydomonas* unrelated to flagellar beating. Video supplement. *Cell Motil. Cytoskeleton* **39**, 347–348.

Kramer-Zucker, A. G., Olale, F., Haycraft, C. J., Yoder, B. K., Schier, A. F., and Drummond, I. A. (2005). Cilia-driven fluid flow in the zebrafish pronephros, brain and Kupffer's vesicle is required for normal organogenesis. *Development* **132**, 1907–1921.

Krock, B. L., and Perkins, B. D. (2008). The intraflagellar transport protein IFT57 is required for cilia maintenance and regulates IFT-particle-kinesin-II dissociation in vertebrate photoreceptors. *J. Cell Sci.* **121**, 1907–1915.

Kulaga, H. M., Leitch, C. C., Eichers, E. R., Badano, J. L., Lesemann, A., Hoskins, B. E., Lupski, J. R., Beales, P. L., Reed, R. R., and Katsanis, N. (2004). Loss of BBS proteins causes anosmia in humans and defects in olfactory cilia structure and function in the mouse. *Nat. Genet.* **36**, 994–998.

Kyttala, M., Tallila, J., Salonen, R., Kopra, O., Kohlschmidt, N., Paavola-Sakki, P., Peltonen, L., and Kestila, M. (2006). MKS1, encoding a component of the flagellar apparatus basal body proteome, is mutated in Meckel syndrome. *Nat. Genet.* **38**, 155–157.

Lansley, A. B., Sanderson, M. J., and Dirksen, E. R. (1992). Control of the beat cycle of respiratory tract cilia by Ca2+ and cAMP. *Am. J. Physiol.* **263**, L232–L242.

Lantinga-van Leeuwen, I. S., Leonhard, W. N., van der Wal, A., Breuning, M. H., de Heer, E., and Peters, D. J. (2007). Kidney-specific inactivation of the Pkd1 gene induces rapid cyst formation in developing kidneys and a slow onset of disease in adult mice. *Hum. Mol. Genet.* **16**, 3188–3196.

Layton, W. M., Jr. (1976). Random determination of a developmental process: Reversal of normal visceral asymmetry in the mouse. *J. Hered.* **67**, 336–338.

Lechtreck, K. F., and Witman, G. B. (2007). *Chlamydomonas reinhardtii* hydin is a central pair protein required for flagellar motility. *J. Cell Biol.* **176**, 473–482.

Lechtreck, K. F., Delmotte, P., Robinson, M. L., Sanderson, M. J., and Witman, G. B. (2008). Mutations in hydin impair ciliary motility in mice. *J. Cell Biol.* **180**, 633–643.

Ledford, A. W., Brantley, J. G., Kemeny, G., Foreman, T. L., Quaggin, S. E., Igarashi, P., Oberhaus, S. M., Rodova, M., Calvet, J. P., and Vanden Heuvel, G. B. (2002). Deregulated expression of the homeobox gene Cux-1 in transgenic mice results in downregulation of p27(kip1) expression during nephrogenesis, glomerular abnormalities, and multiorgan hyperplasia. *Dev. Biol.* **245**, 157–171.

Lee, L., Campagna, D. R., Pinkus, J. L., Mulhern, H., Wyatt, T. A., Sisson, J. H., Pavlik, J. A., Pinkus, G. S., and Fleming, M. D. (2008). Primary ciliary dyskinesia in mice lacking the novel ciliary protein Pcdp1. *Mol. Cell. Biol.* **28**, 949–957.

Lehman, J. M., Michaud, E. J., Schoeb, T. R., Aydin-Son, Y., Miller, M., and Yoder, B. K. (2008). The oak ridge polycystic kidney mouse: Modeling ciliopathies of mice and men. *Dev. Dyn.* **237**, 1960–1971.

Leitch, C. C., Zaghloul, N. A., Davis, E. E., Stoetzel, C., Diaz-Font, A., Rix, S., Alfadhel, M., Lewis, R. A., Eyaid, W., Banin, E., Dollfus, H., Beales, P. L., et al. (2008). Hypomorphic mutations in syndromic encephalocele genes are associated with Bardet–Biedl syndrome. *Nat. Genet.* **40**, 443–448.

Li, J. B., Gerdes, J. M., Haycraft, C. J., Fan, Y., Teslovich, T. M., May-Simera, H., Li, H., Blacque, O. E., Li, L., Leitch, C. C., Lewis, R. A., Green, J. S., et al. (2004). Comparative genomics identifies a flagellar and basal body proteome that includes the BBS5 human disease gene. *Cell* **117**, 541–552.

Li, G., Vega, R., Nelms, K., Gekakis, N., Goodnow, C., McNamara, P., Wu, H., Hong, N. A., and Glynne, R. (2007). A role for Alstrom syndrome protein, alms1, in kidney ciliogenesis and cellular quiescence. *PLoS Genet.* **3**, e8.

Lie, H., and Ferkol, T. (2007). Primary ciliary dyskinesia: recent advances in pathogenesis, diagnosis and treatment. *Drugs* **67**, 1883–1892.

Lin, F., Hiesberger, T., Cordes, K., Sinclair, A. M., Goldstein, L. S., Somlo, S., and Igarashi, P. (2003). Kidney-specific inactivation of the KIF3A subunit of kinesin-II inhibits renal ciliogenesis and produces polycystic kidney disease. *Proc. Natl. Acad. Sci. USA* **100**, 5286–5291.

Liu, A., Wang, B., and Niswander, L. A. (2005a). Mouse intraflagellar transport proteins regulate both the activator and repressor functions of Gli transcription factors. *Development* **132**, 3103–3111.

Liu, W., Murcia, N. S., Duan, Y., Weinbaum, S., Yoder, B. K., Schwiebert, E., and Satlin, L. M. (2005b). Mechanoregulation of intracellular Ca2+ concentration is attenuated in collecting duct of monocilium-impaired orpk mice. *Am. J. Physiol. Renal Physiol.* **289**, F978–F988.

Lyons, R. A., Saridogan, E., and Djahanbakhch, O. (2006). The reproductive significance of human fallopian tube cilia. *Hum. Reprod. Update* **12**, 363–372.

Marszalek, J. R., Ruiz-Lozano, P., Roberts, E., Chien, K. R., and Goldstein, L. S. (1999). *Situs inversus* and embryonic ciliary morphogenesis defects in mouse mutants lacking the KIF3A subunit of kinesin-II. *Proc. Natl. Acad. Sci. USA* **96**, 5043–5048.

Masyuk, A. I., Gradilone, S. A., Banales, J. M., Huang, B. Q., Masyuk, T. V., Lee, S. O., Splinter, P. L., Stroope, A. J., and Larusso, N. F. (2008). Cholangiocyte primary cilia are chemosensory organelles that detect biliary nucleotides via P2Y12 purinergic receptors. *Am. J. Physiol. Gastrointest. Liver Physiol.* **295**, G725–G734.

May, S. R., Ashique, A. M., Karlen, M., Wang, B., Shen, Y., Zarbalis, K., Reiter, J., Ericson, J., and Peterson, A. S. (2005). Loss of the retrograde motor for IFT disrupts localization of Smo to cilia and prevents the expression of both activator and repressor functions of Gli. *Dev. Biol.* **287**, 378–389.

McGrath, J., Somlo, S., Makova, S., Tian, X., and Brueckner, M. (2003). Two populations of node monocilia initiate left–right asymmetry in the mouse. *Cell* **114**, 61–73.

Menini, A. (1999). Calcium signalling and regulation in olfactory neurons. *Curr. Opin. Neurobiol.* **9**, 419–426.

Mikule, K., Delaval, B., Kaldis, P., Jurcyzk, A., Hergert, P., and Doxsey, S. (2007). Loss of centrosome integrity induces p38-p53-p21-dependent G1-S arrest. *Nat. Cell Biol.* **9**, 160–170.

Mitchell, D. R. (2007). The evolution of eukaryotic cilia and flagella as motile and sensory organelles. *Adv. Exp. Med. Biol.* **607**, 130–140.

Mitchell, D. R., and Kang, Y. (1991). Identification of oda6 as a *Chlamydomonas* dynein mutant by rescue with the wild-type gene. *J. Cell Biol.* **113**, 835–842.

Mochizuki, T., Wu, G., Hayashi, T., Xenophontos, S. L., Veldhuisen, B., Saris, J. J., Reynolds, D. M., Cai, Y., Gabow, P. A., Pierides, A., Kimberling, W. J., Breuning, M. H., et al. (1996). PKD2, a gene for polycystic kidney disease that encodes an integral membrane protein. *Science* **272**, 1339–1342.

Mokrzan, E. M., Lewis, J. S., and Mykytyn, K. (2007). Differences in renal tubule primary cilia length in a mouse model of Bardet–Biedl syndrome. *Nephron Exp. Nephrol.* **106**, e88–e96.

Mollet, G., Salomon, R., Gribouval, O., Silbermann, F., Bacq, D., Landthaler, G., Milford, D., Nayir, A., Rizzoni, G., Antignac, C., and Saunier, S. (2002). The gene mutated in juvenile nephronophthisis type 4 encodes a novel protein that interacts with nephrocystin. *Nat. Genet.* **32**, 300–305.

Morgan, D., Turnpenny, L., Goodship, J., Dai, W., Majumder, K., Matthews, L., Gardner, A., Schuster, G., Vien, L., Harrison, W., Elder, F. F., Penman-Splitt, M., et al. (1998). Inversin, a novel gene in the vertebrate left–right axis pathway, is partially deleted in the inv mouse. *Nat. Genet.* **20**, 149–156.

Moyer, J. H., Lee-Tischler, M. J., Kwon, H. Y., Schrick, J. J., Avner, E. D., Sweeney, W. E., Godfrey, V. L., Cacheiro, N. L., Wilkinson, J. E., and Woychik, R. P. (1994). Candidate gene associated with a mutation causing recessive polycystic kidney disease in mice. *Science* **264**, 1329–1333.

Murcia, N. S., Richards, W. G., Yoder, B. K., Mucenski, M. L., Dunlap, J. R., and Woychik, R. P. (2000). The Oak Ridge Polycystic Kidney (orpk) disease gene is required for left–right axis determination. *Development* **127**, 2347–2355.

Mykytyn, K., Mullins, R. F., Andrews, M., Chiang, A. P., Swiderski, R. E., Yang, B., Braun, T., Casavant, T., Stone, E. M., and Sheffield, V. C. (2004). Bardet–Biedl syndrome type 4 (BBS4)-null mice implicate Bbs4 in flagella formation but not global cilia assembly. *Proc. Natl. Acad. Sci. USA* **101**, 8664–8669.

Nachury, M. V. (2008). Tandem affinity purification of the BBSome, a critical regulator of Rab8 in ciliogenesis. *Methods Enzymol.* **439**, 501–513.

Nachury, M. V., Loktev, A. V., Zhang, Q., Westlake, C. J., Peranen, J., Merdes, A., Slusarski, D. C., Scheller, R. H., Bazan, J. F., Sheffield, V. C., and Jackson, P. K. (2007). A core complex of BBS proteins cooperates with the GTPase Rab8 to promote ciliary membrane biogenesis. *Cell* **129**, 1201–1213.

Nauli, S. M., Alenghat, F. J., Luo, Y., Williams, E., Vassilev, P., Li, X., Elia, A. E., Lu, W., Brown, E. M., Quinn, S. J., Ingber, D. E., and Zhou, J. (2003). Polycystins 1 and 2 mediate mechanosensation in the primary cilium of kidney cells. *Nat. Genet.* **33**, 129–137.

Nauli, S. M., Rossetti, S., Kolb, R. J., Alenghat, F. J., Consugar, M. B., Harris, P. C., Ingber, D. E., Loghman-Adham, M., and Zhou, J. (2006). Loss of polycystin-1 in human cyst-lining epithelia leads to ciliary dysfunction. *J. Am. Soc. Nephrol.* **17**, 1015–1025.

Nauta, J., Sweeney, W. E., Rutledge, J. C., and Avner, E. D. (1995). Biliary epithelial cells from mice with congenital polycystic kidney disease are hyperresponsive to epidermal growth factor. *Pediatr. Res.* **37**, 755–763.

Nishimura, D. Y., Fath, M., Mullins, R. F., Searby, C., Andrews, M., Davis, R., Andorf, J. L., Mykytyn, K., Swiderski, R. E., Yang, B., Carmi, R., Stone, E. M., et al. (2004). Bbs2-null mice have neurosensory deficits, a defect in social dominance, and retinopathy associated with mislocalization of rhodopsin. *Proc. Natl. Acad. Sci. USA* **101**, 16588–16593.

Nonaka, S., Tanaka, Y., Okada, Y., Takeda, S., Harada, A., Kanai, Y., Kido, M., and Hirokawa, N. (1998). Randomization of left–right asymmetry due to loss of nodal cilia generating leftward flow of extraembryonic fluid in mice lacking KIF3B motor protein. *Cell* **95**, 829–837.

Nonaka, S., Shiratori, H., Saijoh, Y., and Hamada, H. (2002). Determination of left–right patterning of the mouse embryo by artificial nodal flow. *Nature* **418**, 96–99.

Ocbina, P. J., and Anderson, K. V. (2008). Intraflagellar transport, cilia, and mammalian Hedgehog signaling: Analysis in mouse embryonic fibroblasts. *Dev. Dyn.* **237**, 2030–2038.

Okada, Y., Takeda, S., Tanaka, Y., Belmonte, J. C., and Hirokawa, N. (2005). Mechanism of nodal flow: A conserved symmetry breaking event in left–right axis determination. *Cell* **121**, 633–644.

Olbrich, H., Haffner, K., Kispert, A., Volkel, A., Volz, A., Sasmaz, G., Reinhardt, R., Hennig, S., Lehrach, H., Konietzko, N., Zariwala, M., Noone, P. G., et al. (2002). Mutations in DNAH5 cause primary ciliary dyskinesia and randomization of left–right asymmetry. *Nat. Genet.* **30**, 143–144.

Olsen, B. (2005). Nearly all cells in vertebrates and many cells in invertebrates contain primary cilia. *Matrix Biol.* **24**, 449–450.

Ong, A. C., Ward, C. J., Butler, R. J., Biddolph, S., Bowker, C., Torra, R., Pei, Y., and Harris, P. C. (1999). Coordinate expression of the autosomal dominant polycystic kidney disease proteins, polycystin-2 and polycystin-1, in normal and cystic tissue. *Am. J. Pathol.* **154**, 1721–1729.

Onuchic, L. F., Furu, L., Nagasawa, Y., Hou, X., Eggermann, T., Ren, Z., Bergmann, C., Senderek, J., Esquivel, E., Zeltner, R., Rudnik-Schoneborn, S., Mrug, M., et al. (2002). PKHD1, the polycystic kidney and hepatic disease 1 gene, encodes a novel large protein containing multiple immunoglobulin-like plexin-transcription-factor domains and parallel beta-helix 1 repeats. *Am. J. Hum. Genet.* **70**, 1305–1317.

Oro, A. E., Higgins, K. M., Hu, Z., Bonifas, J. M., Epstein, E. H., Jr., and Scott, M. P. (1997). Basal cell carcinomas in mice overexpressing sonic hedgehog. *Science* **276**, 817–821.

Ostrowski, L. E., Blackburn, K., Radde, K. M., Moyer, M. B., Schlatzer, D. M., Moseley, A., and Boucher, R. C. (2002). A proteomic analysis of human cilia: Identification of novel components. *Mol. Cell. Proteomics* **1**, 451–465.

Ozgul, R. K., Satman, I., Collin, G. B., Hinman, E. G., Marshall, J. D., Kocaman, O., Tutuncu, Y., Yilmaz, T., and Naggert, J. K. (2007). Molecular analysis and long-term clinical evaluation of three siblings with Alstrom syndrome. *Clin. Genet.* **72**, 351–356.

Pan, J., and Snell, W. J. (2002). Kinesin-II is required for flagellar sensory transduction during fertilization in Chlamydomonas. *Mol. Biol. Cell* **13**, 1417–1426.

Pan, J., and Snell, W. J. (2005). Chlamydomonas shortens its flagella by activating axonemal disassembly, stimulating IFT particle trafficking, and blocking anterograde cargo loading. *Dev. Cell* **9**, 431–438.

Pan, X., Ou, G., Civelekoglu-Scholey, G., Blacque, O. E., Endres, N. F., Tao, L., Mogilner, A., Leroux, M. R., Vale, R. D., and Scholey, J. M. (2006). Mechanism of transport of IFT particles in C. elegans cilia by the concerted action of kinesin-II and OSM-3 motors. *J. Cell Biol.* **174**, 1035–1045.

Parisi, M. A., Doherty, D., Chance, P. F., and Glass, I. A. (2007). Joubert syndrome (and related disorders) (OMIM 213300). *Eur. J. Hum. Genet.* **15,** 511–521.

Patel, V., Li, L., Cobo-Stark, P., Shao, X., Somlo, S., Lin, F., and Igarashi, P. (2008). Acute kidney injury and aberrant planar cell polarity induce cyst formation in mice lacking renal cilia. *Hum. Mol. Genet.* **17,** 1578–1590.

Pazour, G. J. (2004). Comparative genomics: prediction of the ciliary and basal body proteome. *Curr. Biol.* **14,** R575–R577.

Pazour, G. J., Dickert, B. L., Vucica, Y., Seeley, E. S., Rosenbaum, J. L., Witman, G. B., and Cole, D. G. (2000). *Chlamydomonas* IFT88 and its mouse homologue, polycystic kidney disease gene tg737, are required for assembly of cilia and flagella. *J. Cell Biol.* **151,** 709–718.

Pazour, G. J., Baker, S. A., Deane, J. A., Cole, D. G., Dickert, B. L., Rosenbaum, J. L., Witman, G. B., and Besharse, J. C. (2002a). The intraflagellar transport protein, IFT88, is essential for vertebrate photoreceptor assembly and maintenance. *J. Cell Biol.* **157,** 103–113.

Pazour, G. J., San Agustin, J. T., Follit, J. A., Rosenbaum, J. L., and Witman, G. B. (2002b). Polycystin-2 localizes to kidney cilia and the ciliary level is elevated in orpk mice with polycystic kidney disease. *Curr. Biol.* **12,** R378–R380.

Pazour, G. J., Agrin, N., Leszyk, J., and Witman, G. B. (2005). Proteomic analysis of a eukaryotic cilium. *J. Cell Biol.* **170**(1)**,** 103–113.

Pedersen, H., and Mygind, N. (1976). Absence of axonemal arms in nasal mucosa cilia in Kartagener's syndrome. *Nature* **262,** 494–495.

Pedersen, H., and Rebbe, H. (1975). Absence of arms in the axoneme of immobile human spermatozoa. *Biol. Reprod.* **12,** 541–544.

Pedersen, L. B., Miller, M. S., Geimer, S., Leitch, J. M., Rosenbaum, J. L., and Cole, D. G. (2005). *Chlamydomonas* IFT172 is encoded by FLA11, interacts with CrEB1, and regulates IFT at the flagellar tip. *Curr. Biol.* **15,** 262–266.

Pennarun, G., Escudier, E., Chapelin, C., Bridoux, A. M., Cacheux, V., Roger, G., Clement, A., Goossens, M., Amselem, S., and Duriez, B. (1999). Loss-of-function mutations in a human gene related to *Chlamydomonas reinhardtii* dynein IC78 result in primary ciliary dyskinesia. *Am. J. Hum. Genet.* **65,** 1508–1519.

Piontek, K., Menezes, L. F., Garcia-Gonzalez, M. A., Huso, D. L., and Germino, G. G. (2007). A critical developmental switch defines the kinetics of kidney cyst formation after loss of Pkd1. *Nat. Med.* **13,** 1490–1495.

Pissios, P., Bradley, R. L., and Maratos-Flier, E. (2006). Expanding the scales: The multiple roles of MCH in regulating energy balance and other biological functions. *Endocr. Rev.* **27,** 606–620.

Plotnikova, O. V., Golemis, E. A., and Pugacheva, E. N. (2008). Cell cycle-dependent ciliogenesis and cancer. *Cancer Res.* **68,** 2058–2061.

Praetorius, H. A., and Spring, K. R. (2001). Bending the MDCK cell primary cilium increases intracellular calcium. *J. Membr. Biol.* **184,** 71–79.

Praetorius, H. A., and Spring, K. R. (2003). Removal of the MDCK cell primary cilium abolishes flow sensing. *J. Membr. Biol.* **191,** 69–76.

Pugacheva, E. N., Jablonski, S. A., Hartman, T. R., Henske, E. P., and Golemis, E. A. (2007). HEF1-dependent Aurora A activation induces disassembly of the primary cilium. *Cell* **129,** 1351–1363.

Qian, F., Germino, F. J., Cai, Y., Zhang, X., Somlo, S., and Germino, G. G. (1997). PKD1 interacts with PKD2 through a probable coiled-coil domain. *Nat. Genet.* **16,** 179–183.

Qian, F., Boletta, A., Bhunia, A. K., Xu, H., Liu, L., Ahrabi, A. K., Watnick, T. J., Zhou, F., and Germino, G. G. (2002). Cleavage of polycystin-1 requires the receptor for egg jelly domain and is disrupted by human autosomal-dominant polycystic kidney disease 1-associated mutations. *Proc. Natl. Acad. Sci. USA* **99,** 16981–16986.

Qian, C. N., Knol, J., Igarashi, P., Lin, F., Zylstra, U., Teh, B. T., and Williams, B. O. (2005). Cystic renal neoplasia following conditional inactivation of apc in mouse renal tubular epithelium. *J. Biol. Chem.* **280**, 3938–3945.

Qin, H., Rosenbaum, J. L., and Barr, M. M. (2001). An autosomal recessive polycystic kidney disease gene homolog is involved in intraflagellar transport in *C. elegans* ciliated sensory neurons. *Curr. Biol.* **11**, 457–461.

Quarmby, L. M., and Parker, J. D. (2005). Cilia and the cell cycle? *J. Cell Biol.* **169**, 707–710.

Robert, A., Margall-Ducos, G., Guidotti, J. E., Bregerie, O., Celati, C., Brechot, C., and Desdouets, C. (2007). The intraflagellar transport component IFT88/polaris is a centrosomal protein regulating G1-S transition in non-ciliated cells. *J. Cell Sci.* **120**, 628–637.

Roepman, R., Letteboer, S. J., Arts, H. H., van Beersum, S. E., Lu, X., Krieger, E., Ferreira, P. A., and Cremers, F. P. (2005). Interaction of nephrocystin-4 and RPGRIP1 is disrupted by nephronophthisis or Leber congenital amaurosis-associated mutations. *Proc. Natl. Acad. Sci. USA* **102**, 18520–18525.

Ross, A. J., May-Simera, H., Eichers, E. R., Kai, M., Hill, J., Jagger, D. J., Leitch, C. C., Chapple, J. P., Munro, P. M., Fisher, S., Tan, P. L., Phillips, H. M., *et al.* (2005). Disruption of Bardet–Biedl syndrome ciliary proteins perturbs planar cell polarity in vertebrates. *Nat. Genet.* **37**, 1135–1140.

Roth, K. E., Rieder, C. L., and Bowser, S. S. (1988). Flexible-substratum technique for viewing cells from the side: Some *in vivo* properties of primary (9 + 0) cilia in cultured kidney epithelia. *J. Cell Sci.* **89**(Pt. 4), 457–466.

Rutland, J., and de Iongh, R. U. (1990). Random ciliary orientation. A cause of respiratory tract disease. *N. Engl. J. Med.* **323**, 1681–1684.

Saadi-Kheddouci, S., Berrebi, D., Romagnolo, B., Cluzeaud, F., Peuchmaur, M., Kahn, A., Vandewalle, A., and Perret, C. (2001). Early development of polycystic kidney disease in transgenic mice expressing an activated mutant of the beta-catenin gene. *Oncogene* **20**, 5972–5981.

Saburi, S., Hester, I., Fischer, E., Pontoglio, M., Eremina, V., Gessler, M., Quaggin, S. E., Harrison, R., Mount, R., and McNeill, H. (2008). Loss of Fat4 disrupts PCP signaling and oriented cell division and leads to cystic kidney disease. *Nat. Genet.* **40**, 1010–1015.

Saccone, V., Palmieri, M., Passamano, L., Piluso, G., Meroni, G., Politano, L., and Nigro, V. (2008). Mutations that impair interaction properties of TRIM32 associated with limb-girdle muscular dystrophy 2H. *Hum. Mutat.* **29**, 240–247.

Salonen, R. (1984). The Meckel syndrome: Clinicopathological findings in 67 patients. *Am. J. Med. Genet.* **18**, 671–689.

Salonen, R., and Paavola, P. (1998). Meckel syndrome. *J. Med. Genet.* **35**, 497–501.

Sanzen, T., Harada, K., Yasoshima, M., Kawamura, Y., Ishibashi, M., and Nakanuma, Y. (2001). Polycystic kidney rat is a novel animal model of Caroli's disease associated with congenital hepatic fibrosis. *Am. J. Pathol.* **158**, 1605–1612.

Sapiro, R., Kostetskii, I., Olds-Clarke, P., Gerton, G. L., Radice, G. L., and Strauss, I. J. (2002). Male infertility, impaired sperm motility, and hydrocephalus in mice deficient in sperm-associated antigen 6. *Mol. Cell Biol.* **22**, 6298–6305.

Satir, P., and Christensen, S. T. (2008). Structure and function of mammalian cilia. *Histochem. Cell Biol.* **129**, 687–693.

Sawamoto, K., Wichterle, H., Gonzalez-Perez, O., Cholfin, J. A., Yamada, M., Spassky, N., Murcia, N. S., Garcia-Verdugo, J. M., Marin, O., Rubenstein, J. L., Tessier-Lavigne, M., Okano, H., *et al.* (2006). New neurons follow the flow of cerebrospinal fluid in the adult brain. *Science* **311**, 629–632.

Sayer, J. A., Otto, E. A., O'Toole, J. F., Nurnberg, G., Kennedy, M. A., Becker, C., Hennies, H. C., Helou, J., Attanasio, M., Fausett, B. V., Utsch, B., Khanna, H., *et al.*

(2006). The centrosomal protein nephrocystin-6 is mutated in Joubert syndrome and activates transcription factor ATF4. *Nat. Genet.* **38,** 674–681.
Schafer, P. W. (1969). Centrioles of a human cancer: intercellular order and intracellular disorder. *Science* **164,** 1300–1303.
Scholey, J. M. (2003). Intraflagellar transport. *Annu. Rev. Cell Dev. Biol.* **19,** 423–443.
Schwartz, E. A., Leonard, M. L., Bizios, R., and Bowser, S. S. (1997). Analysis and modeling of the primary cilium bending response to fluid shear. *Am. J. Physiol.* **272,** F132–F138.
Shah, A. S., Farmen, S. L., Moninger, T. O., Businga, T. R., Andrews, M. P., Bugge, K., Searby, C. C., Nishimura, D., Brogden, K. A., Kline, J. N., Sheffield, V. C., and Welsh, M. J. (2008). Loss of Bardet–Biedl syndrome proteins alters the morphology and function of motile cilia in airway epithelia. *Proc. Natl. Acad. Sci. USA* **105,** 3380–3385.
Sillitoe, R. V., and Joyner, A. L. (2007). Morphology, molecular codes, and circuitry produce the three-dimensional complexity of the cerebellum. *Annu. Rev. Cell Dev. Biol.* **23,** 549–577.
Simons, M., and Mlodzik, M. (2008). Planar cell polarity signaling: From fly development to human disease. *Annu. Rev. Genet.* **42.**
Simons, M., Gloy, J., Ganner, A., Bullerkotte, A., Bashkurov, M., Kronig, C., Schermer, B., Benzing, T., Cabello, O. A., Jenny, A., Mlodzik, M., Polok, B., *et al.* (2005). Inversin, the gene product mutated in nephronophthisis type II, functions as a molecular switch between Wnt signaling pathways. *Nat. Genet.* **37,** 537–543.
Smith, U. M., Consugar, M., Tee, L. J., McKee, B. M., Maina, E. N., Whelan, S., Morgan, N. V., Goranson, E., Gissen, P., Lilliquist, S., Aligianis, I. A., Ward, C. J., *et al.* (2006). The transmembrane protein meckelin (MKS3) is mutated in Meckel–Gruber syndrome and the wpk rat. *Nat. Genet.* **38,** 191–196.
Stolc, V., Samanta, M. P., Tongprasit, W., and Marshal, W. F. (2005). Genome-wide transcriptional analysis of flagellar regeneration in *Chlamydomonas reinhardtii* identifies orthologs of ciliary disease genes. *Proc. Natl. Acada. Sci. USA* **102,** 3703-3707.
Stratigopoulos, G., Padilla, S. L., LeDuc, C. A., Watson, E., Hattersley, A. T., McCarthy, M. I., Zeltser, L. M., Chung, W. K., and Leibel, R. L. (2008). Regulation of Fto/Ftm gene expression in mice and humans. *Am. J. Physiol. Regul. Integr. Comp. Physiol.* **294,** R1185–R1196.
Sturgess, J. M., Chao, J., Wong, J., Aspin, N., and Turner, J. A. (1979). Cilia with defective radial spokes: A cause of human respiratory disease. *N. Engl. J. Med.* **300,** 53–56.
Sturgess, J. M., Chao, J., and Turner, J. A. (1980). Transposition of ciliary microtubules: Another cause of impaired ciliary motility. *N. Engl. J. Med.* **303,** 318–322.
Supp, D. M., Witte, D. P., Potter, S. S., and Brueckner, M. (1997). Mutation of an axonemal dynein affects left–right asymmetry in inversus viscerum mice. *Nature* **389,** 963–966.
Swoboda, P., Adler, H. T., and Thomas, J. H. (2000). The RFX-type transcription factor DAF-19 regulates sensory neuron cilium formation in *C. elegans. Mol. Cell* **5,** 411–421.
Takehara, Y., Takahashi, M., Naito, M., Kato, T., Nishimura, T., Isoda, H., and Kaneko, M. (1989). Caroli's disease associated with polycystic kidney: Its noninvasive diagnosis. *Radiat. Med.* **7,** 13–15.
Tallila, J., Jakkula, E., Peltonen, L., Salonen, R., and Kestila, M. (2008). Identification of CC2D2A as a Meckel syndrome gene adds an important piece to the ciliopathy puzzle. *Am. J. Hum. Genet.* **82,** 1361–1367.
Tan, P. L., Barr, T., Inglis, P. N., Mitsuma, N., Huang, S. M., Garcia-Gonzalez, M. A., Bradley, B. A., Coforio, S., Albrecht, P. J., Watnick, T., Germino, G. G., Beales, P. L., *et al.* (2007). Loss of Bardet–Biedl syndrome proteins causes defects in peripheral sensory innervation and function. *Proc. Natl. Acad. Sci. USA* **104,** 17524–17529.

Tanaka, M., Cohn, M. J., Ashby, P., Davey, M., Martin, P., and Tickle, C. (2000). Distribution of polarizing activity and potential for limb formation in mouse and chick embryos and possible relationships to polydactyly. *Development* **127,** 4011–4021.

Tanaka, Y., Okada, Y., and Hirokawa, N. (2005). FGF-induced vesicular release of Sonic hedgehog and retinoic acid in leftward nodal flow is critical for left–right determination. *Nature* **435,** 172–177.

Tayeh, M. K., Yen, H. J., Beck, J. S., Searby, C. C., Westfall, T. A., Griesbach, H., Sheffield, V. C., and Slusarski, D. C. (2008). Genetic interaction between Bardet–Biedl syndrome genes and implications for limb patterning. *Hum. Mol. Genet.* **17,** 1956–1967.

Taylor, H. C., Satir, P., and Holwill, M. E. (1999). Assessment of inner dynein arm structure and possible function in ciliary and flagellar axonemes. *Cell Motil. Cytoskeleton* **43,** 167–177.

Thauvin-Robinet, C., Cossee, M., Cormier-Daire, V., Van Maldergem, L., Toutain, A., Alembik, Y., Bieth, E., Layet, V., Parent, P., David, A., Goldenberg, A., Mortier, G., et al. (2006). Clinical, molecular, and genotype-phenotype correlation studies from 25 cases of oral-facial-digital syndrome type 1: A French and Belgian collaborative study. *J. Med. Genet.* **43,** 54–61.

Tickle, C. (2003). Patterning systems—from one end of the limb to the other. *Dev. Cell* **4,** 449–458.

Tobin, J. L., and Beales, P. L. (2007). Bardet–Biedl syndrome: Beyond the cilium. *Pediatr. Nephrol.* **22,** 926–936.

Torres, V. E., and Harris, P. C. (2006). Mechanisms of disease: Autosomal dominant and recessive polycystic kidney diseases. *Nat. Clin. Pract. Nephrol.* **2,** 40–55; quiz 55.

Town, T., Breunig, J. J., Sarkisian, M. R., Spilianakis, C., Ayoub, A. E., Liu, X., Ferrandino, A. F., Gallagher, A. R., Li, M. O., Rakic, P., and Flavell, R. A. (2008). The stumpy gene is required for mammalian ciliogenesis. *Proc. Natl. Acad. Sci. USA* **105,** 2853–2858.

Tran, P. V., Haycraft, C. J., Besschetnova, T. Y., Turbe-Doan, A., Stottmann, R. W., Herron, B. J., Chesebro, A. L., Qiu, H., Scherz, P. J., Shah, J. V., Yoder, B. K., and Beier, D. R. (2008). THM1 negatively modulates mouse sonic hedgehog signal transduction and affects retrograde intraflagellar transport in cilia. *Nat. Genet.* **40,** 403–410.

Tritos, N. A., Mastaitis, J. W., Kokkotou, E., and Maratos-Flier, E. (2001). Characterization of melanin concentrating hormone and preproorexin expression in the murine hypothalamus. *Brain Res.* **895,** 160–166.

Valente, E. M., Brancati, F., and Dallapiccola, B. (2008). Genotypes and phenotypes of Joubert syndrome and related disorders. *Eur. J. Med. Genet.* **51,** 1–23.

Verhage, H. G., Bareither, M. L., Jaffe, R. C., and Akbar, M. (1979). Cyclic changes in ciliation, secretion and cell height of the oviductal epithelium in women. *Am. J. Anat.* **156,** 505–521.

Waldherr, R., Lennert, T., Weber, H. P., Fodisch, H. J., and Scharer, K. (1982). The nephronophthisis complex. A clinicopathologic study in children. *Virchows Arch. A Pathol. Anat. Histol.* **394,** 235–254.

Wanner, A., Salathe, M., and O'Riordan, T. G. (1996). Mucociliary clearance in the airways. *Am. J. Respir. Crit. Care Med.* **154,** 1868–1902.

Ward, C. J., Hogan, M. C., Rossetti, S., Walker, D., Sneddon, T., Wang, X., Kubly, V., Cunningham, J. M., Bacallao, R., Ishibashi, M., Milliner, D. S., Torres, V. E., et al. (2002). The gene mutated in autosomal recessive polycystic kidney disease encodes a large, receptor-like protein. *Nat. Genet.* **30,** 259–269.

Ward, C. J., Yuan, D., Masyuk, T. V., Wang, X., Punyashthiti, R., Whelan, S., Bacallao, R., Torra, R., LaRusso, N. F., Torres, V. E., and Harris, P. C. (2003). Cellular and subcellular localization of the ARPKD protein; fibrocystin is expressed on primary cilia. *Hum. Mol. Genet.* **12,** 2703–2710.

Wheatley, D. N., Wang, A. M., and Strugnell, G. E. (1996). Expression of primary cilia in mammalian cells. *Cell Biol. Int.* **20,** 73–81.

Wilkerson, C. G., King, S. M., and Witman, G. B. (1994). Molecular analysis of the gamma heavy chain of *Chlamydomonas* flagellar outer-arm dynein. *J. Cell Sci.* **107**(Pt. 3), 497–506.

Williams, C. L., Winkelbauer, M. E., Schafer, J. C., Michaud, E. J., and Yoder, B. K. (2008). Functional redundancy of the B9 proteins and nephrocystins in *Caenorhabditis elegans* ciliogenesis. *Mol. Biol. Cell* **19,** 2154–2168.

Winkelbauer, M. E., Schafer, J. C., Haycraft, C. J., Swoboda, P., and Yoder, B. K. (2005). The *C. elegans* homologs of nephrocystin-1 and nephrocystin-4 are cilia transition zone proteins involved in chemosensory perception. *J. Cell Sci.* **118,** 5575–5587.

Wood, C. R., Hard, R., and Hennessey, T. M. (2007). Targeted gene disruption of dynein heavy chain 7 of *Tetrahymena thermophila* results in altered ciliary waveform and reduced swim speed. *J. Cell Sci.* **120,** 3075–3085.

Worthington, W. C., Jr., and Cathcart, R. S., III (1963). Ependymal cilia: Distribution and activity in the adult human brain. *Science* **139,** 221–222.

Yoder, B. K., Richards, W. G., Sweeney, W. E., Wilkinson, J. E., Avener, E. D., and Woychik, R. P. (1995). Insertional mutagenesis and molecular analysis of a new gene associated with polycystic kidney disease. *Proc. Assoc. Am. Physicians* **107,** 314–323.

Yoder, B. K., Hou, X., and Guay-Woodford, L. M. (2002). The polycystic kidney disease proteins, polycystin-1, polycystin-2, polaris, and cystin, are co-localized in renal cilia. *J. Am. Soc. Nephrol.* **13,** 2508–2516.

Yu, S., Hackmann, K., Gao, J., He, X., Piontek, K., Garcia-Gonzalez, M. A., Menezes, L. F., Xu, H., Germino, G. G., Zuo, J., and Qian, F. (2007). Essential role of cleavage of Polycystin-1 at G protein-coupled receptor proteolytic site for kidney tubular structure. *Proc. Natl. Acad. Sci. USA* **104,** 18688–18693.

Zhang, Y. J., O'Neal, W. K., Randell, S. H., Blackburn, K., Moyer, M. B., Boucher, R. C., and Ostrowski, L. E. (2002). Identification of dynein heavy chain 7 as an inner arm component of human cilia that is synthesized but not assembled in a case of primary ciliary dyskinesia. *J. Biol. Chem.* **277,** 17906–17915.

Zhang, Q., Davenport, J. R., Croyle, M. J., Haycraft, C. J., and Yoder, B. K. (2005). Disruption of IFT results in both exocrine and endocrine abnormalities in the pancreas of Tg737(orpk) mutant mice. *Lab. Invest.* **85,** 45–64.

Zollinger, H. U., Mihatsch, M. J., Edefonti, A., Gaboardi, F., Imbasciati, E., and Lennert, T. (1980). Nephronophthisis (medullary cystic disease of the kidney). A study using electron microscopy, immunofluorescence, and a review of the morphological findings. *Helv. Paediatr. Acta* **35,** 509–530.

Index

A

Abnormal motile cilia
   ATP hydrolysis, 376
   in brain, 378
   characteristic phenotypes, 375
   cilia and flagella in reproductive system, 381
   consists of, 376
   and left-right asymmetry defects, 379–381
   and neurodevelopment, 383
   PCD phenotypes
      clinical features, 378
      genome-wide linkage analysis, 379
      ultrastructural studies, 378
Acetylation, α-tubulin
   aurora kinase, 92
   axonemal microtubules, 90
   HDAC6 and SIRT2 enzymes, 92–93
   microtubules marker, 91
   protists and invertebrates, 91–92
Adenylyl cyclase, 350
Alstrom syndrome (ALMS), 408
Arl13b mutants, 244
Autogenous hypothesis
   basal body and axonene, 74–76
   MTOC
      axonemal microtubules, 72–74
      modification, 72
   ninefold symmetry
      motility and regulation, 75–76
      mutation, 74–75
      triplet microtubules, 74
Autosomal recessive disorders
   Alstrom syndrome (ALMS), 408
   Bardet-Biedl syndrome (BBS), 401–405
   Joubert syndrome (JBTS), 405–406
   Meckel-Gruber syndrome (MKS), 398–401
   nephronophthisis (NPHP), 396–398
   Orofaciodigital type 1 syndrome (OFD), 406–408
   Senior-Loken syndrome (SLSN), 406
Axonemal dyneins, in cilia motility, 78
Axoneme
   electron micrographs study, 7
   genetic experiments, 8
   motif, 93
   ninefold symmetry, 8–9

B

Bardet–Biedl syndrome (BBS), 240, 310
   genes involved, 130
   mutations involved, 129
Basal body
   and centrosome, 2
   and cilia
      autogenous model, 71–74
      viral hypothesis, 68–69
   and ciliogenesis
      axoneme formation, 7–9
      cell cortex integration, 9–11
      protein composition, 11–12
      spindle pole formation, 12–13
   comparative genomic analysis, 14–15
   complex ultra structure, 3–4
   de novo assembly pathway, 4–5
   deuterosome and blepharoplast, 5–6
   electron tomography image, 16–17
   genetic analysis, 15–16
   genome association, 6–7
   microtubule structure, 3
   positioning, 212
   proteomic analysis, 14
   RNAi-mediated knockdown, 15
   suppression of Wnt signalling
      BBS protein, 180
      cascade components, 180–181
      inversin over expression, 179–180
      nonciliary gene, 181
   transmission of Wnt signalling
      BBS protein, 183
      inversin, 181
      PCP protein, 182–183
      USH protein, 217
Basal cell carcinoma (BCC), 229, 246–247
Basal foot protrudes, 3
BBS. *See* Bardet–Biedl syndrome
BBS genes, 386
BCC. *See* Basal cell carcinoma
Blepharoplast, 5
β-transducin repeat-containing protein (β-TrCP), 231

C

Cadherin, 216–217
Caenorhabditis elegans, 226

CALK proteins, 266
Cell cycle control regulation
  PDGFRαα signaling, 267–272
  Wnt signaling, 282–286
Cell-matrix adhesion complexes, 280
Cell migration
  in fibroblasts, 273–276
  regulated by
    PDGFRα, 272–279
    Wnt signaling, 282–286
Cellular signaling center, 239–240
Chemoreception in cilia, 126–127
Chemosensory systems, 334
Chlamydomonas, 236, 245, 266
Chlamydomonas flagella axoneme, 373, 378
Chronic rhinosinusitus, 357
Ci. *See* Cubitus interruptus
Cilia
  abnormal motile cilia
    ATP hydrolysis, 376
    in brain, 378
    characteristic phenotypes, 375
    consists of, 376
    and left-right asymmetry defects, 379–378
    and neurodevelopment, 383
    PCD phenotypes, 378–379
    role of cilia and flagella in reproductive system, 381
  anterior-posterior limb patterning
    apical ectodermal ridge (AER), 305
    Bardet-Biedl syndrome (BBS), 310
    Gli3 repressor (Gli3R), 309
    hedgehog (Hh) pathway regulation, 306–308
    Meckel-Gruber syndrome, 311
    oralfacial-digital syndrome I (OFD1), 310–311
    Shh signaling, 310
    zone of polarizing activity (ZPA), 306
  in articular cartilage
    articular chondrocyte, 318
    osteoarthritis, 317
  autogenous hypothesis
    basal body and axonene, 74–76
    MTOC, 72–74
    ninefold symmetry, 75–76
  bone collar development
    Ihh signaling, 321
    perichondrial cells, 320–321
    Wnt signaling, 320
  bone maintenance, 321–333
  chondrocytes and perichondrial cells, 304–305
  consists of, 304
  craniofacial development
    craniofacial dymorphology, 333–334
    defects, 334
    human ciliopathies, 333

endochondral bone formation
  Ihh regulators, 312
  perichondrial IFT/cilia, 314
  Pkd1 mutants, 313
  polycycstic kidney disease (PKD), 316
  polycystin 1 (Pc1) and polycystin 2 (Pc2), 316–317
  Prx1-Cre and Col2a-Cre mutants, 314–316
and eukaryotic phylogeny
  gene duplication, 67
  phyletic clade, 66
IFT and sensory function
  multiprotein complex, 77
  vesicle coats and core proteins, 77
immotile primary cilium
  in mammals, 384
  neuronal cilia and obesity phenotypes, 387–390
  primary cilia and cell cycle, 384–386
  sensory abnormalities with cilia mutants, 386–387
intra flagellar transport
  assembly and maintenance, 30–32
  canonical motor, 32, 35
  cilia-mediated signaling, 47–49
  polypeptides, 40
  regulation process, 42
intrinsic polarity in ear
  cadherin and protocadherin, 216–217
  in PCP mutants, 215–216
  Sans protein, 217
  Usher syndrome, 214–215
in kidney and liver diseases, 27
kinocilium, 199–200
LR development
  bending movement, 157–158
  dynein isoforms, 157
  electron microscopy, 156
  nonflow and flow pathways, 168
  ultrastructural arrangements, 158
mechanotransduction, 205–206
membrane proteins
  BBS proteins, 45
  Golgi-derived vesicles, 44–45
  MKS proteins, 47
  nephrocystins, 46–47
  synthesis, 44
  transition fibers, 45–46
motility mechanism, 78–79
mouse mutations, 156, biogenesis
  motility, 158
  polycystin-2, 160
nodal flow and LR organizer
  fluid properties, 163
  Kif3A and Kif3B proteins, 162–163
  leftward flow, 163
  posterior positioning, 164
origin and hypothesis, 67–68

PCP regulation, vertebrates
  BBS proteins, 210
  gastrulation, 209
  noncanonical Wnt signaling activity, 210
  seahorse gene, 211
photoreceptor, 66
PKD-related phenotypes, 390–396
postnatal growth, 317–318
proteins
  axoneme and membrane, 64–66
  mutations, 66
  nuclear genome, 64
  radial spokes, 65
in sensory organs, 26
signaling function, 198
suppression of Wnt signal
  BBS protein, 180
  cascade components, 180–181
  inversin over expression, 179–180
  nonciliary gene, 181
tooth development, 335–336
transmission of Wnt signal
  BBS protein, 183
  inversin, 181
  PCP protein, 182–183
types
  motile, 24–26
  non motile, 26
in vertebrate embryo
  cellular antennae, 156
  monocilia and classic, 155
viral hypothesis
  capsid and mutation, 69
  disadvantages, 71–72
  morphogenesis, 69–70
  ninefold symmetry, 71
  predictions, 70
Wnt phenotypes
  kidney degeneration, 186–187
  noncanonical signal, 185
  pancreatic abnormality, 188
  proteins involved, 185
  retinal and ear degeneration, 186
  tissue and time specific, 188–197
in wound healing, 276–277
Ciliary axoneme, 240
  composition, 339
  post-translational modifications, 340
Ciliary components function, 243–246
Ciliary function, 242–246
Ciliary membrane
  adhesive properties, 122
  BBS proteins and sorting
    genes involved, 130
    mutations involved, 129
  cell signaling, 127
  chemoreception, 126–127

IFT system, protein sorting
  Elipsa gene and β-arrestins, 129
  GFP-tagged IFT20, 128
mechanoreception
  fluid flow and sensing, 124
  nodal flow, 125
  statocysts, 123
photoreception
  opsin and transport, 125–126
  rod and cone cells, 125
polarity proteins and sorting
  Crumbs3-CLPI isoform, 131
  in zebrafish, 131–132
regulation and calcium ions, 121
small GTPases and protein sorting
  genetic analysis, 133–134
  proteomic analysis, 133
  Rab regulators, 132–134
structure
  diffusional barrier, 119
  endocytosis and exocytosis, 119–120
  lipid composition, 118
  membrane proteins, 118–119
  periciliary ridge complex, 121
  shapes, 118
targeting sequences
  cis acting motifs, 135
  KKxx sequences, 134
  mutation of residues, 137
  opsin residue, 135–137, 139
  polycystins, 136
  residue comparison, 137–138
  rhodopsin, 135–136
Ciliary membrane lipids, 342
Ciliary necklace, 24
Ciliary proteome
  canonical G protein-coupled pathway
    adenylyl cyclase and CNG channels, 350
    Ca^{2+}-activated Cl^- channel, 350–351
    odorant receptors, (ORs), 349
    olfactory G protein, 349–350
  ciliary protein entry regulation of, 354
  ciliary proteomics, 353–354
  expression profiling, 352–353
  noncanonical pathways, 351–352
Ciliary proteomics, 353
Ciliary tubulin
  acetylation, 92, aurora kinase
    axonemal microtubules, 90
    HDAC6 and SIRT2 enzymes, 92–93
    microtubules marker, 91
    protists and invertebrates, 91–92
  axonemal microtubule lattice, 87
  CCT subunits, 85
  cilia-specific functions, 86
  CTTs
    acidity nature, 94

Ciliary tubulin (cont.)
  axoneme motif, 93
  detyrosination
    axoneme assembly, 95
    tubulin-tyrosine ligase, 95
    tyrosinated isoforms, 96
    zinc metallocarboxypeptidase, 94
  glutamylation
    axoneme assembly, 97
    enzyme involves, 96
    GT335 antibody, 98
    TTLL1 protein, 96–97
  glycylation, 98–99, glycyl side chains
    axoneme motif, 99–100
    katanin interactions, 101–102
    LM bundles, 101
    mutational effects, 100
  kinesin-13 subfamily, 87–88
  microtubule depolymerization, 88–89
  prefoldin chaperone complex, 84
  recycling, 89
Ciliogenesis
  and cell cycle
    centriole formation, 28–29
    golgi-derived vesicles, 28
  disruption of, 237
  kinesin motors, 35–37
  Wnt signaling pathway, 184
Ciliogenesis and basal body
  axoneme formation
    electron micrographs study, 7
    genetic experiments, 8
    ninefold symmetry, 8–9
  cell cortex integration
    mouse node, 10
    rotational orientation, 10–11
    vesicle complex, 9
  protein composition, 11–12
  spindle pole formation, 12–13
Ciliopathies, 27
  human
    cilia proteome and genome databases and the identification, 410–412
    oak ridge polycystic kidney (orpk) mouse model, 374–376
    oligogenic inheritance and clinical variability, 409–410
Classic cilia, 155
Cochlea, 200
Convergent extension (CE), 206
Craniofacial dymorphology, 333–334
Crk proteins, 271
Crumbs proteins, 131–132
C-terminal tail domains (CTTs)
  acidity nature, 94
  glutamic acids and axoneme motif, 93
Cubitus interruptus (Ci), 230
Cytoplasmic dynein 2, 37

D

Deflagellation, 88
De novo assembly pathway, 4–5
Detyrosination, α-tubulin
  axoneme assembly, 95
  tubulin-tyrosine ligase, 95
  tyrosinated isoforms, 96
  zinc metallocarboxypeptidase, 94
Deuterosome, 5
Dorsal forerunner cells (DFCs), 161
Downstream mechanisms, 279
Drosophila melanogaster, 226, 230
DVL proteins, 177
Dyneins, 157

E

Elipsa gene, 129
Ellis-van Creveld syndrome, 311
Extracellular matrix (ECM), 280–282
Ezrin/radixin/moesin (ERM) proteins, 273, 277–279

F

FABP. See Flagellar and basal body proteome
FK506-binding protein 8 (FKBP8), 246
Flagella
  intra flagellar transport
    assembly and maintenance, 40
    canonical motor, 35
  structure, 24–25
Flagellar and basal body proteome (FABP), 14
F-molecule, 153–154
Ftm-deficient fibroblasts, 244

G

Gastrocoel roof plate (GRP), 160, 162
Gli proteins
  functions of, 233–235
  processing of, 242–243
Gli3 repressor (Gli3R), 309
Gli transcription factors, 231
Glutamylation, tubulin
  axoneme assembly, 97
  enzyme involves, 96
  GT335 antibody, 98
  TTLL1 protein, 96–97
Glycylation, tubulin
  axoneme motif, 99–100
  glycyl side chains, 98–99
  katanin interactions, 101–102
  longitudinal microtubule bundles, 101
  mutational effects, 100
G protein-coupled receptor kinase 2 (GRK2), 242
Groucho protein, 185
GRP. See Gastrocoel roof plate

H

Haptocilia, 122
Head trauma, 356
Hedgehog (Hh) signal
    intraflagellar transport disrupts, 236–239
    mediated tumorigenesis, 246–248
    pattern diverse developing tissues, 226–229
    at primary cilium, 241–242
    proteins, Bhanu P. Jena
    transduction, 229–233
Hensen's node, 166
Heterotaxy (Htx), 152
Human ciliopathies, 333
human Embryonic stem cells (hESCs), 272
Hypertrophic chondrocytes, 312

I

IFT. *See* Intraflagellar transport
$Ift88^{orpk}$ mutation, 318
IFT88 protein, 211–212
Immotile primary cilium, PKD–related phenotypes
    ALMS (omim 203800), 408
    BBS, 394–397
    JBTS (omim 213300), 405–406
    in mammals, 378
    MKS, 398–340
    neuronal cilia and obesity phenotypes
        feeding behavior, 387–390
        proopiomelanocortin (POMC) neurons, 388–390
        sensory neurons, 391
    NPHP, 396–398
    OFD (omim 311200), 406–408
    PKD-related phenotypes
        ADPKD patients, 395–396
        cystic kidney disease phenotypes, 396
        ischemic reperfusion (IR), 393
        planar cell polarity (PCP) pathway, 393–394
        polycystic kidney, 374–376
        proteolytic peptides, 391
        renal cysts, 390–396
    primary cilia and cell cycle, 384–386
    sensory abnormalities with cilia mutants, 386–387
    SLSN (omim 266900), 406
Intraflagellar transport (IFT), 347
    canonical motor
        flagella maintenance, 35
        kinesin-2 family, 32, 35
        subunits and components, 33–34
    cilia and sensory function
        multiprotein complex, 77
        vesicle coats and core proteins, 77
    cilia-mediated signaling
        environmental signals, 48

        protein targeting, Q47–48
        transcriptional effects, 48–49
    in ciliary assembly, 32
    ciliary protein sorting
        Elipsa gene and β-arrestins, 129
        GFP-tagged IFT20, 128
    complex A polypeptides
        flagella assembly, 40
        retrograde process, 41
    complex B polypeptides
        functions, 40
        types, 39
    components, 242
    DIC microscopy observation, 30–32
    functions, 30
    motility process, 29
    mutants, 242, 244–245
    particle polypeptides
        gene cloning and sequencing, 39
        isolation, 38
    proteins, 236, 239
    regulation
        ciliary length, 42
        flagellar and ciliary tip, 43
        immunolocalization study, 41
    retrograde motor
        heavy chain, 37
        light chain, 38
        light intermediate chain, 37–38
    role, 236
Intrinsic cellular polarity and cilia
    PCP protein
        cellular alignment, 07–216
        ciliary mutants, 215
        extracellular directional cue, 08
    Usher syndrome, 208–209, cadherin and protocadherin
        genes, 214
        hair bundle polarity, 216–217
        proteins involved, 215
        Sans protein, 217

J

JBTS genes, 405
Jeune asphyxiating thoracic dystrophy (JATD), 409
Joubert syndrome (JBTS), 240, 405–406

K

Kartagener syndrome, 155
Katanin, 102
Kinocilium
    ablation, 212
    and hair cells, 201
    mechanotransduction, 205–206
    single primary cilium, 199
    stereociliary bundles
        hair cell regeneration, 204

Kinocilium (cont.)
  maturation and development, 202
  PCP regulation, 204–205
  polarization, 204
Kupffer's vesicle (KV), 162

L

Last eukaryotic common ancestor (LECA), 66–67
Lathosterolosis, 241
LECA. See Last eukaryotic common ancestor
Left-right (LR) asymmetry, vertebrates
  AP and DV axes, 154
  cilia
    mouse mutations, 158–160
    nonflow and flow pathways, 168
  DFC and embryonic shield, 161–162
  molecular asymmetry, 152–154
  motile primary cilia
    bending movement, 157–158
    dynein isoforms, 157
    electron microscopy, 156
    ultrastructural arrangements, 158
  mouse node, 161
  three classes, 152
Longitudinal microtubule bundles (LM), 101
LR organizer, 161–162
  asymmetric calcium signals
    avian embryo, 166
    chemical hypothesis, 167
    combined NVP and mechanosensation, 168
    mechanosensory hypothesis, 167–168
    mouse node, 164–166
    zebrafish, 166
  nodal flow and cilia
    Kif3A and Kif3B proteins, 58–163
    leftward flow, 163
    luid properties, 163
    posterior positioning, 164

M

Mammalian Smo activates, 241–242
Mastigonemes projection, 11
Mechanoreception, cilia
  fluid flow and sensing, 124
  nodal flow, 125
  statocysts, 123
Meckel–Gruber syndrome (MKS), 187, 240, 311
Microtubule-organizing center (MTOC)
Microvilli, 198
MKS genes, 398–401
MKS. See Meckel–Gruber syndrome
Molar tooth malformation (MTM), 397
Monociliaq, 155
Monosiga brevicollis, 226
MTOC. See Microtubule-organizing center
Mutants

Arl13b, 244
Dnchc2, 245
IFT, 242, 244–245
Ift 172, 237
Shh, 227
THM1, 245

N

Na+/H+ exchanger (NHE1), 277–279
Nephronopthisis (NPHP) genes, 365, 396–398
Neuronal cilia and obesity phenotypes
  feeding behavior, 388
  proopiomelanocortin (POMC) neurons, 387–390
  sensory neurons, 386
NHE1. See Na+/H+ exchanger
Nima A-related kinase (NRK), 88–89
Nodal flow
  fluid properties, 59
  Kif3A and Kif3B proteins, 58–163
  leftward flow, 59
  posterior positioning, 60
Nodal vesicular particles (NVPs), 167–168
Noncanonical Wnt signaling activity, 210
NRK. See Nima A-related kinase

O

Oak ridge polycystic kidney (orpk) mouse model, 374–376
Odorant-binding proteins (OBPs), 338
Odorant receptors, (ORs), 349
Ofd1. See Oral-facial-digital type I
Olfaction, sensory modality, 334
Olfactory cilia
  disease
    chronic rhinosinusitis, 357
    head trauma, 356–357
    olfactory ciliopathies, 355–356
    pathogenic target, 358
  formation
    dendrites and dendritic knobs, 344–345
    neuronal differentiation, 344
    olfaction process, 346
    protein expression, 346–347
  intraflagellar transport
    Bardet-Biedl syndrome (BBS) proteins, 348
    Caenorhabditis elegans, 347
  structure of
    basal body, 343
    cilia axoneme, 339
    ciliary necklace/transition zone, 342
    ciliary rootlet, 343
    lipid composition, 340–342
Olfactory ciliopathies, 355
Olfactory epithelium (OE)
  cell types and ultrastructure

Index

basal cell types, 337
  olfactory sensory neurons, 337–338
gross anatomy
  inspired odorants, 335
  main olfactory bulb (MOB), 335–336
regeneration, 338
Olfactory G protein, 349–350
Olfactory sensory neurons (OSNs), 337
Opsin, 135–137
Oralfacial-digital syndrome I (OFD1), 243–244, 310–311
Organ of Corti, 200

P

PCD. See Primary ciliary dyskinesia
PCP. See Planar cell polarity
PDGF isoforms, 269–270, 273–274, 276
PDGFR-α. See Platelet-derived growth factor receptor alpha
Photoreception, cilia
  opsin and transport, 125–126
  rod and cone cells, 125
Photoreceptors, 386–387
PKD genes, 392
pkd1 mutant, 313
PKD-related phenotypes
  ADPKD patients, 391
  Oral-facial-digital type I (OFD1) syndrome, 406–408
  cystic kidney disease phenotypes, 393
  ischemic reperfusion (IR), 393
  planar cell polarity (PCP) pathway, 393
  polycystic kidney, 374
  proteolytic peptides, 391
  renal cysts, 390–396
PKD. See Polycystic kidney disease
Planar cell polarity (PCP)
  cilia and regulation in vertebrates
    bbs mouse mutants study, 211
    BBS proteins, 210
    Cre-mediated recombination, 211–212
    gastrulation, 209
    noncanonical Wnt signal, 210
    seahorse gene, 211
  inner ear sensory organs
    CE regulation, 207
    molecules involved, 208
    multiple regulatory components, 206
    patterning and growth, 209
    protein localization, 207–208
  mammalian inner ear
    cochlea and vestibule, 201
    phalangeal processes, 202
    sensory organs, 200
    stereocilia and kinocilium, 199–201
    stereociliary bundles and kinocilia, 204–205
Platelet-derived growth factor receptor alpha (PDGFR-α), 262, 265

cell migration regulated by, 272–279
PNC. See Posterior notochord
Polarity proteins, ciliary membrane
  Crumbs3-CLPI isoform, 131
  in zebrafish, 131–132
Polarization, by Wnt signaling, 282–286
Polycystic kidney disease (PKD), 237, 316
Polycystin (Pc), 136, 316–317
Postaxial polydactyly. See Ellis-van Creveld syndrome
Posterior notochord (PNC), 160, 162
Prefoldin, 84–85
Primary ciliary dyskinesia (PCD)
  clinical features, 372
  genome-wide linkage analysis, 373–374
  ultrastructural studies, 373
Primary cilium, 64
Proopiomelanocortin (POMC) neurons, 388–389
Proteasomes, 243
Protocadherin, 216–217

R

Rabin8 and Rab8 proteins, 130
Resistance nodulation-cell division (RND), 241
Rfx3, 245
Rhodopsin, 135–136
RNAi-mediated knockdown, 15
RND. See Resistance nodulation-cell division

S

Sans protein, 216–217
Scale reservoir, 119
seahorse gene, 211
Senior-Loken syndrome (SLSN), 406
Sensory membrane patch, 77
Shh signaling, 310
Situs inversus (SI), 152
Situs solitus (SS), 152
SLSN genes, 369
Small GTPases, ciliary membrane
  genetic analysis, 133–134
  proteomic analysis, 133
  Rab regulators, 132–134
Smith-Lemli-Opitz syndrome, 241
Sonic hedgehog (Shh), 226
  mutants, 227
S-opsin, 136
Statocysts, 123
Stereocilia
  and hair cells, 201
  microvilli, 198–199
  and Usher syndrome, 214–215
Striated rootlet protrudes, 3

T

Thm1^{aln} mutants, 310
Tubulin
   acetylation
      aurora kinase, 92
      axonemal microtubules, 90
      HDAC6 and SIRT2 enzymes, 92–93
      microtubules marker, 91
      protists and invertebrates, 91–92
   axonemal microtubule lattice, 87
   CCT subunits, 85
   cilia-specific functions, 86
   CTTs
      acidity nature, 94
      axoneme motif, 93
   detyrosination
      axoneme assembly, 95
      tubulin-tyrosine ligase, 95
      tyrosinated isoforms, 96
      zinc metallocarboxypeptidase, 94
   glutamylation
      axoneme assembly, 97
      enzyme involves, 96
      GT335 antibody, 98
      TTLL1 protein, 96–97
   glycylation, 101, LM bundles
      axoneme motif, 99–100
      glycyl side chains, 98–99
      katanin interactions, 101–102
      mutational effects, 100
   kinesin-13 subfamily, 87–88
   microtubule depolymerization, 88–89
   prefoldin chaperone complex, 84
   recycling, 89
Tubulin-tyrosine ligase (TTL), 95
Tumorigenesis, Hh pathway-mediated, 246–248

U

UNI
   linkage group, 7
   phenotypes, 16
Usher syndrome (USH)
   and basal body, 217
   genes in, 214
   hair bundle polarity, 216–217
   proteins involved, 215
USH. *See* Usher syndrome

V

Vestibule, 200
Viral hypothesis, cilia
   capsid and mutation, 69
   disadvantages, 71–72
   morphogenesis, 69–70
   ninefold symmetry, 71
   predictions, 70

W

Wiscott-Aldrich syndrome protein (WASP), 275
Wnt signaling pathways, 320
   and ciliogenesis, 184
   DVL proteins, 177
   genes coding, 176
   in human health and disease, 282–283
   ligands role, 179
   overview, 178
   PCP and ear sensory organs, 208
   phenotypes and cilia dysfunction
      Groucho protein, 185
      kidney degeneration, 186–187
      noncanonical signal, 185
      pancreatic abnormality, 188
      retinal and ear degeneration, 186
      tissue and time specific, 188–197
   and primary cilium, 283–286
   suppression in cilia and basal body
      BBS protein, 180
      cascade components, 180–181
      inversin, 179–180
      nonciliary gene, 181
   transcriptional targets, 177
   transmission, cilia and basal body
      BBS protein, 183
      inversin, 181
      PCP protein, 182–183
Wound healing
   primary cilia in, 276–277

Z

Zone of polarizing activity (ZPA), 229

Contents of Previous Volumes

Volume 47

1. **Early Events of Somitogenesis in Higher Vertebrates: Allocation of Precursor Cells during Gastrulation and the Organization of a Moristic Pattern in the Paraxial Mesoderm**
   Patrick P. L. Tam, Devorah Goldman, Anne Camus, and Gary C. Shoenwolf

2. **Retrospective Tracing of the Developmental Lineage of the Mouse Myotome**
   Sophie Eloy-Trinquet, Luc Mathis, and Jean-François Nicolas

3. **Segmentation of the Paraxial Mesoderm and Vertebrate Somitogenesis**
   Olivier Pourqulé

4. **Segmentation: A View from the Border**
   Claudio D. Stern and Daniel Vasiliauskas

5. **Genetic Regulation of Somite Formation**
   Alan Rawls, Jeanne Wilson-Rawls, and Eric N. Olsen

6. **Hox Genes and the Global Patterning of the Somitic Mesoderm**
   Ann Campbell Burke

7. **The Origin and Morphogenesis of Amphibian Somites**
   Ray Keller

8. **Somitogenesis in Zebrafish**
   Scott A. Halley and Christiana Nüsslain-Volhard

9. **Rostrocaudal Differences within the Somites Confer Segmental Pattern to Trunk Neural Crest Migration**
   Marianne Bronner-Fraser

Volume 48

1. **Evolution and Development of Distinct Cell Lineages Derived from Somites**
   Beate Brand-Saberi and Bodo Christ

2. **Duality of Molecular Signaling Involved in Vertebral Chondrogenesis**
   Anne-Hélène Monsoro-Burq and Nicole Le Douarin

3. **Sclerotome Induction and Differentiation**
   Jennifer L. Docker

4. **Genetics of Muscle Determination and Development**
   Hans-Henning Arnold and Thomas Braun

5. **Multiple Tissue Interactions and Signal Transduction Pathways Control Somite Myogenesis**
   Anne-Gaëlle Borycki and Charles P. Emerson, Jr.

6. **The Birth of Muscle Progenitor Cells in the Mouse: Spatiotemporal Considerations**
   Shahragim Tajbakhsh and Margaret Buckingham

7. **Mouse–Chick Chimera: An Experimental System for Study of Somite Development**
   Josiane Fontaine-Pérus

8. **Transcriptional Regulation during Somitogenesis**
   Dennis Summerbell and Peter W. J. Rigby

9. **Determination and Morphogenesis in Myogenic Progenitor Cells: An Experimental Embryological Approach**
   Charles P. Ordahl, Brian A. Williams, and Wilfred Denetclaw

Volume 49

1. **The Centrosome and Parthenogenesis**
   Thomas Küntziger and Michel Bornens

2. **γ-Tubulin**
   Berl R. Oakley

3. **γ-Tubulin Complexes and Their Role in Microtubule Nucleation**
   Ruwanthi N. Gunawardane, Sofia B. Lizarraga, Christiane Wiese, Andrew Wilde, and Yixian Zheng

4. **γ-Tubulin of Budding Yeast**
   Jackie Vogel and Michael Snyder

5. **The Spindle Pole Body of *Saccharomyces cerevisiae*: Architecture and Assembly of the Core Components**
   Susan E. Francis and Trisha N. Davis

6. The Microtubule Organizing Centers of *Schizosaccharomyces pombe*
   Iain M. Hagan and Janni Petersen

7. Comparative Structural, Molecular, and Functional Aspects of the *Dictyostelium discoideum* Centrosome
   Ralph Gräf, Nicole Brusis, Christine Daunderer, Ursula Euteneuer, Andrea Hestermann, Manfred Schliwa, and Masahiro Ueda

8. Are There Nucleic Acids in the Centrosome?
   Wallace F. Marshall and Joel L. Rosenbaum

9. Basal Bodies and Centrioles: Their Function and Structure
   Andrea M. Preble, Thomas M. Giddings, Jr., and Susan K. Dutcher

10. Centriole Duplication and Maturation in Animal Cells
    B. M. H. Lange, A. J. Faragher, P. March, and K. Gull

11. Centrosome Replication in Somatic Cells: The Significance of the G_1 Phase
    Ron Balczon

12. The Coordination of Centrosome Reproduction with Nuclear Events during the Cell Cycle
    Greenfield Sluder and Edward H. Hinchcliffe

13. Regulating Centrosomes by Protein Phosphorylation
    Andrew M. Fry, Thibault Mayor, and Erich A. Nigg

14. The Role of the Centrosome in the Development of Malignant Tumors
    Wilma L. Lingle and Jeffrey L. Salisbury

15. The Centrosome-Associated Aurora/Ipl-like Kinase Family
    T. M. Goepfert and B. R. Brinkley

16 Centrosome Reduction during Mammalian Spermiogenesis
    G. Manandhar, C. Simerly, and G. Schatten

17. The Centrosome of the Early *C. elegans* Embryo: Inheritance, Assembly, Replication, and Developmental Roles
    Kevin F. O'Connell

18. The Centrosome in *Drosophila* Oocyte Development
    Timothy L. Megraw and Thomas C. Kaufman

19. The Centrosome in Early *Drosophila* Embryogenesis
    W. F. Rothwell and W. Sullivan

20. **Centrosome Maturation**

   Robert E. Palazzo, Jacalyn M. Vogel, Bradley J. Schnackenberg, Dawn R. Hull, and Xingyong Wu

Volume 50

1. **Patterning the Early Sea Urchin Embryo**

   Charles A. Ettensohn and Hyla C. Sweet

2. **Turning Mesoderm into Blood: The Formation of Hematopoietic Stem Cells during Embryogenesis**

   Alan J. Davidson and Leonard I. Zon

3. **Mechanisms of Plant Embryo Development**

   Shunong Bai, Lingjing Chen, Mary Alice Yund, and Zinmay Rence Sung

4. **Sperm-Mediated Gene Transfer**

   Anthony W. S. Chan, C. Marc Luetjens, and Gerald P. Schatten

5. **Gonocyte–Sertoli Cell Interactions during Development of the Neonatal
   Rodent Testis**

   Joanne M. Orth, William F. Jester, Ling-Hong Li, and Andrew L. Laslett

6. **Attributes and Dynamics of the Endoplasmic Reticulum in Mammalian Eggs**

   Douglas Kline

7. **Germ Plasm and Molecular Determinants of Germ Cell Fate**

   Douglas W. Houston and Mary Lou King

Volume 51

1. **Patterning and Lineage Specification in the Amphibian Embryo**

   Agnes P. Chan and Laurence D. Etkin

2. **Transcriptional Programs Regulating Vascular Smooth Muscle Cell Development and Differentiation**

   Michael S. Parmacek

3. **Myofibroblasts: Molecular Crossdressers**

   Gennyne A. Walker, Ivan A. Guerrero, and Leslie A. Leinwand

4. **Checkpoint and DNA-Repair Proteins Are Associated with the Cores of Mammalian Meiotic Chromosomes**
   Madalena Tarsounas and Peter B. Moens

5. **Cytoskeletal and Ca^{2+} Regulation of Hyphal Tip Growth and Initiation**
   Sara Torralba and I. Brent Heath

6. **Pattern Formation during *C. elegans* Vulval Induction**
   Minqin Wang and Paul W. Sternberg

7. **A Molecular Clock Involved in Somite Segmentation**
   Miguel Maroto and Olivier Pourquié

Volume 52

1. **Mechanism and Control of Meiotic Recombination Initiation**
   Scott Keeney

2. **Osmoregulation and Cell Volume Regulation in the Preimplantation Embryo**
   Jay M. Baltz

3. **Cell–Cell Interactions in Vascular Development**
   Diane C. Darland and Patricia A. D'Amore

4. **Genetic Regulation of Preimplantation Embryo Survival**
   Carol M. Warner and Carol A. Brenner

Volume 53

1. **Developmental Roles and Clinical Significance of Hedgehog Signaling**
   Andrew P. McMahon, Philip W. Ingham, and Clifford J. Tabin

2. **Genomic Imprinting: Could the Chromatin Structure Be the Driving Force?**
   Andras Paldi

3. **Ontogeny of Hematopoiesis: Examining the Emergence of Hematopoietic Cells in the Vertebrate Embryo**
   Jenna L. Galloway and Leonard I. Zon

4. **Patterning the Sea Urchin Embryo: Gene Regulatory Networks, Signaling Pathways, and Cellular Interactions**
   Lynne M. Angerer and Robert C. Angerer

Volume 54

1. **Membrane Type-Matrix Metalloproteinases (MT-MMP)**
   Stanley Zucker, Duanqing Pei, Jian Cao, and Carlos Lopez-Otin

2. **Surface Association of Secreted Matrix Metalloproteinases**
   Rafael Fridman

3. **Biochemical Properties and Functions of Membrane-Anchored Metalloprotease-Disintegrin Proteins (ADAMs)**
   J. David Becherer and Carl P. Blobel

4. **Shedding of Plasma Membrane Proteins**
   Joaquín Arribas and Anna Merlos-Suárez

5. **Expression of Meprins in Health and Disease**
   Lourdes P. Norman, Gail L. Matters, Jacqueline M. Crisman, and Judith S. Bond

6. **Type II Transmembrane Serine Proteases**
   Qingyu Wu

7. **DPPIV, Seprase, and Related Serine Peptidases in Multiple Cellular Functions**
   Wen-Tien Chen, Thomas Kelly, and Giulio Ghersi

8. **The Secretases of Alzheimer's Disease**
   Michael S. Wolfe

9. **Plasminogen Activation at the Cell Surface**
   Vincent Ellis

10. **Cell-Surface Cathepsin B: Understanding Its Functional Significance**
    Dora Cavallo-Medved and Bonnie F. Sloane

11. **Protease-Activated Receptors**
    Wadie F. Bahou

12. **Emmprin (CD147), a Cell Surface Regulator of Matrix Metalloproteinase Production and Function**
    Bryan P. Toole

13. **The Evolving Roles of Cell Surface Proteases in Health and Disease: Implications for Developmental, Adaptive, Inflammatory, and Neoplastic Processes**
    Joseph A. Madri

6. **Directions in Cell Migration Along the Rostral Migratory Stream: The Pathway for Migration in the Brain**
   Shin-ichi Murase and Alan F. Horwitz

7. **Retinoids in Lung Development and Regeneration**
   Malcolm Maden

8. **Structural Organization and Functions of the Nucleus in Development, Aging, and Disease**
   Leslie Mounkes and Colin L. Stewart

Volume 62

1. **Blood Vessel Signals During Development and Beyond**
   Ondine Cleaver

2. **HIFs, Hypoxia, and Vascular Development**
   Kelly L. Covello and M. Celeste Simon

3. **Blood Vessel Patterning at the Embryonic Midline**
   Kelly A. Hogan and Victoria L. Bautch

4. **Wiring the Vascular Circuitry: From Growth Factors to Guidance Cues**
   Lisa D. Urness and Dean Y. Li

5. **Vascular Endothelial Growth Factor and Its Receptors in Embryonic Zebrafish Blood Vessel Development**
   Katsutoshi Goishi and Michael Klagsbrun

6. **Vascular Extracellular Matrix and Aortic Development**
   Cassandra M. Kelleher, Sean E. McLean, and Robert P. Mecham

7. **Genetics in Zebrafish, Mice, and Humans to Dissect Congenital Heart Disease: Insights in the Role of VEGF**
   Diether Lambrechts and Peter Carmeliet

8. **Development of Coronary Vessels**
   Mark W. Majesky

9. **Identifying Early Vascular Genes Through Gene Trapping in Mouse Embryonic Stem Cells**
   Frank Kuhnert and Heidi Stuhlmann

Volume 63

1. **Early Events in the DNA Damage Response**
   Irene Ward and Junjie Chen

2. **Afrotherian Origins and Interrelationships: New Views and Future Prospects**
   Terence J. Robinson and Erik R. Seiffert

3. **The Role of Antisense Transcription in the Regulation of X-Inactivation**
   Claire Rougeulle and Philip Avner

4. **The Genetics of Hiding the Corpse: Engulfment and Degradation of Apoptotic Cells in *C. elegans* and *D. melanogaster***
   Zheng Zhou, Paolo M. Mangahas, and Xiaomeng Yu

5. **Beginning and Ending an Actin Filament: Control at the Barbed End**
   Sally H. Zigmond

6. **Life Extension in the Dwarf Mouse**
   Andrzej Bartke and Holly Brown-Borg

Volume 64

1. **Stem/Progenitor Cells in Lung Morphogenesis, Repair, and Regeneration**
   David Warburton, Mary Anne Berberich, and Barbara Driscoll

2. **Lessons from a Canine Model of Compensatory Lung Growth**
   Connie C. W. Hsia

3. **Airway Glandular Development and Stem Cells**
   Xiaoming Liu, Ryan R. Driskell, and John F. Engelhardt

4. **Gene Expression Studies in Lung Development and Lung Stem Cell Biology**
   Thomas J. Mariani and Naftali Kaminski

5. **Mechanisms and Regulation of Lung Vascular Development**
   Michelle Haynes Pauling and Thiennu H. Vu

6. **The Engineering of Tissues Using Progenitor Cells**
   Nancy L. Parenteau, Lawrence Rosenberg, and Janet Hardin-Young

7. **Adult Bone Marrow-Derived Hemangioblasts, Endothelial Cell Progenitors, and EPCs**
   Gina C. Schatteman

8. **Synthetic Extracellular Matrices for Tissue Engineering and Regeneration**
   Eduardo A. Silva and David J. Mooney

9. **Integrins and Angiogenesis**
   D. G. Stupack and D. A. Cheresh

Volume 65

1. **Tales of Cannibalism, Suicide, and Murder: Programmed Cell Death in *C. elegans***
   Jason M. Kinchen and Michael O. Hengartner

2. **From Guts to Brains: Using Zebrafish Genetics to Understand the Innards of Organogenesis**
   Carsten Stuckenholz, Paul E. Ulanch, and Nathan Bahary

3. **Synaptic Vesicle Docking: A Putative Role for the Munc18/Sec1 Protein Family**
   Robby M. Weimer and Janet E. Richmond

4. **ATP-Dependent Chromatin Remodeling**
   Corey L. Smith and Craig L. Peterson

5. **Self-Destruct Programs in the Processes of Developing Neurons**
   David Shepherd and V. Hugh Perry

6. **Multiple Roles of Vascular Endothelial Growth Factor (VEGF) in Skeletal Development, Growth, and Repair**
   Elazar Zelzer and Bjorn R. Olsen

7. **G-Protein Coupled Receptors and Calcium Signaling in Development**
   Geoffrey E. Woodard and Juan A. Rosado

8. **Differential Functions of 14-3-3 Isoforms in Vertebrate Development**
   Anthony J. Muslin and Jeffrey M. C. Lau

9. **Zebrafish Notochordal Basement Membrane: Signaling and Structure**
   Annabelle Scott and Derek L. Stemple

10. **Sonic Hedgehog Signaling and the Developing Tooth**
    Martyn T. Cobourne and Paul T. Sharpe

Volume 66

1. **Stepwise Commitment from Embryonic Stem to Hematopoietic and Endothelial Cells**
   Changwon Park, Jesse J. Lugus, and Kyunghee Choi

2. **Fibroblast Growth Factor Signaling and the Function and Assembly of Basement Membranes**
   Peter Lonai

3. **TGF-β Superfamily and Mouse Craniofacial Development: Interplay of Morphogenetic Proteins and Receptor Signaling Controls Normal Formation of the Face**
   Marek Dudas and Vesa Kaartinen

4. **The Colors of Autumn Leaves as Symptoms of Cellular Recycling and Defenses Against Environmental Stresses**
   Helen J. Ougham, Phillip Morris, and Howard Thomas

5. **Extracellular Proteases: Biological and Behavioral Roles in the Mammalian Central Nervous System**
   Yan Zhang, Kostas Pothakos, and Styliana-Anna (Stella) Tsirka

6. **The Genetic Architecture of House Fly Mating Behavior**
   Lisa M. Meffert and Kara L. Hagenbuch

7. **Phototropins, Other Photoreceptors, and Associated Signaling: The Lead and Supporting Cast in the Control of Plant Movement Responses**
   Bethany B. Stone, C. Alex Esmon, and Emmanuel Liscum

8. **Evolving Concepts in Bone Tissue Engineering**
   Catherine M. Cowan, Chia Soo, Kang Ting, and Benjamin Wu

9. **Cranial Suture Biology**
   Kelly A Lenton, Randall P. Nacamuli, Derrick C. Wan, Jill A. Helms, and Michael T. Longaker

Volume 67

1. **Deer Antlers as a Model of Mammalian Regeneration**
   Joanna Price, Corrine Faucheux, and Steve Allen

2. **The Molecular and Genetic Control of Leaf Senescence and Longevity in *Arabidopsis***
   Pyung Ok Lim and Hong Gil Nam

3. **Cripto-1: An Oncofetal Gene with Many Faces**
   Caterina Bianco, Luigi Strizzi, Nicola Normanno, Nadia Khan, and David S. Salomon

4. **Programmed Cell Death in Plant Embryogenesis**
   Peter V. Bozhkov, Lada H. Filonova, and Maria F. Suarez

5. **Physiological Roles of Aquaporins in the Choroid Plexus**
   Daniela Boassa and Andrea J. Yool

6. **Control of Food Intake Through Regulation of cAMP**
   Allan Z. Zhao

7. **Factors Affecting Male Song Evolution in *Drosophila montana***
   Anneli Hoikkala, Kirsten Klappert, and Dominique Mazzi

8. **Prostanoids and Phosphodiesterase Inhibitors in Experimental Pulmonary Hypertension**
   Ralph Theo Schermuly, Hossein Ardeschir Ghofrani, and Norbert Weissmann

9. **14-3-3 Protein Signaling in Development and Growth Factor Responses**
   Daniel Thomas, Mark Guthridge, Jo Woodcock, and Angel Lopez

10. **Skeletal Stem Cells in Regenerative Medicine**
    Wataru Sonoyama, Carolyn Coppe, Stan Gronthos, and Songtao Shi

Volume 68

1. **Prolactin and Growth Hormone Signaling**
   Beverly Chilton and Aveline Hewetson

2. **Alterations in cAMP-Mediated Signaling and Their Role in the Pathophysiology of Dilated Cardiomyopathy**
   Matthew A. Movsesian and Michael R. Bristow

3. **Corpus Luteum Development: Lessons from Genetic Models in Mice**
   Anne Bachelot and Nadine Binart

4. **Comparative Developmental Biology of the Mammalian Uterus**
   Thomas E. Spencer, Kanako Hayashi, Jianbo Hu, and Karen D. Carpenter

5. Sarcopenia of Aging and Its Metabolic Impact
   Helen Karakelides and K. Sreekumaran Nair

6. Chemokine Receptor CXCR3: An Unexpected Enigma
   Liping Liu, Melissa K. Callahan, DeRen Huang, and Richard M. Ransohoff

7. Assembly and Signaling of Adhesion Complexes
   Jorge L. Sepulveda, Vasiliki Gkretsi, and Chuanyue Wu

8. Signaling Mechanisms of Higher Plant Photoreceptors: A Structure-Function Perspective
   Haiyang Wang

9. Initial Failure in Myoblast Transplantation Therapy Has Led the Way Toward the Isolation of Muscle Stem Cells: Potential for Tissue Regeneration
   Kenneth Urish, Yasunari Kanda, and Johnny Huard

10. Role of 14-3-3 Proteins in Eukaryotic Signaling and Development
    Dawn L. Darling, Jessica Yingling, and Anthony Wynshaw-Boris

Volume 69

1. Flipping Coins in the Fly Retina
   Tamara Mikeladze-Dvali, Claude Desplan, and Daniela Pistillo

2. Unraveling the Molecular Pathways That Regulate Early Telencephalon Development
   Jean M. Hébert

3. Glia–Neuron Interactions in Nervous System Function and Development
   Shai Shaham

4. The Novel Roles of Glial Cells Revisited: The Contribution of Radial Glia and Astrocytes to Neurogenesis
   Tetsuji Mori, Annalisa Buffo, and Magdalena Götz

5. Classical Embryological Studies and Modern Genetic Analysis of Midbrain and Cerebellum Development
   Mark Zervas, Sandra Blaess, and Alexandra L. Joyner

6. Brain Development and Susceptibility to Damage; Ion Levels and Movements
   Maria Erecinska, Shobha Cherian, and Ian A. Silver

7. **Thinking about Visual Behavior; Learning about Photoreceptor Function**
   Kwang-Min Choe and Thomas R. Clandinin

8. **Critical Period Mechanisms in Developing Visual Cortex**
   Takao K. Hensch

9. **Brawn for Brains: The Role of MEF2 Proteins in the Developing Nervous System**
   Aryaman K. Shalizi and Azad Bonni

10. **Mechanisms of Axon Guidance in the Developing Nervous System**
    Céline Plachez and Linda J. Richards

Volume 70

1. **Magnetic Resonance Imaging: Utility as a Molecular Imaging Modality**
   James P. Basilion, Susan Yeon, and René Botnar

2. **Magnetic Resonance Imaging Contrast Agents in the Study of Development**
   Angelique Louie

3. **$^1H/^{19}F$ Magnetic Resonance Molecular Imaging with Perfluorocarbon Nanoparticles**
   Gregory M. Lanza, Patrick M. Winter, Anne M. Neubauer, Shelton D. Caruthers, Franklin D. Hockett, and Samuel A. Wickline

4. **Loss of Cell Ion Homeostasis and Cell Viability in the Brain: What Sodium MRI Can Tell Us**
   Fernando E. Boada, George LaVerde, Charles Jungreis, Edwin Nemoto, Costin Tanase, and Ileana Hancu

5. **Quantum Dot Surfaces for Use *In Vivo* and *In Vitro***
   Byron Ballou

6. ***In Vivo* Cell Biology of Cancer Cells Visualized with Fluorescent Proteins**
   Robert M. Hoffman

7. **Modulation of Tracer Accumulation in Malignant Tumors: Gene Expression, Gene Transfer, and Phage Display**
   Uwe Haberkorn

8. **Amyloid Imaging: From Benchtop to Bedside**
   Chungying Wu, Victor W. Pike, and Yanming Wang

9. ***In Vivo* Imaging of Autoimmune Disease in Model Systems**
   Eric T. Ahrens and Penelope A. Morel

Volume 71

1. **The Choroid Plexus-Cerebrospinal Fluid System: From Development to Aging**
   Zoran B. Redzic, Jane E. Preston, John A. Duncan, Adam Chodobski, and Joanna Szmydynger-Chodobska

2. **Zebrafish Genetics and Formation of Embryonic Vasculature**
   Tao P. Zhong

3. **Leaf Senescence: Signals, Execution, and Regulation**
   Yongfeng Guo and Susheng Gan

4. **Muscle Stem Cells and Regenerative Myogenesis**
   Iain W. McKinnell, Gianni Parise, and Michael A. Rudnicki

5. **Gene Regulation in Spermatogenesis**
   James A. MacLean II and Miles F. Wilkinson

6. **Modeling Age-Related Diseases in *Drosophila*: Can this Fly?**
   Kinga Michno, Diana van de Hoef, Hong Wu, and Gabrielle L. Boulianne

7. **Cell Death and Organ Development in Plants**
   Hilary J. Rogers

8. **The Blood-Testis Barrier: Its Biology, Regulation, and Physiological Role in Spermatogenesis**
   Ching-Hang Wong and C. Yan Cheng

9. **Angiogenic Factors in the Pathogenesis of Preeclampsia**
   Hai-Tao Yuan, David Haig, and S. Ananth Karumanchi

Volume 72

1. **Defending the Zygote: Search for the Ancestral Animal Block to Polyspermy**
   Julian L. Wong and Gary M. Wessel

2. **Dishevelled: A Mobile Scaffold Catalyzing Development**
   Craig C. Malbon and Hsien-yu Wang

3. **Sensory Organs: Making and Breaking the Pre-Placodal Region**
   Andrew P. Bailey and Andrea Streit

4. **Regulation of Hepatocyte Cell Cycle Progression and Differentiation by Type I Collagen Structure**
   Linda K. Hansen, Joshua Wilhelm, and John T. Fassett

5. **Engineering Stem Cells into Organs: Topobiological Transformations Demonstrated by Beak, Feather, and Other Ectodermal Organ Morphogenesis**
   Cheng-Ming Chuong, Ping Wu, Maksim Plikus, Ting-Xin Jiang, and Randall Bruce Widelitz

6. **Fur Seal Adaptations to Lactation: Insights into Mammary Gland Function**
   Julie A. Sharp, Kylie N. Cane, Christophe Lefevre, John P. Y. Arnould, and Kevin R. Nicholas

Volume 73

1. **The Molecular Origins of Species-Specific Facial Pattern**
   Samantha A. Brugmann, Minal D. Tapadia, and Jill A. Helms

2. **Molecular Bases of the Regulation of Bone Remodeling by the Canonical Wnt Signaling Pathway**
   Donald A. Glass II and Gerard Karsenty

3. **Calcium Sensing Receptors and Calcium Oscillations: Calcium as a First Messenger**
   Gerda E. Breitwieser

4. **Signal Relay During the Life Cycle of *Dictyostelium***
   Dana C. Mahadeo and Carole A. Parent

5. **Biological Principles for *Ex Vivo* Adult Stem Cell Expansion**
   Jean-François Paré and James L. Sherley

6. **Histone Deacetylation as a Target for Radiosensitization**
   David Cerna, Kevin Camphausen, and Philip J. Tofilon

7. **Chaperone-Mediated Autophagy in Aging and Disease**
   Ashish C. Massey, Cong Zhang, and Ana Maria Cuervo

8. Extracellular Matrix Macroassembly Dynamics in Early Vertebrate Embryos

   Andras Czirok, Evan A. Zamir, Michael B. Filla, Charles D. Little, and Brenda J. Rongish

Volume 74

1. Membrane Origin for Autophagy

   Fulvio Reggiori

2. Chromatin Assembly with H3 Histones: Full Throttle Down Multiple Pathways

   Brian E. Schwartz and Kami Ahmad

3. Protein–Protein Interactions of the Developing Enamel Matrix

   John D. Bartlett, Bernhard Ganss, Michel Goldberg, Janet Moradian-Oldak, Michael L. Paine, Malcolm L. Snead, Xin Wen, Shane N. White, and Yan L. Zhou

4. Stem and Progenitor Cells in the Formation of the Pulmonary Vasculature

   Kimberly A. Fisher and Ross S. Summer

5. Mechanisms of Disordered Granulopoiesis in Congenital Neutropenia

   David S. Grenda and Daniel C. Link

6. Social Dominance and Serotonin Receptor Genes in Crayfish

   Donald H. Edwards and Nadja Spitzer

7. Transplantation of Undifferentiated, Bone Marrow-Derived Stem Cells

   Karen Ann Pauwelyn and Catherine M. Verfaillie

8. The Development and Evolution of Division of Labor and Foraging Specialization in a Social Insect (*Apis mellifera* L.)

   Robert E. Page Jr., Ricarda Scheiner, Joachim Erber, and Gro V. Amdam

Volume 75

1. Dynamics of Assembly and Reorganization of Extracellular Matrix Proteins

   Sarah L. Dallas, Qian Chen, and Pitchumani Sivakumar

2. Selective Neuronal Degeneration in Huntington's Disease

   Catherine M. Cowan and Lynn A. Raymond

3. RNAi Therapy for Neurodegenerative Diseases
   Ryan L. Boudreau and Beverly L. Davidson

4. Fibrillins: From Biogenesis of Microfibrils to Signaling Functions
   Dirk Hubmacher, Kerstin Tiedemann, and Dieter P. Reinhardt

5. Proteasomes from Structure to Function: Perspectives from Archaea
   Julie A. Maupin-Furlow, Matthew A. Humbard, P. Aaron Kirkland,
   Wei Li, Christopher J. Reuter, Amy J. Wright, and G. Zhou

6. The Cytomatrix as a Cooperative System of Macromolecular and Water Networks
   V. A. Shepherd

7. Intracellular Targeting of Phosphodiesterase-4 Underpins Compartmentalized cAMP Signaling
   Martin J. Lynch, Elaine V. Hill, and Miles D. Houslay

Volume 76

1. BMP Signaling in the Cartilage Growth Plate
   Robert Pogue and Karen Lyons

2. The CLIP-170 Orthologue Bik1p and Positioning the Mitotic Spindle in Yeast
   Rita K. Miller, Sonia D'Silva, Jeffrey K. Moore, and Holly V. Goodson

3. Aggregate-Prone Proteins Are Cleared from the Cytosol by Autophagy: Therapeutic Implications
   Andrea Williams, Luca Jahreiss, Sovan Sarkar, Shinji Saiki,
   Fiona M. Menzies, Brinda Ravikumar, and David C. Rubinsztein

4. Wnt Signaling: A Key Regulator of Bone Mass
   Roland Baron, Georges Rawadi, and Sergio Roman-Roman

5. Eukaryotic DNA Replication in a Chromatin Context
   Angel P. Tabancay, Jr. and Susan L. Forsburg

6. The Regulatory Network Controlling the Proliferation–Meiotic Entry Decision in the *Caenorhabditis elegans* Germ Line
   Dave Hansen and Tim Schedl

7. Regulation of Angiogenesis by Hypoxia and Hypoxia-Inducible Factors
   Michele M. Hickey and M. Celeste Simon

Volume 77

1. **The Role of the Mitochondrion in Sperm Function: Is There a Place for Oxidative Phosphorylation or Is this a Purely Glycolytic Process?**
   Eduardo Ruiz-Pesini, Carmen Díez-Sánchez, Manuel José López-Pérez, and José Antonio Enríquez

2. **The Role of Mitochondrial Function in the Oocyte and Embryo**
   Rémi Dumollard, Michael Duchen, and John Carroll

3. **Mitochondrial DNA in the Oocyte and the Developing Embryo**
   Pascale May-Panloup, Marie-Françoise Chretien, Yves Malthiery, and Pascal Reynier

4. **Mitochondrial DNA and the Mammalian Oocyte**
   Eric A. Shoubridge and Timothy Wai

5. **Mitochondrial Disease—Its Impact, Etiology, and Pathology**
   R. McFarland, R. W. Taylor, and D. M. Turnbull

6. **Cybrid Models of mtDNA Disease and Transmission, from Cells to Mice**
   Ian A. Trounce and Carl A. Pinkert

7. **The Use of Micromanipulation Methods as a Tool to Prevention of Transmission of Mutated Mitochondrial DNA**
   Helena Fulka and Josef Fulka, Jr.

8. **Difficulties and Possible Solutions in the Genetic Management of mtDNA Disease in the Preimplantation Embryo**
   J. Poulton, P. Oakeshott, and S. Kennedy

9. **Impact of Assisted Reproductive Techniques: A Mitochondrial Perspective from the Cytoplasmic Transplantation**
   A. J. Harvey, T. C. Gibson, T. M. Quebedeaux, and C. A. Brenner

10. **Nuclear Transfer: Preservation of a Nuclear Genome at the Expense of Its Associated mtDNA Genome(s)**
    Emma J. Bowles, Keith H. S. Campbell, and Justin C. St. John

Volume 78

1. **Contribution of Membrane Mucins to Tumor Progression Through Modulation of Cellular Growth Signaling Pathways**
   Kermit L. Carraway III, Melanie Funes, Heather C. Workman, and Colleen Sweeney

2. **Regulation of the Epithelial Na^+ Channel by Peptidases**
   Carole Planès and George H. Caughey

3. **Advances in Defining Regulators of Cementum Development and Periodontal Regeneration**
   Brian L. Foster, Tracy E. Popowics, Hanson K. Fong, and Martha J. Somerman

4. **Anabolic Agents and the Bone Morphogenetic Protein Pathway**
   I. R. Garrett

5. **The Role of Mammalian Circadian Proteins in Normal Physiology and Genotoxic Stress Responses**
   Roman V. Kondratov, Victoria Y. Gorbacheva, and Marina P. Antoch

6. **Autophagy and Cell Death**
   Devrim Gozuacik and Adi Kimchi

Volume 79

1. **The Development of Synovial Joints**
   I. M. Khan, S. N. Redman, R. Williams, G. P. Dowthwaite, S. F. Oldfield, and C. W. Archer

2. **Development of a Sexually Differentiated Behavior and Its Underlying CNS Arousal Functions**
   Lee-Ming Kow, Cristina Florea, Marlene Schwanzel-Fukuda, Nino Devidze, Hosein Kami Kia, Anna Lee, Jin Zhou, David MacLaughlin, Patricia Donahoe, and Donald Pfaff

3. **Phosphodiesterases Regulate Airway Smooth Muscle Function in Health and Disease**
   Vera P. Krymskaya and Reynold A. Panettieri, Jr.

4. Role of Astrocytes in Matching Blood Flow to Neuronal Activity
   Danica Jakovcevic and David R. Harder

5. Elastin-Elastases and Inflamm-Aging
   Frank Antonicelli, Georges Bellon, Laurent Debelle, and William Hornebeck

6. A Phylogenetic Approach to Mapping Cell Fate
   Stephen J. Salipante and Marshall S. Horwitz

Volume 80

1. Similarities Between Angiogenesis and Neural Development: What Small Animal Models Can Tell Us
   Serena Zacchigna, Carmen Ruiz de Almodovar, and Peter Carmeliet

2. Junction Restructuring and Spermatogenesis: The Biology, Regulation, and Implication in Male Contraceptive Development
   Helen H. N. Yan, Dolores D. Mruk, and C. Yan Cheng

3. Substrates of the Methionine Sulfoxide Reductase System and Their Physiological Relevance
   Derek B. Oien and Jackob Moskovitz

4. Organic Anion-Transporting Polypeptides at the Blood–Brain and Blood–Cerebrospinal Fluid Barriers
   Daniel E. Westholm, Jon N. Rumbley, David R. Salo, Timothy P. Rich, and Grant W. Anderson

5. Mechanisms and Evolution of Environmental Responses in *Caenorhabditis elegans*
   Christian Braendle, Josselin Milloz, and Marie-Anne Félix

6. Molluscan Shell Proteins: Primary Structure, Origin, and Evolution
   Frédéric Marin, Gilles Luquet, Benjamin Marie, and Davorin Medakovic

7. Pathophysiology of the Blood–Brain Barrier: Animal Models and Methods
   Brian T. Hawkins and Richard D. Egleton

8. Genetic Manipulation of Megakaryocytes to Study Platelet Function
   Jun Liu, Jan DeNofrio, Weiping Yuan, Zhengyan Wang, Andrew W. McFadden, and Leslie V. Parise

9. Genetics and Epigenetics of the Multifunctional Protein CTCF
   Galina N. Filippova

Volume 81

1. **Models of Biological Pattern Formation: From Elementary Steps to the Organization of Embryonic Axes**
   Hans Meinhardt

2. **Robustness of Embryonic Spatial Patterning in *Drosophila Melanogaster***
   David Umulis, Michael B. O'Connor, and Hans G. Othmer

3. **Integrating Morphogenesis with Underlying Mechanics and Cell Biology**
   Lance A. Davidson

4. **The Mechanisms Underlying Primitive Streak Formation in the Chick Embryo**
   Manli Chuai and Cornelis J. Weijer

5. **Grid-Free Models of Multicellular Systems, with an Application to Large-Scale Vortices Accompanying Primitive Streak Formation**
   T. J. Newman

6. **Mathematical Models for Somite Formation**
   Ruth E. Baker, Santiago Schnell, and Philip K. Maini

7. **Coordinated Action of N-CAM, N-cadherin, EphA4, and ephrinB2 Translates Genetic Prepatterns into Structure during Somitogenesis in Chick**
   James A. Glazier, Ying Zhang, Maciej Swat, Benjamin Zaitlen, and Santiago Schnell

8. **Branched Organs: Mechanics of Morphogenesis by Multiple Mechanisms**
   Sharon R. Lubkin

9. **Multicellular Sprouting during Vasculogenesis**
   Andras Czirok, Evan A. Zamir, Andras Szabo, and Charles D. Little

10. **Modelling Lung Branching Morphogenesis**
    Takashi Miura

11. **Multiscale Models for Vertebrate Limb Development**
    Stuart A. Newman, Scott Christley, Tilmann Glimm, H. G. E. Hentschel, Bogdan Kazmierczak, Yong-Tao Zhang, Jianfeng Zhu, and Mark Alber

12. Tooth Morphogenesis *in vivo, in vitro* and *in silico*
    Isaac Salazar-Ciudad
13. Cell Mechanics with a 3D Kinetic and Dynamic Weighted Delaunay-Triangulation
    Michael Meyer-Hermann
14. Cellular Automata as Microscopic Models of Cell Migration in Heterogeneous Environments
    H. Hatzikirou and A. Deutsch
15. Multiscale Modeling of Biological Pattern Formation
    Ramon Grima
16. Relating Biophysical Properties Across Scales
    Elijah Flenner, Francoise Marga, Adrian Neagu, Ioan Kosztin, and Gabor Forgacs
17. Complex Multicellular Systems and Immune Competition: New Paradigms Looking for a Mathematical Theory
    N. Bellomo and G. Forni

Volume 82

1. Ontogeny of Erythropoiesis in the Mammalian Embryo
   Kathleen McGrath and James Palis
2. The Erythroblastic Island
   Deepa Manwani and James J. Bieker
3. Epigenetic Control of Complex Loci During Erythropoiesis
   Ryan J. Wozniak and Emery H. Bresnick
4. The Role of the Epigenetic Signal, DNA Methylation, in Gene Regulation During Erythroid Development
   Gordon D. Ginder, Merlin N. Gnanapragasam, and Omar Y. Mian
5. Three-Dimensional Organization of Gene Expression in Erythroid Cells
   Wouter de Laat, Petra Klous, Jurgen Kooren, Daan Noordermeer, Robert-Jan Palstra, Marieke Simonis, Erik Splinter, and Frank Grosveld
6. Iron Homeostasis and Erythropoiesis
   Diedra M. Wrighting and Nancy C. Andrews

7. **Effects of Nitric Oxide on Red Blood Cell Development and Phenotype**
   Vladan P. Čokić and Alan N. Schechter

8. **Diamond Blackfan Anemia: A Disorder of Red Blood Cell Development**
   Steven R. Ellis and Jeffrey M. Lipton

Volume 83

1. **Somatic Sexual Differentiation in *Caenorhabditis elegans***
   Jennifer Ross Wolff and David Zarkower

2. **Sex Determination in the *Caenorhabditis elegans* Germ Line**
   Ronald E. Ellis

3. **The Creation of Sexual Dimorphism in the *Drosophila* Soma**
   Nicole Camara, Cale Whitworth, and Mark Van Doren

4. ***Drosophila* Germline Sex Determination: Integration of Germline Autonomous Cues and Somatic Signals**
   Leonie U. Hempel, Rasika Kalamegham, John E. Smith III, and Brian Oliver

5. **Sexual Development of the Soma in the Mouse**
   Danielle M. Maatouk and Blanche Capel

6. **Development of Germ Cells in the Mouse**
   Gabriela Durcova-Hills and Blanche Capel

7. **The Neuroendocrine Control of Sex-Specific Behavior in Vertebrates: Lessons from Mammals and Birds**
   Margaret M. McCarthy and Gregory F. Ball

Volume 84

1. **Modeling Neural Tube Defects in the Mouse**
   Irene E. Zohn and Anjali A. Sarkar

2. **The Etiopathogenesis of Cleft Lip and Cleft Palate: Usefulness and Caveats of Mouse Models**
   Amel Gritli-Linde

3. **Murine Models of Holoprosencephaly**
   Karen A. Schachter and Robert S. Krauss

4. **Mouse Models of Congenital Cardiovascular Disease**
   Anne Moon

5. **Modeling Ciliopathies: Primary Cilia in Development and Disease**
   Robyn J. Quinlan, Jonathan L. Tobin, and Philip L. Beales

6. **Mouse Models of Polycystic Kidney Disease**
   Patricia D. Wilson

7. **Fraying at the Edge: Mouse Models of Diseases Resulting from Defects at the Nuclear Periphery**
   Tatiana V. Cohen and Colin L. Stewart

8. **Mouse Models for Human Hereditary Deafness**
   Michel Leibovici, Saaid Safieddine, and Christine Petit

9. **The Value of Mammalian Models for Duchenne Muscular Dystrophy in Developing Therapeutic Strategies**
   Glen B. Banks and Jeffrey S. Chamberlain

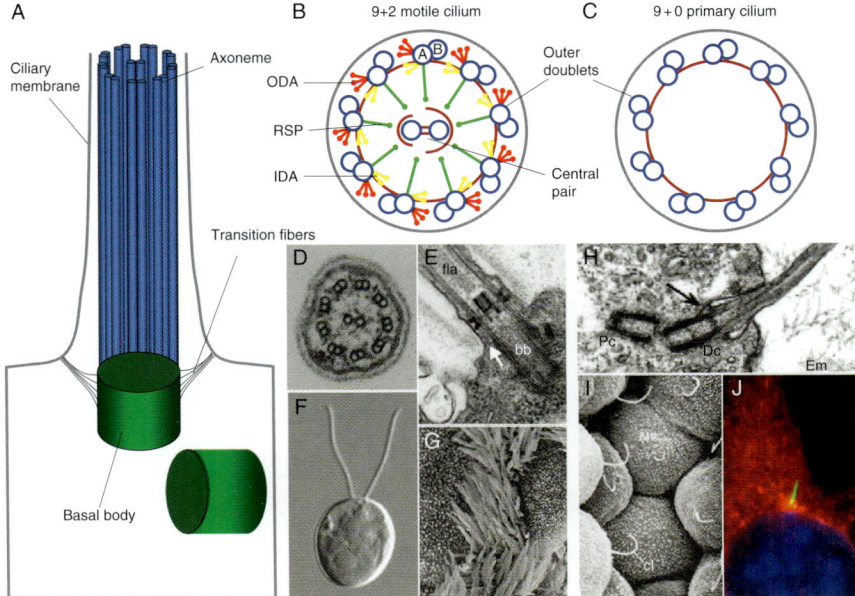

Lotte B. Pedersen and Joel L. Rosenbaum, Figure 2.1 Diversity and structure of cilia and flagella. (A) Schematic showing the overall structure of a 9 + 2 cilium. (B) Schematic of a cross section of a 9 + 2 cilium highlighting the main axonemal components (ODA, outer dynein arm; IDA, inner dynein arm; RSP, radial spoke). (C) Schematic of a cross section of a 9 + 0 cilium that lacks motility-related structures. (D) TEM of a cross section of a *Chlamydomonas* flagellum (courtesy of Stefan Geimer, University of Bayreuth, Germany). (E) Longitudinal section of the base of a *Chlamydomonas* flagellum visualized by TEM. The arrow points to the flagellar base. *Abbreviations*: bb, basal body; fla, flagella. From Mitchell *et al.* (2005) with permission. (F) DIC of a *Chlamydomonas* cell with two 9 + 2 flagella. (G) Scanning EM of motile (9 + 2) tracheal cilia from a 4-week-old mouse (courtesy of Karl F. Lechtreck and George B. Witman, UMass Medical School, Worcester, MA). (H) Longitudinal section (TEM) of a primary cilium from a chicken chondrocyte showing the mother/distal centriole (Dc) and the daughter/proximal centriole (Pc). Arrow points to the transition fibers. *Abbreviation*: Em, extracellular matrix. Reprinted from Jensen *et al.* (2004) with permission. (I) SEM of ciliated renal epithelial cells from a kidney collecting tubule. *Abbreviations*: Ci, cilium; Mv, microvillus. From Kessel and Kardon (1979) with permission from Randy H. Kardon. (J) Immunofluorescence micrograph of a mouse NIH3T3 fibroblast stained with an antibody against detyrosinated α-tubulin to label the primary cilium (green) and with a p150^{Glued} antibody that labels the ciliary base (red). The nucleus is stained with DAPI (blue). Courtesy of Jacob M. Schrøder, University of Copenhagen, Denmark, who also generated the completed figure.

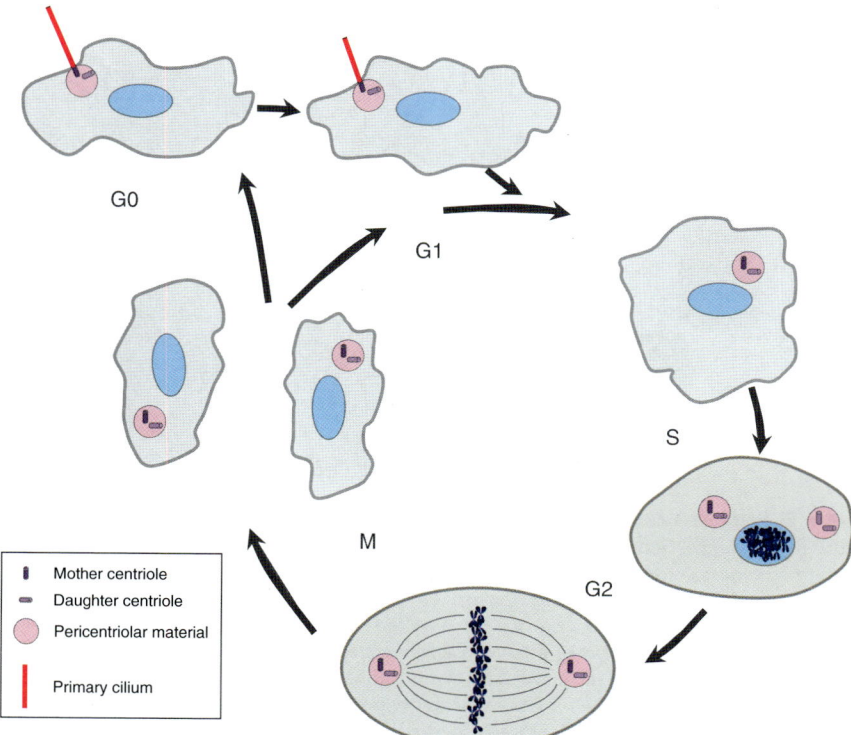

Lotte B. Pedersen and Joel L. Rosenbaum, Figure 2.2 Assembly and disassembly of primary cilia are coordinated with the cell cycle. The primary cilium is formed in G1/G0 following docking of the mother centriole at the ciliary assembly site at the apical plasma membrane. Cilium elongation is mediated by IFT-dependent addition of ciliary precursors to the distal end of the mother centriole, which at this point is termed a basal body. Disassembly of the cilium prior to mitosis allows both centriole pairs to function in mitotic spindle formation.

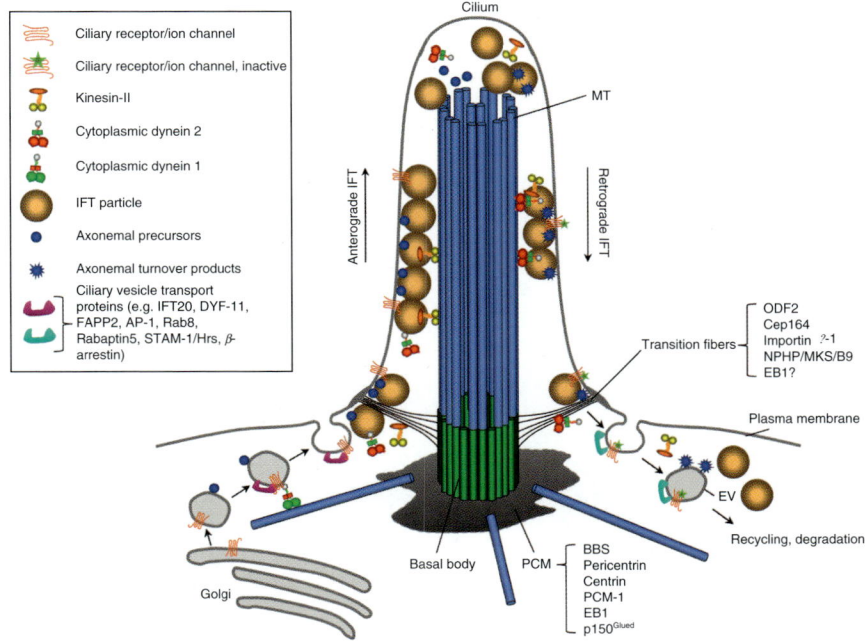

Lotte B. Pedersen and Joel L. Rosenbaum, Figure 2.3 IFT and targeting of proteins to the ciliary compartment. Proteins destined for the ciliary compartment (membrane proteins as well as axonemal components) are transported in Golgi-derived vesicles to the base of the cilium where the vesicles are exocytosed and the ciliary proteins associate with IFT particles. This Golgi-to-cilium-mediated vesicle transport, which is proposed to involve cytoplasmic dynein 1 MT-based movement, depends on the IFT complex B proteins IFT20 and DYF-11, the small G protein Rab8 and associated GEFs (e.g., Rabin 8, Rabaptin 5), FAPP2, and adapter proteins such as AP-1. BBS proteins and other proteins localized in the pericentrosomal region (e.g., PCM-1, EB1, p150^{Glued}) may provide a link between the Golgi-derived vesicles and the transition fibers at the ciliary base and may also serve to anchor MTs at the basal body. At the ciliary base, only proteins (or protein complexes) containing specific ciliary targeting motifs are allowed access through the zone defined by the transition fibers. Selective entry of proteins into the ciliary compartment probably involves specific G proteins and GEFs that are associated with NPHPs, MKS, and B9 domain-containing proteins. Following entry into the ciliary compartment, these proteins, along with inactive cytoplasmic dynein 2, are transported anterogradely along the axoneme by kinesin-II-mediated IFT. At the ciliary tip IFT particles are remodeled, kinesin-II inactivated, and cytoplasmic dynein 2 activated. Ciliary turnover products (e.g., inactive receptors) are, in turn, transported retrogradely along ciliary axonemes by cytoplasmic dynein 2 for recycling or degradation in the cytoplasm. Recycling or turnover of ciliary membrane receptors may involve ubiquitination (e.g., via BBS proteins) and/or dephosphorylation of the receptors as well as binding to endosomal vesicle adapter proteins such as STAM-1/Hrs (Bae and Barr, 2008) or β-arrestin (Kovacs et al., 2008). Figure based on references (Azimzadeh and Bornens, 2007; Leroux, 2007; Rosenbaum and Witman, 2002) as well as references cited in the text. *Abbreviations*: EV, endocytic vesicle; MT, microtubule; PCM, pericentriolar material. Figure generated by Jacob M. Schrøder, University of Copenhagen.

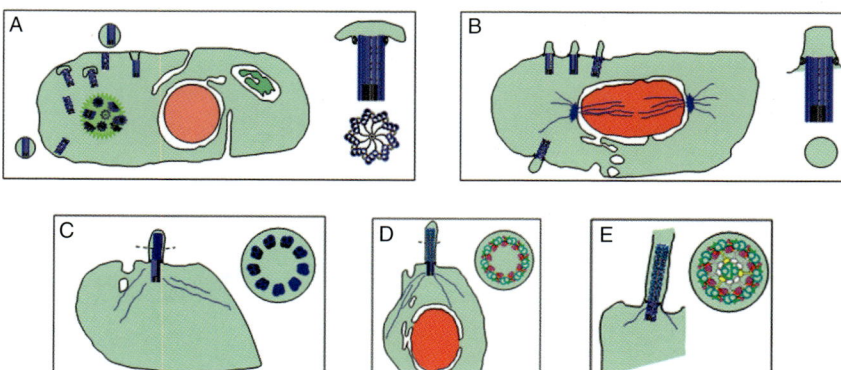

Peter Satir et al., Figure 3.2 The viral hypothesis of ciliary evolution. (A) Membrane trafficking in the protoeukaryotic bacterial cytoplasm leads to endocytosis of cell organelles. An RNA enveloped virus with ninefold symmetry invades this cytoplasm. Cross section shows ninefold symmetry at the cartwheel. (B) The centriolar virus remains attached to the cell membrane and a sensory bulge capable of positional signaling develops. GPS function leads to restriction in organelle number and development of cell polarity. Cross section shows area of attachment at the ciliary necklace. (C) Elongation of the centriolar capsid, utilizing α- and β-tubulin, IFT and motor proteins of the host cell produces the 9 + 0 axoneme (cross section) in a sensory organelle. (D) Cytoplasmic dynein diversifies to give axonemal isoforms—arms between doublets in cross section—motility begins. (E) Misplaced cytoplasmic MTs give rise to the central pair leading to an increase in efficient motility in a stabilized 9 + 2 axoneme (cross section). Efficient signaling and motility in the 9 + 2 axoneme are advantageous for protistan survival and outcompete earlier forms. Images from Satir *et al.* (2007). Courtesy *Cell Motil. Cytoskeleton*.

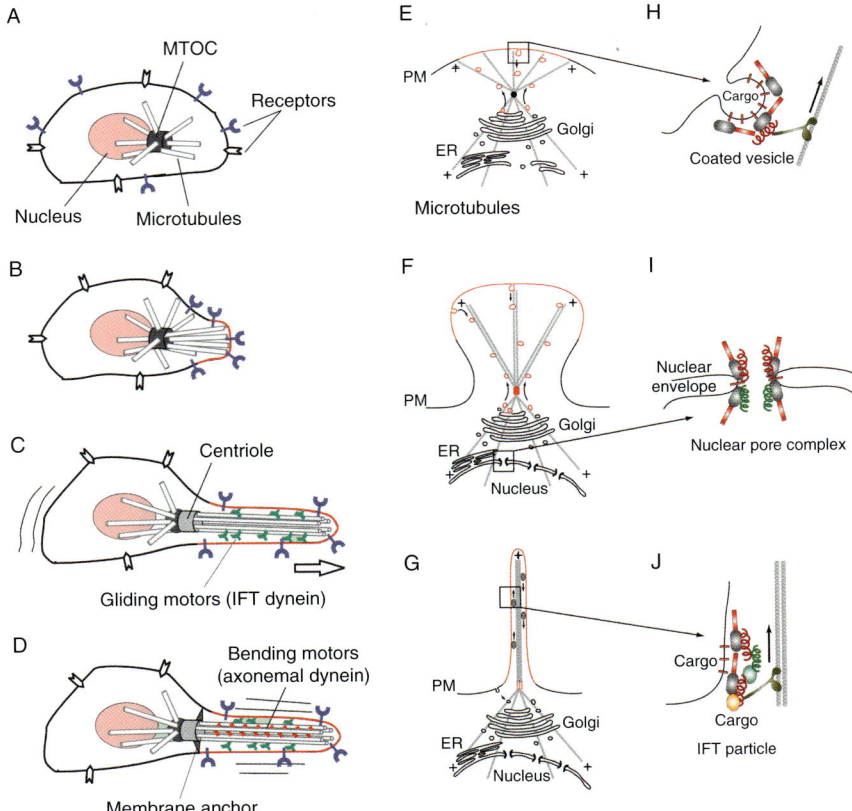

Peter Satir *et al.*, Figure 3.3 The autogenous hypothesis of the evolution of motile ciliary functions and IFT from coated vesicle transport. (A) The evolving eukaryotic cell possesses microtubules emanating from an organizing center (MTOC) and receptors randomly distributed on the cell surface. (B) Sensory receptor capping and directional transport lead to MT elongation and bundling. A specialized membrane patch (red) appears at the ends of elongated MT. (C) The bundled MTs form the primitive basal body (centriole) and a primitive axoneme. Vesicle transport motors evolve into IFT kinesin and dynein (green). The axoneme uses IFT motors for gliding motility, promoting selection of the polarized MT bundle/protoaxoneme. (D) Axonemal dynein isoforms evolve (red) that permit bending movement of the axoneme, which drives selection of doublet microtubules and of a centriole with membrane anchors. The 9 + 2 pattern became fixed at a later time based on its superior regulation of bending parameters. (E) The proto-IFT complex in a protoeukaryote mediated transport of transmembrane proteins (endocytosis and/or recycling) at a specialized region of the plasma membrane (shown in red). (F) Microtubule growth exposed this membrane domain to the environment. Transmembrane proteins were transported in vesicles, proteins of the developing axoneme arrived by diffusion. Nuclear pore complexes also diverged from a protocoatomer complex. (G) Bidirectional IFT in modern cilia with kinesin-II as the anterograde and dynein 1b/2 as the retrograde motor. The motors are thought to have evolved from the ones that moved coated vesicles. IFT cargo can be either transmembrane or cytoplasmic. During the subsequent stages of cilia evolution basal body structure coevolved with axonemal structure. (H), (I), and (J) Organization of coated vesicle complexes, nuclear pore complexes, and the IFT complex. Adapted from Mitchell (2004) and Jékely and Arendt (2006).

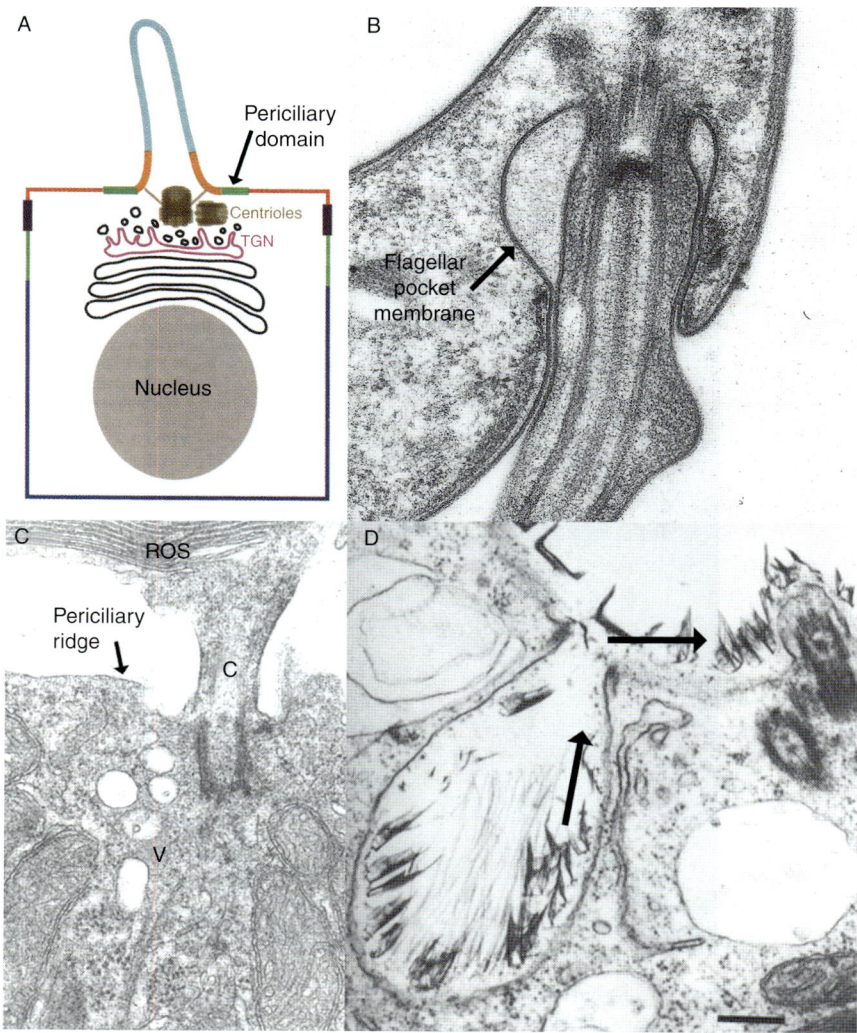

Gregory J. Pazour and Robert A. Bloodgood, Figure 5.2 Four examples of putative periciliary sorting domains. (A) Diagram from Reiter and Mostov (2006) showing location of periciliary domains in canine kidney cells. (B) Transmission electron micrograph (courtesy of Y.-D. Stierhof) showing the location of the flagellar membrane pocket in *Trypanosoma brucei*. (C) Transmission electron micrograph from Papermaster *et al*. (1985) showing the apical end of a frog rod photoreceptor inner segment and connecting cilium, with the periciliary ridge domain indicated by the arrow. (D) Transmission electron micrograph from McFadden and Wetherbee (1985) showing a portion of the cell surface of the algal cell *Pyramimonas* showing the scale reservoir located near the base of a flagellum. It is presumed that the flagellar membrane scales migrate from the scale reservoir membrane onto the flagellar membrane.

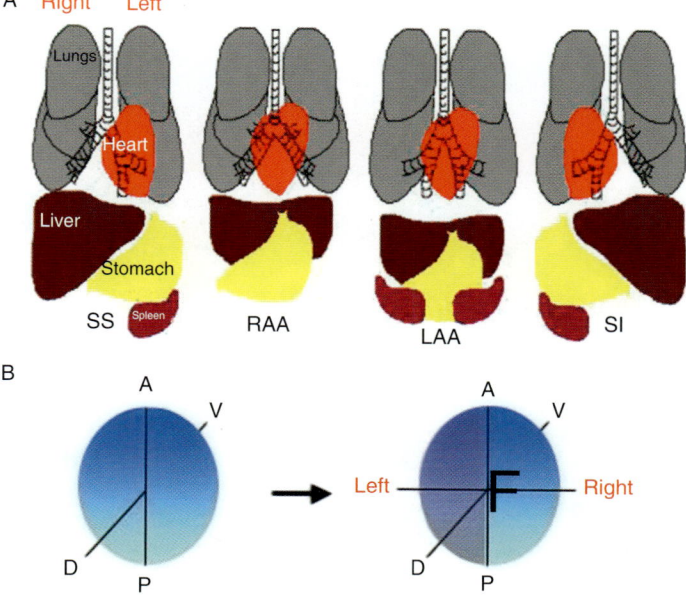

(Adapted from: Brown and Wolpert, development 1990)

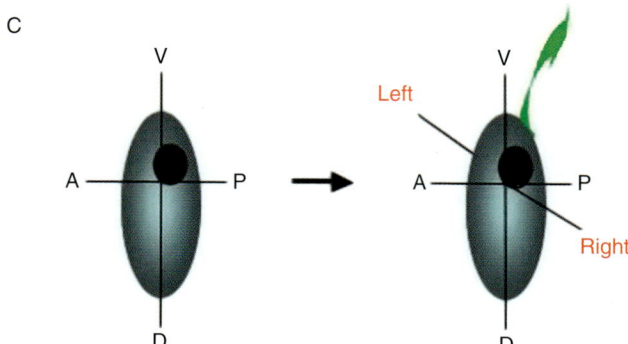

Basudha Basu and Martina Brueckner, Figure 6.1 (A) The anatomic spectrum of organ laterality (SS, situs solitus; RAA, right atrial isomerism). The liver is midline, there are two eparterial bronchi, the position of the stomach and cardiac apex is indeterminate and there is asplenia (LAA, left atrial isomerism). The liver is midline, there are two hyparterial bronchi, the position of the stomach and cardiac apex is indeterminate and there are multiple spleens (SI, situs inversus). (B) Orientation of the hypothetical "F" chiral molecule or macromolecular structure along the AP and DV axes creates obligate chiral asymmetry. (C) The cilium shown as a green arrow is the chiral macromolecular structure that creates chiral asymmetry of the cells at the mouse node.

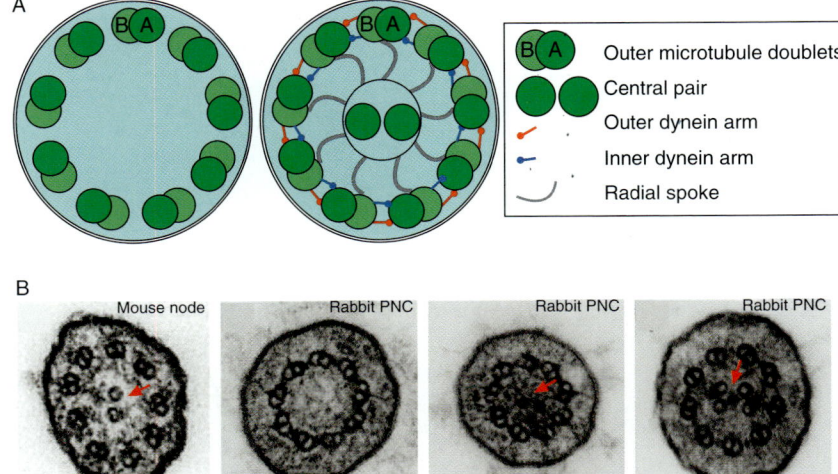

Basudha Basu and Martina Brueckner, Figure 6.2 (A) Outline of the ciliary axoneme. The axoneme shown at the left can be adapted to a variety of purposes, including the full array of motility components diagrammed on the right. (B) TEM of cilia at the mouse node (Caspary *et al.*, 2007) and rabbit PNC (Feistel and Blum, 2006) showing a wide spectrum of ciliary architecture including $9 + 2$, $9 + 0$, and $9 + 4$ microtubule configurations. Red arrow indicates the location of the central pair(s) of microtubules.

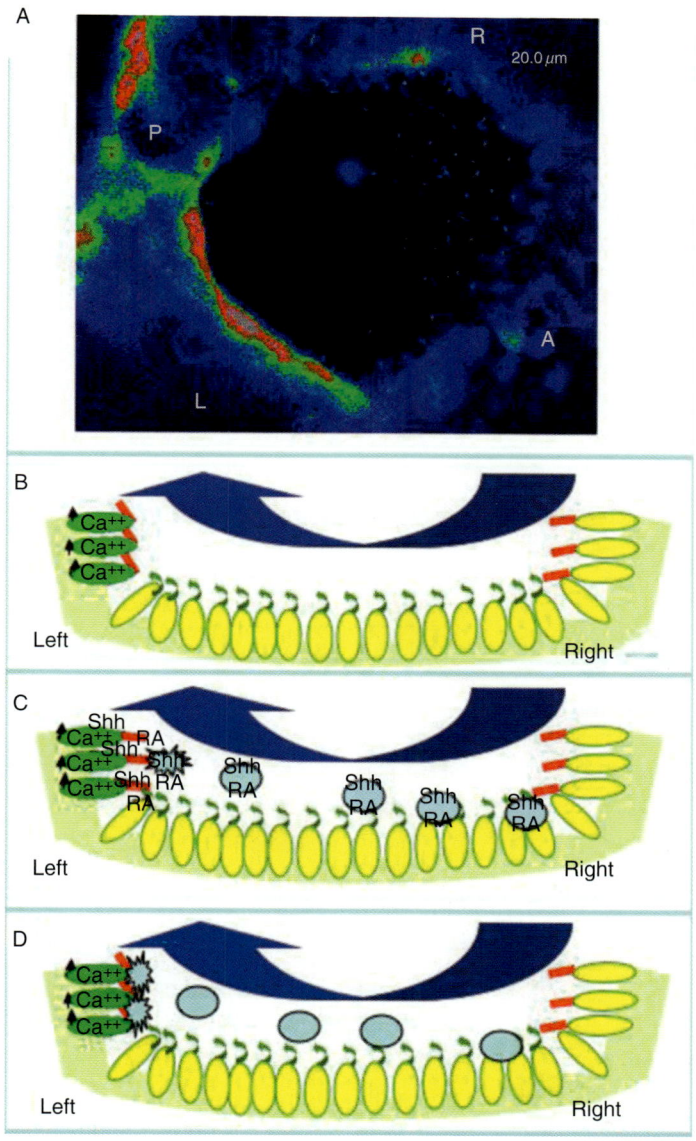

CTDBCH-06fig-03

Basudha Basu and Martina Brueckner, Figure 6.3 (A) Calcium signal at the mouse node. e7.8 embryo showing fluo3 fluorescence at left and posterior margins of the node. Node cilia fluoresce secondary to the GFP-tagged lrd. The cilia are motile, as is indicated by the multiple signals emanating from a single cilium during the acquisition of a single slow scan. (B) The mechanosensory model for left-sided calcium signal. Motile cilia are shown in green, polycystin-containing nonmotile cilia are shown in red. Cilia at the left side of the node are affected differently than on the right to the geometric configuration of the node coupled with mediolateral asymmetry of the cilia themselves. (C) The nodal vesicular parcel model. Motile cilia are shown in green. NVPs loaded with putative morphogens ''shatter'' preferentially on cilia and cell wall at the left. (D) The ''combined'' model. Flow-driven interaction of NVPs with mechanosensory cilia leads to an intracellular calcium signal on the left.

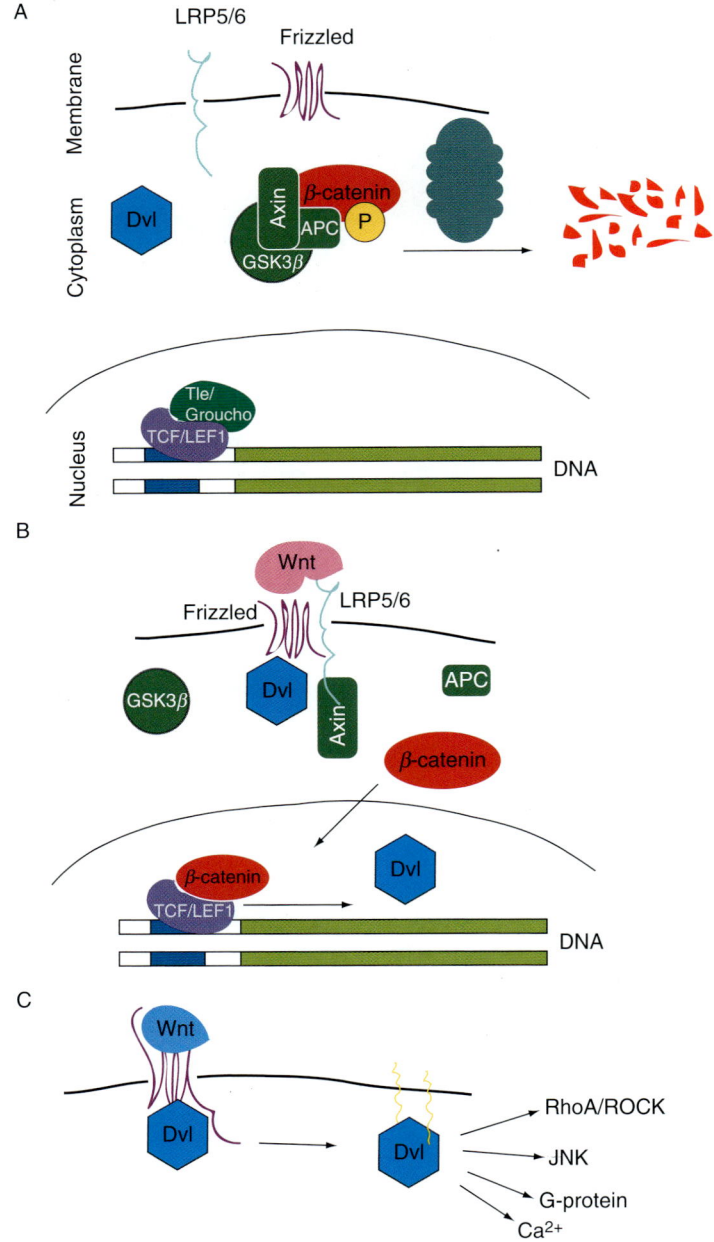

Jantje M. Gerdes and Nicholas Katsanis, Figure 7.1 Wnt signaling overview. (A) In absence of extracellular Wnt ligand, β-catenin is constitutively phosphorylated by the β-catenin destruction complex and thus targeted for proteasomal degradation. Transcription factors TCF/LEF1 are complexed by Tle/Groucho protein to repress transcription of β-catenin/TCF/LEF1 targets. (B) In canonical β-catenin-dependent Wnt

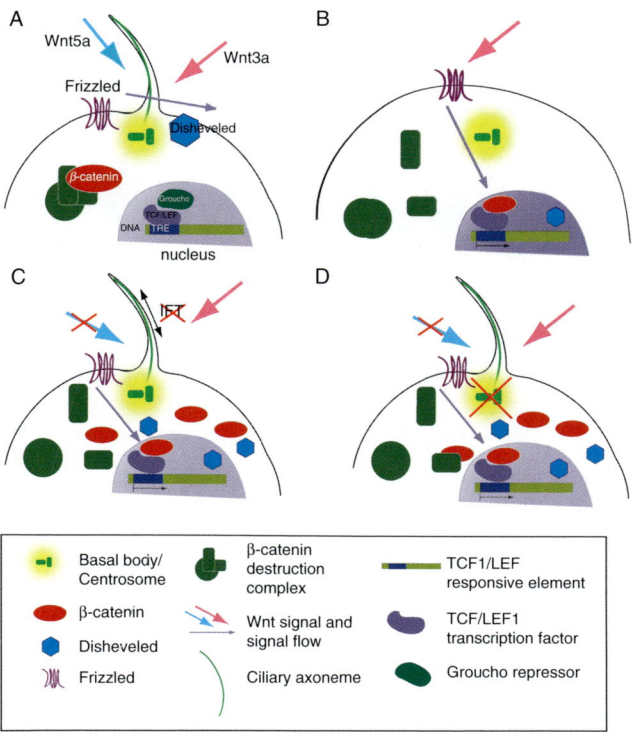

Jantje M. Gerdes and Nicholas Katsanis, Figure 7.2 The role of the primary cilium and basal body in Wnt signaling. (A) In the presence of a primary cilium, noncanonical Wnt signals (such as Wnt5a) antagonize canonical signals (such as Wnt3a) and the Wnt signal is transmitted through membrane-associated DVL to downstream targets. (B) In unciliated cells, the external Wnt stimulus is transmitted through Fz receptors and leads to stabilization of β-catenin and subsequent transcriptional activation of β-catenin targets. (C) and (D) If the ciliary axoneme is disrupted (such as by suppression of KIF3a or IFT88), or the basal body is compromised (by suppression of BBS4 or Ofd1), antagonism of noncanonical and canonical Wnt signals and thus inhibition of β-catenin-dependent signaling is lost; both cytoplasmic β-catenin and DVL are stabilized and accumulate in the cytoplasm and nucleus, leading to transcriptional activation of β-catenin responsive elements.

signaling, external Wnt ligand binds to a heterodimer of Fz receptor and LRP5/6 coreceptor, leading to dissociation of the β-catenin destruction complex. Both DVL and β-catenin are localized to the nucleus in the context of canonical Wnt signaling, where β-catenin displaces Tle/Groucho and leads to transcriptional activation of β-catenin targets. (C) In β-catenin-independent signaling, an external Wnt ligand binds to Fz receptor and activates DVL, which is commonly associated with the membrane in the context of noncanonical Wnt signaling. DVL then activates putative noncanonical Wnt signaling components such as RhoA/ROCK, JNK, G protein or Ca^{2+} signaling.

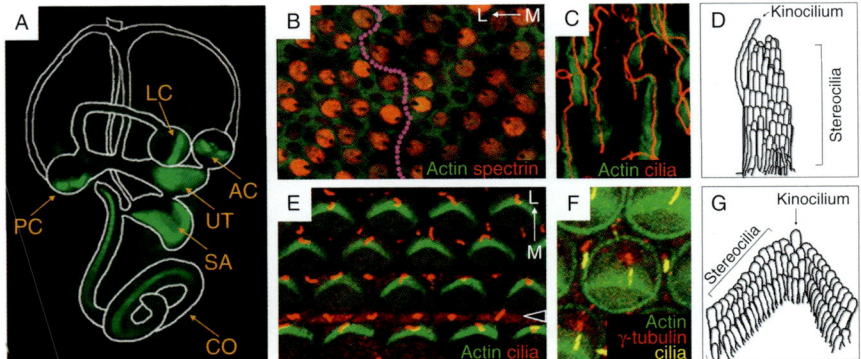

Chonnettia Jones and Ping Chen, Figure 8.1 Planar cell polarity of inner ear sensory epithelia. (A) Mouse inner ear isolated at E18.5. The white tracing outlines the fluid-filled labyrinth that connects the vestibule and the cochlea of the inner ear. A Math1-GFP transgene (green) is expressed in hair cells and highlights the sensory epithelia of the six sensory organs: the organ of Corti of the cochlea (CO); the maculae of the utricle (UT) and the saccule (SA); and the three perpendicular cristae: the posterior crista (PC), anterior crista (AC), and lateral crista (LC). (B)–(G) The planar cell polarity of vestibular (B)–(D) and cochlear (E)–(G) sensory organs isolated from a mouse embryo at E18.5 viewed by immunohistochemistry and confocal microscopy (B), (C), (E), (F), and the intrinsic polarity of vestibular (D) and cochlear (G) hair cells illustrated by schematic diagrams. The hair cells are polarized across the sensory epithelium along the medial-to-lateral axis (M, medial; L, lateral). (B) The utricle viewed at a level just beneath the cell cortex. Actin (green) is enriched at the cell membranes. Spectrin (red) accumulates in the cuticular plates of hair cells, is excluded from the pericentriolar region surrounding the basal body and serves as a convenient marker of hair cell polarity. The dotted line (purple) demarcates the line of polarity reversal, where the polarities of the hair cells are reversed in the utricle. (E) The surface of the organ of Corti. Actin (green) accentuates the microvilli that make up the stereociliary bundles. Tubulin (red) marks both the cilia associated with stereociliary bundles atop the sensory hair cells and the cilia of intervening nonsensory supporting cells. The black arrowhead marks the separation of the inner hair cells from the outer hair cells, which are uniformly oriented toward the abneural, or lateral, edge of the sensory epithelium. (C) and (F) Confocal projections of vestibular (C) and cochlear (F) hair cell bundles at high magnification. The kinocilia (C, red; F, yellow) are assembled from the eldest of two centrioles (C, not shown; F, red) that make up the basal bodies, and are closely apposed to the hair cell bundles (C, F, green). Kinocilia and stereocilia of the vestibule (C) are considerably longer than those of the cochlea (F). (D) and (G) Diagrams of a typical vestibular (D) and cochlear (G) hair cell bundle showing the array of stereocilia rows increasing in height and the eccentrically placed kinocilium which protrudes from behind the tallest row of stereocilia. Note the round morphology of the vestibular hair cell bundle in comparison to the ''V''-shaped morphology of the cochlear hair cell bundle.

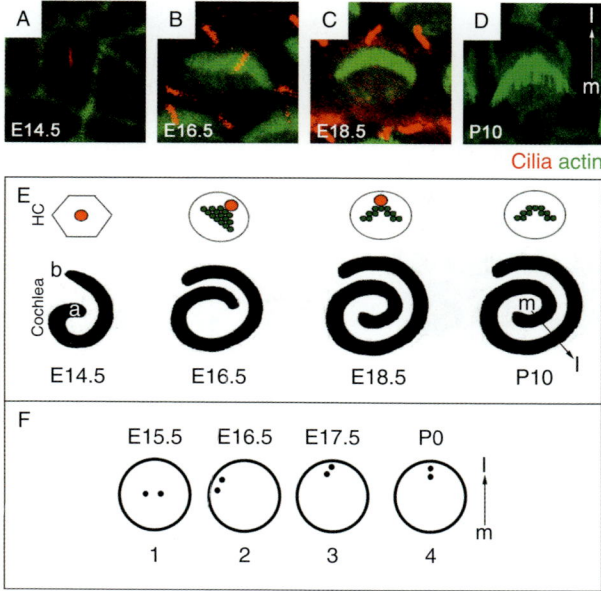

Chonnettia Jones and Ping Chen, Figure 8.2 Kinocilium and basal body polarization. Confocal images of cochlear whole mounts (A)–(D) and a schematic diagram (E) focusing on a single cochlear hair cell bundle at successive stages in mouse ear development. Extension of the cochlear spiral unidirectionally from the base (b) to the distal tip, or apex (a), occurs concurrently with hair cell (HC) differentiation in the organ of Corti. In an immature hair cell, the kinocilium (red) projects centrally from the apical cell membrane. The kinocilium proceeds to polarize the growing microvilli (green) toward the lateral edge of the cell surface. By the end of embryogenesis, the polarized V-shaped morphology of the stereociliary bundle becomes apparent. The orientation of the hair cell bundle is refined postnatally until the kinocilium associated with the vertex of the stereociliary bundle points distally along the medial-to-lateral axis. The medial (neural) and the lateral (abneural) sides of the sensory epithelium face the interior and periphery of the cochlea, respectively. The kinocilium regenerates postnatally prior to the onset of hearing in mammalian cochleae. (F) Diagram illustrating the positioning of the ciliary basal body during cochlear hair cell bundle development. The pair of centrioles that make up the basal body undergoes similar polarization movements during hair cell maturation. The movements can be described as a four-step process: (1) the pair of centrioles is positioned centrally beneath the cell cortex in newly differentiating hair cells; (2) the pair of centrioles migrates to the lateral cell surface; (3) one of the centrioles leads the other toward the distal edge of the sensory epithelium; (4) the pair of centrioles go through a presumed period of refinement as they align themselves along the medial-to-lateral axis, such that the daughter centriole is ultimately positioned distally (laterally), while the elder maternal centriole, from which the kinocilium is assembled, is positioned proximally (medially) and is closely apposed to the stereociliary bundle (not shown) (M, medial; L, lateral).

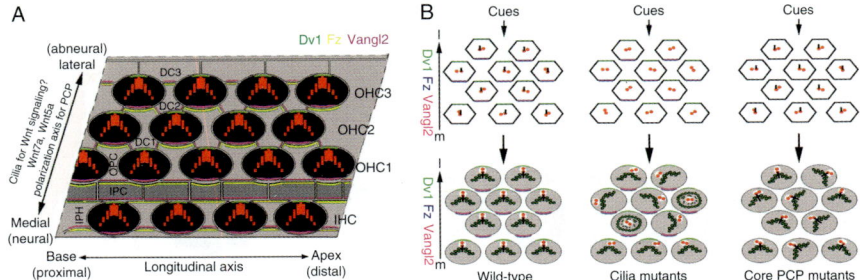

Chonnettia Jones and Ping Chen, Figure 8.3 Models of PCP regulation in the ear. (A) Schematic drawing of the organ of Corti viewed from its surface. Hair cells (black) are arranged into four parallel rows and are separated from each other by nonsensory supporting cells (gray). Hair cells are synchronously aligned across the sensory epithelium, displaying planar cell polarity. The apical and asymmetric distribution of conserved PCP proteins, Dvl (green), Fz (yellow), and Vangl2 (magenta) along the planar polarity axis is required to direct planar cell polarity of the hair cells in the cochlea. Wnt7a and Wnt5a have been implicated in regulating planar cell polarity of hair cells. The role of primary cilia in planar cell polarity has recently been confirmed, and cilia may act as receivers for Wnts or an unknown polarizing signal for PCP regulation (OHC, outer hair cell; IPC, inner pillar cell; IPHC, inner phalangeal cell; DC, Deiters cell). (B) Model of the role of cilia in the morphological polarization of hair cells in the organ of Corti. An unknown polarization signal directs the asymmetric membrane sorting of core PCP proteins Dvl (green), Fz (blue), and Vangl2 (magenta) in wild-type animals and ciliary mutants. Cell–cell interactions reinforce the distribution of PCP proteins along the planar polarity axis. In contrast, the PCP proteins fail to organize properly in core PCP mutants. Consequently, in wild-type cells, the kinocilia (black lines) and hair cell bundles (green) are polarized and all the hair cells are uniformly aligned along the medial-to-lateral, or planar polarization, axis. Like wild-type animals, PCP mutants exhibit normally polarized kinocilia and hair cell bundles, but the bundles are randomly oriented. Ciliary mutants lack kinocilia and exhibit both randomly polarized hair cell bundles and bundles that have lost their intrinsic polarity. Therefore, the distribution of core PCP proteins is necessary but not sufficient to direct polarization of the kinocilia and the stereociliary bundle. Basal bodies, through the positioning and alignment of the centrioles (red), may direct the organization of the cytoskeletal network to build the polarized structure of the stereociliary bundles. It is likely that the interactions between core PCP proteins and the basal bodies, together with their associated cytoskeletal components, cooperate to direct the coordination of stereociliary bundles across the sensory epithelium.

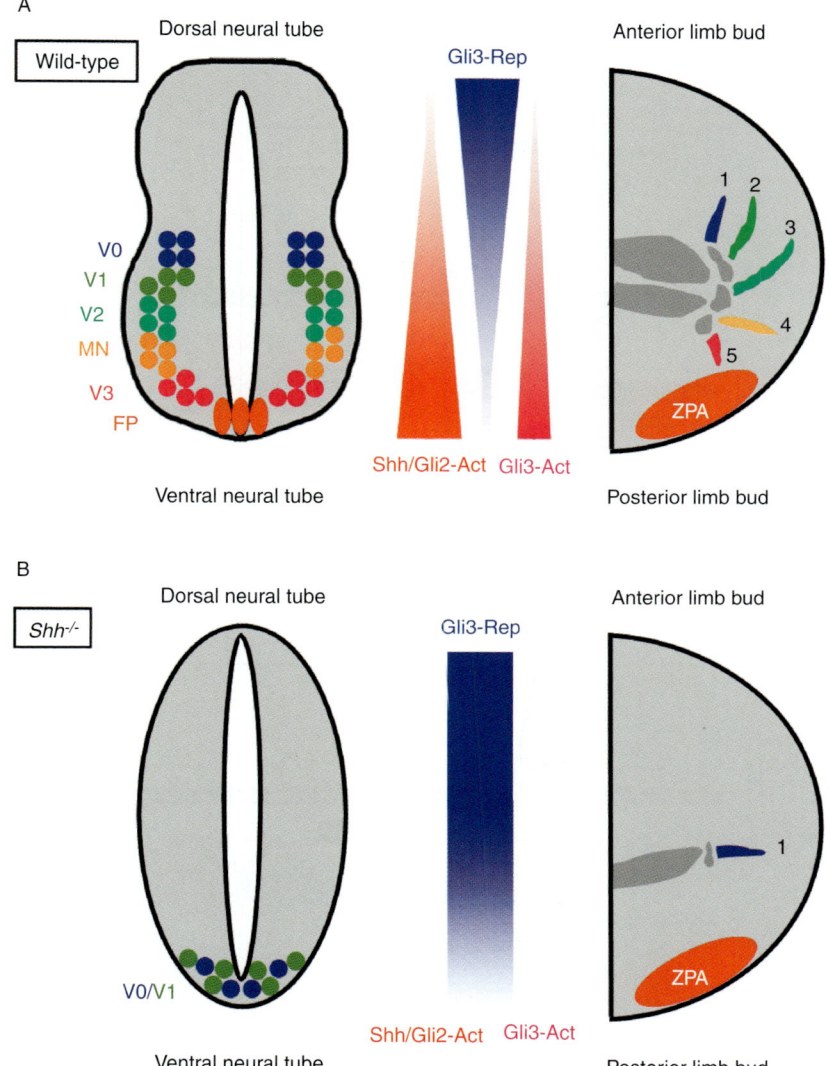

Sunny Y. Wong and Jeremy F. Reiter, Figure 9.1 Proper patterning of the neural tube and limb buds is mediated by a gradient of Shh. (A) In the neural tubes of wild-type embryos, Shh produced by the notochord, floor plate (FP) and possibly gut forms a gradient along the dorsal–ventral axis. Similarly, Shh released by cells in the zone of polarizing activity (ZPA) in the limb buds forms a gradient along the anterior–posterior axis. Increased Shh concentration and duration of action are associated with increased Gli2 activator (Gli2-Act) and Gli3 activator (Gli3-Act) functions, and reduced Gli3 repressor (Gli3-Rep) formation. High-level Hh signaling is critical for specifying the ventral-most cell types in the neural tube, including FP and V3 interneurons (V3). More dorsal cell types (V0, V1, V2, and motor neurons (MN)) are specified by lower levels of Hh signaling. In limb buds, the Shh and Gli3-Rep gradients are interpreted for the specification of digit identity (digits 1–5). (B) In *Shh* mutant embryos, the gradient of Hh pathway activation is absent. Gli3 is mostly converted into Gli3-Rep along the dorsal–ventral axis of the neural tube and the anterior–posterior axis of the limb buds, although Ihh may still activate the pathway to some degree. In mutant neural tubes, only the V0 and V1 ventral neuronal subtypes are specified, and in mutant limb buds, only the anterior-most digit (digit 1) is formed.

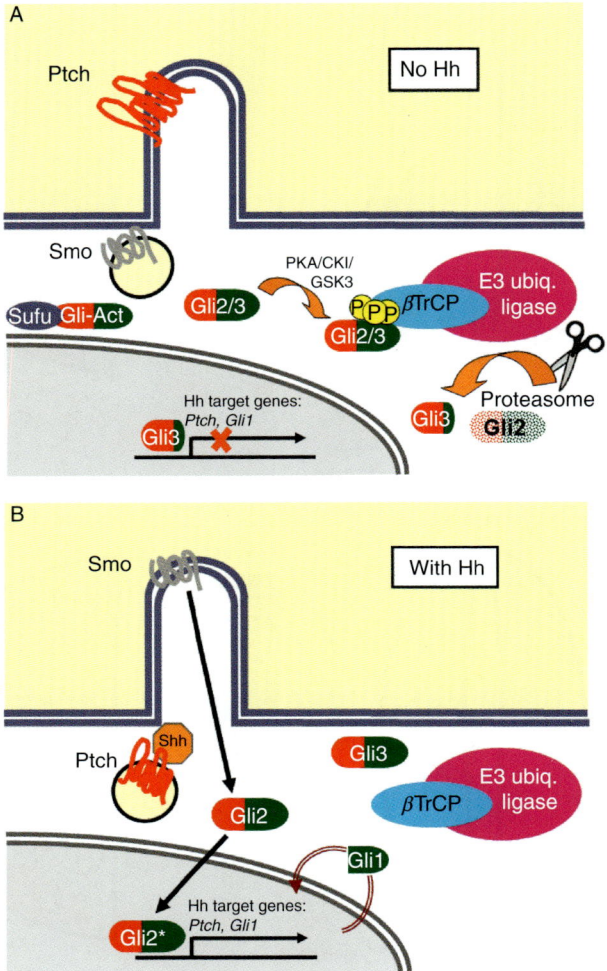

Sunny Y. Wong and Jeremy F. Reiter, Figure 9.2 A model of mammalian Hh signal transduction. (A) In the absence of Shh, Ptch localizes to the primary cilium and, through an unknown mechanism, prevents Smo from entering the cilium. Gli2 and Gli3 are phosphorylated by kinases, including PKA, CKI, and GSK3, to generate phosphopeptide-binding motifs for β-TrCP, an important substrate recognition component of the E3 SCF ubiquitin ligase complex. Ubiquitination of Gli2 and Gli3 targets these proteins to the proteasome, which processes Gli3 into a carboxy-terminal-truncated repressor form and degrades Gli2. Gli3 subsequently enters the nucleus and inhibits transcription of downstream Hh target genes such as *Ptch1* and *Gli1*. (B) In the presence of Shh, Ptch is internalized, allowing Smo to traffic into the primary cilium. It is unclear whether Smo traffics directly to the cilium, or accumulates first in the plasma membrane. At the cilium, Smo inhibits the formation of Gli3 repressor and activates Gli2, which enters the nucleus to promote transcription of *Ptch1* and *Gli1*, whose protein products negatively and positively feed back on this pathway, respectively.

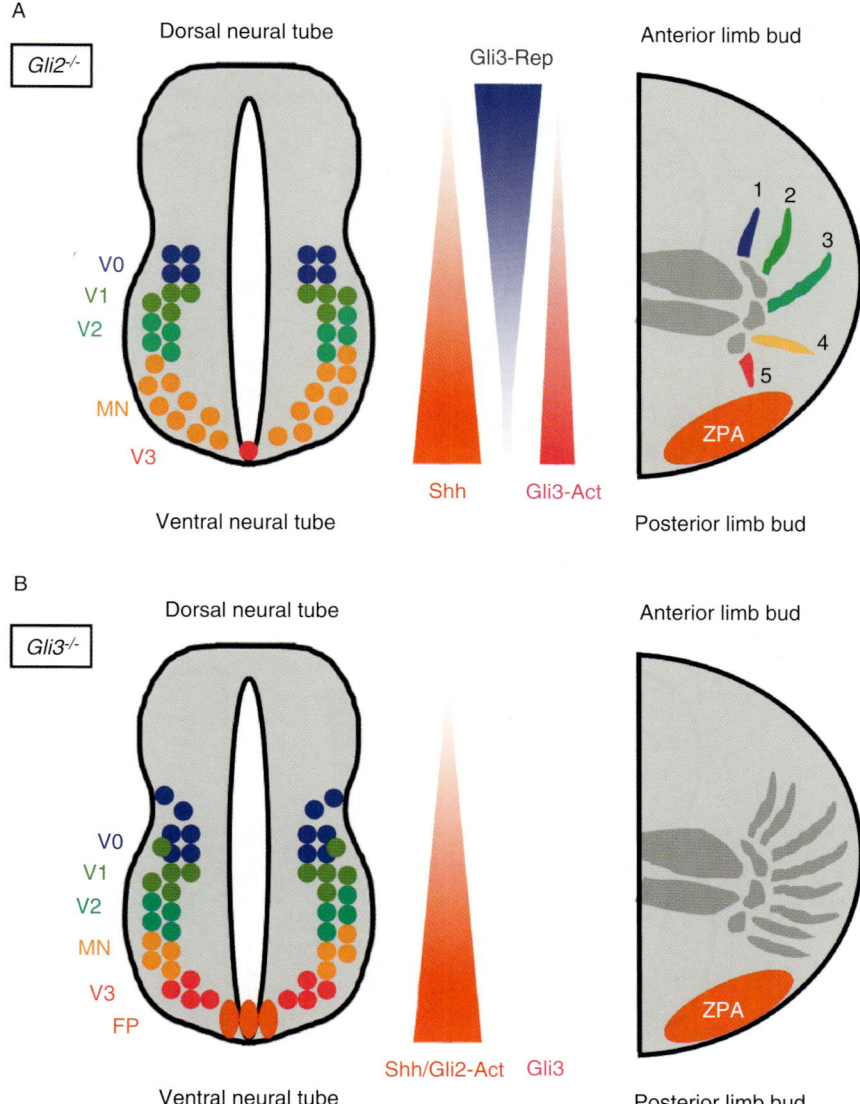

Sunny Y. Wong and Jeremy F. Reiter, Figure 9.3 The neural tube and limb buds differ in their reliance on Gli2 and Gli3 for proper patterning. (A) In *Gli2* mutant embryos, gradients of Shh and Gli3-Rep/Act are established, but downstream signaling is not properly activated. In the neural tube, which relies predominantly on Gli2 activator function, specification of neuronal subtypes that require the highest levels of Hh pathway activity (FP and V3) is deficient. Because proper patterning of the limb bud depends upon a gradient of Gli3-Act/-Rep, and not on Gli2-Act, digit specification is normal in *Gli2* mutants. (B) In *Gli3* mutant embryos, gradients of Shh and Gli2-Act are established, but formation of Gli3-Rep is absent. *Gli3*-deficient neural tubes display only a subtle expansion of V0 and V1 neuronal subtypes. Because proper limb patterning depends on a gradient of Gli3-Rep, *Gli3* mutant limb buds upregulate genes normally repressed by Gli3-Rep, including *Gremlin* and *Hoxd13*, and exhibit polydactyly.

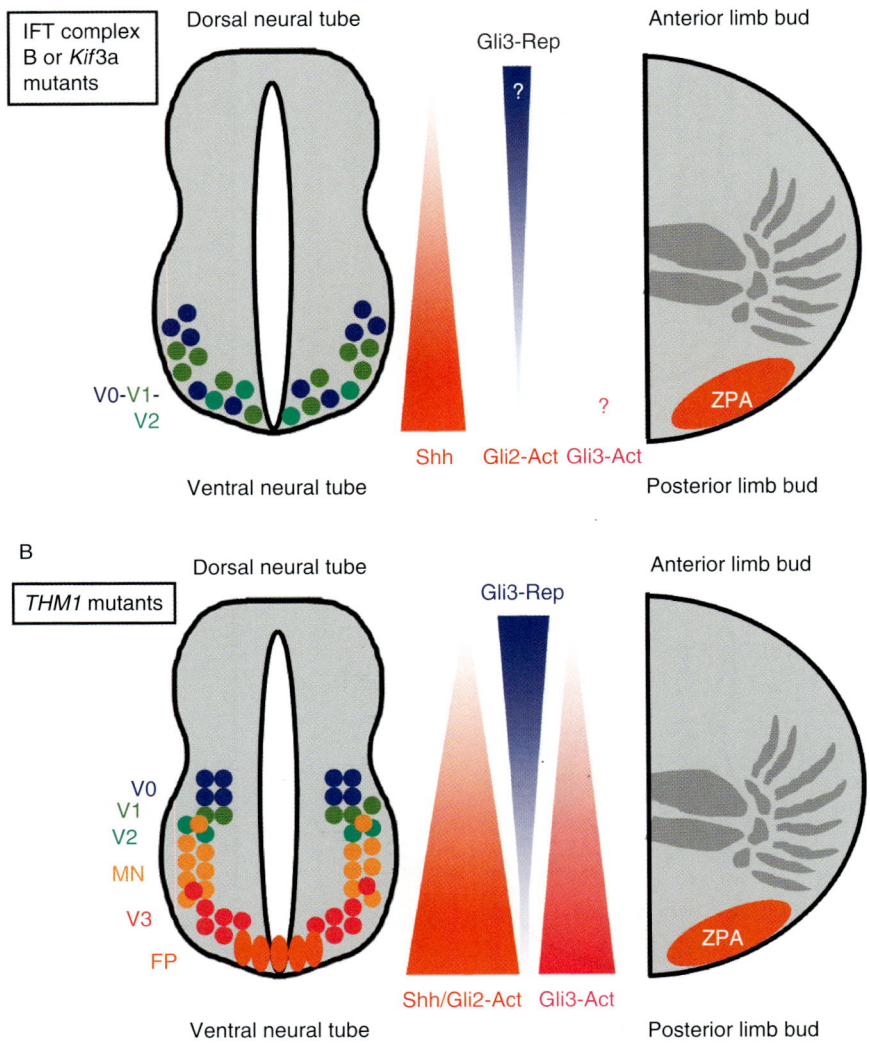

Sunny Y. Wong and Jeremy F. Reiter, Figure 9.4 Many IFT mutations cause polydactyly and loss of ventral neural tube fates. (A) In embryos lacking IFT components such as IFT88, IFT172, and Kif3a, a gradient of Shh is established, but signaling through Gli2-Act and formation of Gli3-Rep are disrupted. Consequently, these mutants do not properly activate the Hh transcriptional targets, and display loss of ventral cell types (FP, V3, MN) and expansion of more dorsal subtypes (V0, V1, V2) in the neural tube. In IFT mutant limb buds, disruption of Gli3-Rep formation leads to polydactyly. (B) In contrast to other IFT mutants, *THM1* mutants display increased Hh signaling mediated by Gli2-Act or Gli3-Act, leading to an increase in ventral subtypes (FP, V3, MN) in the neural tube. THM1 is also essential for Gli3 processing, defects in which result in polydactyly.

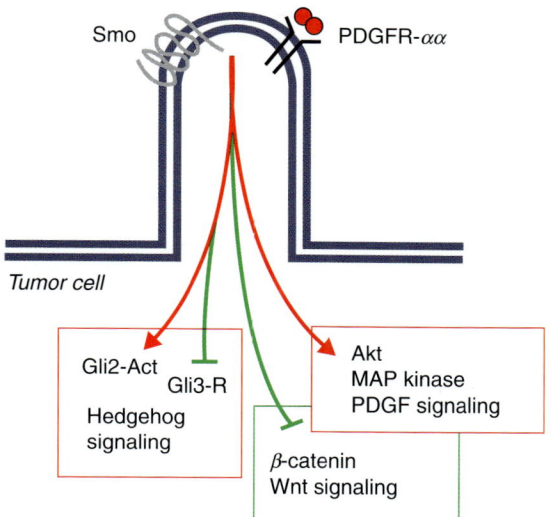

Sunny Y. Wong and Jeremy F. Reiter, Figure 9.5 The primary cilium likely coordinates a variety of signaling pathways important for cancer. In Hh-dependent cancers such as BCC and medulloblastoma, the cilium may modulate the activity of signaling pathways in addition to Hh. For instance, localization of PDGF receptor-α to the cilium is important for transducing downstream signals mediated by Akt and MAP kinases. The cilium may also normally restrain Wnt pathway activity by preventing the accumulation of β-catenin. The balance of the downstream effects of these pathways likely governs cellular decisions for proliferation and apoptosis.

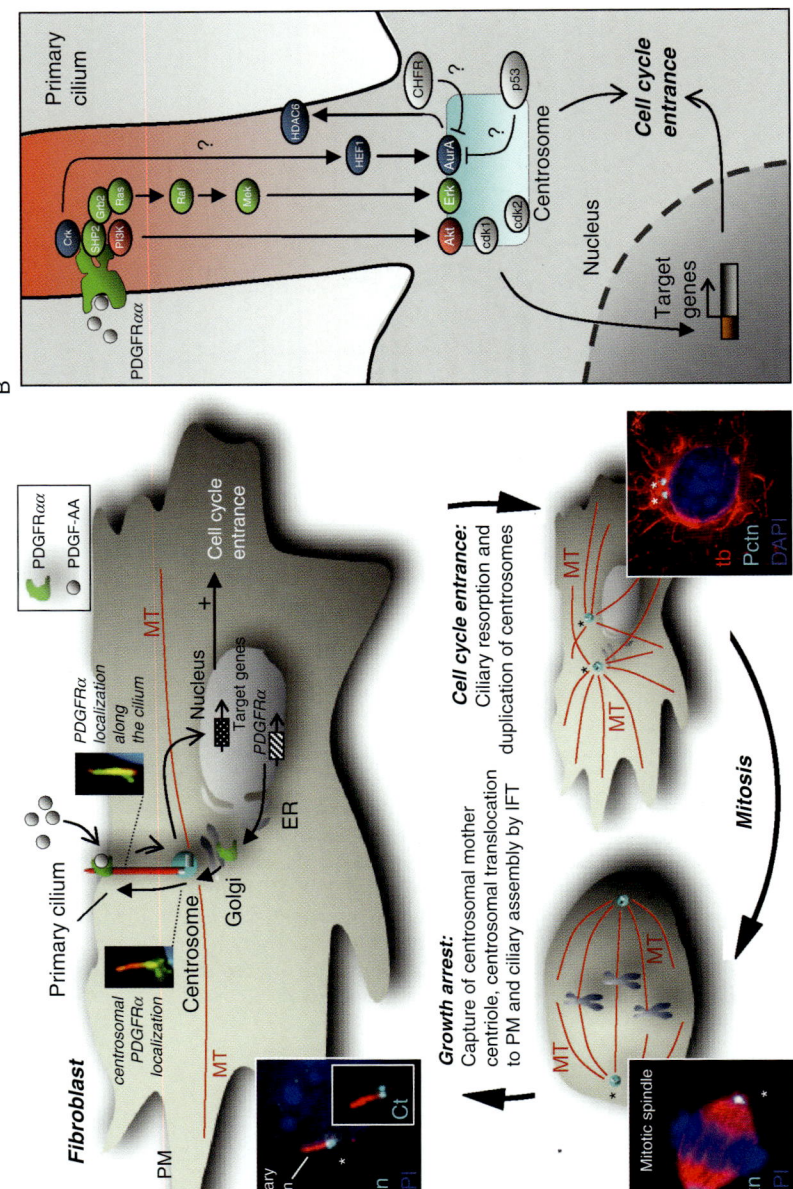

Søren T. Christensen et al., Figure 10.1 *Cell cycle control regulated by PDGFRαα in the primary cilium of fibroblasts.* (A) Upon growth arrest PDGFRα expression is upregulated and the receptor is translocated to the centrosome/ciliary base and then into the cilium. Activation of ciliary PDGFRαα by, e.g., PDGF-AA turns on a series of signal transduction pathways in the cilium and to the cell. Ciliary localization of PDGFRα in NIH3T3 fibroblasts is shown by immunofluorescence microscopy analysis (IF) with antiacetylated a-tubulin (tb, cilium, red) and anti-PDGFRα (green). The nucleus is stained with DAPI (blue). From Schneider et al. (2005) with permission. PDGFRαα activation causes ciliary resorption, duplication of the centrosome and cell cycle entrance followed by mitosis where the two centrosomes form the mitotic spindle. After cytokinesis the daughter cells may reenter growth arrest by capturing of the centrosomal mother centriole and translocation of the centrosome to the plasma membrane (PM) at which the primary cilium is formed by intraflagellar transport (IFT). The three major inset images show IF of NIH3T3 fibroblasts in interphase growth (lower right image), mitosis (lower image left) and growth arrest (upper left image). The primary cilium, cytoskeletal microtubules (MT) networks, and mitotic spindle are stained with acetylated a-tubulin (tb, red), centrosomes are stained with antipericentrin (Pctn, light blue, asterisks) and the nucleus/chromosomes is stained with DAPI (dark blue). The centrioles at the base of the primary cilium were further stained with anticentrin (Ct, light blue) (upper left image). (B) Current model on some of the signal transduction pathways in the ciliary/centrosome axis that controls growth arrest and cell cycle entrance. Activation of PDGFRαα in the primary cilium leads to ciliary resorption and cell cycle entrance via activation of the PI3K-Akt, Mek1/2-Erk1/2 and hypothetically Crk/Hef1 pathways in the cilium and/or at the centrosome. This results in altered gene expression and induces ciliary resorption via Aurora A kinase (AurA) and tubulin deacetylase (HDAC6). Other hypothetical players in cell cycle control at the ciliary/centrosome axis may include cyclin-dependent kinases, cdk1 and cdk2, and tumor suppressor proteins, p53 and CHFR, which we suggest to inhibit ciliary resorption and cell cycle entrance by blocking the activity of AurA.

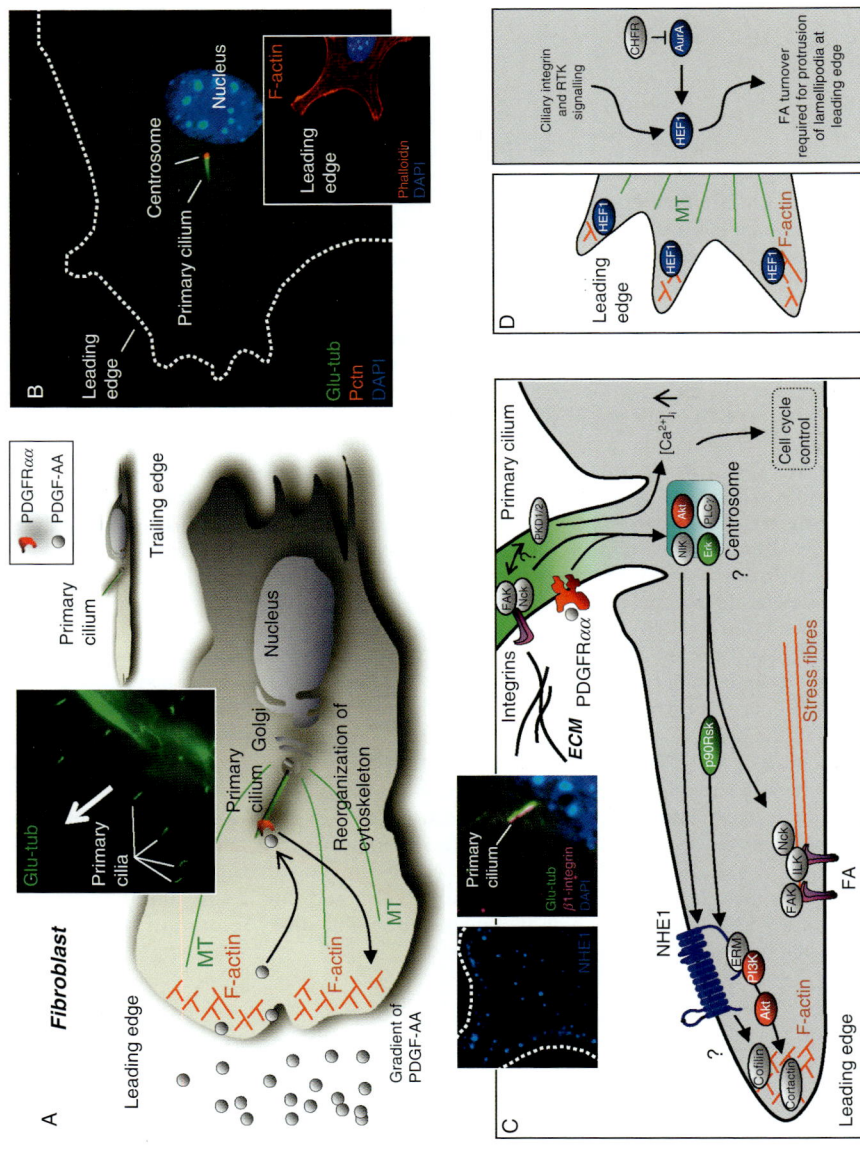

Søren T. Christensen et al., Figure 10.2 *Cell migration regulated by PDGFRαα and integrins in the primary cilium.* (A) In migrating cells the primary cilium functions as a cellular GPS that locates in front of the nucleus and orients in the direction of the leading edge to coordinate signaling (e.g., PDGFRαα signaling) in reorganization of microtubules (MT) and F-actin in lamellipodia. The inset shows IF of primary cilia stained with antidetyrosinated tubulin (Glu-tub, green) orienting to the leading edge of growth-arrested NIH3T3 cells in an *in vitro* scratch assay. (B) IF of a single migrating fibroblast with the primary cilium (Glu-tub, green) orienting toward the leading edge and emerging from the centrosome stained with antipericentrin (Pctn, red) and in front of the nucleus (DAPI, blue). (C) Our model of some of the signal transduction pathways in the ciliary/centrosome axis that coordinate cell migration via PDGF-AA-mediated activation of PDGFRαα and ECM-mediated activation of integrins in the primary cilium. Activation of the ciliary receptors leads to activation of signal transduction molecules at the centrosome (e.g., Akt, Erk1/2, phospholipase C-gamma, PLCγ, and Nck-interacting kinase, NIK). In integrin signaling these molecules are activated by focal adhesion kinase (FAK) and Nck. Also, polycystin-2 (PKD2) may be involved in both cell migration and cell cycle control by interaction with integrin signaling and mechanical deflection of the cilium to increase $[Ca^{2+}]_i$. Downstream of the ciliary/centrosome axis, a series of critical molecules in cell motility are activated to control formation of focal adhesions (FA) and lamellipodia at the leading edge of migrating cells. These include FAK, Nck, and NIK at FA and the plasma membrane Na^+/H^+ exchanger (NHE1) at leading edge lamellipodia. As examples, NHE1 directly interacts with ezrin/radixin/moesin (ERM) proteins to regulate cortactin via the PI3K-Akt pathway, and NHE1 regulates cofilin activity at the leading edge. IF insets show NHE1 (blue) localizing to the leading edge lamellipodia (dashed line, upper left IF) and β1-integrin (magenta) localizing to the primary cilium (anti-Glu-tub, green) (upper right IF) in NIH3T3 fibroblasts. While many of these signaling pathways are well described in the literature, little is known on how signaling from the cilium/centrosome axis is transmitted to NHE1 and FA. Some of the hypothetical pathways are also indicated with a question mark. (D) shows in two panels our hypothesis on the role of ciliary integrin and receptor tyrosine kinase (e.g., PDGFRαα) signaling in activation of human enhancer of filamentation 1 (HEF1) in control of protrusion of lamellipodia at the leading edge. In this pathway tumor suppressor, CHFR, may restrain excessive cell migration (as seen in invasive cancers) by inhibiting the activity of AurA, which activates HEF1.

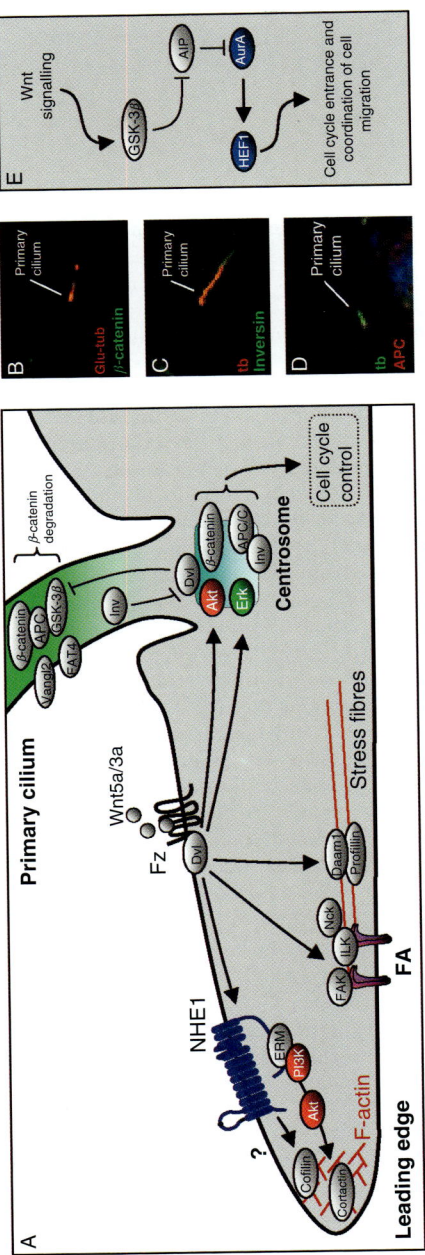

Søren T. Christensen et al., Figure 10.3 *Cell migration and cell cycle control regulated by Wnt signaling.* (A) Hypothesis on some of the Wnt signaling components at the plasma membrane and in the ciliary/centrosome axis that coordinate cell migration and cell cycle control. Ciliary Inversin impairs the canonical Wnt pathway by restricting Dvl-mediated impairment of β-catenin degradation by Adenomatous poliposis coli (ACP) and glycogen synthase kinase 3 beta (GSK-3β) in the primary cilium and potentiates degradation Wnt signaling through membrane-bound dishevelled (Dvl), which is activated by Wnt5a binding to Frizzled (Fz) 3. In turn, Dvl activates NHE1 and F-actin reorganization at the leading edge in addition to stress fiber formation and focal adhesion (FA) turnover. Dvl also mediates Wnt5a-induced activations of integrins via the PI3K-Akt pathway, in addition to Src- and PKC-mediated activation of focal adhesion kinase (FAK). Besides β-catenin-mediated transcription of genes in cell proliferation Wnt3a activates the Raf-Mek1/2-Erk1/2 pathway at the centrosome for cell cycle progression. Moreover, Inversin interacts with Anaphase Promoting Complex/Cyclosome (APC/C), which controls mitosis. (B–D) IF localization of β-catenin (green) (Corbit et al., 2008; with permission) (B), Inversin (green) (Courtesy Prof. J. Goodship) (C) and APC (red) (Corbit et al., 2008; with permission) (D) to the primary cilium stained with either detyrosinated tubulin (Glu-tub) or acetylated α-tubulin (tb). (E) Our hypothesis on the potential overlap between Wnt signaling and AurA-induced cell cycle control and migration. In addition to its role in β-catenin degradation complex, ciliary GSK-3β releases the inhibition of centrosomal AurA through AurA-Interacting protein (AIP), resulting in activation of human enhancer of filamentation 1 (HEF1).

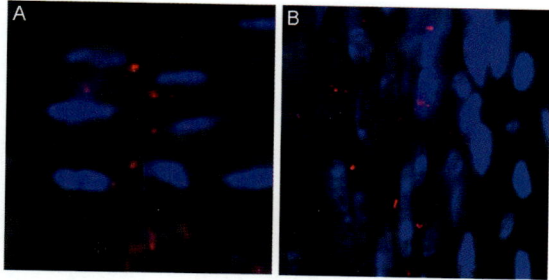

Courtney J. Haycraft and Rosa Serra, Figure 11.1 *Localization of cilia on chondrocytes and perichondrial cells.* (A) Cilia, visualized by immunostaining with anti-acetylated α-tubulin antibodies (red), are aligned in the center of columns of proliferating chondrocytes in the growth plate of endochondral bones. Nuclei (blue) are located on alternate sides of the cells within a column. (B) In the perichondrium, cilia are present on both elongated and cuboidal cells adjacent to the cartilage anlagen (right side of image). Cilia were visualized by immunostaining with anti-Ift88 antibodies (red). Nuclei are blue.

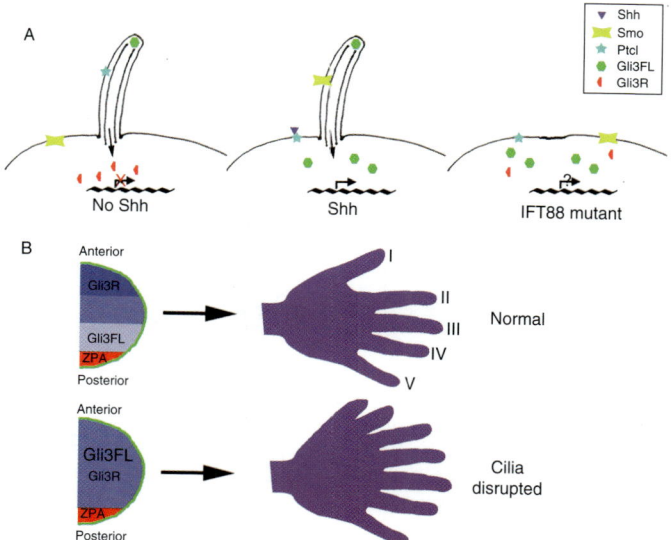

Courtney J. Haycraft and Rosa Serra, Figure 11.2 *Mammalian Hh signal transduction and limb patterning.* (A) *Hh signal transduction.* In the absence of Hh ligand, Ptc1 localizes to the ciliary membrane. Full length Gli3 (Gli3FL) is found at the distal tip of the cilium and subsequent proteolytic processing results in increased amounts of Gli3 repressor (Gli3R). Gli3R is a potent transcriptional repressor resulting in repression of target gene transcription including Ptc1 and Gli1. Upon ligand binding, Ptc1 translocates out of the cilium and Smo is localized in the cilium membrane. Smo translocation to the cilium inhibits Gli3 processing leading to an increased Gli3FL:Gli3R ratio allowing transcription of target genes. In Ift88 mutants, the cilium is not formed and Gli3 processing is inefficient. As a result, target genes such as Gremlin and Hand2, which are repressed by Gli3R, are expressed while transcription of Gli1 and Ptc1 is not activated likely due to disruption of Gli2 function by an unknown mechanism. (B) *Role of Shh in development of the mammalian limb.* In the developing limb, Shh is expressed by cells in the ZPA (red). Activation of Shh signaling leads to decreased Gli3R levels in the posterior limb bud while the level of Gli3R remains high in the anterior limb bud. In cilia mutants, such as those lacking Ift88 or Kif3a, disruption of the normal gradient of Gli3FL:Gli3R in conjunction with loss of Gli2 activity across the limb bud results in polydactyly.

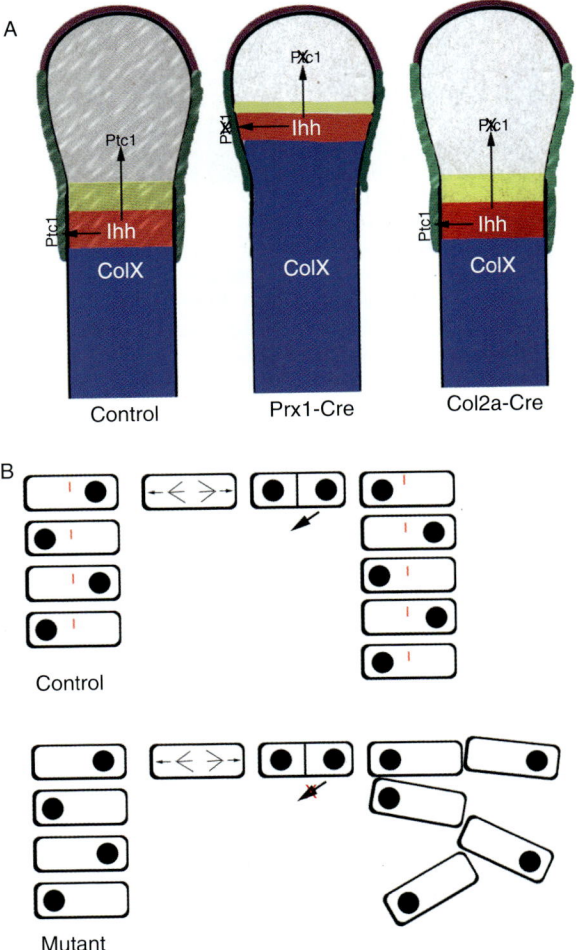

Courtney J. Haycraft and Rosa Serra, Figure 11.3 *A role for primary cilia in endochondral bone formation and the postnatal growth plate.* (A) *Primary cilia regulate Ihh signaling during embryonic endochondral bone formation.* Ihh is expressed in chondrocytes that are committed to hypertrophic differentiation (red area). Ihh normally acts on both prehypertrophic chondrocytes (gray and yellow areas) and the perichondrium (green). Ptc1, a downstream target of Ihh signaling, was used to determine how depletion of cilia affects Ihh signaling during endochondral bone formation. In normal control mice, cilia are present on chondrocytes as well as cells in the perichondrium and Ptc1 expression is detected in both cell types. In mice in which ciliogenesis is disrupted in Prx1-Cre expressing cells, cilia are absent from both chondrocytes and cells in the perichondrium. Disruption of ciliogenesis in the cartilage and in the perichondrium resulted in accelerated hypertrophic differentiation (blue area) similar to that seen in the Ihh-null mice. Ptc1 expression was dramatically reduced in both chondrocytes and cells of the perichondrium confirming that Ihh signaling was disrupted. In contrast, when ciliogenesis was disrupted using Col2a-Cre, which only targets the chondrocytes, hypertrophic differentiation was normal. Ptc1 expression was reduced in chondrocytes but maintained in the perichondrium. (B) *Primary cilia mediate chondrocyte rotation in the postnatal growth plate.* In normal control mice, flat cells in the growth plate have polarity and are aligned in columns parallel to the long axis of the bone. Cell division occurs perpendicular to the long axis. One cell then migrates under the other to form the columns. This process is called chondrocyte rotation. Cilia are present on chondrocytes in the columns and are required for normal chondrocyte rotation. In cilia depleted growth plates, chondrocytes still divide perpendicular to the long axis of the bone and polarity is maintained; however, the orientation of the cells one to another is altered and the cells are not maintained in columns.

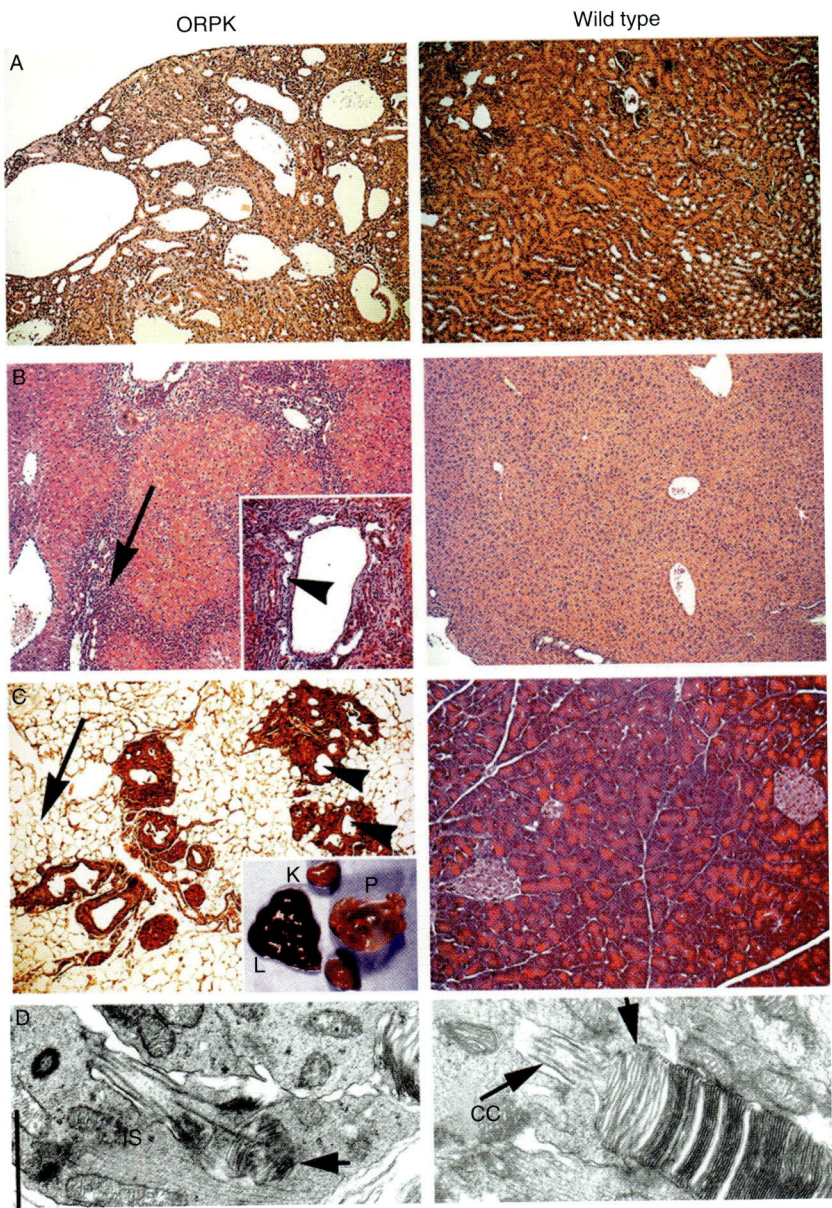

Neeraj Sharma et al., Figure 13.1 Characteristic phenotypes in the ORPK mouse. Left panels show various disease phenotypes observed in different tissue types in ORPK (FVB/N background) mutants and right panels show wild-type controls. (A) Enlarged cysts present in the kidney. (B) Biliary (arrow) and bile duct (arrow head) hyperplasia developed in the liver. Inset shows higher magnification view of a central vein displaying multiple dysplastic bile ductules. (C) Dilated ducts (arrowhead) and acini atrophy

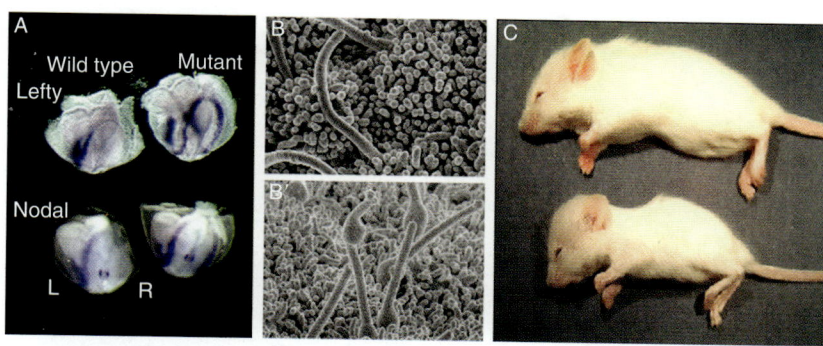

Neeraj Sharma et al., Figure 13.3 Motile cilia defects. Left–right asymmetry defects in day 7.5 mouse embryos. *In situ* hybridization for two genes expressed asymmetrically in the embryonic node (Lefty and Nodal). In Wild-type embryos, both are expressed on the left while in cilia mutant embryos, this expression pattern is lost (Panel A, reproduced with permission of the Company of Biologists, adapted from Murcia *et al.* (2000) Copyright © 2000, The company of Biologists). Scanning electron microscopy of the embryonic node showing wild-type primary cilia (Panel B) and Dync2h1 mutant cilia with atypical bulges (Panel B' Images courtesy of Danwei Huangfu and Kathryn Anderson). Mice mutant for the gene *Hydin* (*Hy*drocephalus *In*ducing) develop perinatal hydrocephalus due to ependymal cell motile cilia defects (Panel C image adapted from Robinson *et al.* (2002), Copyright © 2002, Springer, New York).

(arrow) in the OPRK mutant pancreas. Inset show the individual affected organs (K, kidney; L, liver; P, pancreas). (D) P10. The ORPK mutants develop disrupted discs and misshaped OSs filled with amorphous material. (IS, inner segment; CC, connecting cilium). Panel D images are reproduced from Pazour *et al.* (2002) *J. Cell Biol.* (Copyright © 2002, The Rockefeller University Press). All panels were reproduced from Lehman *et al.* (2008), *Dev. Dyn.* (Copyright © 2008, Wiley-Liss).

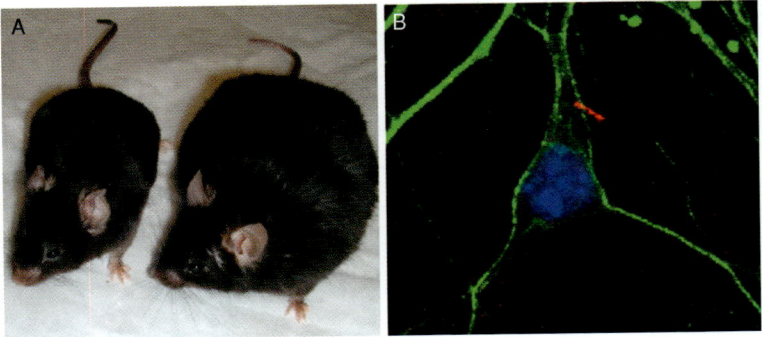

Neeraj Sharma et al., Figure 13.5 Neuronal cilia have a role in feeding behavior. The loss of neuronal cilia results in hyperphagia induced obesity in mice. Panel A shows an obese mouse that has the conditional allele of IFT88 and a neuronal specific Cre transgene (SynI-Cre), next to its Cre negative littermate (Image courtesy of Mandy Croyle). Panel B shows a primary hypothalamic cultured neuron (neuronal marker β-Tubulin III in green) with a Mchr1 positive cilium in red (Image Courtesy of Kirk Mykytyn).

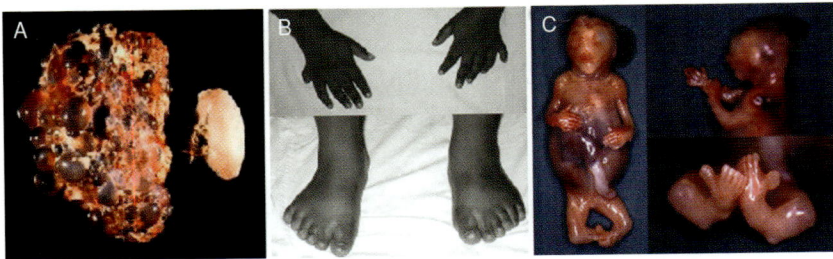

Neeraj Sharma et al., Figure 13.6 Examples of clinical features of ciliopathies. A polycystic kidney in shown next to a normal kidney in panel A. Polydactyly is a cardinal feature of BBS, panel B. Embryonic lethality with exencephaly and polydactyly is observed in MKS, panel C.